Residential Exposure Assessment

Residential Exposure Assessment
A Sourcebook

Edited by

Scott Baker
International Copper Association
New York, New York

Jeffrey Driver
infoscientific.com, Inc. and risksciences.net
Manassas, Virginia

and

David McCallum
FOCUS GROUP
Tilghman Island, Maryland

Springer Science+Business Media, LLC

Library of Congress Cataloging-in-Publication Data

Residential exposure assessment: a sourcebook/edited by Scott Baker, Jeffrey Driver and David McCallum.

p. cm.

Includes bibliographical references and index.

ISBN 978-1-4613-5475-8 ISBN 978-1-4615-1279-0 (eBook)

DOI 10.1007/978-1-4615-1279-0

1. Health risk assessment. 2. Environmental health. 3. Pollution—Measurement. 4. Environmental monitoring. I. Baker, Scott, 1950 Jan. 13– II. Driver, Jeffrey. III. McCallum, David, 1943–

RA566.27 .R465 2000
615.0'02—dc21

00-049770

Sponsored by the Society for Risk Analysis and International Society for Exposure Assessment.

This project is a joint effort by the Society for Risk Analysis and International Society for Exposure Assessment to encourage the development and dissemination of improved methods and data for determining exposures in environments outside the work place. The editors and the project team would like to acknowledge the financial and technical support provided by the following: U.S. Environmental Protection Agency, Society for Risk Analysis, International Society of Exposure Analysis, American Chemical Council, Chemical Specialties Manufacturers Association, American Crop Protection Association, Responsible Industry for a Sound Environment, The Procter & Gamble Company

The Society for Risk Analysis is an international professional association which promotes research and education on all aspects of risk. The Society supervises grants from governmental and nongovernmental sources to qualified experts for the preparation of reports on important subjects relating to risk, but each report represents solely the work of the authors and not the views of the Society.

ISBN 978-1-4613-5475-8

Originally published by Kluwer Academic / Plenum Publishers, New York in 2001
Softcover reprint of the hardcover 1st edition 2001
http://www.wkap.nl/

10 9 8 7 6 5 4 3 2 1

A C.I.P. record for this book is available from the Library of Congress

PREFACE

Residential Exposure Assessment: A Source Book is the result of a multiyear effort known as the Residential Exposure Assessment Project (REAP) which was initiated by the Society for Risk Analysis and the International Society of Exposure Analysis. This textbook is the primary product of the REAP and it contains contributions from over 30 professionals from a variety of disciplines such as chemistry, biology, physics, engineering, industrial hygiene, toxicology, pharmacology, and environmental law, reflecting the diverse knowledge and resources necessary to assess and manage potential exposures occurring in and around the home. Expert working groups were organized for each of the 13 chapters to address such issues as U.S. legislation relevant to products used in and around the residence, methods for measuring and modeling exposures across multiple pathways and routes, and distributional data available for key residential exposure factors.

This volume is a compendium of information about predictive methods and tools, monitoring methods, data sources, and key variables that characterize exposures in the residential setting. It presents approaches for doing exposure assessments in and around all types of residences. The purpose of the Source Book is to provide a resource for use in educational programs and for "practitioners" of residential exposure assessment. Accordingly, this book is intended for risk assessors, exposure assessors, students, initiates new to the concept of risk assessment, industrial hygienists assessing health hazards in the home, engineers, and monitoring specialists. Residential exposure assessment includes measurement of agents that are present in and around the home, as well as the use of predictive models to estimate potential fate, transport, exposure, and bioavailability via multiple pathways (*e.g.*, direct inhalation of agents in indoor air, ingestion of contaminated food, surface-to-skin transfer of dislodgeable residues from contaminated surfaces) and routes (inhalation, ingestion, dermal contact). This volume also provides guidance for identifying and developing key data, whether through measurement or simulation, used as input to residential exposure assessments. References are cited to facilitate finding additional information from the extensive literature on these subjects. This volume is not intended as a guide for determining which agents should be the subjects of an exposure assessment: that decision is taken in another phase of the risk assessment process. The chapters making up the body of this volume vary considerably volume in depth and detail. The authors' intent is to provide those undertaking an evaluation of exposure in some residential environment with an introduction and means to frame and conduct analyses, as well as an introduction to some of the relevant literature. The authors do not intend this to be a comprehensive treatise on the subject.

This Source Book is a reference document that builds on other primary references on the topic of exposure. Some of these references are U.S. Environmental Protection

Agency's Exposure Factors Handbook, Standard Operating Procedures for Residential Exposure Assessment, Dermal Exposure Assessment: Principles and Applications, Guidelines for Exposure Assessment, Superfund Human Health Evaluation Manual, the American Industrial Health Council's Exposure Factors Source book, consumer product surveys (*e.g.*, the the National Home and arden Pesticide Useage Survey, the USEPA's Total Exposure Assessment Methodology studies, the USEPA's National Usage Survey of Household Cleaning Products), and the European Center for Ecotoxicology and Toxicology of Chemicals (ECETOC). The reader is encouraged to obtain and utilize other key publications referenced herein. Further, studies and data relevant to residential exposure assessment are being continuously developed. Thus, as with many scientific disciplines, it is important to update and supplement the information presented in this volume. To facilitate this process, definitions and key terms, a software bibliography, and residential assessment practices in Europe are provided in Section 6, *Supplemental Data and Information*, to facilitate the identification and exchange of new data as they become available.

Funding for the REAP effort was provided by the U.S. Environmental Protection Agency, the Chemical Manufacturers Association, the Chemical Specialties Manufacturers Association, American Crop Protection Association, Responsible Industry for a Sound Environment, and the Procter & Gamble Company.

This Source Book contains six sections. Section 1 chapters contain an overview of principles and practices in residential exposure assessment. Section 2 explores methodologies for assessing residential exposures, including the inhalation, dermal, and ingestion routes of exposure, and for total exposure from all sources via all routes. Section 3 addresses indoor and outdoor consumer products, and biological agents as sources and for their fate processes in the residential environment. Section 4 identifies the various exposure factors involved in conducting a residential exposure assessment, including activity patterns, specific human factors, and environmental factors. Section 5 addresses issues concerning the presentation, interpretation, and communication of residential exposure assessments, including uncertainty analysis. Section 6 contains supplemental information on residential exposure, including a glossary of terms, a bibliography of relevant literature, and a synopsis of international activities on this topic.

Editorial Committee
Residential Exposure Assessment Project (REAP)

PARTICIPANTS

Principal Investigator

David McCallum
FOCUS Group

Management Committee

James Wilson, *Chairman*
Resources for the Future

Paul Price
Ogden Environmental and Energy Services

Kenneth Sexton
School of Public Health, University of Minnesota

Jerald Schroy
Monsanto Company

Harlee Strauss
Silent Spring Institute

William Wood
U.S. Environmental Protection Agency

Editorial Committee

Jeffrey Driver, *Chairman*
infoscientific.com, Inc. and risksciences. net

Annette Guiseppi-Elie
DuPont Haskell Laboratory

P.J. (Bert) Hakkinen
The Procter & Gamble Company

David McCallum
FOCUS Group

Judith Schreiber
New York State Department of Health

John Spengler
Harvard School of Public Health

Joan Tell
Exxon Biomedical Sciences, Inc.

William Wood
U.S. Environmental Protection Agency

Susan Youngren
Novigen Sciences, Inc.

CONTENTS

II Methodologies for Assessing Residential Exposures

III Evaluating Sources in the Residential Environment

IV Residential Exposure Factors

V Presenting, Interpreting, and Communicating the Results of Residential Exposure Assessments

I

Residential Exposure Assessment

An Overview of Principles and Practices

1

INTRODUCTION

This Source Book is intended for use by risk analysts whose experience with assessing risks in residences is limited, who want to sharpen their skills, and for advanced practitioners of risk assessment. This Source Book is a compendium of information about predictive methods and tools, monitoring methods, data sources, and key variables that characterize exposures in the residential setting. The document presents approaches for doing exposure assessments in and around all types of residences. The Source Book addresses exposure to releases from sources on the property and in the residence and agents that enter the residential environment from off site sources. It provides an overview of data on key residential exposure variables related to human physiological characteristics, activity patterns and such residential "building factors" as air exchange rates. Data for these variables are not provided as single recommended "values" or a single "recommended distributions"; nor are the data presented considered be an exhaustive presentation of the results of all studies on the variables discussed. However, key references are cited, so that the user is provided with sources for more detailed information. Finally, the emphasis of the REAP effort is on North American residential environments. Thus, it is important to note that application of these data to residential environments outside North America should take into account sociological and geographical differences. A supplemental section is provided which provides sources for information relevant to residential exposure assessment in Europe and we hope that continued interest in this area will result in similar information for other countries and geographical areas.

This book is intended as an introduction and guide to the art of assessing exposures in a particular setting, the residential environment. For the purposes of this Source Book, the residential environment and its microenvironment are collectively defined as indoor—inside the residence—as well as outdoor—for instance "backyard"—areas.

The book also provides guidance for identifying and developing key data, whether through measurement or simulation, used as input to residential exposure assessments. References are cited to facilitate finding additional information from the extensive literature on these subjects. This book is not intended as a guide for determining which agents should be the subjects of an exposure assessment: that decision is taken in another phase of the risk assessment process.

The art of evaluating exposures in the residential environment is early in its development. Practices for evaluating exposures from some kinds of sources, for some kinds of agents, and for some exposure pathways are more advanced than for others. For some, such as transport of agents in air, the basic science is well understood and the application in specific instances quite sophisticated. For others, such as transport of agents from particles adsorbed to skin into the body, the grounding sciences are poorly understood, the art relatively primitive.

Residential Exposure Assessment, edited by Baker *et al.*
Kluwer Academic/Plenum Publishers, New York, 2000.

As the young profession of health risk analysis has evolved, two complementary subdisciplines have developed. One deals with assessment of exposures; residential exposure assessment constitutes a speciality within it. The second attempts to answer the question, "What harm can the agent bring about, and under what conditions?" At this writing, this subdiscipline comprises primarily people who evaluate information about chemical hazards inferred from animal experiments, "toxicologists", and those who evaluate information from human experience, "epidemiologists". Those in the profession who deal mainly with harm potentially caused by microorganisms are yet distinct, mainly because of the different scientific roots from which their analyses spring. We attempt, in this volume, to begin the integration of analysis of exposure to microorganisms within the exposure assessment subdiscipline. However, readers should note that exposure analysis originally was concerned with risks posed by exposure to chemical and physical hazards, especially in the workplace, and that much of the material we have to present will reflect the bias this origin causes.

Because both toxicologists and microbiologists rely on results from exposure analysis, exposure assessors need to use vocabulary that communicates to both these groups, as well as to those not trained in either discipline. Of particular importance is the term used to refer to hazards potentially of concern. Toxicologists commonly use "toxin" or "chemical" for this purpose, even if the hazard may actually derive from a complex mixture of individual chemical substances. Microbiologists commonly use "microbe" as a generic term encompassing both individual kinds of organisms and mixtures of them, and also "agent." Because this latter term is general enough and broad enough to include hazardous materials both living and dead, we will use "agent" for this purpose throughout this book.

WHY DO EXPOSURE (AND RISK) ANALYSES?

Decisions on managing risks to people or the environment are made every day by managers in government and industry, by public interest groups, and by informed consumers. Decisions may often be based on personal experience or subjective values. The "exposure assessment" provides information relevant to the risk assessment and directly useful to decisions about risk.

The decisions themselves may be concerned with a variety of issues, including:

- The safety of products either used in or being considered for use in residences
- The nature of warning labels on product containers, as derived from an evaluation of the consequences of misuse
- The kinds of protective measures needed in normal use of a product
- Effective strategies for minimizing harm from unavoidable, natural agents, and
- The kinds of remediation/action taken for contaminated media impacting the home and its residents.

Exposure assessments may also be conducted as part of retrospective (epidemiologic) studies of human response to various agents, or to inform design of prospective clinical studies.

Human health risk assessment entails the evaluation of scientific information on the hazardous properties of chemical, biological and physical agents and on the extent of human exposure to those agents.

The risk assessment process is generally viewed as three discrete components: risk assessment, risk management, and risk communication (NRC 1983) (see Figure 1-1).

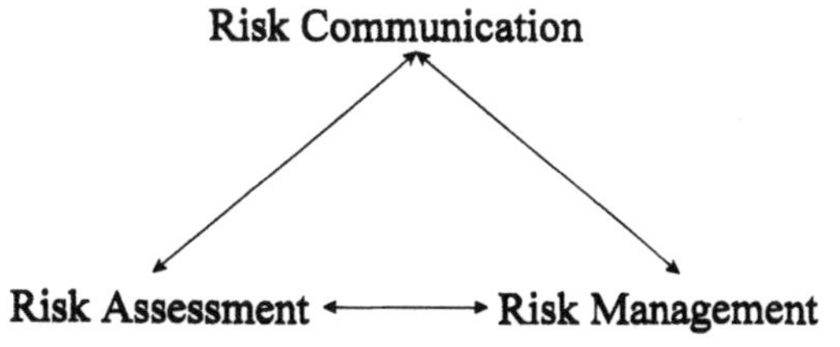

Figure 1-1. Three components of risk analysis.

However, it is useful to consider these three components as a triangle, each with its own unique character but inseparable from the others.

The product of a risk assessment is a statement regarding the probability that exposed populations will be harmed, and to what degree. The probability may be expressed quantitatively or in relatively qualitative ways. There are other types of risk assessment that use similar processes, such as the relative safety of a nuclear power plant, and travel via commercial airplanes. In 1983 the National Research Council (NRC 1983) outlined a four-step analytic process for human health risk assessment (see Figure 1-2). This process was developed to facilitate the use of a systematic approach to quantifying heath risks.

In the context of residential exposure assessment, chemical, biological and physical agents are emitted from sources in and around the home, and then partition and move through various media (*e.g.*, dust, air, clothing) and may result in exposures (*e.g.*, residents inhale airborne agents). The external exposure (amount present at biological exchange membranes—lungs, skin, gastrointestinal tract) may results in a locally or systemically absorbed dose. The magnitude, duration and time of exposure/absorbed dose determine the extent to which the toxic properties of a given agent are realized in exposed individuals (*i.e.*, health risks). This model is captured in the NRC paradigm as follows:

Step 1. Hazard Identification entails identification of the agents that are suspected to pose health hazards, quantification of the concentrations at which they are present in the environment, a description of the specific forms of toxicity (*e.g.*, neurotoxicity, developmental toxicity,

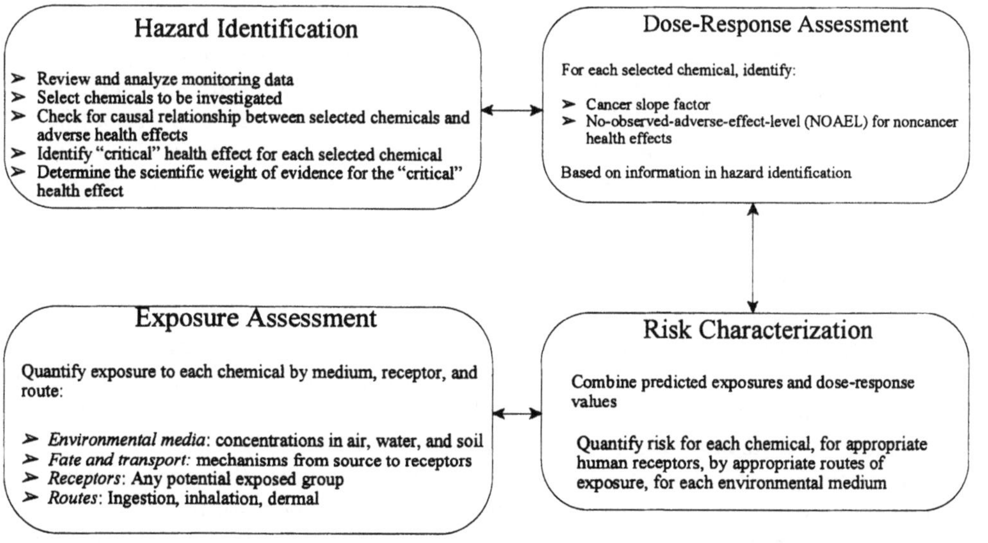

Figure 1-2. Conceptual framework for risk assessment: how the analysis is conducted.

carcinogenicity) that can be caused by the agent(s) of concern, and an evaluation of the conditions under which these forms of toxicity might be expressed in exposed humans. Information for this step is typically derived from studies in laboratory animals, humans (epidemiology studies) and other specialized studies (*e.g.*, in vitro assays, mechanistic studies).

Step 2. Dose-response assessment entails a further evaluation of the conditions under which the toxic properties of a given agent might be manifested in exposed individuals, with particular emphasis on the quantitative relation between the dose that the toxic response. The development of this relationship may involve the use of mathematical models. This step may include and assessment of variations in response, for example, differences in susceptibility between young and old people.

Step 3. Exposure Assessment involves specifying the subpopulations that might be expose to a given agent of concern, identifying the routes through which exposures can occur, and measuring and/or estimating the magnitude, duration and time-course of exposure. This step also often includes the measurement and/or estimation of absorbed dose.

Step 4. Risk Characterization involves the integration of information from the first three steps to develop a quantitative or qualitative estimate of the likelihood that any of the hazards associated with the agent(s) of concern will be realized in exposed people. This is the step in which risk assessment results are expressed. Risk characterization should also include a characterization of the uncertainties associated with the estimates of exposure and risk.

In 1983, the NRC recognized that completion of the four steps of the risk assessment process includes the use of professional judgements, some of which are not based on scientific consensus. Thus, in many cases, exposure/risk assessors may be faced with several scientifically plausible approaches [*e.g.*, choosing the most reliable dose-response model for extrapolation beyond the range of observable effects, selection of the most appropriate indoor air model (and input parameters, such as air exchange rates) for residences in a specific geographical area during a specific season of the year] with no definitive basis for distinguishing among them. In response to this reality, guidelines have been developed by Federal and State Agencies to improve the consistency of the general elements of human health risk assessment. Further, quantitative methods of uncertainty analysis (see Chapter 12) are now viewed as being critically important for understanding the results of a given risk assessment and there relative significance. Further, guidelines are continuously updated and improved as better scientific information becomes available.

One exposure/risk assessment strategy component that deserves emphasis is the need for "iteration" or "tiers" in the development of an appropriate approach to conducing a residential assessment. Neither the resources, nor the necessary scientific data exist to perform a full-scale exposure/risk analysis for most agents in the residential setting. Nor, in many cases, is an extremely rigorous assessment needed. An iterative approach begins with relatively inexpensive screening-level methods and moves to more resource-intensive levels of data gathering, model construction and model application, as the particular situation warrants. To guard against the possibility of underestimating exposure, screening-level methods must be constructed such that the results can be considered conservative (it is desirable to quantify the degree of conservatism using uncertainty analysis techniques). Use of an iterative approach can improve the scientific basis of risk assessment decisions and more appropriately address risk management goals, such as the degree of protection and resource constraints (NRC 1994).

The paradigm of risk management also assumes that either the agents to which residential exposure occurs are valuable to people, and used indoors for that purpose, or are difficult, perhaps impossible—certainly costly—to avoid. That is, risk management is not needed if a substance is not wanted or needed in the residential environment and can be readily eliminated. In that case, no amount of exposure need be tolerated. Estimate of exposure may not be necessary. However, if removing the agent from the residential environment causes harm or inconvenience, or imposes large financial or other costs, then it becomes necessary to know about the relation between the duration and intensity in use and that causing harm. Lacking knowledge of the relation between exposure and harm, one cannot make sensible decisions about how to balance the costs and benefits of a particular use or exposure. Thus analysis of the exposure to potential hazards is necessary to good risk-management decision making.

As noted above, actual management of these risks is often done by those who live in the residential space; evaluation of the degree of exposure caused by different kinds of potential hazards is often done in order to inform householders of the need to take appropriate precautions. Thus analysts need to anticipate that the recipients of the information they generate will not necessarily be attuned to the nuances of the analyst's art, and think about how communication of the results will most usefully be accomplished. Sometimes the risks are managed by more global regulatory actions (*e.g.*, indoor-use pesticide approval). In such cases, it may be that the consumer of the information generated will be technically sophisticated enough to understand a nuanced report. In any case, the exposure characterization must be appropriate to the needs of the users (USEPA 1996).

Effective risk communication is vital for successful implementation of risk management decisions, involving clear identification of stakeholder concerns to risk decision-makers in all sectors of society, and transfer of important relevant information and decision-making options.

REFERENCES

NRC (National Research Council). 1983. Risk assessment in the federal government: Managing the process. National Academy Press. Washington DC. 196 pages.

NRC (National Research Council). 1994. Science and judgement in risk assessment. National Academy Press. Washington DC. 652 pages.

USEPA (U.S. Environmental Protection Agency). 1996. Exposure factors handbook (Draft). EPA Publication No. Volume I 600/P-95/002Ba, Volume II 600/P-95/002Bb, and Volume III 600/P-95/002Bc.

2

RESIDENTIAL EXPOSURE ASSESSMENT

Review of Scientific and Regulatory History

Every day people come in contact with time-varying amounts of chemical, biological, and physical agents in air, water, food, soil and on various surfaces. Exposures to these agents occur principally because humans engage in normal activities in various microenvironments (defined areas in the home and its surroundings) that bring them into relatively close proximity. These activities and concurrent sources of chemicals occur in outdoor air (*i.e.*, ambient levels of air pollutants such as nitrogen oxides, carbon monoxide and particulates), in the work setting (*e.g.*, exposure to industrial chemicals in manufacturing facilities), as pollutant exposures in vehicles while in transit or refueling (*e.g.*, passenger-compartment benzene levels), and as exposures in the residence (*e.g.*, influenza aerosols). In the case of the residential environment (in and around the home), there are numerous potential sources of exposures including:

- Consumer products such as cleaners, waxes, paints, pesticides, adhesives, paper products/printing ink, and clothing and furnishings that can release volatile organic compounds
- Building sources, such as combustion products from household appliances and vehicle motors running in attached garages, building materials, and ventilation and humidification systems
- Personal sources such as tobacco smoke and biological agents of human, animal, and plant origin
- Outdoor sources of chemicals, such as garden and lawn care products, combustion pollutants, contaminated soil particles that can infiltrate or be tracked into the home, drinking water (which can release volatile organics during showering or other use in the home), and contaminated subsurface water (*e.g.*, infiltration of VOCs into basement areas), and
- Dietary consumption of chemical residues in food.

Various environmental and public health issues over the past two decades have recognized the importance of residential environments to personal exposures. These issues included (Gammage and Berven 1996, USEPA 1997):

- The energy crisis of the 1970's and the resulting demand for changes in residential construction, including reduction in air infiltration (*e.g.*, "air-tight" homes to reduce heat loss)
- General concerns regarding ambient and indoor air quality, and

- The importance of measuring personal exposures to a variety of agents associated with public health concerns, including:
  nitrogen dioxide
  carbon monoxide
  environmental tobacco smoke
  volatile organic compounds
  ozone
  radon, and indoor allergens
- Concerns regarding potential exposures to infants and children in and around the home.

Over 75 years ago, Carrier devised mechanical means of controlling and conditioning air for the seasonal ventilation of buildings. For over 50 years, guidance for building ventilation codes was predicated on the acceptance of air quality by human perception. Yagloo's chamber studies during the1930s and 1940s found that odor quality of occupied spaces would become increasingly unacceptable (below 80 percent acceptance) as the ventilation rates dropped below 10 cfm/person. Homes and apartments, not identified as unique building structures *per se*, were considered "leaky" enough such that minimal consideration was given to adequate ventilation.

Several profound changes have occurred over the last half of the 20th century that have important consequences for indoor environments, including residences:

- Proliferation of synthetic compounds used in consumer products and all sorts of building materials
- Changes in construction techniques and design that has increased dependency on air conditioning (HVAC systems), and
- Escalation of energy cost as a fraction of operational expenses leading to increased recirculations of previously conditioned air.

Exploratory studies during the 1960s and 1970s looked at the obvious products of combustion (particles, NO_2, SO_2, CO). These early studies have recounted situations in homes, ice skating rinks, garages, and other locations where indoor levels of air contaminants exceed ambient levels. Transferring experience from occupational settings, public and government organizations became concerned about asbestos in schools and offices, and about radon in homes. Meanwhile, the National Institute of Occupational Safety and Health was receiving an increased number of requests to conduct health hazard evaluations in hospitals, offices, and schools. By the early 1980s, when the National Research Council (NRC) was completing a comprehensive evaluation of indoor air pollution, there was growing concern for the implications of indoor exposures.

Today, exposure analysis is considered an essential activity in support of epidemiology, hazardous waste site evaluation, consumer product registration, and other research and development and regulatory activities. A variety of federal and state regulatory programs and legislation require quantitative exposure assessment (see this chapter, *Overview of Federal and State Regulations Relating to Products Used In and Around the Home*). These requirements have continued to foster the development of exposure monitoring methods, measurements, and predictive mathematical models. In the last two decades, new understanding has developed regarding the relationship between exposure, absorption through biological membranes, and the kinetics of *in vivo* transport, fate, and elimination. Further, exposure (*i.e.*, external concentrations), biological monitoring, and the development of validated pharmacokinetic models (models predicting the absorption, distribution, metabolism, and excretion of chemicals in/from the body) has, in some cases, supported estimation of target-tissue dose concentrations. In combination with knowledge of a given agent's toxicological mechanism-of-action, advanced risk analysis models have been developed. New information about exposure, pharmacokinetics, toxic mechanisms, and exposure and risk factors related to contaminants in food, water, soil, and air will continue to advance our understanding of exposures in the residential environment. The U.S. Environmental Protection Agency (USEPA), a major benefactor of exposure information, supports numerous laboratory, field, and survey studies to increase the general knowledge about time-activity behavior, food consumption patterns, consumer product use, and potential exposures to such agents as volatile organic chemicals (VOCs), particles, and pesticides. The U.S. Food and Drug Administration (USFDA) routinely surveys pesticide and metal residues in the American food supply. In the U.S. and elsewhere, research by the private sector, nongovernmental organizations, academic researchers, and many governmental organizations has greatly expanded information relevant to exposure assessment. There are now several professional organizations, including the Society for Risk Analysis (SRA) and International Society for Exposure Assessment (ISEA), that publish articles and organize symposia on subjects related to residential exposure assessment. This specialized field has matured to the extent where mathematical models are constructed to predict exposures to a broad class of chemical compounds or physical and biological agents. Statistical sophistication has been realized in commercially available software that allows prediction of exposure pattern distributions (see **Appendix B**).

As noted above, in response to the need to measure and evaluate multiple chemical exposures from multiple media (*e.g.*, soil, air, food, water, contaminated surfaces) in and around homes, research and development efforts and scientific studies have been undertaken over the past two decades to develop and validate residential monitoring and assessment methods. In a somewhat parallel manner, various federal and state laws have been enacted regarding the regulation of household substances and products.

WHAT WE HAVE LEARNED FROM PREVIOUS SCIENTIFIC STUDIES

Active scientific research programs have elucidated important information about the relationship between Indoor and Outdoor Air Exposures, and assessment of personal exposure in the residential setting.

The Relationship between Indoor and Outdoor Air Exposures

Numerous studies relating to the measurement and characterization of exposures to agents found in the "indoor environment" have been published in the U.S. and worldwide over the past several decades. These studies have been conducted for a wide variety of reasons. However, they are all related to public health concerns and, in the U.S., many of these concerns were initially focused on the "healthfulness" of indoor air quality. Atmospheric pollution has been defined as "the presence in the ambient air space of one or more air contaminants or combinations thereof in such quantities and of such duration as to cause a nuisance, be injurious, or, on the basis of current information, be potentially injurious to human or animal life, to vegetation, or to property, or unreasonably interfere with the comfortable enjoyment of life and property or the conduct of business (Samet and Spengler 1991). This definition, developed by the Commonwealth of Massachusetts' Department of Public Health in 1961 is appropriate for both indoor and outdoor air pollution.

Indoor air quality is generally related to ambient air quality. Thus, human exposure to air pollutants is determined by concentrations outdoors and indoors. For example, in 1970 the global environmental monitoring system (GEMS), a project sponsored by the World Health Organization (WHO) and the United Nations Environment Program (UNEP), was created to monitor air quality in countries all over the world. Through this program and other research efforts, it was observed that many urbanized areas in developing countries have ambient SO_2 and particulate pollution levels that are ten to one-hundred times the concentrations currently experienced in the U.S. and western Europe. In this and other situations, human personal exposures will be more proportionally influenced by ambient conditions.

In the U.S., it has become increasingly apparent since the early 1970's that indoor exposures in environments such as residences, office buildings, and public modes of transportation represent a significant fraction of "total exposure" on a time-weighted basis, given that most people spend 90 percent or more of their time indoors (NRC 1981, Spengler and Sexton 1983, Sexton and Ryan 1988, Samet and Spengler 1991, Sexton 1993, Tsang and Klepeis 1996). Further, it had been previously recognized that indoor exposures can, under certain conditions, be harmful to human health (NRC 1981a,b, 1984, 1985, 1986, Spengler and Sexton 1983, Samet and Spengler 1991, Gold 1992, Sexton 1993). Indoor air contaminants that have been studied historically have included organic vapors, combustion products, aeroallergens, microorganisms, radon, particles, and fibers (NRC 1981a,b, 1984, 1986, 1991, 1994). In modern indoor environments, a complex array of potential sources of indoor air exposure exists, including synthetic building materials, consumer products, and furnishings. Airborne emissions also occur because of the people, pets, and plants inhabiting these spaces (Sexton and Dyer 1996). Further, as noted previously, efforts to lower energy costs by reducing ventilation rates have increased the likelihood that pollutants generated indoors can accumulate (Spengler and Sexton 1983, Gold 1992).

Sexton and Dyer (1996) identified five major factors regarding concerns related to the evaluation of indoor air exposures and potential health risks: people's time-activity patterns and the contribution of indoor air pollutants to total exposure, presence of sources indoors, measured indoor concentrations, health-related complaints associated with indoor environments, and formal assessments of indoor air-related health risks. Human time-activity patterns, for example, have a dramatic influence regarding potential indoor air exposures. Some individuals, such as infants, the infirm and the elderly, are inside virtually all the time. In general, because people spend so much time indoors, indoor air concentrations (and residential exposures in general), even if they are uniformly lower than outdoor levels, make a significant contribution to time-weighted, average exposures to most pollutants. The evidence is compelling that realistic residential exposure estimates depend on a comprehensive understanding of pollutant concentrations in and around (outside) the home, as well as quantification of the time that people spend in different microenvironments (NRC 1981a, Spengler and Sexton 1983, Samet and Spengler 1991, Sexton 1993). Numerous monitoring studies inside residences and buildings have consistently found that the concentrations of many air pollutants tend to be higher indoors than outdoors. Indoor air has been shown to be a complex mixture of chemical, biological, and physical agents, only a small fraction of which have been characterized adequately (NRC 1981a, 1986, 1991, Samet and Spengler 1991, Sexton and Dyer 1996).

Assessing Personal Exposures In and Around the Residence

In recent assessments of the human health impact of airborne pollutants, there has been increasing focus on the contribution of various microenvironments (*e.g.*, indoors, outdoors, in-transit) and sources (*e.g.*, consumer products, combustion appliances, outdoor sources) to total human exposure to a given chemical. During the past decade and a half, a number of studies, most notably the Total Exposure Assessment Methodology (TEAM) studies sponsored by the USEPA, have demonstrated that, for a variety of contaminants, residential indoor air is often a more significant source of exposure than outdoor air (Thomas *et al.* 1993; Wallace 1993, Pellizzari *et al.* 1987; Curry 1994). Some of the studies conducted in the past have found elevated indoor concentrations of certain pollutants, raising questions about the types, sources, levels, and human health implications of indoor exposures (Spengler *et al.* 1983; Melia *et al.* 1978; Dockery and Spengler 1981). Assessment of potential consumer exposures has also been recognized by industry as a key part of the overall risk evaluation process for consumer products (Hakkinen *et al.* 1991). For example, several studies of potential indoor air exposures from use of consumer products have been conducted and published by industry and trade associations to support and confirm the safety of these particular products (Wooley 1990, Gibson *et al.* 1991).

Since 1980, the USEPA's Office of Research and Development has conducted a series of studies, referred to as TEAM studies, on human exposure to different classes of pollutants. In the TEAM studies, when total personal exposures to VOCs in individual breathing zones, measured as VOC concentrations in exhaled breath, personal exposures most often exceeded outdoor air concentrations. Median personal concentrations of VOCs were on the order of 2 to 5 times outdoor levels; maximum personal concentrations were roughly 5 to 70 times the highest outdoor levels (Wallace 1993). This observed variability in exposures indicates (1) the role of various human activities in bringing individuals into contact with chemicals indoors, and (2) the importance of specific sources of exposures that may not be present in residential settings for all individuals. Personal exposures to particles in the daytime are 50 percent greater than either indoor or outdoor concentrations. The recent Valdez Air Health Study in Valdez, Alaska (Goldstein *et al.* 1992) generally supports the findings of the TEAM studies in terms of the importance of personal sources of exposure relative to outdoor sources, where mean personal concentrations of benzene were roughly 3 to 4 times higher than outdoor concentrations.

FEDERAL AND STATE REGULATIONS RELATING TO PRODUCTS USED IN AND AROUND THE HOME

As noted above, indoor air pollution research represented the initial focus of residential exposure assessment and in a somewhat parallel manner, federal regulations evolved in response. By the end of the 1940s, air pollution in the U.S. obtained national recognition. President Truman directed the Secretary of the Interior to head the interdepartmental committee that organized the first U.S. Technical Conference on Air Pollution, held in May 1950. Five years later Congress enacted the first national legislation, the Air Pollution Control Act and in 1963 the Clean Air Act was passed, establishing the federal governments right over the states to legislate air pollution on the constitutional basis of interstate commerce. The Clean Air Act has been amended multiple times during the 1970s. The Clean Air Act Amendments of 1970 and the establishment of the USEPA significantly changed the course of pollution control. Once such control measure was the establishment of National Ambient Air Quality Standards (NAAQS).

Residential exposure assessments are conducted for many purposes, including the evaluation of indoor air quality, compliance with U.S. statues such as the Federal Insecticide, Fungicide and Rodenticide Act (FIFRA), the Federal Food, Drug and Cosmetic Act (FFDCA), the Resource Conservation and Recovery Act (RCRA), California's Proposition 65, and evaluating consumer product safety as part of industry research and development efforts and establishing appropriate product labeling requirements required by the Consumer Products Safety Commission (CPSC) and other state and federal agencies. The importance of residential exposure assessment has historically been emphasized by the need to define and maintain adequate indoor air quality and consumer product safety. Further, human activity pattern surveys have revealed that in California, for example, the average person spends approximately 60 percent of his or her time indoors at home. In a broader perspective, residential exposure assessment has been an important part of understanding public health issues associated with drinking water, food safety (*e.g.*, microbial hazards, pesticide residues in agricultural commodities and processed foods), allergens, and waste management (*e.g.*, septic and sewer systems maintenance, garbage disposal). Residential exposure assessment was emphasized by amendments to FIFRA and FFDCA (*i.e.*, the Food Quality Protection Act (FQPA) of 1996), which requires the USEPA to make a "determination of safety" with respect to potential aggregate exposures to adults and children resulting from the use of pesticides in the agricultural and residential settings. This determination must include consideration of anticipated dietary and all other exposures (*e.g.*, drinking water, consumer products) for which there is reliable information. Further, legislative activities and scientific interest in Europe and Japan have also resulted in an increasing demand for the development of scientific methods for measuring and modeling potential exposures in and around the home.

REFERENCES

Curry, K.K. 1994. Personal exposures to toluene during use of nail lacquers in residences: Description of the results of a preliminary study. J. Exposure Anal. Environ. Epidemiol. 4(4):443–456.

Dockery, D.W. and J.D. Spengler. 1981. Indoor-outdoor relationships of respirable sulfates and particles. Atmos. Environ. 15:335–343. As cited in Wilson *et al.* 1993.

Gammage, R.B. and B.A. Berven. 1996. Indoor air and human health. Lewis Publishers. Boca Raton.

Gibson, W.S., F.R. Keller, D.J. Foltz, and G.J. Harvey. 1991. Diethylene glycol monobutyl ether concentrations in room air from application of cleaner formulations to hard surfaces. J. Expos. Anal. Environ. Epidemiol. 1(3):369–383.

Gold, D.R. 1992. Indoor air pollution. In: E.D. Epler (ed.). Clinics in chest medicine. Occup. Lung Dis. 13:215–229.

Goldstein, B.D., R.G. Tardiff, S.R. Baker, G.F. Hoffnagle, D.R. Murray, P.A. Catizone, R.A. Kester, and D.G. Caniparoli. 1992. Valdez air health study. Anchorage, Alaska: Alyeska Pipeline Service Co. As cited in Wallace (1993).

Hakkinen, P.J., C.K. Kelling, and J.C. Callender. 1991. Exposure assessment of consumer products: Human body weights and total body surface areas to use, and sources of data for specific products. Vet. Hum. Toxicol. 33(1):61–65.

Melia, R.J.W., C. duV. Florey, S.C. Darby, E.D. Palmes, and B.D. Goldstein. 1978. Differences in NO_2 levels in kitchens with gas or electric cookers. Atmos. Environ. 12:1379–1381.

NRC (National Research Council). 1981a. Indoor pollutants. National Academy Press. Washington, D.C.

NRC (National Research Council). 1981b. Formaldehyde and other aldehydes. National Academy Press. Washington, D.C.

NRC (National Research Council). 1984. Asbestiform fibers: Nonoccupational health risks. National Academy Press. Washington, D.C.

NRC (National Research Council). 1985. Epidemiology and air pollution. National Academy Press. Washington, D.C.

NRC (National Research Council). 1986. Environmental tobacco smoke: Measuring and assessing health effects. National Academy Press. Washington, D.C.

NRC (National Research Council). 1991. Human exposure for airborne pollutants: Advances and opportunities. National Academy Press. Washington, D.C.

NRC (National Research Council). 1994. Science and Judgment in Risk Assessment. National Academy Press. Washington, D.C. 652 pages.

Pellizzari, E.D., T.D. Hartwell, R.L. Perritt, C.M. Sparacino, L.S. Sheldon, R.W. Whitmore, and L.A. Wallace. 1987. Comparison of indoor and outdoor residential levels of volatile organic chemicals in five U.S. geographic areas. Environ. Internat. 12:619–623.

Samet, J.M. and J.D. Spengler. 1991. Indoor air pollution: A health perspective. Johns Hopkins University Press. Baltimore, Maryland.

Sexton, K. 1993. Air pollution health risks: Do class and race matter? Toxicol. Indust. Health. 9:843–878.

Sexton, K. and R. Dyer. 1996. Effects of indoor air quality on human health: Setting strategic research directions and priorities. In: Gammage, R.B. and B.A. Berven (eds.). Indoor air and human health. Lewis Publishers. Boca Raton, Florida.

Sexton, K. and B. Ryan. 1988. Human exposure to air pollution: Methods, measurements, and models. In: A. Watson, R.R. Bates, and D. Kennedy (eds.). Air pollution, the automobile, and public health. National Academy Press. Washington, D.C.

Spengler, J.D. and K. Sexton. 1983. Indoor air pollution: A public health perspective. Science 221:9–17.

Spengler, J.D., C.P. Duffy, R. Letz, T.W. Tibbets, and B.G. Ferris Jr. 1983. Nitrogen dioxide inside and outside 137 homes and implications for ambient air quality standards and health effects research. Environ. Sci. Technol. 17(3):164–168.

Thomas, K.W., E.D. Pellizzari, C.A. Clayton, R.L. Perritt, R.N. Dietz, R.W. Goodrich, W.C. Nelson, and L.A. Wallace. 1993. Temporal variability of benzene exposures for residents in several New Jersey homes with attached garages or tobacco smoke. J. Expos. Anal. Environ. Epidem. 3(1):49–73.

Tsang, A.M. and N.E. Klepeis. 1996. Results tables from a detailed analysis of the national human activity pattern survey (NHAPS) response. Draft report prepared for the U.S. Environmental Protection Agency by Lockheed Martin. Contract No. 68-W6-001, Delivery Order No. 13.

USEPA (U.S. Environmental Protection Agency). 1996. Exposure factors handbook (Draft). EPA Publication No. Volume I 600/P-95/002Ba, Volume II 600/P-95/002Bb, and Volume III 600/P-95/002Bc.

Wallace, L. 1993. A decade of studies of human exposure: What have we learned? Risk Analysis 13:135–139.

Wooley, J., W.W. Nazaroff, and A.T. Hodgson. 1990. Release of ethanol to the atmosphere during use of consumer cleaning products. JAWMA 40:1114–1120.

3

GENERAL FRAMEWORK FOR ASSESSING POTENTIAL HUMAN EXPOSURES TO CHEMICAL, BIOLOGICAL, AND PHYSICAL AGENTS IN THE RESIDENTIAL ENVIRONMENT

Exposure assessments are used to relate contaminant sources into quantitative estimates of the amount of chemical that comes in contact with the visible exterior of an individual of potentially exposed population. The potentially exposed population is the one that does or plausibly could contact the source of contamination. This contact is the basis for estimating a potential dose used in the characterization of potential health risks. Defining exposure pathways is an important component of the exposure assessment. An exposure pathway is the course a chemical or physical agent takes from a source to an exposed individual. An exposure pathway describes a unique mechanism by which an individual or population is exposed to a chemicals or physical agents at or originating from a source. Each exposure pathway includes a source or release from a source, a transport/exposure medium (such as air) or media (in cases of intermedia transport, such as water to air) also is included.

Exposure assessments often rely implicitly on the assumption that exposure can be linked by simple parameters to ambient concentrations in air, water, soil, and raw food products. However, for human populations, total exposure assessments that include time and activity patterns and microenvironmental data reveal than an exposure assessment is most valuable when it provides a comprehensive view of exposure pathways and identifies major sources of uncertainty. Thus we see the need to address many types of "multiples" in the quantification of exposure and dose, such as multiple media (air, water, soil); multiple exposure pathways or multiple routes (ingestion, inhalation, dermal), and multiple target tissues for dose and effect.

An assessment of intake requires that we determine how much crosses the exchange boundaries of the organism. Thus, exposure is the condition of a chemical contacting the outer boundary of an organism and exposure over a period of time can be represented by a time-dependent profile of the exposure concentration. Intake is the process by which a chemical is physically moved through an opening in the outer boundary of an exposed individual. Applied dose is the amount of a contaminant that comes in contact with the living tissues of an individual by entering the lungs, the gastrointestinal tract, or the outer skin layer. In some experimental designs, the applied dose is referred to as the adminis-

Residential Exposure Assessment, edited by Baker *et al.*
Kluwer Academic/Plenum Publishers, New York, 2000.

tered dose. Potential dose is an approximation to applied doe that is simply the amount of chemical in the food or water ingested, air inhaled, or material applied to skin surface. Absorbed dose is the amount of contaminant penetrating the exchange boundaries of an organism after contact. Absorbed dose is calculated from the intake and absorption efficiency. For human populations, it is typically expressed as the mass of contaminant absorbed into the body per unit mass per unit time such as mg/kg-day.

Accurate exposure-related data are crucial to valid risk analysis. For example, exposure data must match up temporally with the health end points of concern. Key issues in the evaluation of exposure include:

- Characterization of emissions (*e.g.*, offgassing of volatile organic chemicals from consumer products)
- Consideration of transport and fate of a given agent across time (*e.g.*, chemical and physical properties of the agent, source and sinks for the agent, and rates of generation and degradation of the agent)
- The toxicological endpoint of interest (*e.g.*, acute irritation versus developmental toxicity vs. cancer)
- The subpopulations at risk (*e.g.*, adult consumer product applicators, infants and children post-application, pregnant females, integrated, long-term exposures to residents)
- Their activity patterns (*e.g.*, where people spend their time, how people use products)
- Their routes of exposure (*e.g.*, incidental ingestion associated with hand-to-mouth behavior in infants and children, inhalation exposure during application of a hand-held aerosol product)
- The durations of their exposure (*e.g.*, several minutes during application and several hours post-application versus long-term—entire period spent in a given residence), and
- The nature and degree of multiroute, multipathway or aggregate exposure to one or more agents.

Rarely are all of these issues resolved by the exposure data available for a given assessment. Therefore, efforts to collect the data should focus on the minimum needed to conduct a scientifically credible analysis and meet the goals of the assessment in its risk management context.

CONDUCTING RESIDENTIAL EXPOSURE ASSESSMENTS

There are a number of sources of residential exposures including:

- Consumer products such as cleaners, waxes, paints, pesticides, adhesives, paper products/printing ink, clothing and furnishings that can offgas VOCs
- Building sources, such as combustion products from appliances and attached garages, building materials that can release VOCs like formaldehyde and HVAC systems
- Personal sources such as tobacco smoke, and biological contaminants such as allergens of human, animal, and plant origin, and
- Outdoor sources of chemicals leading to infiltration of the residential environment. These sources include ambient combustion pollutants, contaminated

soil particles that can infiltrate or be tracked into the home, drinking water (which can release volatile organics during showering or other use in the home), and contaminated subsurface water leading to infiltration of VOCs into basement areas.

The residential environment should be thought of in dynamic terms. VOCs that enter the residential environment can be absorbed to surfaces, or "sinks," and then later be released as airborne levels are depleted by various mechanisms, including air exchange with other rooms of the house and with outdoor air, and with chemical/physical transformations in residential air (see Figure 3-1). As illustrated in Figure 3-2, there is evidence that particulate contaminants, whether generated inside the residence, tracked in, or infiltrated from the outdoor environment are resuspended and recycled within the house by walking on floors and rugs, sweeping and dusting, and vacuuming. In short, the residence is the exposure unit.

QUANTITATIVE EXPOSURE ASSESSMENT

Quantitative estimation of exposure, which is often the central feature of exposure evaluation activities, can be accomplished in three general ways: point-of-contact measurements, reconstruction of internal dose to calculate associated exposure, or creation of an exposure scenario (USEPA 1992). The point-of-contact approach requires measurements of actual exposure as it occurs at the point of contact with the human body,

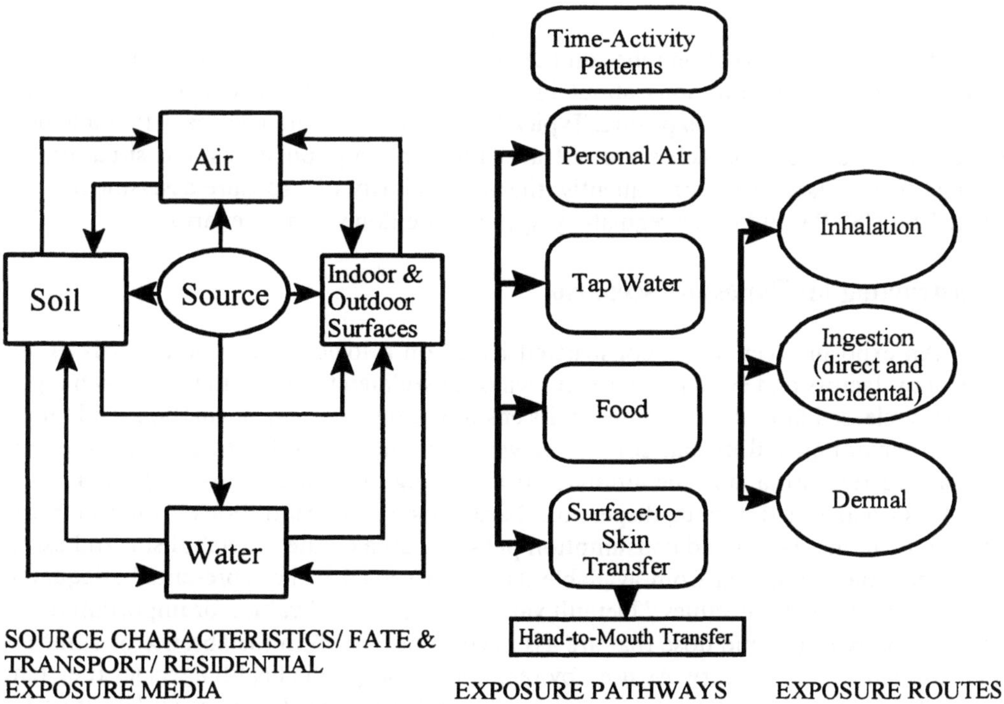

Figure 3-1. A multimedia, multiple pathway residential exposure model.

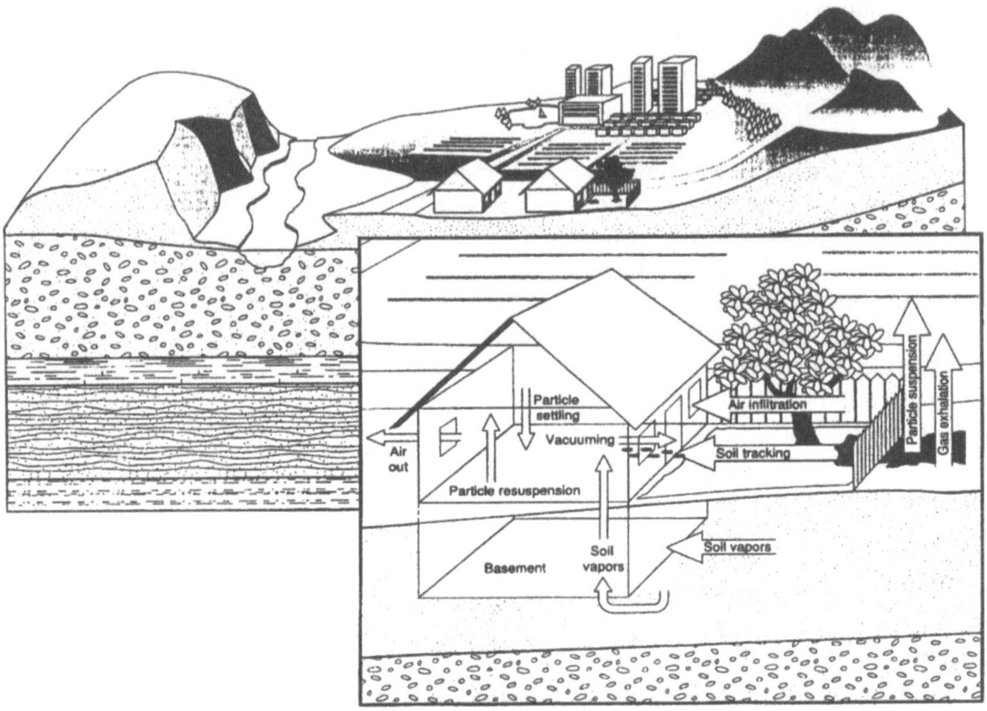

Figure 3-2. Potential residential exposure pathways associated with contaminants in soil and dust (adapted from McKone 1992).

while exposure reconstruction relies on human tissue measurements, knowledge of pharmacokinetics, and assumptions about intake and uptake parameters to "reconstruct" internal dose and calculate exposure. Typically, however, the data necessary to implement either of these two approaches are not available for the populations and situations of interest to risk assessors. Consequently, the vast majority of exposure assessments conducted by USEPA and other regulatory agencies are done using scenarios.

Constructing an Exposure Scenario

An exposure scenario is constructed based on a logical, stepwise analysis of the important factors and events that are known or postulated to affect how and when a pollutant is released into the environment, its ensuing transport, transformation, and fate in various environmental media, contact between people and pollutant concentrations in relevant carrier media, and the ultimate dose entering the human body. Important parameters, such as emission rates, dispersion characteristics, concentrations in air, water, and food, human inhalation and consumption patterns, uptake rates, metabolism, and excretion, are either estimated from available data or assumed to be represented adequately by appropriate "default values." Default values are postulated values for important determinants of exposure and dose that are assumed to be correct in the absence of sufficient information to the contrary. As used by regulatory and public health agencies they tend to be "conservative" or "prudent" premises intended to guard against underestimating risks.

Because of the scarcity of hard facts, and the associated need to make assumptions, some of which are necessarily heroic, scenarios do not usually include a complete description of the population distribution for exposure or dose. Instead, scenarios typically emphasize estimation of several points on the distribution, including: average exposure (either arithmetic or geometric), "high-end" exposure (90th percentile and above), and most exposed individual (highest point on the distribution).

Although scenarios have obvious limitations, such as lack of data, unvalidated default assumptions, and reliance on point estimates, which contribute significantly to uncertainties in risk assessment, they remain the only viable method to estimate exposure or dose in the absence of direct measurements. The primary advantage of the scenario approach is that it enables assessors to estimate exposure with very limited data; conversely, the uncertainty introduced by the need to make assumptions and inferences in the face of limited information is also its major disadvantage. The scenario approach is most useful when the assessor has some insight into the completeness, soundness, validity, and appropriateness of the underlying assumptions and inferences, and understands their overall contribution to the uncertainty of the derived estimates for exposure and dose.

Sources

An initial step in any exposure assessment is the characterization of it sources in different components of the residential environment. Knowledge of the source-term characteristics of a toxic agent is important in order to develop measurement or modeling efforts to determine the substance's concentrations in contact media. Pertinent information includes the physical and chemical properties of the substance(s) of concern and attributes of the source (*e.g.*, emission rate of gas or particles to indoor and outdoor air). Sources can be categorized in terms of space (*e.g.*, area source versus point source), time (*e.g.*, transient versus chronic release), and mode of formation (endogenous or exogenous substance). Source-term analyses are typically designed to determine key physical and chemical factors that produce the observed level of contact from some consumer product or from a given process and to determine release to contact media.

Once the source or sources of a toxic substance have been identified, a followon task is to quantify the emission rate of the substance(s) to various components of the indoor environment. One technique for determining emissions of gases and particles from indoor sources is to solve an indoor-transport model for the emission term (*i.e.*, release rate) that provides the best agreement between measured and predicted concentrations of the contaminant.

Transport and Transformation

For the purposes of exposure assessment, a toxic substance that has been identified in one component of the environment must be characterized in terms of its transport and transformation within that compartment and its transport to other components of the environment. We are particularly interested in those components of the environment with which humans are most likely to have contact. In order to carry out this characterization, we view the environment as a series of interacting compartments. In this framework, one must determine whether a substance will (a) remain or accumulate within the compartment of its origin, (b) be transported by dispersion or advection within the compartment of its origin, (c) be physically, chemically, or biologically transformed within the compartment of its origin (*i.e.*, by hydrolysis, oxidation, *etc.*), or (d) be transported to another compartment by cross-media transfer (*i.e.*, volatilization, precipitation, *etc.*).

Several approaches have been used to estimate expected indoor air pollutant concentrations (for a review). These include deterministic models based on a pollutant mass balance around a particular indoor air volume, a variety of empirical approaches based on a statistical evaluation of test data and (usually) a least-squares regression analysis, or a combination of both approaches—empirically fitting the parameters of a mass-balance model with values statistically derived from experimental measurements. All three approaches have advantages and weaknesses. The mass-balance models provide more generality in their application, but the results lack accuracy and precision. The empirical models, when applied within the range of measured conditions for which they were fitted, provide more accurate information. Mass-balance models include single and multiple compartment models. Often the component of the indoor-air mass balance models that is most difficult to represent is the role of indoor surfaces as sources or sinks for contaminants.

Using Human Exposure Factors to Evaluate Scenarios

Evaluation of an exposure scenario typically involves developing a mathematical equation that expresses dose as a function of pollutant concentration and other important parameters. The form of the equation is affected by qualitative assumptions used in constructing the scenario, while quantitative assumptions affect the values of the parameters in the equation. Individual parameters in the dose equation, with the exception of pollutant concentration, are referred to as exposure factors (USEPA 1992, 1996). In the context of this discussion, the term human exposure factors refers specifically to a) human characteristics, such as body weight, surface area, life expectancy, inhalation rates for air, and consumption rates for food, drinking water, and soil, and b) human behaviors, such as activity patterns, occupational and residential mobility, and consumer product use, which are used by exposure assessors to calculate potential dose.

The process of an environmental agent entering the body can be described as a series of steps:

- Exposure: Contact between the agent and a person for a defined period of time
- Intake: The agent crosses an outer boundary of the body by bulk transport from the mouth or nose (intake is relevant to inhalation and ingestion but not dermal contact), and
- Uptake: The agent crosses an absorption barrier, such as the skin, gastrointestinal tract, or respiratory tract.

Once the agent enters the body by either intake or uptake, it is described in terms of "dose." Several different types of dose are distinguished for purposes of exposure assessment (USEPA 1992). Potential (Administered) dose is the amount of the agent that is actually ingested, inhaled, or applied to the skin. Applied dose is the amount of the agent directly in contact with the body's absorption barriers. Internal (Absorbed) dose is the amount of the agent crossing the absorption barriers and, therefore, available to undergo metabolism, transport, storage, or elimination. Delivered dose (body burden) is that portion of the internal dose that reaches a tissue of interest. Biologically effective (target) dose is that portion of the delivered dose that reaches the site or sites of toxic action.

Starting with a general integral form of the equation for exposure, several dose equations can be derived depending on boundary assumptions (USEPA 1992). One of the most important is the Average Daily Potential Dose (ADDpot) equation, which calculates an average dose over the time period of interest and is widely used by USEPA for many noncancer health effects (USEPA 1996). The ADDpot can be calculated by dividing the total potential dose by the product of body weight and an averaging time, where total potential dose is equal to the product of four factors: concentration of the agent in an environmental medium that is in contact with the outer boundary of the body; intake rate, which is the rate of inhalation for air pollutants or rate of ingestion for water, food, and soil contaminants; exposure duration, which is the length of time contact occurs between the agent and the person; and exposure frequency, which is how often contact occurs (USEPA 1992, 1996).

$$ADDpot = \frac{C \times IR \times ED \times EF}{BW \times AT} \qquad (1)$$

where

ADDpot = average daily potential dose (*e.g.*, mg/kg-day)
C = concentration of the agent in a carrier medium (*e.g.*, mg/L, mg/g, mg/m^3, mg/cm^2)
IR = intake rate:
- inhalation rate, amount of air breathed per unit time, *e.g.*, m^3/day)
- ingestion rate, amount of water or food consumed per unit time (*e.g.*, L/day, g/day)
- dermal absorption (not applicable)

ED = exposure duration, time of contact at specified values for C and IR (*e.g.*, years)
EF = exposure frequency, if intermittent, number of times exposures occur (*e.g.*, days/year)
BW = body weight for exposed person or average for exposed population (*e.g.*, kg)
AT = averaging time, period over which dose is averaged (*e.g.*, days)

For cancer risk assessment, which often focuses on estimating lifetime risks, the USEPA typically uses the Lifetime Average Daily Potential Dose (LADDpot) instead of the ADDpot, and normal life expectancy (*e.g.*, 70 years) replaces averaging time (AT) in equation 1 above. The LADDpot is used even though exposure and dose do not necessarily occur over the entire lifetime (USEPA 1996).

The general ADDpot equation described above is applicable when intake rate, exposure duration, exposure frequency, body weight, and pollutant concentration remain constant over the time period of interest. If they change over time, then it is necessary to use either a summation or integration approach to calculate potential dose (USEPA 1992, 1996).

A major challenge confronting exposure assessors is deciding how best to construct a plausible scenario and evaluate it quantitatively to obtain a realistic estimate of potential dose. Decisions about which values to use for critical human exposure factors are central to resolving key exposure- and dose-related questions successfully. Depending on the complexity and comprehensiveness of a particular exposure assessment, literally hundreds of variables may need to be considered, as, for example, with multichemical, multipathway assessments. Although typically only a relatively few human exposure factors cause most of the variability and uncertainty in the final estimate, it is not always clear at the outset which are most important and which have minimal or negligible effects.

Quantified values for human exposure factors are best determined on a case-by-case basis, with site-specific and source-specific circumstances driving choices about appropriate values for intake rates, exposure duration and frequency, body weight, averaging time, and other related variables affecting calculation of ADDpot. Each exposure assessment is unique and the assessor must construct a scenario and tailor related human exposure factors to fit the conditions at hand. Consequently, it is difficult, and potentially misleading, to make generic statements about which scenarios and human exposure factors are most appropriate.

One type of exposure assessment that does lend itself to general discussion is a "screening-level" assessment. A screening-level assessment refers to a preliminary, evaluation that is usually cursory and relatively quick and inexpensive. Moreover, it is general in nature rather than either site- or situation-specific. Typically, screening-level assessments are performed using available, off-the-shelf values for human exposure factors to give the assessor a sense of (1) whether preliminary evaluation suggests the possibility of adverse health risks, thereby indicating the need for more indepth analysis, or (2) the approximate range of exposures that might be expected in the general population under normal, everyday conditions, which is helpful for putting site- and situation-specific values into perspective.

Uncertainty and Variability

There are five factors that determine the precision or reliability of an exposure assessment. They are:

- Specification of the problem (scenario development)
- Formulation of the conceptual model (the influence diagram)
- Formulation of the computational model
- Estimation of parameter values, and
- Calculation and documentation of results including uncertainties.

An uncertainty analysis involves the determination of the variation of imprecision in an output function based on the collective variance of model inputs.

One of the five issues in uncertainty analysis that must be confronted is how to distinguish between the relative contribution of variability (*i.e.*, heterogeneity) versus true certainty (measurement precision) to the characterization of predicted outcome. Variability refers to quantities that are distributed empirically—such factors as soil characteristics, weather patterns, and human characteristics—that come about through processes that we expect to be stochastic because they reflect actual variations in nature. These processes are inherently random or variable, and cannot be represented by a single value, so that we can determine only their moments (mean, variance, skewness, *etc.*) with precision. In contrast, true uncertainty or model specification error (*e.g.*, statistical estimation error) refers to an input that, in theory, has a single value, which cannot be known with precision due to measurement or estimation error.

ASSESSMENT OF INHALATION EXPOSURES IN THE RESIDENCE

Factors considered in assessing inhalation exposures to chemicals in the residence include:

- *Source characteristics.* Important factors in determining the impact of sources in the residence on inhalation exposures are the nature of the source (*e.g.*, consumer product or residential construction material such as floor or wall surface), how it is released (fine respirable aerosols, nonrespirable coarse aerosols, or vapor release), and the source strength (roughly proportional to the concentration of the chemical in the source or product).
- *Human exposure factors.* These include body weight, which varies between and within age and gender categories, and inhalation rates, which vary primarily by age, gender, and activity level.
- *Physical, chemical and biological properties.* In the case of chemical agents, these include factors such as molecular weight and vapor pressure that determine the rate of evaporation into air of a chemical in an applied material such as paint, or the release from aqueous solution (for instance, the role of the Henry's law constant in determining the release of volatile organic compounds from tap water used in the home). In the case of biological agents these include, for example, pathogenicity to humans, infectious dose levels, aerosol particle size distribution.
- *Residential building factors.* The basic characteristics of the rooms and building in which residential exposures occur, as well as the ventilation configuration (*i.e.*, number of windows and doors open, the rate of mechanical ventilation and air mixing, rate of infiltration of outside air) will determine the extent and rate of dilution of the agent of interest in a specific indoor air setting.
- *Exposure frequency and duration.* The exposure frequency (the number of days per year, years per lifetime) and duration of exposure (minutes or hours of exposure to an agent for a given day on which exposure occurs) are critical variables for estimating residential exposures to agents. These are a function of product use patterns, human activities that bring individuals in contact with areas that may contain an agent, and the nature of the population's mobility, all of which limit the total number of years an individual may be exposed to a site-specific contaminant like radon in the residence.

As discussed in Whitmyre *et al.* (1992a,b), a number of these factors are associated with a wide range of variability across an affected population, resulting in a wide band of uncertainties. Thus, the true distribution of exposures across the population would likely span several orders of magnitude.

A number of indoor air modeling tools are available for use in assessing inhalation exposures to a variety of contaminants from a variety of sources. Some, like the Screening-Level Consumer Inhalation Exposure Software (SCIES) developed by the Exposure Assessment Branch of the U.S. Environmental Protection Agency's (USEPA's) Office of Pollution Prevention and Toxics (USEPA 1994), are more oriented toward assessment of exposures to chemical agents when the specific emission term is not known. MAVRIQ is a model that can be used to estimate indoor inhalation exposures to organic chemicals due to volatilization from indoor uses of water (Wilkes and Small 1992).

A number of validated USEPA modeling tools exist to address indoor airborne levels of agents from many types of emission sources. The Multi-Chamber Concentration and Exposure Model (MCCEM), developed for the USEPA's Environmental Monitoring Systems Laboratory in Las Vegas (USEPA 1991a), is an example of an indoor air model that can be used when the emission term, such as an aerosol product released at a rate of 1.5 g/sec for 3 minutes), is known. MCCEM is a user-friendly computer program that estimates indoor concentrations for, and inhalation exposures to, agents released from products or materials used indoors. Concentrations can be modeled in as many as four zones in a building. The user provides values for emission rates, the zone where the source is located, the zone where exposure occurs, duration of exposure, air exchange rates, the nature of the building, and whether a short-term model (including average and maximum peak values) or long-term model is desired. The model contains room volume data and measured air flow rate data between different rooms for different building configurations and different geographic locations, or the user may build a hypothetical house or building, assigning the desired room (zone) volume and air exchange rates. Other examples of similar modeling tools include several USEPA models, as well as the CONTAM model developed and updated regularly by the National Institute of Standards and Technology (NIST 1994).

A new database/model management tool developed by the University of Nevada at Las Vegas for the USEPA is anticipated to revolutionize the modeling of indoor air exposures. This software tool is called the Total Human Exposure Risk Database and Advanced Simulation Environment (THERdbASE). This software integrates a number of indoor air models with distributional data on variables such as demographics, time-activity, food consumption, and physiological parameter data that can be subset according to the needs of the assessment (www.epa.gov/nerlpage/heasd/therdbase.htm. 1998, ISEA 1995). THERdbASE can also be used for estimating dermal and ingestion exposures and total human exposure via multiple agents and pathways (*i.e.*, multiple agents present in more than one media and coming into contact with humans via multiple exposure pathways and routes).

Exposure pathways other than inhalation that need to be addressed in characterizing and quantifying human residential exposures to agents include:

- Dermal exposure to dislogdeable residues on surfaces, such as pesticides on floors and carpeting and chemicals resulting from use of hard surface cleaners, and
- Incidental ingestion exposure to surface contaminants, such as that due to hand-to-mouth activity, particularly in infants and toddlers.

There are several examples of studies and reviews that have addressed and provide examples of noninhalation residential exposures (Calvin 1992, CTFA 1983, ECETOC 1994, Turnbull and Rodricks 1989, Vermiere *et al.* 1993).

ASSESSMENT OF DERMAL EXPOSURES IN THE RESIDENCE

There are numerous opportunities for dermal exposure to chemicals, allergens, and other household pollutants in the residential environment. These include, but are not limited to:

- Direct contact with cleaning and laundry products (*e.g.*, cleanser, laundry detergent) during use
- Indirect contact with cleaning product residues (*e.g.*, laundry detergent residues in washed clothing)
- Contact with dislodgeable residues of a chemical after use (*e.g.*, crawling infant contact with pesticide residues on rug, and
- Direct contact with materials that are intentionally applied to the skin (*e.g.*, soap, cosmetics).

It is important to distinguish between external dose and absorbed dose. External dose is the amount of material in contact with the outer surface of the skin and potentially available for absorption. Absorbed dose is the amount of the material that actually crosses the outer skin layer, known as the stratum corneum, during a given time period and is thus, considered to be systemically absorbed into the body.

There are valid methods available for measuring both external exposure (skin rinse techniques, dermal dosimetry patches) and internal or absorbed dose biomonitoring (USEPA 1996). Special considerations are necessary in most efforts to either measure or estimate external or absorbed dose. For example, exposure media, like soils, have characteristic chemical binding affinities depending on the chemical and soil composition. This may significantly impact bioavailability and absorbed dose (USEPA 1992).

There are basically two types of approaches to estimating a dermally absorbed dose in the case of chemical agents: (1) the film-thickness approach, and (2) dermal-permeability based approaches (USEPA 1992). The film-thickness approach assumes that a uniform layer of a material (*e.g.*, liquid consumer product) is present on a certain area of the skin, and that all of the material in that layer is available for absorption. Default film thickness data, in the absence of data on the actual product of interest, is available in USEPA (1987). Other variables that are unique to the film thickness approach are the density of the product (g/cm^3) and the percent dermal absorption anticipated during the per-event exposure period. Absorption can be assumed to be 100 percent for screening-level assessments, but large overestimation of dermal exposure is likely to occur.

In contrast, dermal permeability-based methods recognize the fact that dermal absorption is a time-dependent process, and that under controlled conditions, the dermal penetration can be expressed as a time-dependent parameter known as the dermal permeability coefficient (K_p). Measured and estimated dermal flux (μg/cm^2/hr) and/or permeability coefficients (cm/hr) have been published for various substances (USEPA 1992, Driver *et al.* 1993). Additional information regarding dermal exposure assessment and percutaneous absorption kinetics can be found in USEPA (1992), Kasting and Robinson (1993), and Wilschut *et al.* (1995).

Regardless of which general approach is taken, several factors must be taken into account to quantify human dermal exposure:

- *Human exposure factors.* Apart from body weight, which varies between and within age and gender categories, it is necessary to build an exposure scenario that specifies the amount of skin surface area exposed. One can use total surface area statistics and take a fraction representing the exposed area, or specify body parts that are exposed (*e.g.*, hands) and use body part surface area data (USEPA 1996, AIHC 1995). Because skin surface area is closely correlated with body weight, data on the ratio of surface area : body weight should ideally be used in calculating the dermal exposure (Phillips *et al.* 1993).

- *Frequency and duration of exposure*. Duration of exposure should represent the anticipated pollutant contact time with the skin prior to washing or removal.
- *Concentration of the agent on the skin*. Estimation or measurement of vapor-phase or aqueous-phase concentration of a given agent in contact with the skin. For example, aqueous-phase exposures are usually expressed as μg of agent per cm^3 of aqueous solution.
- *Surface area of skin exposed*. The amount of surface area exposed is proportional to the amount of a given substance that may be percutaneously absorbed.

ASSESSMENT OF INCIDENTAL INGESTION EXPOSURES IN THE RESIDENCE

Ingestion of agents can occur in the home beyond chemical residues and microbial agents consumed in food derived from nominally contaminated raw agricultural commodities. Primary examples of incidental chemical residues include ingestion of cleaning agent and pesticide residues on plates and silverware following product use, and ingestion of trace levels of organics (*e.g.*, haloforms) in drinking water entering the home. Another important pathway for incidental ingestion exposure is hand-to-mouth behavior in infants and toddlers in particular. Vacarro (1996) has shown this to be actually the predominant exposure pathway (for this age group) for exposure to pesticide residues applied to carpets either directly or incidentally (*e.g.*, through insecticide fogger use, such as a flea bomb), more so than inhalation or dermal contact through crawling on or touching contaminated surfaces. For food-related incidental contact, it will be necessary to consider the nature of the toxicological endpoint (*e.g.*, short-term versus long-term health effects) to determine which type of dietary consumption data are most appropriate (*e.g.*, an upper-bound on the amount eaten during one day in which the commodity is consumed, or long-term averages, which would include days on which the commodity is not consumed).

GENERAL CONSIDERATIONS FOR MULTIPATHWAY, MULTIROUTE RESIDENTIAL EXPOSURE ASSESSMENTS

During the past 15 years, a number of studies, most notably the USEPA Total Exposure Assessment Methodology (TEAM) studies, have demonstrated that residential air is often a more significant source of exposure to various chemicals, such as volatile organic compounds (VOCs), than outdoor air. Many of the compounds of interest in residential air are present in consumer products that are used in and around the residence. Recent studies have investigated the relationship between use period/post-use period activities and exposures to a variety of chemicals in consumer products. While the resulting residential exposures are likely to be low in some cases, nonetheless there is a need to characterize these exposures. For certain chemicals such as pesticides, post-application exposures in particular may require characterization of various exposure pathways/routes and subpopulations to fully understand the magnitude of exposure associated with consumer uses of these chemicals. In performing such assessments, it is necessary to consider the range of approaches that can be taken, including use of body-burden modeling for intermittent exposures, use of indoor air modeling tools, incorporation of time-activity data, consideration of the form of the airborne concentration dissipation curve in determining post-application exposures, and use and adjustment of emissions/concentration data for surrogate compounds to obtain an emission rate/airborne level for the compound of interest.

A number of studies have been made of exposures to outdoor-use chemicals, most notably lawn chemicals, since opportunities exist for the general public to become exposed to lawn chemicals. Likely exposure pathways include dermal exposure to liquids during mixing/loader of formulation, inhalation of aerosols and vapors, inhalation of dust, and accidental or incidental spills onto body extremities. Other factors that affect residential exposure include the use of protective equipment or additional layers of clothing, the frequency and duration of applications, and the use rate and percent active ingredient of the product used. Significant post-application exposures may also occur from contact with dislodgeable residues of lawn chemicals during normal backyard activities.

Monitoring has been performed to collect compound-specific data, with the intention of also being able to use them more generally to characterize exposures for specific application scenarios and human use patterns. Studies characterizing post-application consumer exposures to lawn chemicals have used passive dosimeters (*e.g.*, patch and partial or whole body covers) and fluorescent tracers to characterize and quantify dermal exposures. There are some significant method differences, in that the mean dermal exposures measured by dosimeter-based methods (*e.g.*, fabric patches or whole-body covers) are about an order of magnitude higher than that quantified using fluorescent tracer techniques. Thus, dosimeter-based methods may significantly overestimate dermal exposures to lawn chemicals (Eberhart 1994). In addition, attempts to remove and quantify dislodgeable residues from treated turf using methods such as polyurethane foam (PUF) rollers have allowed researchers to estimate transfer coefficients.

A residential exposure task force for turf chemicals known as the Outdoor Residential Exposure Task Force (ORETF) has been convened recently, comprised of approximately 33 member companies. It will focus on reviewing existing data, as well as conducting new studies that will provide the basis for development of a generic data base for exposure assessment. This generic database will allow risk assessments to be conducted on both new and existing lawn care products.

DATA SOURCES FOR RESIDENTIAL EXPOSURE ASSESSMENT

A number of important data sources exist for performing residential exposure assessment:

- Data on human exposure factors (*e.g.*, distributions of body weights and skin surface areas, inhalation rates) can be obtained from the USEPA's Exposure Factors Handbook (USEPA 1996). This source is updated periodically as new information emerges.
- Residential air exchange rate data have been summarized in Pandian *et al.* (1993), and further refined in Murray and Burmaster (1995).
- Human time-activity data in the U.S. have been summarized in USEPA (1991b), compiled in the THERdbASE software (Pandian *et al.* 1995), and updated by John Robinson of the University of Maryland. These data will be published in the USEPA's future revision to the Exposure Factors Handbook.

- Dermal exposure assessment methods and dermal permeability coefficients for some organic chemicals are contained in the USEPA's dermal exposure assessment guidance document (USEPA 1992).
- Because skin surface area and body weight are closely correlated, total skin surface area to body weight ratios for use in residential exposure assessments are available from Phillips *et al.* (1993).
- Sources of food commodity consumption rate data for food-related incidental ingestion exposure analyses include software, the USEPA's Dietary Risk Evaluation System (DRES), the 1977–1994 United States Department of Agriculture (USDA) U.S. food consumption survey data, and specialty databases from various institutes and trade associations (*e.g.*, National Food Processors Association pesticide residue databases, the National Institute on Alcoholism and Alcohol Abuse database on wine consumption).
- Data relevant to consumer product exposure assessments are contained in ECETOC (1994).

Some of these data sources are discussed in more detail in subsequent chapters.

In the future, there will be advances in multipollutant exposure monitoring surveys, such as the the USEPA's National Human Exposure Assessment Survey, which will provide empirical data to further refine models. There will also be further development of biomarkers and pharmacokinetic models, advances in sensor technology to expand the exposure database derived from personal exposures.

Increased knowledge is needed about exposures to many environmental contaminants and agents in consumer products, in particular those agents that may persist in the residental environment and which may pose a risk to human health risk.

When performing an exposure assessment, it is important to be thorough by including key elements such as:

- A clear problem formulation (risk management context)
- General approach to be taken (*e.g.*, measurement, modeling)
- Definition of source or sources
- Characterization of specific receptor populations or subpopulations of concern
- Determination of fate and transport mechanisms by which the receptors come in contact with the agents
- Identification of appropriate exposure routes, and
- Consideration of factors (*e.g.*, contact rate, time activity pattern) than influence exposure.

However, key variables in determining exposure in some cases may be less obvious, and hence there is a need to be insightful.

For example, toluene occurs widely in consumer products, such as nail lacquers. Therefore, one might anticipate that toluene would be found in the residential indoor air environment associated with use of particular consumer products. A monitoring study on inhalation exposures to toluene associated with use of nail lacquers in California residences was conducted by Curry *et al.* (1994). The study involved collection of personal, area, and background samples (charcoal tubes) before, during, and after application of nail lacquer products by five subjects in their homes. Airborne levels of toluene declined relatively rapidly after application. For the poorest-ventilated residence, a log-linear dissipation curve was observed, with a post-application half-life of 67 minutes in the subject's breathing zone. Although the study captured a variety of residential housing stock types, room volumes, and ventilation conditions, the highest observed exposure was not associated with the residence having the poorest ventilation conditions. Rather, it was experienced by a woman who held her hand closest to her face (due to poor vision) during application of the nail lacquer. This case study illustrates the importance of personal behavior on exposures. Also, it illustrates the need to continually evaluate whether all relevant dependent variables have been considered in the design, conduct, and analysis phases of an exposure assessment study.

REFERENCES

AIHC (American Industrial Health Council). 1995. Exposure factors sourcebook. Washington, D.C. 118 pages

Calvin, G. 1992. Risk Management Case Histories—Detergents. Chapter 10 in: Richards, M.L. (ed.). Risk management of chemicals. The Royal Society of Chemistry. London, United Kingdom.

CTFA (Cosmetic, Toiletry and Fragrance Association). 1983. Summary of the results of surveys of the amount and frequency of use of cosmetic products by women. Report prepared by ENVIRON Corporation.

Curry, K.K., D.J. Brookman, G.K. Whitmyre, J.H. Driver, R.J. Hackman, P.J. Hakkinen, and M.E. Ginevan. 1994. Personal exposures to toluene during the use of nail lacquers in residences: Description of the results of a preliminary study. J. Exposure Anal. Environ. Epidemiol. 4(4):443–456.

Driver, J.H., R.G. Tardiff, L. Sedik, R.C. Wester, and H.I. Maibach. 1993. In vitro percutaneous absorption of [^{14}C] ethylene glycol. J. Exposure Anal. Environ. Epidemiol. 3(3):277–284.

Eberhart, D.C. 1994. Current activities in assessing human exposures to lawn chemicals. Presented at the Workshop on Residential Exposure Assessment, Annual Meeting of the International Society for Exposure Analysis and the International Society for Environmental Epidemiology. September 18, 1994. Research Triangle Park, North Carolina.

ECETOC (European Center for Ecotoxicology and Toxicology of Chemicals). 1994. Assessment of non-occupational exposure to chemicals. Technical Report Number 58. Brussels, Belgium.

ISEA (International Society of Exposure Analysis). 1995. ISEA Newsletter, Summer 1995 Issue. Argonne, Illinois: Argonne National Laboratory.

Kasting, G.B. and P.J. Robinson. 1993. Can we assign an upper limit to skin permeability? Pharmaceutical Res. 10:930–931.

McKone, T. 1992. Multipathway residential exposures: An overview. In: B. Hakkinen, J. Driver, and G. Whitmyre (eds.). Reference house workshop: Assessment of residential exposures. Annual meeting of the Society for Risk Analysis. December 6, 1992. 29 pages.

Murray, D.M. and D.E. Burmaster. 1995. Residential air exchange rates in the United States: Empirical and estimated parametric distributions by season and climatic region. Risk Anal. 15(4):459–465.

NIST (National Institute of Standards and Technology). 1994. CONTAM93. User manual. Building and Fire Research Laboratory, NIST, U.S. Department of Commerce. Gaithersburg, Maryland.

Pandian, M.D., W.R. Ott, and J.V. Behar. 1993. Residential air exchange rates for use in indoor air and exposure modeling studies. J. Exposure Anal. Environ. Epidemiol. 3(4):407–416.

Phillips, L.J., R.J. Fares, and L.G. Schweer. 1993. Distributions of total skin surface area to body weight ratios for use in dermal exposure assessments. J. Exposure Anal. Environ. Epidemiol. 3(3):331–338.

Turnbull, D. and J.V. Rodricks. 1989. A comprehensive risk assessment of DEHP as a component of baby pacifiers, teethers and toys. In: Paustenbach, D.J. (ed.). The risk assessment of environmental and human health hazards: A textbook of case studies. John Wiley and Sons. New York, New York.

USEPA (U.S. Environmental Protection Agency). 1987. Methods for assessing exposure to chemical substances. Volume 7. Methods for assessing consumer exposure to chemical substances. EPA Publication No. 560/5-85-007. Exposure Assessment Branch, Office of Pollution Prevention and Toxic Substances. Washington, D.C.

USEPA (U.S. Environmental Protection Agency). 1991a. MCCEM. Multi-chamber concentration and exposure model. User's guide. Version 2.3. Environmental Monitoring Systems Laboratory, Office of Research and Development. Washington, D.C.

USEPA (U.S. Environmental Protection Agency). 1991b. Time spent in activities, locations, and microenvironments: A California-national comparison. EPA Publication Number 600/4-91/006. Office of Research and Development, Environmental Monitoring Systems Laboratory. Las Vegas, Nevada.

USEPA (U.S. Environmental Protection Agency). 1992. Dermal exposure assessment: Principles and applications. EPA Publication Number 600/8-91-011. Exposure Assessment Group, Office of Health and Environmental Assessment, Office of Research and Development. Washington, D.C.

USEPA (U.S. Environmental Protection Agency). 1994. Screening-level consumer inhalation exposure software (SCIES): Description and user's manual. Version 3.0. Exposure Assessment Branch, Office of Pollution Prevention and Toxic Substances. Washington, D.C.

USEPA (U.S. Environmental Protection Agency). 1996. Exposure factors handbook (Draft). EPA Publication Number Volume I 600/P-95/002Ba, Volume II 600/P-95/002Bb, and Volume III 600/P-95/002Bc.

Vacarro, J.R. 1996. The use of unique study design to estimate exposure of adults and children to surface and airborne chemicals. American Society of Testing and Materials. West Conshohocken, Pennsylvania.

Vermiere, T.G., P. van der Poel, R.T.H. van de Laar, and H. Roelfzema. 1993. Estimation of consumer exposures to chemicals: applications of simple models. Sci. Total Environ. 136:155–176.

Whitmyre, G.K., J.H. Driver, M.E. Ginevan, R.G. Tardiff, and S.R. Baker. 1992a. Human exposure assessment I: Understanding the uncertainties. Toxicol. Indust. Health 8(5):297–320.

Whitmyre, G.K., J.H. Driver, M.E. Ginevan, R.G. Tardiff, and S.R. Baker. 1992b. Human exposure assessment II: Quantifying and reducing the uncertainties. Toxicol. Indust. Health 8(5):321–342.

Wilkes, C.R. and M.J. Small. 1992. Inhalation exposure model for volatile chemicals from indoor uses of water. Atmos. Environ. 26A:2227–2236.

Wilschut, A., W.F. ten Barge, P.J. Robinson, and T.E. McKone. 1995. Estimating skin permeation: The validation of five mathematical skin permeation models. Chemosphere 30:1275–1296.

www.epa.gov/nerlpage/heasd/therdbase.htm. 1998. THERdbASE: Total human exposure risk database and advanced simulation environment. Version 1.2—September 1998.

II

Methodologies for Assessing Residential Exposure

4

INHALATION EXPOSURES IN RESIDENCES

We are in our homes about 60 percent of the time (USEPA 1996), and in other homes and buildings an additional 25 percent of the time. Clearly, then, the air we breathe in residences makes up the majority of our total intake. If the air in our homes were "clean" (unpolluted), this perhaps would not matter greatly—but the fact is, our home air is generally "dirtier" than the outdoor air, because our consumer products, building materials, and personal activities such as smoking add to the pollutants entering our homes from outside. This is true for scores and probably hundreds of agents that may be harmful to our health.

In this chapter, we will identify the major agents of interest, briefly describe the most current sampling and analysis methods for these agents, and provide a summary of the personal exposures and indoor air concentrations that have been measured in major studies of exposure. We shall also include some discussion of sources, source strengths, and ancillary variables that affect exposures (such as the air exchange rate in a home). Finally, we include a short section on modeling personal exposures and indoor air concentrations.

The major agents include biological agents, volatile organic compounds (VOCs), formaldehyde, particles, carbon monoxide, pesticides, radon, environmental tobacco smoke (ETS), and combustion particles such as polycyclic aromatic hydrocarbons (PAHs). It is instructive to rank these in terms of their indoor : outdoor ratios. Those agents with very high indoor : outdoor ratios are often those that are easiest to reduce or eliminate, because most of the time they are due to sources that the resident has introduced or allowed to accumulate in the home, and which can often be removed.

The highest indoor : outdoor ratios are observed for bioaerosols. Airborne infectious agents commonly occur in dangerous concentrations only indoors. Dust mite, cockroach, and pet-source allergens also present little, if any, risk outdoors, but are major contributors to inhalation-related disease indoors. In homes contaminated with microorganisms, concentrations can be thousands of times higher indoors than out. For users of air fresheners, mothballs, pot pourri, and other products designed to raise indoor concentrations for long periods of time, the mean concentrations (of naphthalene, paradichlorobenzene, limonene, and many fragrances) are typically 10 to 100 times outdoor levels, and individual homes reach 1,000 times outdoor levels (Wallace *et al.* 1989, 1991a). For homes with smokers who smoke indoors, ETS (including benzene, fine particles, PAHs, and nitrosamines) is a major contaminant (Spengler and Dockery 1981). Persons using indoor pesticides will raise concentrations greatly during use, remaining slightly elevated over long periods of time due to the slow release from indoor sinks. In

Residential Exposure Assessment, edited by Baker *et al.*
Kluwer Academic/Plenum Publishers, New York, 2000.

many of these examples, except for human-source agents, some allergen sources, and residual pesticides, the resident can nearly eliminate exposures by removing the source from the house.

For some agents in a very large number of homes, the indoor concentrations may be 10 to 20 times greater than outdoors. This is the case for radon and formaldehyde. Homes treated with chlordane or chlorpyrifos for termites may draw pesticide-contaminated soil gas into the home for years or even decades following home treatment (Lewis and Bond 1987). Since the foundation soil is heavily contaminated, and can be tracked in on shoes, it may add to the total loading of pesticide in household dust (Roberts *et al.* 1992).

For a number of common VOCs, indoor air concentrations in homes are 2 to 5 times outdoor air concentrations (Wallace 1987a). These include benzene, chloroform, tetrachloroethylene, and trichloroethylene.

Inhalable particles ($\leq 10\,\mu m$ in aerodynamic diameter) are elevated in homes where there are smokers or where food is often fried or grilled. However, in homes without these two major sources, indoor concentrations are generally 50 to 80 percent of outdoor levels due to aerosol deposition on walls and floors.

Finally, there are a few agents whose indoor concentrations are *lower* than outdoors. This is true of ozone, which is created outdoors but is highly reactive, and does not last long indoors. However, even for this chemical, because of the long time people spend indoors and the short time outdoors, total indoor exposure indoors may exceed total outdoor exposure. Another agent that declines indoors is acid aerosols, due to reactions with the ammonia that is common in homes. The acid concentrations decline even further as they approach the breathing zone, due to the endogenous ammonia we all exhale. Fungal spore levels are often lower indoors than out during the growing season unless fungi are actively growing in indoor reservoirs.

Table 4-1 lists the most common volatile organic indoor air agents and their sources. Table 9-1 lists common sources for indoor biological agents causing hay fever and asthma. The information in Table 4-1 has been obtained largely in the last two decades.

Table 4-1. Common volatile organic chemicals and their sources

Chemicals	Major sources of exposure
Acetone	Cosmetics
Alcohols (ethanol, isopropanol)	Sprits, cleansers
Aromatic hydrocarbons (toluene, xylenes, ethylbenzene, trimethylbenzenes)	Paints, adhesives, gasoline, combustion sources
Aliphatic hydrocabons (octane, decane, undecane)	Paints, adhesives, gasoline, combustion sources
Benzene	Smoking, auto exhaust, passive smoking, driving, pumping gas
Carbon tetrachloride	Fungicides, global background
Chloroform	Showering (10 min average), washing clothes, dishes
p-Dichlorobenzene	Room deodorizers, moth cakes
Formaldehyde	Pressed wood products
Methylene chloride	Paint stripping, solvent use
Styrene	Smoking
Tetrachloroethylene	Wearing and storing dry-cleaned clothes, visiting dry cleaners
1,1,1-Trichloroethane	Wearing and storing dry-cleaned clothes, aerosol sprays
Trichloroethylene	Unknown (cosmetics, electronic parts, correction fluid)
Terpenes (limonene, α-pinene)	Scented deodorizers, polishes, fabrics, fabric softeners, cigarettes, food, beverages

It has depended first on the development of measurement methods capable of measuring the exquisitely low environmental concentrations of many of these agents. Secondly, large-scale monitoring studies using probability-selected populations have been necessary to determine the likely distribution of concentrations and indoor : outdoor ratios. Finally, the results of the monitoring studies have sometimes been incorporated into indoor air models to allow us to extend our understanding to the point of predicting results under a large number of scenarios. These three important pillars of our understanding—measurement methods, monitoring, and modeling—make up the substance of this chapter.

MEASUREMENT METHODS

In most cases, methods suitable for outdoor air sampling cannot simply be transferred to indoor settings without considerable modification or, in some cases, completely different configurations. For example, high-volume air samplers cannot be used indoors because they would suck in (and filter) much more air than would normally enter the home. Thus, low-flow air monitors must be employed indoors. Noise levels must also be reduced to be acceptable to persons living in the home. Bulkiness must also be reduced, particularly if several instruments of various types are to be deployed in one area of the home. For personal sampling, even more stringent requirements pertain. Low weight, on the order of a pound or less, is generally required, and noise and bulkiness must be even further reduced beyond acceptable indoor values.

Living Organisms

Living organisms in air are sampled to estimate risk of infectious disease and as an indicator of exposure to microbial allergens and toxins.

Air Sampling Methods. The most important concern for sampling living organisms is to protect their viability (ability to grow) during the sampling process. Airborne living organisms can be considered particles, and they are most commonly collected by inertial impaction onto semi-solid nutrient medium or into liquid, or by electrostatic precipitation. Airborne organisms may travel through air as single cells, groups of cells, or as one or more cells attached to other particles. Counts derived from impaction onto agar represent organism/bearing particles, and are expressed as colony forming units (CFU) per cubic meter of air. Collection of samples into a liquid for subsequent culture probably releases single cells, and data could be expressed as culturable cells per cubic meter of air. However, CFU is the generally used term for all cultural sampling.

In general, air sampling methods for bioaerosols leave a great deal to be desired. Sampling times are short—often 5 to 10 minutes. Several air sampling apparatuses are available:

- *Andersen sieve plate impactors.* Sieve plate impactors accelerate air through a multi-hole plate and impact particles onto an agar-filled Petri dish. The 400-hole, 6-stage Andersen cascade impactor (Grasby-Andersen, Atlanta) was described in the late 1950s (Andersen 1958) and has become somewhat of a standard among microbial air samplers. The sampler impacts particles onto agar-filled 100 mm Petri dishes. Particles are separated into six size categories as the air passes through progressively smaller holes. The largest (greater than 7 μm aerodynamic diameter) are collected below the first or top stage and the smallest (0.65 to 1.1 μm aerodynamic diameter) below the sixth stage. The Andersen microbial sampler is now most typically used for indoor air quality investigations with only the sixth stage in place, thereby collecting all particles above about 0.65 μm onto a single Petri dish. In this configuration it is often called the Andersen N6 (Jones *et al.* 1985). Graseby-Andersen also produces a two-stage sieve-plate impactor that separates particles into "nonrespirable" (above about 8 μm diameter) and "respirable" (about 1 to 8 μm) size fractions. The sampler has 200 holes per stage instead of the 400 holes in the original Andersen sampler.

The Andersen samplers are not portable, requiring a vacuum pump that will draw one cubic foot per minute (28.3 liters per minute) through the device. The two-stage model is provided with a critical orifice. If one assumes that a minimum sampling time is about 30 seconds and the maximum about 10 minutes (the agar dries excessively soon afterward), and that (at least for fungi) a colony density of about 10 to 50 colonies per plate is optimal, the air concentrations for which the sampler is useful can be calculated. Calculations based on these assumptions indicate that the N6 could reliably detect levels of 35 to 5,500 colony forming units per cubic meter of air. The use of the two-stage sampler increases the upper limit to some degree, althought the majority of bioaerosol particles penetrate to the second stage.

- *Other sieve plate impactors.* The Burkard Portable Sampler for Culture Plates (Burkard Manufacturing Co., Ltd., Rickmansworth, UK) is a single-stage sieve impactor that is small (15 × 12 cm), with a built-in air mover, timer, and rechargeable battery. It draws air through 100 × 1 mm diameter holes. At 20 liters/min, this results in a calculated d50 of about 2.9 μm. While flow rates are stated by the manufacturer to be 10 or 20 liters per minute depending on the model, measured flow rates are often much higher (between 28 and 55 liters per minute).

  The Spiral Air Systems sampler (SAS, Spiral Systems Instruments, Bethesda, Maryland) operates similarly to the Andersen samplers but is relatively portable with a rechargeable battery pack. The samplers operate at flow rates of 90 or 180 liters per minute, depending on the model, and the d50s are 2 and 1.45 μm, respectively. Particles are collected onto 65 mm diameter Rodac contact plates. At 180 liters per minute, assuming about 25 fungal colonies on the plate is maximum, the upper concentration limit is a little over 400 CFU per cubic meter of air, and the sampler is best used in relatively clean environments.
- *Slit impactors.* A sampler that pulls air through a narrow slit and impacts particles onto a rotating agar surface was described in the early 1940s (Bourdillon *et al.* 1941). The currently available New Brunswick (New Brunswick Scientific Co., Inc., Edison, New Jersey) and the Matson-Garvin (Barramundi Corp., Homosassa Springs, Florida) slit-to-agar (STA) samplers rotate agar-filled Petri dishes (either 100 or 150 mm) beneath a narrow slit over a period that can be adjusted from minutes to one hour depending on the model. They allow time discrimination of culturable fungal and bacterial levels (as CFU) and can be very useful in making connections between changes in concentrations and potentially causative events. STA samplers generally use flow rates between 28

and 50 liters per minute and have a very high collection efficiency for bacterial and fungal-sized particles (d50 near 0.5 µm). Because of the relatively high flow rates and the slow plate rotation speeds, these samplers can be used only in fairly clean environments.

- *Centrifugal impaction*. One popular sampler for culturable organisms collects particles by centrifugal impaction. The RCS Biotest sampler (Biotest Diagnostics Corporation, Denville, New Jersey) is a portable, battery operated device with a built-in air mover that draws air into a chamber and accelerates particles toward a peripheral agar strip. The flow rate through the sampler is nominally 40 liters per minute, although there is evidence that this theoretical calculation is inaccurate. The d50 for this sampler is not readily calculated and has not been experimentally determined. In comparative studies, it has performed poorly in relation to the Andersen impactors.
- *Liquid impingers*. Liquid impingers were developed in the early 1920s and used primarily to sample airborne bacteria. The Greenberg-Smith impinger pulls particles through a glass tube, accelerates them through a narrow opening, and impacts them onto a glass platform which is submerged in liquid. The particles are washed from the platform and suspended in the fluid by the liquid turbulence created by the air jet. A number of variations on this early impinger are in use today, including the midget impinger, and the all glass impingers (AGI 4 and AGI 30, Ace Glass, Inc., Vineland, New Jersey). The AGIs use a critical orifice (flow rate about 12.5 liters per minute) and impact particles onto the bottom of the glass reservoir (AGI 4) or into the fluid (AGI 30).

  While the d50 of the liquid impingers can be very low (<1 µm), other factors can have a major impact on recoveries. Hydrophobic fungal spores (and other hydrophobic particles) are generally considered to be poorly retained in the fluid reservoir, although this has not been well studied. Due to aerosolization from bursting bubbles and liquid turbulence, bacterial cells can also be lost from the fluid reservoir. In addition, violent physical forces can damage, if not kill, some microbes, especially vegetative bacterial cells (Buttner and Stetzenbach 1991).
- *Electrostatic precipitators*. The Litton high volume electrostatic precipitator has been used to collect viruses and bacteria from aerosols primarily for research purposes (Crook 1997). The device results in concentration of aerosol particles into a small volume of liquid, and is especially useful where very low aerosol concentrations are expected.
- *Filtration sampling for culturable cells*. Filtration sampling, often used for other types of particle collection (see below), is rarely used for recovery of cells for culture. Losses due to desiccation can be high, depending on the resistance of the organism, and are rarely predictable. In spite of this major drawback, filtration is sometimes useful in environments such as those encountered in agricultural situations where very concentrated aerosols are present, and where the goal is to estimate the order of magnitude of the exposure.

Dust Sampling Methods. Dust samples, collected by vacuuming dust from floors, bedding, and furniture (see below), have also been analyzed for living cells as a surrogate for fungal allergen exposure. Relationships between levels in dust and air remain to be documented, although preliminary indications are that many organisms present in dust do not routinely become airborne, at least in a culturable state.

Other Kinds of Reservoir Sampling. Samples for estimating exposure to living cells are also collected from water reservoirs, and from bulk solids such as building materials, fabrics, and foods. Sampling water from reservoirs from which aerosols are routinely dispersed (*e.g.*, machining coolant or humidifier reservoirs, cooling coil drip pans, cooling towers, shower heads) is the method of choice for estimating the potential for exposure to agents of Legionnaires' disease (*Legionella pneumophila* and other virulent strains).

Analysis of Samples. In order to be sure that a cell is living, it must be cultured, although some attempts have been made to estimate the prevalence of living cells using "vital" staining (Madelin and Madelin 1997). Culture involves providing an environment (including nutrients, water, an appropriate atmosphere, and temperature) that encourages an organism to reproduce and develop a recognizable colony (or effect in the case of viruses). No single set of conditions will allow the growth of all (or even most) organisms, and cultural analysis is always highly selective, and always underestimates actual concentrations of viable cells.

- *Viruses.* Viruses are obligate intracellular parasites, and will reproduce only inside of a living cell. Some viruses can be induced to grow in cultured cells, and counts are made by examining the cell culture for clear spots indicating death of the cells.
- *Bacteria.* Most bacteria are free-living and grow readily in culture. Most bacteria will grow over a temperature range of 25°C to 40°C, with the range between 30°C and 37°C often optimum. Bacteria that can survive the airborne state are always aerobic, although the growth of some is stimulated by increased levels of CO2. Culture media for bacteria are usually near neutral (pH $\simeq$ 7), have a water activity in excess of 0.9, and (often) contain nitrogen in the form of amino acids or proteins, and relatively simple sugars as a carbon source. Bacteria that are recovered from air are often damaged, and need time for repair before growth begins. Minimal media such as R2A allow time for such repair and often maximize the number of cells that are recovered. Although bacteria recovered directly from sources (*e.g.*, from clinical specimens) may grow rapidly and produce countable colonies after 24 hours, bacteria collected from air often require up to 5 days of incubation, especially when grown at 30°C on minimal medium. Some bacteria (including *Mycobacterium tuberculosis*) require complex media components for growth and several weeks to produce visible colonies.

  Once colonies have been produced, further identification requires staining with microscopic examination (to determine the Gram reaction that divides the bacteria into two large groupings: Gram positive and Gram negative. Further identification relies on specific enzyme patterns, the ability to utilize a range of carbohydrates, and other biochemical and physiological tests.

  For some kinds of bacteria, immunological tests that rely on the antigenic specificity of bacterial species are available. *Legionella pneumophila* is routinely identified using a fluorescent antibody stain combined with microscopy. The direct fluorescent antibody (DFA) method for bulk water samples involves concentrating cells or cell fragments by filtration or centrifugation, then staining and counting microscopically. The method allows detection of a single cell/ml of liquid, or even fewer if larger volumes are filtered before staining. Although the method as most often used detects both living and dead cells, the approach has been used to detect actively respiring cells (Rodriguez *et al.* 1992).

Recently, the development of techniques for amplifying DNA has provided a new tool for the identification of some bacteria. The polymerase chain reaction (PCR) allows the DNA in a cell to be replicated many times, so that its structure can readily be determined and compared with a library of structures for identified microorganisms.

- *Fungi.* Fungi, like bacteria, are free-living and most can be cultured. All fungi of interest in indoor air are aerobic, and most grow optimally at room temperature (18–25°C). Fungi grow over a range of water activities from <0.7 to >0.9. The most commonly used fungal culture medium is malt extract agar (MEA) which contains extract of malted barley, dextrose, peptone, and agar. This is a relatively rich medium, and the use of a simpler formulation containing 2 percent malt extract and agar has been suggested for maximizing recoveries. Another medium which better supports fungi that require low water activity for optimal growth (xerophilic fungi) is DG18 (Hocking and Pitt 1980). This medium restricts colony diameters through the use of dichloran (a pesticide) and restricts water activity with 18 percent glycerol. A disadvantage is it's apparent inability to support the growth of some fungi and reduced sporulation of others. The fungi are noted for their ability to degrade complex carbohydrates. Some produce cellulases, ligninases, and keratinases and can degrade cellulose (cotton, paper), lignin (wood), and keratin (human skin). The organisms producing these enzymes can be selected (enriched) by using culture medium containing the appropriate complex carbohydrate as the sole source for carbon.
- *Other microorganisms.* Living algal cells and protozoa may also be present in aerosols. No known risk is associated with algal cells that depends on their viability, although they do carry allergens. Protozoa rarely cause infections following exposure in relatively large droplets. Air sampling has not been used for estimating such exposures.

Allergens and Antigens

Allergens and antigens are (generally) glycoproteins that are produced by living organisms. Allergens have been isolated from fungi, plants (*e.g.*, pollen, peanuts, rubber tree latex), arthropods (*e.g.*, fecal material from mites and cockroaches), mammals (*e.g.*, cats, dogs, mice, rats), and birds (*e.g.*, pigeons). Exposure to allergens and other antigens is estimated by measuring indicators (*i.e.*, allergen-bearing particles or sources) or by direct measurement of the allergens themselves.

Sampling Culturable Particles. The most common method for estimating indoor exposure to fungal allergens is sampling either dust or air for culturable (living) cells (see above).

Sampling Microscopically Identifiable Particles or Sources (Fungi, Pollen, Whole Dust Mites)

- *Air sampling.* Fungal spores and pollen can be counted and, to some extent, identified microscopically from particle collections. World wide, the most commonly used sampler for counting spores and pollen in outdoor air (and thereby estimating exposure to allergens) is the Burkard modification of the Hirst (1952) spore trap (Recording Spore Sampler, Burkard Manufacturing Co., Ltd., Rickmansworth, United Kingdom). The sampler, which is wind-oriented using a large wind vane, pulls air at 10 Lpm through a side-facing slit protected by a rain shield onto a greased transparent tape that moves at 2 mm/hour. The tape is mounted on a drum that rotates once in 7 days. Following exposure, the tape is cut into 7 × 24-hour pieces, and mounted on microscope slides.

For indoor use, the Burkard Recording Spore Trap is available with an upward-facing slit, and without the wind vane. This model is available with interchangeable heads that sample for 24 hours onto glass slides or for 7 days onto the tape-wrapped drum. Similar, though less-well characterized is the Samplair (AllergenCo, San Antonio), which incrementally moves a greased microscope slide beneath an upward facing slit.

Also available from the Burkard company is the "personal" spore trap, which is small (11.5 × 9 cm), battery powered device that impacts particles onto a stationary microscope slide.

Filtration might be one of the earliest types of bioaerosol sampling as Pasteur used a plug of gun cotton in glass tubing to collect airborne particles. He dissolved the cotton "filter" in an alcohol and ether mixture, let the particles settle, and viewed the recovery microscopically.

Filtration is now commonly used for particle sampling where analysis is to be by gravimetrics, or elution and subsequent assay (see below). However, it can also be used for microscopic analysis, provided particles can be readily washed from the filter material or the material itself is optically transparent. For particle removal, smooth surface filters such as track-etched polycarbonate or teflon are usually preferred. Although filters can be "cleared" using oils of matching refractive index, such treatment collapses biological particles, and filter material must transmit light in water based mounting media. Alternatively, epi-illumination can be used following staining with fluorescent dyes (see below).

- *Dust sampling*. Although commonly used to estimate allergen exposure, microscopy is rarely used for analysis, although microscopic examination of dust samples may serve to document (or contradict) culture data.

  Dust mite counts are sometimes used as an indicator for exposure to dust mite allergens, although dust mite populations reach a peak somewhat before allergen levels. Dust is sieved, washed, and counter-stained leaving dust-mite bodies white.

Sampling Indicator Components. Ergosterol is the principle fungal steroid, and its primary source in dust and air is fungi. Ergosterol analysis is commonly used to assess fungal biomass in foods, but has also been used for dust samples. Likewise, measurement of 1–3 β-d glucan has been used as a surrogate for the presence of fungi and fungal allergens in dust. The presence of the amino acid guanine has been used as a surrogate measure for dust mite levels, since guanine is excreted in abundance only by acarids (mites) and arachnids (spiders). It is assumed that dust mite populations far exceed those of spiders in house dust.

Sampling for Direct Allergen Analysis

- *Dust sampling.* Exposure to amorphous allergen-bearing particles is almost exclusively estimated using analysis of dust samples collected using vacuum devices from reservoirs such as bedding, carpets and smooth floors, and furniture. Vacuum flow rate is rarely either controlled or monitored. Dust is collected until there is enough (by visual inspection) to run the desired assays. Sometimes some standardization of collection is attempted by collecting dust from a measured area of surface over a specified length of time. Dust samples are usually sieved to remove large particles (*e.g.*, hairs, paper clips, pieces of food, cockroach bodies). Results are expressed as amounts of assayed material per gram of fine dust, usually with no indication of the amount of dust, and therefore, the amount of allergen, that might be present in the environment. Collection media for the dust have included 180 thread count bed sheets, cellulose extraction thimbles, and teflon-faced filter material (*i.e.*, Goretex®).
- *Air sampling.* Air sampling for amorphous allergens is rarely a part of survey studies. It appears that the important dust mite allergens are borne on large particles and air levels are low except during very active disturbance of reservoirs. Thus epidemiological studies have relied on dust measurements for all allergens, although some (*e.g.*, cat allergen) is abundant in air, and could routinely be measured using air sampling. Fungal allergen exposure has also been estimated from air samples.

  Collection of air samples for allergen assay has been done using high volume filtration devices using glass fiber filters or Teflon membranes (Swanson *et al.* 1985, 1989), and filter cassettes (Sakaguchi *et al.* 1995).

Analysis

- *Culture.* Culture is the most commonly used analytical method to estimate exposure to microbial allergens in spite of the fact that no data have been collected that demonstrate a good relationship between culturability and allergen content. Unfortunately, good immunoassays have not been developed for the majority of fungal allergens. In addition, specific identification of fungi is essential for estimation of allergen exposure, and no better method than culture is currently available for the vast majority of organisms likely to be encountered in indoor environments.
- *Microscopy.* Microscopy is the method most commonly used to assess exposure to outdoor allergen-bearing particles including pollen and fungal spores, and many published surveys of outdoor levels of these particles are available (*e.g.*, Solomon 1976, Delfino *et al.* 1996). Limited data support a strong relationship between, for example, birch pollen counts and the major birch allergens as measured by immunoassay (Schappi *et al.* 1997). However, there is also evidence that some allergen is borne on particles much smaller than pollen (Solomon *et al.* 1983).
- *Immunoassay.* Immunoassays can be used for any material against which antibodies can be induced. Allergen immunoassays measure the actual agent that causes the allergic disease, and therefore are the assay methods of choice for allergens. They are sensitive, specific, reproducible, and quantify allergen in

absolute units. Immunoassays rely on the production of specific antibodies against each individual allergen to be measured. Such antibodies are usually raised in mice (*i.e.*, monoclonal antibodies) or (often) in rabbits (polyclonal antibodies) and are of the IgG class. The RadioAllergoSorbent Test (RAST) uses pooled human serum containing IgE to specific allergens. Good IgG immunoassays that use either monoclonal antibodies, or high-quality polyclonal systems are available for the most common dust mite allergens (Der f 1, Der p 1), cat (Fel d 1), cockroach (Bla g 1, Bla g 2), dog (Can f 1), mouse (Mus m 1), and a few others. While a few fungal allergens have also been purified (*e.g.*, Alt a 1, Asp f 1, Cla h 1), assays for measuring allergen in environmental samples are not widely available. Data from immunoassays are usually reported as ng (or units) of allergen/gram of dust.
Immunoassays have also been used to measure 1–3 β d glucans as a surrogate for total fungi in dust (including allergens).

- *Chemical analyses*. Ergosterol is measured using GCMS and levels have been shown to be related to levels of culturable fungi in organic dust (Saraf *et al.* 1997), and in house dust.
- *Surrogate measures for amorphous allergens*. The most commonly used surrogate measure for allergen exposure is questionnaire data on the presence or absence of specific sources. Thus, the presence of a cat, dog, visible cockroaches, or mice is assumed to indicate exposure to associated allergens. Likewise, visible mold growth, or even visible signs of dampness is considered to indicate exposure to fungal allergens.

Microbial Toxins

Sampling Methods. Air samples for analysis of bacterial endotoxin are collected using endotoxin-free filter material. There is considerable controversy in the literature regarding the specific type of material that is most useful, and recoveries from different types appear to depend on surface area available that irretrievably binds the endotoxin, and on the methods used for elution. However, track-etched polycarbonate membranes appear to work well for most applications providing the entire protocol described for their use is followed. Endotoxin analysis is also readily performed on dust and fluid samples (Milton *et al.* 1990).

Exposure to fungal toxins (glucans, mycotoxins) is most often assessed by cultural air sampling using the Andersen or other culture plate impactor, or analysis of dust or other bulk materials using culture, assay for indicator chemicals, or direct analysis for the toxins.

Analysis

- *Culture*. Culture as been used as a surrogate measure for exposure to endotoxin and to mycotoxins. While the presence of Gram negative bacteria in a sample surely indicates the presence of endotoxin (since endotoxin is an inevitable part of the cell wall of these organisms), the biological activity of the endotoxin is not necessarily represented by the number of culturable bacterial cells. The composition of the endotoxin (*i.e.*, the nature of the 3-OH fatty acids) determine its biological activity. In addition, much endotoxin can be present as fragments of cells.

Culture is also often used as a surrogate for exposure to mycotoxins. However, there is mounting evidence that the production of mycotoxins is strain and substrate dependent, and that the presence of a species known to produce toxin is not sufficient evidence of the presence of the toxin itself. Also, culturability is unlikely to be related to the amount of mycotoxin contained in a fungal spore.

- *Microscopy*. Microscopy has also been used to assess levels of bacteria (combined with limited cultural analysis to verify types) as a surrogate for endotoxin exposure. Likewise, culture is often used to indicate the presence (or absence) of mycotoxin-producing fungi.
- *Bioassay*. The most commonly used assay method for estimating endotoxin exposure is the Limulus amoebocyte assay, which depends on the dose-response clotting of horseshoe crab blood in response to endotoxin exposure. The assay is sensitive, but difficult to perform consistently, requiring stringent quality control (Milton *et al.* 1992).

  Limulus amoebocyes have also been used to measure glucans. Batches of lysate have been recovered that are specific for the 1–3 β d glucans in preference to endotoxin. However, the lysate, and therefore the assay, is not readily available.

  Bioassays are also used to estimate toxin content of samples assumed to contain mycotoxins. The brine shrimp assay, which uses as an endpoint the death of *Artemia salina*, is an estimate of total cytotoxicity in a sample, and promises to be useful in estimating mycotoxin exposure. The Ames assay, used to assess mutagenicity of many toxic substances, is also used for mycotoxins, although not usually in exposure assessment studies.
- *Chemical assay*. As a measure for endotoxin GCMS is less sensitive than Limulus, and measures chemical content of 3-OH fatty acids rather than the biological activity of the toxin. However, GCMS is more reproducible than the Limulus assays. GCMS has served to elucidate relationships between the kinds of bacteria (and therefore the kinds of fatty acids in their endotoxin) and the biological response measured by the Limulus assay.

  GCMS and HPLC are the methods usually used to measure mycotoxins in bulk samples. Neither method is sensitive enough to measure levels of mycotoxins that are likely to be found in air in any but the most severely contaminated residential, industrial, or agricultural environment where moldy materials are being handled during sampling.
- *Immunoassay*. An immunoaffinity column assay is commercially available for the measurement of aflatoxin. Its sensitivity is adequate for bulk samples supporting active growth of toxin-producing fungi, and the assay could be useful for air sample analysis in environments where contaminated materials are being actively disturbed.

Volatile Organic Chemicals (VOCs, Including Microbial VOCs)

Sampling with Activated Charcoal. Historically, the measurement method of choice for *occupational* exposures (generally 1 to 100 parts per million, or ppm, for a given VOC) has been to pump a sample of air across a sorbent (usually activated charcoal) in order to concentrate the VOCs. They are then recovered by a solvent such as carbon disulfide.

In the early 1980s, passive badges employing activated charcoal were developed for use in occupational sampling. The badges operate on the principle of diffusion, and are often operated over an 8-hour workday to provide an integrated average exposure for comparison to the occupational standards (*e.g.*, the threshold limit value, or TLV). However, the manufacturing process for these badges leaves residues of VOCs on the activated carbon that make the badges unsuitable for short-term sampling at *environmental* concentrations, which are usually at part-per-billion (ppb) levels. However, the high background contamination on the badges can be overcome by extending the time of sampling to a week or more, and several studies of indoor air pollution have adopted this technique (Mailahn *et al.* 1987, Seifert and Abraham, 1983).

Sampling with Tenax. The background problems associated with activated charcoal, as well as problems in obtaining reliable recoveries of sorbed chemicals, led to a search for a more suitable sorbent. A polymer known as Tenax was widely adopted during the 1970s as a more reliable sorbent than charcoal for ppb levels (Barkley *et al.* 1980, Krost *et al.* 1982). Tenax, properly cleaned, has very low background contamination for almost all VOCs of interest. It is also stable at temperatures up to 250°C, allowing thermal desorption instead of solvent desorption. (Solvent desorption involves a redilution of the VOCs, thus partially negating the concentration made possible by the sorbent.) Drawbacks include artifact formation of several chemicals (*e.g.*, benzaldehyde, phenol) and an inability to retain very volatile organic chemicals (*e.g.*, vinyl chloride, methylene chloride). Although most uses of Tenax sorbent have been with active (pumped) samplers, a passive badge containing Tenax has also been developed (Coutant *et al.* 1985, 1986, Lewis *et al.* 1985).

Sampling with Multisorbent Systems. In the late 1980s, attempts were made to combine the best attributes of charcoal and Tenax into a multisorbent system. Newer types of activated charcoal (Spherocarb, Carbosieve) were developed to provide more reliable recoveries. Tandem systems employing Tenax as the first sorbent and activated charcoal as the second, or backup, sorbent were employed. The Tenax collected the bulk of the VOCs and the activated charcoal collected those more volatile VOCs that "broke through" the Tenax. Systems were also developed using three sorbents, such as Tenax, Ambersorb, and Spherocarb or Carbosieve (Hodgson *et al.* 1986). All such systems allow collection of a broader range of chemical types and volatilities.

Direct (Whole Air) Sampling. This method, first developed in the 1970s for upper atmosphere sampling, avoids the sorption-desorption step, which should theoretically allow less chance for contamination. (However, it requires great sensitivity on the part of the detection instruments.) The method may involve real-time sampling in mobile laboratories, with direct injection of the air sample into a cold trap attached to a GC; or sampling in evacuated electropolished aluminum canisters for later laboratory analysis (Oliver *et al.* 1986).

Comparison of Sampling Methods. No single method of sampling VOCs in the atmosphere or indoors has become a standard or reference method. In the U.S., the two preferred methods are Tenax and evacuated canisters. These two methods were compared under controlled conditions in an unoccupied house (Spicer *et al.* 1986). Ten chemicals were injected at nominal levels of about 3, 9, and 27 ug/m^3. The results showed that the two methods were in excellent agreement, with precisions of better than 10 percent for all chemicals at all spiked levels.

In Europe, the two most comon methods are Tenax and activated charcoal. One study employing both methods side by side (Skov *et al.* 1990) found consistently higher levels of total VOC on the charcoal sorbent. The difference may be due to very volatile organics such as pentane and isopentane, which are collected by charcoal but which break through Tenax readily.

The sorbent methods lend themselves to personal monitoring—a small battery-powered pump is worn for an 8-h or 12-h period to provide a time-integrated sample. However, at present, the whole-air methods employ bags or canisters that are too bulky or heavy to be used as personal monitors.

Analysis. Samples are usually analyzed by first separating the components using gas chromatography (GC). Three detection methods in common use are flame ionization (FID), electron capture (ECD) and mass spectrometry (MS). Only GC-MS has the ability to unambiguously identify many chemicals. Neither GC-FID nor GC-ECD is able to separate chemicals that coelute (emerge from the chromatographic column at the same time). Also, GC-FID response is depressed by chlorine and other halogens, so it is not suitable for samples containing halogens. Mass spectrometry, by breaking chemicals into fragments and then identifying these fragments, is often capable of differentiating even among coeluting chemicals. However, since chemicals are identified by comparing these mass fragment spectra to existing libraries, and the libraries are incomplete, even GC-MS identifications are often tentative or mistaken. (One study using known mixtures of chemicals found about 75 percent accuracy of identification for several different GC-MS computerized spectral search systems.)

Particles (Nonspecific)

Particles may be measured by determining their weight on a filter (gravimetric methods) or by counting them (optical methods). Historically, most of the measurement methods employed in environmental field studies have been gravimetric. There are no optical methods accepted by EPA as reference methods. This may change as improved optical methods are developed and made widely available.

Gravimetric Particle Sampling. The traditional particle sampler consists basically of a pump to pull air at a known rate across a filter. The filter collects particles over a period of time and is then weighed. Although this sounds simple and straightforward, in fact it is often brutally difficult to provide an accurate sample. Particularly for residential sampling, the greatest difficulty is the small amount of mass that can be collected, and the resultant requirements for extreme care in handling such a small sample. Larger amounts can be collected by increasing either the flow rate or the sampling time. However, a very large pump flow rate indoors would result in changing the existing indoor conditions of velocity, airflow, pressure, and the results would be meaningless. Therefore as a practical matter, it is not possible to sample at much above 20 Lpm. For a typical indoor PM_{10} concentration of $20\,\mu g/m^3$, an 8-hour sampling period would collect only about $200\,\mu g$ of material. Considering that a typical filter weighs 100 mg, one is measuring the difference between 100 mg and 100.2 mg, a difference of only 0.2 percent. The filter can increase or decrease its weight by more than that due to absorbing water vapor due to changes in relative humidity. Because of this, filters must be conditioned for lengthy periods in humidity and temperature-controlled conditions both before and after sampling. Static charge can also cause spurious weight readings, and therefore must be removed from the filters before weighing. Despite these problems, very careful field work can succeed in weighing filters reproducibly to within 10 μg. For the USEPA's most recent study of particles, the sampling period was 12 hours, the flow rate was 4 Lpm, and the average indoor concentration was 80 μg/m^3, leading to a total amount on the filter of only 230 μg on average, and much less than that in "clean" homes; yet the filters could be weighed to within 4 μg and the resulting precision was within 5 percent.

Various pump and filter samplers are available for personal and indoor sampling in residences. One type employs a 4 Lpm monitor for personal sampling and a 10 Lpm microenvironmental exposure monitor. The personal sampler collects one size fraction (either $PM_{2.5}$ or PM_{10}) on 37 mm filters. The indoor monitor can collect both size fractions simultaneously (Marple *et al.* 1987).

All such pump and filter samplers must be returned to the laboratory for weighing, no earlier than a day or two after the measurements. Therefore, they are unable to provide a real-time reading. One gravimetric instrument that is able to provide a *semi*-real-time reading is the Piezobalance.

The Piezobalance (TSI, Minneapolis) uses an impactor to collect particles below 3.5 μm aerodynamic diameter at 1 Lpm. The particles pass through a chamber with an electrostatic precipitator that forces them to deposit on a piezoelectric crystal. The crystal is kept under forced vibration by an electric current. The vibration frequency changes as a result of the mass deposited (the piezoelectric effect). This change in frequency is monitored by continuous comparison with the frequency of an identical crystal in a closed chamber. The resulting *cumulative* change in frequency is divided by the time (usually two minutes, but the time can be lengthened under cases of very clean environments) to determine the total mass collected in that time. The frequency change is linear with mass up to a total mass of some 30 ng, at which time the monitor must be cleaned, a process that takes about 1 to 2 minutes. Maintenance includes cleaning the needle assembly for the electrostatic precipitator and coating the impactor with an even layer of grease. The instrument is calibrated at the laboratory, and is usually recalibrated once per year. The Piezobalance has never been tested under the USEPA's reference method program, and so has failed to be selected as a reference sampler.

Nonetheless, the Piezobalance is a gravimetric sampler with the ability to make near-real-time measurements, and has therefore been used in a number of studies of indoor environments. One of the most well-known studies was carried out by Repace and Lowry (1980) and resulted in documenting increased concentrations of RSP ($PM_{3.5}$) related to ETS in restaurants, offices, and other areas allowing smoking.

Automated Particle Counters. Recently, particle counters capable of measuring six or eight size ranges simultaneously have become commercially available. One type uses laser diode technology to provide counts of particles in six size ranges, from 0.3 μm to 10 μm. A built-in printer prints out the data at programmable intervals. A version of this monitor uses a manifold allowing automatic sampling from several locations. Portable (hand-held) versions are also available for personal sampling. Count data are stored in memory for transfer to a computer.

Although these monitors provide useful data on numbers of particles, because of widely varying particle densities they can be misleading when used to estimate mass. If the aerosol is of known composition (*e.g.*, ETS), the monitors can be calibrated to the particular aerosol and provide trustworthy mass information. But if the aerosol is of unknown composition, it is presently impossible to obtain accurate mass information from optical counters.

Analysis

- *Elemental analysis*. Particles collected on filters can be analyzed for their elemental content using either x-ray fluorescence (XRF) or proton-induced x-ray emission (PIXE). Teflon filters are often employed to avoid the high elemental background of quartz fiber filters. Depending on air concentrations and volume of air sampled, up to 30 or 40 elements can be analyzed using one of these methods. In the PTEAM Study, which used monitors collecting only 3 m^3 of air, 14 elements were commonly detectable by XRF. Besides providing data directly on toxic metals such as lead and cadmium, metal analysis can also be used for source apportionment (*e.g.*, marker elements, such as Si, for crustal material, and V for home heating oil).
- *Organic Analysis*. Particle-bound organics, such as the heavier polyaromatic hydrocarbons (PAHs), can also be analyzed following collection of a sample. Since the lighter PAHs are often in the vapor phase, and medium-weight PAHs may exist in both aerosol and gaseous states, a combination of a filter and a sorbent is sometimes used to collect all the PAHs in both states, with subsequent extraction of the filter and sorbent together. This method provides an accurate total, but does not allow for identification of the relative amounts in the vapor and particle stage. A method that can determine the phase distribution more exactly employs a diffusion denuder followed by a filter and sorbent (Coutant *et al.* 1985, 1986, Gundel *et al.* 1995).
  A continuous monitor for total PAHs has also been developed and tested in the field (Ott, personal communication). By collecting indoor and outdoor data simultaneously, the impact of woodsmoke, for example, on indoor air quality can be determined.

Carbon Monoxide

A large number of continuous monitors, both active (pumped) and passive (diffusion), are commercially available for CO. Most are electrochemical, depending on counting the electrons produced when CO is oxidized to CO_2. Precision of these monitors is generally very good, with typical errors on the order of 0.1 ppm. Interferences can be a problem in some cases. For example, when the monitors are used for breath analysis, endogenously produced hydrogen (from eating certain foodstuffs such as beans) can be a positive interference.

Pesticides

Although personal and indoor monitors have been used in field studies (Lewis and McLeod 1982), none are commercially available. Also, analysis continues to be extremely expensive, on the order of $1,000 per sample to identify a suite of common pesticides. Cost savings can be achieved if only one or a few pesticides are the targets, but in general pesticides are simply too difficult and expensive to allow for much data to be collected in homes.

NO_x

A useful monitor for nitrogen dioxide (NO_2) is the Palmes Tube (Palmes *et al.* 1986), which consists of a short plastic tube with a filter soaked with a solution of triethanolamine, which reacts with NO_2 and can later be quantitated by a colorimetric method. The system has a sensitivity of about 600 ppb-h, so that for typical environmental concentrations of 10 to 20 ppb, a sampling period of a few days is sufficient to obtain a measurement.

An improvement in some ways over the Palmes Tube is the Yanagisawa Badge (Yanagisawa and Nishimura, 1982), which employs a baffle to reduce sensitivity to wind velocity, and a smaller length-to-area distance to improve diffusion rate; the badge is about 10 times as sensitive as the Palmes Tube, allowing shorter collection periods.

Radon

Nature of the Exposure. Measurement methods for radon are complicated by the fact that the radon gas itself (which is chemically inert) is not the cause of health effects, but is the source of the radioactive decay products that can enter and lodge in the lungs. These decay products include ^{218}Po, ^{214}Bi, ^{214}Pb, and ^{214}Po. Since the radioactive decays themselves include three types of energy (alpha, beta, and gamma decay) which can require different measurement techniques, several different measurement methods have been developed. However, before describing these methods, several concepts need to be discussed.

First, since some methods depend on measuring radon directly, it is important to know the ratio of the decay products to the parent radon gas (the equilibrium factor). This factor has often been quoted as 0.5, but recent measurements indicate that it may be closer to 0.35 to 0.4.

Second, the decay products listed above may stick to larger particles, which can then be breathed into the lung, or they may enter the lung directly (the "unattached" fraction). Since the particles may plate out on indoor surfaces, it is thought that the unattached fraction may have a higher efficiency in entering the lung. Therefore, it has been thought important to determine the size of this fraction, which for indoor air is not well characterized, but may be between 0.01 and 0.1.

Third, since health effects are proportional to total energy, an equation for the total energy of the radon decay products is required. This quantity is called the Potential Alpha Energy Concentration (PAEC). The total airborne potential alpha energy concentration, PAEC, is calculated as follows:

$$C_p J_m^{-3} = (5.79C_1 + 28.6C_2 + 21.0C_3) \times 10^{-10} \tag{1}$$

where

Cp	=	PAEC in Joules per cubic meter (J_m^{-3})
C_1, C_2, C_3	=	activity concentrations of ^{218}Po, ^{214}Pb, and ^{214}Bi in Becquerels per cubic meter (Bq m^{-3}), respectively.

Exposure can then be expressed as the PAEC multiplied by the length of exposure in hours: Joue-hours per cubic meter (Jhm^{-3}). Unfortunately, the use of old units from the U.S. uranium mining days—the working level and the working level month—persist in the literature. One working level is equal to a PAEC of 0.0206 mJm^{-3} while 1 WLM (170 hours) equals 3.5 mJhm^{-3}.

Shorter-Term Measurement of Radon

- *Scintillation cells.* Radon can be measured directly at environmental levels by using the ability of the emitted alpha particle to excite a ZnS(Ag) scintillator to produce measurable emitted light. Both hemispherical (Lucas 1957) and simple right circular cylindrical (George 1976) detector cells with optically clear flat windows and interior walls coated with ZnS(Ag) have been used. For simple grab measurements, the evacuated cell is opened and filled with room air. The cell is then placed on a photomultiplier tube and the count rate of light pulses is measured. However, the concentration of radon in a building is highly variable. Thus, single grab samples of radon rarely reflect the long-term average indoor concentration. Also, in most cases, the concentration of radon is low so that a long counting interval may be needed to obtain sufficient statistical precision in the total alpha count.

  An alternative approach is an active system to pull air through the scintillation chamber. The counts in a given time period (*e.g.*, 15 minutes) can then be converted to an approximate radon concentration. There are also problems in counting statistics at low radon concentrations so that relatively large scintillation chambers are needed for accurate ambient measurements. Scintillation counting is, therefore, not convenient for very long term monitoring.
- *Electrostatic Collection of $^{218}Po^+$.* An alternative method for semi-continuous measurements of radon is to collect the charged decay product, $^{218}Po^+$, on a scintillation or a solid state alpha detector and relate the number of decays of ^{218}Po to the concentration of ^{222}Rn in the air. It has been found that in almost 90 percent of the decays of ^{222}Rn to ^{218}Po, the Po molecule is a singly charged, positive ion at the end of the recoil path (Wellisch 1913, Porstendörfer and Mercer 1979, Dua *et al.* 1983, Chu and Hopke 1985, Hopke 1989). This high percentage of charged Po molecules has led a number of investigators to use the collection of the Po in an electrostatic field as a measure of the Rn content of the air (Albrecht and Kaul 1967, Wrenn *et al.* 1975, Spitz and Wrenn 1977). However, the method may be adversely affected not only by water vapor (George 1977, Frey *et al.* 1981, Goldstein and Hopke 1985, Chu and Hopke 1988), but also by other common atmospheric species (including NO, NO_2, NH_3 and many organic compounds) with an ionization potential low enough to serve as electron donors (Busigin *et al.* 1981, Chu and Hopke 1988).

Longer-Term Measurement of Radon

- *Activated carbon.* There are two methods available for obtaining an integrated measure of the radon concentration. For short intervals (2 to 7 days), canisters containing activated carbon can be set out into a room (Countess 1970). The radon is adsorbed on the carbon and thus accumulates over time. After the sampling period is over, the canister is sealed and the decay products build up in the container. Since equal activities of the shorter lived decay products will develop after about 4 hours, a measurement of the emitted gamma radiations from the progeny can be used to determine the radon concentrations. However, the carbon canisters are only useful for a limited period and can have difficulties because of water vapor adsorption reducing the amount of adsorbed radon. Standard methods have been defined for the use of activated carbon canisters (George 1984) and these procedures have been adopted by the USEPA for use in general home monitoring (Gray and Windham 1987).

In order to reduce the effects of water vapor and to damp the effects of the variations in radon concentrations on the observed activity, the open area of activated carbon can be covered with a semi-permeable membrane (Pritchard and Marien 1985). Cohen and Nason (1986) developed a diffusion barrier canister with an integration time constant of 84 hours and uses a 1 week (7-day) exposure. George and Weber (1990) have extended the integration time constant to 9 to 11 days by using a sintered metal filter covering the sampling port in the lid of the charcoal canister. It appears that with proper care, charcoal canisters can provide adequate precision, integrated measurements of radon concentrations for periods between 2 and 10 days. However, due to the seasonal variability of the indoor radon levels, these short-term measurements may not provide an adequate estimate of the annual average exposure to ^{222}Rn which people encounter in indoor air.

- *Solid state nuclear track detectors.* A better long term, integrated measurement can be made using a track etch detector. This type of measurement is generally considered to be the best estimate of the annual average exposure of individuals to radon. Thus, track etch detectors are commonly used in national surveys to relate geology to potential radon exposure and epidemiological studies of radon and lung cancer in the general population. This type of measurements is also considered to be the most reliable way to verify the performance of any mitigation measures applied in houses with elevated radon levels.

To perform such measurements two kinds of simple passive radon monitors were developed: "bare" or "open" type and "closed" type. The "open" type monitor measured the total airborne alpha activity by exposure of a bare piece of detecting material to investigated atmosphere. The "closed" type monitor is typically a small diffusion chamber with detecting material inside. Radon diffuses into the chamber through a filter that intercepts dust and airborne radon decay products. The signal is therefore proportional directly to radon gas concentration.

As a detecting material, solid state nuclear track detectors (SSNTD) have been used. There are basically two kinds of materials sensitive to alpha radiation: cellulose nitrate (CN) (Fantini and Renard, 1981) and diethylene glycol bis (allyl carbonate) (PADC) (Cartwright and Shirk 1978). The more commonly used material is PADC plastic (known as CR-39), which is produced by several companies, *e.g.*, American Acrylics, Inc. USA) The "tracks" are narrow (3–10 nm) trails of damage by heavily ionizing particles (*e.g.*, alpha particles). The damage in polymers consists in modification of the polymeric chains lengths and possible production of free radicals. Certain chemical reagents ("etchants") dissolve or degrade these damaged regions at much higher rates than the undamaged material. The etching process enlarges the track, until it is visible under an optical microscope, allowing detail analysis of single tracks for determining the nature of incident particles or simply counting numbers of tracks for dosimetric purposes.

The alpha track counting techniques vary from very simple manual counting with microfiche reader or optical microscope, to very sophisticated microscope—video camera—computer, fully automatic units. Estimated track density is proportional to radon gas concentration. The most common configuration of the radon monitor there is a small plastic cup with one or more pieces of CR-39 inside. For such monitors, a 3-month exposure to average radon concentrations is considered to be sufficient to obtain measurement accuracies of ±15 percent.

- *Electrets.* A recent development is the use of an electret for passive radon measurements (Kotrappa *et al.* 1988). The concept is similar to the track-etch detector in that the detector is placed in a cup so that the radon can enter it. However, the measure of radon exposure is the decrease in surface electric charge on an electret made of Teflon FEP. Such a system has a much simpler readout device that only has to measure the residual surface charge on the electret, and may provide a useful alternative measurement method for periods of several weeks to many months depending on the radon concentration. Their performance in intercomparison tests have been very encouraging. One problem with electrets is that the local gamma ray background also produces ionization that reduces the charge on the electret. Thus, additional monitors are needed to determine typical gamma backgrounds so appropriate corrections can be made to the estimated radon concentration.

Measurement of Radon Decay Products. Measurement of radon decay products is more difficult because it is necessary to determine both the concentration and size distribution for the decay products. The decay product behavior is quite complex because of their ability to attach to airborne particles as well as to indoor macro surfaces such as walls, ceilings, and furnishings. The particle size determines the ability of the radioactivity to be deposited both in the room and in the respiratory system. It is the respiratory deposition that provides the dose to the critical tissues. The sizes of most concern are those less than 10 nm. These highly diffusive particles can more easily deposit in the body whereas only about 20 percent of the particles with diameters around 100 nm are retained in the respiratory tract.

- *Potential Alpha Energy Concentration (PAEC).* Measurement of the total concentration of alpha emitting particulate matter is determined by drawing air though a membrane filter. The collected alpha activity can be measured using either with a ZnS(Ag) scintillator for total alpha counting or with a solid-state detector that provides α-spectroscopic determination of the specific decay products. In the case of gross alpha counts, there are two approaches to extracting the concentrations of the three decay products from a sequence of counts. A series of three count rate measurements at various times following the collection of the sample was first proposed by Tsivoglou *et al.* (1953). The method was improved by Thomas (1970, 1972) by measuring the integrated counts obtained during 3 sequential and different length intervals. This method yields 3 equations in the 3 unknown activity concentrations. Nazaroff (1984) suggested somewhat different time intervals beginning sooner after the collection of the sample in order to reduce the uncertainty in the ^{218}Po activity determination.

  The method developed by Raabe and Wrenn (1969) uses a series of many short counting intervals (typically 1 minute) and then performs a non-linear least squares fit to the growth and decay equations to extract the activity concentrations of the 3 decay products. Details of the method are provided by EML (1990) and the computer programs to calculate the results are described by Knutson (1989).

The proper choice of filter can provide both quantitative collection of activity and an adequate radioactive source for accurate measurement of the collected activity. Membrane filters are generally used to provide a good quality alpha spectrum However, a problem can arise with membrane filters because of their high static charge. In areas where there is a high concentration of ^{218}Po, the charged filter can collect the $^{218}Po^{+}$ ions while it is being placed into the filter holder or being removed to place it in a counter. This problem needs to be considered when planning the sampling procedure.

- *Activity-weighted size distributions: penetration of particles through wire screens.* An equation of penetration through wire screens based on the theory of fibrous filtration was derived by Cheng and Yeh (1980), Cheng *et al.* (1980) and Yeh *et al.* (1982). Scheibel and Porstendörfer (1984) have further verified the model for particle sizes $d_p \approx 4\,nm$. Other work with charged and uncharged ^{218}Po clusters in the 0.5–1.5 nm size range has also indicated general agreement with the wire screen model penetration (Ramamurthi *et al.* 1990). A more detailed study of the problems associated with counting the radioactivity that becomes attached to wire screens was performed by Solomon and Ren (1992). Intercomparisons of activity-weighted size distribution methods has been described by Hopke et al. (1992).

  Initially, penetrations through single screens was used to determine the "unattached" fraction (George 1972, James *et al.* 1972). However, a wide variety of definitions have developed (Ramamurthi and Hopke 1989) and the whole rationale for measuring the "unattached" fraction has changed as more has been learned about respiratory deposition particularly in the extrathorasic region (Hopke 1992).

Ancillary Measurements

Air Exchange. Air exchange is one of several crucial ancillary variables in understanding indoor-outdoor relationships. For example, a "tight" (well insulated) house affords more protection against short-term peak episodes of high ambient particles than a poorly insulated house. Tracer gases such as SF_6 or other perfluorinated tracers (PFTs) not found in nature are generally used to measure air exchange rates.

In one approach, a source with a constant emission rate is placed in the home for the duration of the monitoring period. A collector device (often a sorbent such as activated charcoal) collects an amount of tracer that is inversely proportional to the total air flow through the house. A second approach also uses a continuous source but utilizes a continuous direct readout device to provide more detailed temporal record of air exchange. A third approach is to introduce the tracer into the home over a limited time period and then to record its decay. The latter approach is generally good only for shorter time intervals of 8–16 hours, but has the advantages of (a) using less tracer material, and (b) providing a direct indication of the air exchange rate without having to determine the volume of the structure.

In the first two methods, division of the air flow rate (m^3/h) by the volume of the house (m^3) gives the air exchange rate, in inverse hours. That is, an air exchange rate of

$1\ h^{-1}$ indicates that the home exchanges its volume of air with the outdoors every hour. (In fact, under a well-mixed scenario, a fraction (1/e) of the air in the home at the beginning of the hour remains at the end of the hour.)

Use of the perfluorinated tracers (Dietz and Cote 1982) provides the most accurate measurement of air exchange, but less expensive methods can also be employed. Because of the general decline of CO emissions from automobiles, many homes in suburban neighborhoods have very low CO concentrations just outside; for such homes with gas stoves, the decay of CO concentrations in the home just after extensive cooking has been used to measure air exchange rates of the home.

KEY MONITORING STUDIES

Relevant studies are described below in terms of their overall study design, specific methods, results, and implications.

Living Organisms

Legionella. Arnow *et al.* (1985) surveyed 95 homes (8 houses, 87 apartments) in Chicago for culturable Legionella pneumophila. They collected both hot and cold tap water samples (after 1 minute of water flow) and faucet and shower scrapings. *L. pneumophila* was isolated from 30 (32 percent) of the homes. Nearly all isolates were from hot water, and recovery of the organism was strongly associated with water temperatures <60°C, in which 42 percent were positive. Sera from an occupant of each home was tested for *Legionella* antibodies. Only two subjects with positive environmental cultures had high antibody titers. Overall, antibody titers were not related to the recovery of *Legionella* in the water systems of this group of healthy individuals. This overall level of ~30 percent positive systems and the relationship with hot water temperatures <60°C appears consistent across several studies. While the presence of colonization with this opportunistic pathogen appears to present little risk to normal, healthy individuals, it may be appropriate to recommend maintenance of water temperatures above 60°C in homes of especially susceptible people.

Other Bacteria. Reponen *et al.* (1992) collected 6-stage Andersen samples in 71 homes over a 4 year period. Samples were collected over all seasons, resulting in 162 indoor sampling events and 141 outdoor events. Further details of the sampling plan are not given. Geometric mean bacterial levels in these homes were 570 (GSD-1600/570) overall, with little difference between summer and winter. Indoor levels were consistently higher than those outdoors. Homes with mechanical air supply and exchange had lower levels (in summer only) than either naturally ventilated homes, or homes with only mechanical exchange. The authors propose an upper limit of bacteria in homes of $5000/m^3$ based on the 95th percentile. These data provide supporting evidence for the widely held opinion that indoor bacteria are derived from indoor sources, primarily human occupants. However, the data must be interpreted conservatively. The study design does not provide information on numbers of people in the houses, activity levels of the occupants, or evidence of a paired sampling design (indoor/outdoor) that would allow accurate calculation of indoor/outdoor ratios. Such studies have not been reported for North American homes.

Fungi. Nine epidemiological studies have used air or dust sampling to assess exposure to culturable fungi in homes, and have related such exposure to health outcomes. These studies have been reviewed by Verhoeff and Burge (1997). In six of these studies, significant relationships between exposure measures and symptoms were reported. As examples, a summary of one air sampling study and one dust sampling study are presented here.

(1) Su *et al.* (1998, submitted for publication) investigated 150 homes of school-aged children in Topeka, Kansas. Air sampling data were related to questionnaire-derived health outcomes. Specifically, the group assessed respiratory symptoms in one school-aged child in each house using questionnaires that also assessed smoking status of occupants, and home characteristics that might relate to fungal contamination (*i.e.*, dampness indicators). Culturable fungal levels were measured using duplicate 1 minute Andersen culture plate samples in each of three rooms (living, bedroom, basement) and outdoors. Fine particulates and NO_2 were also measured in each home.

Indoor fungal concentrations (CFU/m^3) ranged from 165.2 to 3608.5 with geometric means of 432.9 for winter and 1628.9 for summer. Outdoor levels levels ranged from 181.0 to 7538.3 (geometric mean = 458.5 for winter and 3656.4 for summer). Dominant kinds of fungi (taxa) recovered were *Cladosporium, Alternaria, Penicillium, Aspergillus, and Epicoccum.*

The presence of defined groups of fungi were associated (using factor analysis) with housing factors such as crawl spaces, presence of gas stoves, and water collection.

Indoor winter log levels of *Cladosporium,* yeast, and *Epicoccum* were related to hay fever [OR = 1.4 ($\rho < 0.025$), 1.27 ($\rho < 0.05$), and 1.27 ($\rho < 0.1$)] controlling for sex, maternal smoking, and levels of NO_2 and fine particulates. Log levels of *Aspergillus* were related to the reporting of asthma symptoms, ($\rho < 0.02$) in a nonlinear way.

The presence of significant relationships between respiratory symptoms in children randomly selected with respect to genetic risk (*i.e.*, family history) of allergic disease and fungal levels is significant, and either reflects a very strong role for fungi in the development of allergic rhinitis and (possibly) asthma, or indicates other pathways for symptom production than allergic sensitization.

(2) Verhoeff *et al.* (1994) measured dust levels of culturable fungi in homes of 60 children and related this measure to reported respiratory symptoms. Dust was collected from bedroom floor and mattress, sieved, and cultured without dilution. Approximately 9 percent of samples were too overgrown to count. Overall, the geometric mean fungal concentration was 8990 (GSD = 2.3). Levels were significantly higher in dust from textile floors compared to smooth floors. Dominant taxa recovered were species of *Alternaria, Aspergillus, Aureobasidium, Cladosporium, Eurotium, Mucor, Penicillium, Scopulariopsis, and Wallemia.* The majority of homes in the study reported no signs of dampness. Fungal levels in mattresses were significantly ($\rho < 0.02$) higher in 7 homes with reported damp spots than than in the 46 homes not reporting such a condition. Other comparisons were not significant. Fungal concentrations did not correlate with humidity measures. No differences were seen between cases (children with reported respiratory symptoms) and controls (children without

reported respiratory symptoms) for total fungi or for any taxon except *Wallemia*, which was higher in mattress dust of cases than of controls. These data are often taken as an indication that fungi are not involved in respiratory disease in homes. However, it is perhaps as likely that a one-time dust measurement may poorly represent exposure.

Outdoor Spores and Pollen

(1) The American Academy of Allergy, Asthma, and Immunology operates a network of outdoor pollen and spore counting stations across the U.S., each of which collects data at least 3 days/week during the growing season, and reports all data to a central station yearly for analysis. Some stations also report to an 800 number daily.

Stations use either the rotorod rotating arm impactor or the Burkard recording spore trap placed on a rooftop at least 1 story above ground. Samples are counted daily using 400× (pollen) and 1000× (fungi) light microscopy. Each counter at each station is certified by a central laboratory. Certification consists of correctly counting and identifying pollen and a specific set of fungal spores on standard slides.

Summary data from all 85 stations currently in the network is published by the AAAAI (Milwaukee, Wisconsin) in an annual report. Daily averages are summarized graphically, and yearly medians and quartiles for a few kinds of particles are compared within and between stations to a national average. Using these methods, the 5 year national median ragweed pollen concentration (from only those stations reporting ragweed pollen) is 8 grains/m^3.

For these comparisons (based on all available data from all stations in the network), levels from 0 to the median are considered low. From the median to the 75th percentile are considered moderate. From the 75th to the 90th percentile are considered high. Above the 90th percentile is considered very high. The authors note that the categories are representative of the environment, not of health risk. For ragweed pollen in grains/m^3 (AAAAI 1994) LOW = <8, MODERATE = 8 to 44, HIGH = >44 to138, and VERY HIGH = >138. For grass pollen (AAAAI 1995) LOW = <5, MODERATE = 5 to 14, HIGH = >14 to 54, and VERY HIGH = >54.

(2) Neas *et al.* (1996) monitored biological particles in outdoor air over one summer season in State College Pennsylvania, and assessed daily symptoms in 108 children by peak flow measurements and diary reports. The Burkard spore trap was used to generate 24 hour average concentrations of pollens and fungal spores, identified to genus where possible. An increment of 10,000 *Cladosporium* spores was associated with a 1 liter/min decrement in peak flow the following morning, while an increment of 60 *Epicoccum* spores was associated with a 1.5 liter/min decrement in morning peak flow and increased the odds of cough the next morning (OR = 1.8, 95 percent CI 1.0–3.2). These outdoor data support Su *et al.* (in press) findings for *Cladosporium* and *Epicoccum*, and provide further evidence that exposure to fungi can affect respiratory function in children. Also of note is the fact that much fungal exposure occurs outside, and any studies of indoor exposure/health effects requires rigorous control of this important background exposure.

Animal Allergens

Chew *et al.* (1998) evaluated associations between answers to questions on a technician-administered questionnaire with measures of specific allergens in floor (living room and child's bedroom) and child's bed dust. Concentrations of allergens were categorized for the analysis. Approximately 500 homes were visited. Of these, there was insufficient dust for any analysis in 2.4 percent, 4.6 percent and 40.2 percent of the living room, bedroom, and bed samples respectively. No dust was generally interpreted to mean no allergen. Dust mite allergen levels were >10 µg/g in 25 percent of floor dust samples, and 8 percent of bed samples. Cat allergen was present in 93 percent of homes in levels >8 µg/g when a cat was present, but also in 3 percent of homes without reported cats. Cockroach allergen was present in concentrations >2 U/g in 8 percent of floor dust samples, and ~10 percent of bed dust samples.

Housing characteristics that predicted high levels of mite allergen included type of building (house/duplex *versus* apartment), cloth mattress *versus* plastic, carpeted *versus* smooth floors and sampling season (fall, winter, summer). Factors predicting detectable cockroach allergen in kitchen dust were apartment *versus* single-family home, and use of boric acid *versus* no boric acid. These studies make clear that questionnaires alone provide insufficient information to accurately predict risk of exposure to these important biological agents.

Ingram *et al.* (1995) measured dust mite, cockroach, cat and dog allergens in 109 homes in Los Alamos, New Mexico, a city with little rain and, at a high elevation (7200 feet), very dry. Multiple dust samples were collected over a two-week period in September in each home, and were assayed immunochemically for Der p 1, Der f 1, Bla g 2 (cockroach), Can f 1, and Fel d 1. The highest value for each home was used in the analysis. Skin tests, specific IgE, and a test of bronchial hyperreactivity (histamine challenge), were used to group patients as: (1) reactive with symptoms, (2) reactive without symptoms, or (3) controls. Cockroach allergen was recovered from only 2 homes. Mite allergen levels (Der p 1 + Der f 1) ranged from below detection to about 9 µg/g with 96 percent of homes <2 µg/g. All homes with a cat and/or dog had detectable cat and/or dog allergen. Levels in homes with cats ranged from 1 to >100 and all but three were >10. In homes without cats, levels ranged from undetectable to 100, with most below 8.

Astygma symptoms were related to sensitization but not measured concentrations of allergens. These data indicate that allergen is abundant in homes housing animals, and may be present in high concentrations in the absence of pets. Symptoms are related to sensitization rather than concurrent exposure, at least as measured in dust samples.

Platt-Mills *et al.* (1986) conducted a study in 12 Virginia homes designed to reveal seasonal patterns of variation in dust mite allergen concentrations. They collected dust samples monthly over one year from bedding, carpet, and/or sofas. Dust was analyzed by radioimmunoassay for cross-reacting determinants on the major allergens of *Dermatophagoides pteronyssinus* and *D. Farinae*. They also counted mite bodies and used a RAST assay to measure rye grass pollen allergens in some samples. Overall, dust mite allergen levels peaked in the period from August to December, and geometric mean levels for sofa dust across all homes were higher than those for carpets (significant) and bedding (not significant). In at least one home, mite allergen levels were high throughout the year, and in at least two others, levels were consistently low. Mite body counts reached a peak before the mite allergen peak for the three homes for which data are presented. Grass pollen allergen concentrations peaked sharply in June (as would be expected), and reached baseline before the beginning of August.

However, similar kinds of data collected from 20 homes in Boston do not support a strong late-summer peak for mite allergens. Instead, there appear to be strong differences between types of housing (apartments *versus* single family or duplex dwellings) with levels in apartments consistently low throughout the year, and those in single family or duplexes consistently high (Chew *et al.* 1998).

Microbial Toxins

There are no reported studies that have monitored for microbial toxins in a cross-section of indoor environments. Information on residential exposure to endotoxin is not available, and data on mycotoxins are confined to cross-sectional studies where air or dust samples have been collected and analyzed for culturable fungi, and case study reports. In these, mycotoxin exposure has generally been inferred from presence of fungi known to produce toxins. In some cases, strains have been isolated and toxin production demonstrated in culture. In a few, toxins have been measured in sources (bulk samples) and exposure inferred. No studies have documented airborne levels of mycotoxins in residences.

VOCs

Total Exposure Assessment Methodology (TEAM) Studies. Between 1979 and 1987, the USEPA carried out the TEAM (Total Exposure Assessment Methodology) studies to measure personal exposures of the general public to VOCs in several geographic areas in the United States (Pellizzari *et al.* 1987a,b, 1988, 1990, Wallace 1987a). About 20 target VOC were included in the studies, which involved about 750 persons, representing 750,000 residents of the areas. Each participant carried a personal air quality monitor containing 1.5 g Tenax. A small battery-powered pump pulled about 20 L of air across the sorbent over a 12-hour period. Two consecutive 12-hour personal air samples were collected for each person. Concurrent outdoor air samples were also collected in the participants' back yards. In the studies of 1987, fixed indoor air samplers were also installed in the living room of the homes.

The initial TEAM pilot study (Wallace *et al.* 1982) in Beaumont, Texas and Chapel Hill, North Carolina indicated that personal exposures to about a dozen VOCs exceeded outdoor air levels, even though Beaumont, Texas has major oil producing, refining, and storage facilities. These findings were supported by a second pilot study in Bayonne-Elizabeth, New Jersey (another major chemical manufacturing and petroleum refining area) and Research Triangle Park, North Carolina (Wallace *et al.* 1982). A succeeding major study of 350 persons in Bayonne-Elizabeth (Wallace *et al.* 1984, 1985, 1986) and an additional 50 persons in a nonindustrial city and a rural area (Wallace *et al.* 1987a) reinforced these findings (Table 4-2a). A second major study in Los Angeles and Antioch-Pittsburg, California (Pellizzari *et al.* 1988, Wallace *et al.* 1988), with a follow-up study to Los Angeles in 1987 (Wallace *et al.* 1991a) added a number of VOCs to the list of target chemicals with similar results (Table 4-2b).

Major findings of these TEAM Studies included the following:

- Personal exposures exceeded median outdoor air concentrations by factors of 2 to 5 for nearly all of the 11 prevalent VOCs. The difference was even larger (factors of 10 or 20) when the maximum values were compared (Figure 4-1).

Table 4-2a. Weighted estimates of air and breath concentrations of 11 prevalent compounds for 130,000 Elizabeth-Bayonne residents (fall 1981), 110,000 residents (summer 1982), and 49,000 residents (winter 1983)

	Season I (Fall)			Season II (Summer)			Season III (Winter)		
Agent	Personal air (340)	Outdoor air (86)	Breath (300)	Personal air (150)	Outdoor air (60)	Breath (110)	Personal air (49)	Outdoor air (9)	Breath (49)
1,1,1-Trichloroethane	94[a]	7.0[a]	15[b]	67	12	15	45	1.7	4.0
m,p-Dichlorobenzene	45	1.7	8.1	50	1.3	6.3	71	1.2	6.2
m,p-Xylene	52	11	9.0	37	10	10	36	9.4	4.7
Tetrachloroethylene	45	6.0	13	11	6.2	10	28	4.2	11
Benzene	28	9.1	19	NC[c]	NC	NC	NC	NC	NC
Ethylbenzene	19	4.0	4.6	9.2	3.2	4.7	12	3.8	2.1
o-Xylene	16	4.0	3.4	12	3.6	5.4	13	3.6	1.6
Trichloroethylene	13	2.2	1.8	6.3	7.8	5.9	4.6	0.4	0.6
Chloroform	8.0	1.4	3.1	4.3	13	6.3	4.0	0.3	0.3
Styrene	8.9	0.9	1.2	2.1	0.7	1.6	2.4	0.7	0.7
Carbon tetrachloride	9.3	1.1	1.3	1.0	1.0	0.4	ND[d]	ND	ND
Total (11 agents)	338	48	80	200	59	66	216	25	31

[a] Average of arithmetic means of day and night 12-hour samples (μg/m³).
[b] Arithmetic mean.
[c] Not calculated—high background contamination.
[d] Not detected in most samples.

Table 4-2b. Weighted estimates of air and breath concentrations of 19 prevalent compounds for 360,000 Los Angeles residents (February 1984), 330,000 Los Angeles residents (May 1984), and 91,000 Contra Costa residents (June 1984)

	Los Angeles I (February 1984)			Los Angeles II (May 1984)			Contra Costa (June 1984)		
Agent	Personal air (110)	Outdoor air (24)	Breath (110)	Personal air (50)	Outdoor air (23)	Breath (50)	Personal air (76)	Outdoor air (10)	Breath (67)
1,1,1-Trichloroethane	96[a]	34[a]	39[b]	44	5.9	23	16	2.8	16[b]
m,p-Xylene	28	24	3.5	24	9.4	2.8	11	2.2	2.5
m,p-Dichlorobenzene	18	2.2	5.0	12	0.8	2.9	5.5	0.3	3.7
Benzene	18	16	8.0	9.2	3.6	8.8	7.5	1.9	7.0
Tetrachloroethylene	16	10	12	15	2.0	9.1	5.6	0.6	8.6[b]
o-Xylene	13	11	1.0	7.2	2.7	0.7	4.4	0.7	0.6
Ethylbenzene	11	9.7	1.5	7.4	3.0	1.1	3.7	0.9	1.2
Trichloroethylene	7.8	0.8	1.5	6.4	0.1	1.0	3.8	0.1	0.6
n-Octane	5.8	3.9	1.0	4.3	0.7	1.2	2.3	0.5	0.6
n-Decane	5.8	3.0	0.8	3.5	0.7	0.5	2.0	3.8	1.3
n-Undecane	5.2	2.2	0.6	4.2	1.0	0.7	2.7	0.4	1.2
n-Dodecane	2.5	0.7	0.2	2.1	0.7	0.4	2.1	0.2	0.4
α-Pinene	4.1	0.8	1.5	6.5	0.5	1.7	2.1	0.1	1.3
Styrene	3.6	3.8	0.9	1.8	—	—	1.0	0.4	0.7
Chloroform	1.9	0.7	0.6	1.1	0.3	0.8	0.6	0.3	0.4
Carbon tetrachloride	1.0	0.6	0.2	0.8	0.7	0.2	1.3	0.4	0.2
1,2-Dichloroethane	0.5	0.2	0.1	0.1	0.06	0.05	0.1	0.05	0.04
p-Dioxane	0.5	0.4	0.2	1.8	0.2	0.05	0.2	0.1	0.2
o-Dichlorobenzene	0.4	0.2	0.1	0.3	0.1	0.04	0.6	0.07	0.08
Total (19 agents)	240	120	80	150	33	56	72	16	44

[a] Average of arithmetic means of day and night 12-hour samples (μg/m^3).
[b] One very high value removed.

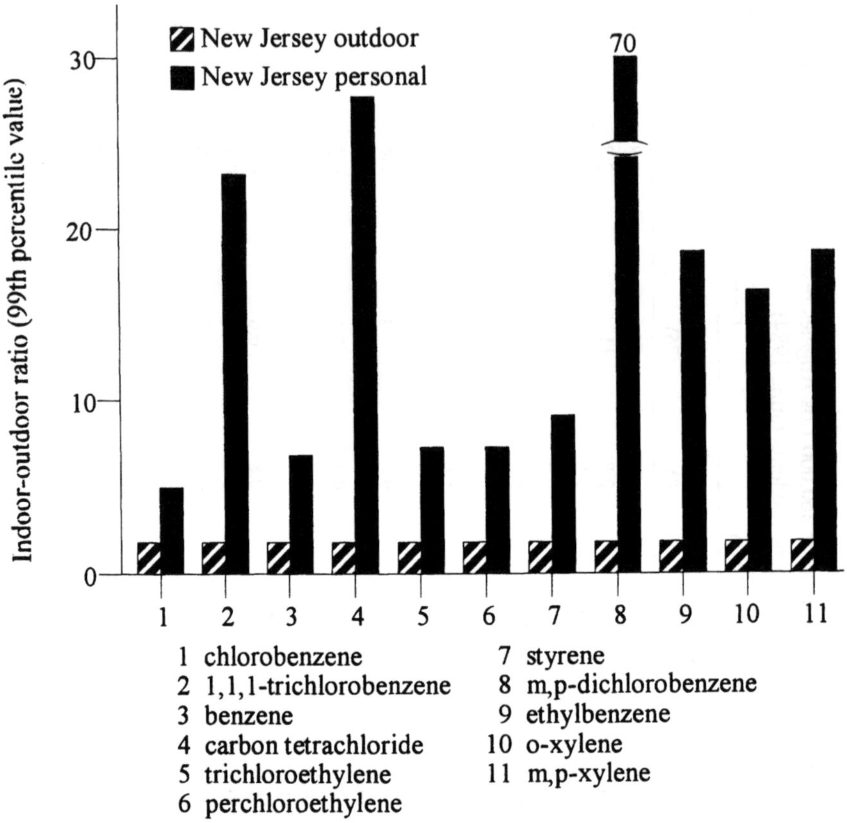

Figure 4-1. Comparison of unweighted 99th percentile concentrations of 11 prevalent chemicals in overnight outdoor air and overnight personal air in New Jersey during the fall of 1981.

This is so despite the fact that most of the outdoor samples were collected in areas with heavy industry (New Jersey) or heavy traffic (Los Angeles).

- Major sources (Table 4-1) are consumer products (bathroom deodorizers, moth repellents); personal activities (smoking, driving); and building materials (paints and adhesives). In the US, one chemical (carbon tetrachloride) has been banned from consumer products and exposure is limited to the global background of about 0.7 $\mu g/m^3$.
- Traditional sources (automobiles, industry, petrochemical plants) contributed only 20–25 percent of total exposure to most of the target VOCs (Wallace 1991). No difference in exposure was noted for persons living close to chemical manufacturing plants or petroleum refineries.

A more recent study of benzene and toluene in 293 California homes (Wilson *et al.*, 1993a,b, 1995, Colome *et al.* 1994) resulted in some interesting differences between the two agents. For benzene, 48-h average indoor concentrations correlated fairly well with outdoor levels, but for toluene, almost no correlation was observed. This is likely due to the much wider use of toluene in consumer products. Major variables associated with higher net indoor benzene levels were presence of a gas furnace and having two cars parked in an attached garage. For toluene, a particular brand of furnace had the highest partial correlation with net indoor toluene concentrations; apartments also had higher concentrations.

Another study of benzene in four New Jersey homes was focused on the extent of contamination from attached garages (Thomas *et al.* 1993). Each home was monitored for either 6 or 10 consecutive 12-hour periods. At all four homes, garage levels of benzene were higher than outdoors, and at three of the four homes the garage levels were higher than in the living area. Air exchange measurements made it possible to calculate the amount of benzene transferred from the garage to the living area in the four homes; in the home without elevated benzene levels in the garage, the total air flow between the garage and the living area was extremely small. Benzene concentrations in the garages ranged from 5–200 $\mu g/m^3$, and the 12-h average source strength estimates ranged from 730 to 26,000 µg/h. The mere presence of an attached garage was not a significant factor in affecting living area benzene concentrations. However, the total number of hours the car was parked in the garage had a significant effect on living area benzene concentrations, as did the mass flow rate of benzene from the garage to the home.

During the most recent National Health and Nutrition Examination Survey (NHANES), the isotope dilution method for blood described above was employed to look for about 20 VOCs in approximately 1000 persons representing the United States population (Ashley *et al.* 1994). Essentially the same dozen or so VOCs observed in the TEAM Studies were prevalent in the blood of the respondents.

A study of personal exposure to benzene was carried out in Valdez, Alaska to determine the contribution of the major petroleum terminal (Goldstein *et al.* 1992). Fifty-eight residents wore Tenax personal samplers for one day in summer and winter, while simultaneous indoor and outdoor measurements were made in or near their homes. To estimate the impact of the terminal, a tracer gas (SF_6) was emitted continuously from the terminal. The amount of SF_6 on the residents' charcoal collectors could be used to estimate the proportion of benzene exposure contributed by the terminal. This proportion was found to be 11 percent, even though the terminal accounted for at least 90 percent of the total atmospheric emissions of benzene.

Indoor Air: Concentrations. As illustrated in Table 4-3, three large studies of VOCs, involving 300 to 800 homes, have been carried out in the Netherlands (Lebret *et al.* 1986), West Germany (Krause *et al.* 1987) and the U.S. (Wallace 1987a). A smaller study of 15 homes was carried out in Northern Italy (De Bortoli *et al.* 1986). A recent study was carried out in a number of homes in Canada (Fellin and Otson 1993). Observed concentrations were remarkably similar for most chemicals, indicating similar sources in these countries. One exception is chloroform, present at typical levels of 1–4 $\mu g/m^3$ in the U.S. but not found in European homes. This is to be expected, since the likely source is volatilization from chlorinated water (Dowty *et al.* 1975, Wallace *et al.* 1982, Andelman 1985a,b, 1990, McKone 1987, Giardino *et al.* 1990, Jo *et al.* 1990a,b). The two European countries do not chlorinate their water.

Major findings of these indoor air studies include:

- Indoor levels in homes and older buildings (>1 year) are typically several times higher than outdoor levels. Sources include dry-cleaned clothes, cosmetics, air fresheners, and cleaning materials.
- New buildings (<1 month) have levels of some VOC (aliphatics and aromatics) 100 times higher than outdoor, falling to 10 times outdoor about 2–3 months later. Major sources include paints and adhesives.
- About half of 750 homes in the U.S. had total VOC levels (obtained by integrating the total ion current response curve of the mass spectrometer) greater

Table 4-3. Volatile organic concentrations in indoor air in Germany and The Netherlands compared with personal exposures in the United States. NM = not measured

	Concentration (μg/m³)					
	Arithmetic mean		Median in	Maximum		
Compound/Class	West Germany[a]	United States[b]	The Netherlands[c]	West Germany	United States[d]	The Netherlands
Chlorinated						
Chloroform	NM	3	NM	NM	210	NM
1,1,1-Trichloroethane	9	52	NM	260	8,300	NM
Trichloroethylene	11	6	<2	120	350	106
Tetrachloroethylene	14	16	<2	810	250	205
p-Dichlorobenzene	14	25	1	1,260	1,600	299
Aromatic						
Benzene	10	16	6	90	510	148
Styrene	2	3	NM	41	76	NM
Ethylbenzene	10	9	2	160	380	138
o-Xylene	7	9	10[e]	45	750	753
m + *p*-Xylene	23	26	—	—	300	3,100
Toluene	84	NM	35	1,710	NM	2,252
Aliphatic						
Octane	5	4[f]	1	92	122	533
Nonane	10	12[f]	4	140	177	407
Decane	15	6[f]	10	240	161	905
Undecane	10	8[f]	6	120	385	445
Dodecane	6	4[f]	2	72	72	118
Terpenes						
α-Pinene	10	4[f]	NM	120	208	NM
Limonene	28	43[f]	30	320	2,530	773

[a] Kraus *et al.* 1987.
[b] Wallace and Clayton (1987), 24-hour averages, 526 persons in New Jersey and California.
[c] Lebret *et al.* (1986), 1-week averages, 319 homes in The Netherlands.
[d] Wallace (unpublished), 12-hour averages, overnight maxima in 666 homes in New Jersey, California, Maryland.
[e] *o* + *m* + *p* isomers.
[f] Wallace (unpublished), 24-hour averages, 315 persons in California and Maryland.

than 1 mg/m³, compared to only 10 percent of outdoor samples (Wallace *et al.* 1991a,b)

- More than 500 VOCs were identified in four buildings in Washington, D.C., and Research Triangle Park, North Carolina (Sheldon *et al.* 1988a).

One study (Wallace *et al.* 1989) involved seven volunteers undertaking about 25 activities suspected of causing increased VOC exposures; a number of these activities (using bathroom deodorizers, washing dishes, cleaning an auto carburetor) resulted in 10–1000-fold increases in 8-h exposures to specific VOCs. A recent study of 170 homes in Avon, England found mean indoor levels of benzene to be 8 μg/m³, compared to outdoor concentrations of 5 μg/m³ (Brown and Crump 1996). The study employed passive Tenax tubes to collect 28-day indoor and outdoor samples. These results were in good agreement with the levels of 10 μg/m³ indoors and 6 μg/m³ outdoors at 50 homes in Los Angeles measured over two seasons in 1987 (Wallace *et al.* 1991a).

A transport mode of particular interest in assessing residential exposures is the movement of soil gas into basements. This is of particular interest in situations involving plumes of hydrocarbons emanating from gasoline spills, hazardous waste sites, and leaking underground storage tanks. Fundamental physical theory and experiments have been developed, notably by the Indoor Air program at Lawrence Berkeley Laboratories, to determine the important variables affecting soil-gas transport, as well as corrective actions that may be taken (Hodgson *et al.* 1988). It appears that the same forces (*e.g.*, negative pressure) that allow radon to enter buildings in soil gas can also result in VOCs and perhaps some termiticides to enter. Some of the same protective actions against radon (*e.g.*, rerouting soil gas into the atmosphere) may also reduce entry of VOCs.

Indoor Air: Sources. Early studies of organics indoors were carried out in the 1970s in the Scandinavian countries (Johansson 1978, Mølhave and Møller 1979, Berglund *et al.* 1982a,b). Mølhave (1982) showed that many common building materials used in Scandinavian buildings emitted organic gases. Seifert and Abraham (1982) found higher levels of benzene and toluene associated with storage of magazines and newspapers in German homes. Early U.S. measurements were made in nine Love Canal residences (Pellizzari *et al.* 1979), 34 Chicago homes (Jarke and Gordon 1981), and in several buildings (Hollowell and Miksch 1981).

Hundreds of VOCs have been identified in environmental tobacco smoke (Higgins 1987, Guerin *et al.* 1987, Jermini *et al.* 1976, Löfroth *et al.* 1989), which contaminates about 60 percent of all U.S. homes and workplaces (Repace and Lowrey 1980, 1985). Among these are several human carcinogens, including benzene. Benzene was elevated in the breath of smokers by a factor of 10 above that in the breath of nonsmokers (Wallace *et al.* 1987a,b.) The amount of benzene in mainstream smoke appears to be directly related to the amount of tar and nicotine in the cigarette (Higgins *et al.* 1983). In the U.S., it is calculated that the 50 million smokers are exposed to about half of the total nationwide "exposure budget" for benzene (Wallace 1989, 1990). Other indoor combustion sources such as kerosene heaters (Traynor *et al.* 1990) and woodstoves (Highsmith *et al.* 1988) may emit both volatile and semivolatile organic compounds.

Later studies also investigated building materials (Sheldon *et al.* 1988a,b) but added cleaning materials and activities such as scrubbing with chlorine bleach or spraying insecticides (Wallace *et al.* 1987c) and using adhesives (Girman *et al.* 1986) or paint removers (Girman *et al.* 1987a,b). Knoppel and Schauenburg (1987) studied VOC emissions from 10 household products (waxes, polishes, detergents). Nineteen alkanes, alkenes, alcohols, esters and terpenes were among the chemicals emitted at the highest rates from the ten products. All of these studies employed either head-space analysis or chambers to measure emission rates.

Other studies estimated emission rates from measurements in homes or buildings. For example, Wallace (1987a) estimated emissions from a number of personal activities (visiting dry cleaners, pumping gas) by regressing measurements of exposure or breath levels against the specified activities. Girman and Hodgson (1987) extended their chamber studies of paint removers to a residence, finding similar (and very high—ppm) concentrations of methylene chloride in this more realistic situation. Thomas *et al.* (1991) estimated emissions of tetrachloroethylene from dry cleaned garments in homes.

The U.S. National Aeronautics and Space Agency (NASA) has measured organic emissions from about 5000 materials used in space missions (Nuchia 1986). Perhaps 3000 of these materials are in use in general commerce (Ozkaynak *et al.* 1987). The chemicals emitted from the largest number of materials included toluene (1896 materials), methyl ethyl ketone (1261), and xylenes (1111).

A 41-day chamber study (Berglund *et al.* 1987) of aged building materials taken from a "sick" preschool indicated that the materials had absorbed about 30 VOCs, which they re-emitted to the chamber during the first 30 days of the study. Only 13 of the VOCs originally present in the first days of the study continued to be emitted in the final days, indicating that these 13 were the only true components of the materials. This finding has significant implications for remediating "sick buildings". Even if the source material is identified and removed, weeks may be needed before re-emission of organics from sinks in the building stops.

Emission rates of most chemicals in most materials are greatest when the materials are new. For "wet" materials such as paints and adhesives, most of the total volatile mass may be emitted in the first few hours or days following application (Tichenor and Mason 1987, Tichenor *et al.* 1990). USEPA studies of new buildings indicated that eight of 32 target chemicals measured within days after completion of the building were elevated 100-fold compared to outdoor levels: xylenes, ethylbenzene, ethyltoluene, trimethylbenzenes, decane, and undecane (Sheldon *et al.* 1988b). The half-lives of these chemicals varied from two to six weeks; presumably some other nontarget chemicals, such as toluene, would have shown similar behavior. The main sources were likely to be paints and adhesives. Thus new buildings would be expected to require about six months to a year to decline to the VOC levels of older buildings.

For dry building materials such as carpets and pressed wood products, emissions are likely to continue at low levels for longer periods. Formaldehyde from pressed wood products may be slowly emitted with a half-life of several years (Breysse 1984). According to several recent studies, 4-phenylcyclohexene (4-PC), a reaction product occurring in the styrene-butadiene backing of carpets, is the main VOC emitted from carpets after the first few days. 4-PC is likely to be largely responsible for the new carpet odor.

A major category of human exposure to toxic and carcinogenic VOCs is room air fresheners and bathroom deodorants. Since the function of these products is to maintain an elevated indoor air concentration in the home or the office over periods of weeks (years with regular replacement), extended exposures to the associated VOCs are often the highest likely to be encountered by most (nonsmoking) persons. The main VOCs used in these products are ρ-dichlorobenzene (widely used in public restrooms), limonene, and α-pinene. The first is carcinogenic to two species (NTP 1986), the second to one (NTP 1988), and the third is mutagenic. Limonene (lemon scent) and α-pinene (pine scent) are also used in many cleaning and polishing products, causing short-term peak exposures during use, but which might not provide as much total exposure as the freshener.

Awareness is growing that most exposure comes from these small nearby sources. In California, Proposition 65 focuses on consumer products, requiring makers to list carcinogenic ingredients. Bills focusing on indoor air were introduced in both the U.S. House and the Senate in 1989 and 1990. ETS was declared a known human carcinogen by the USEPA in 1991. Smoking has been banned from many public places and many private workplaces during the last few years.

Outdoor Air. Outdoor air levels of many of the most common VOCs, even in heavily industrialized areas or areas with high densities of vehicles, are usually considerably lower than indoor levels (Wallace 1987a). This fact has not been fully recognized or incorporated into regulations. For example, in the U.S., the recent (1990) reauthorization of the Clean Air Act continues to deal only with outdoor air ("air external to buildings"), while adding 189 toxic chemicals to the list of those to be regulated. Many of these chemicals, which include common solvents and household pesticides, have been shown to be far more prevalent and at higher concentrations in homes than outdoors.

For many hydrocarbons, the major source of outdoor air levels is gasoline vapor or auto exhaust (Sigsby *et al.* 1987, Zweidinger *et al.* 1988). For example, about 85 percent of outdoor air benzene levels in the U.S. is from mobile sources and only about 15 percent from stationary sources. Exposure to certain aromatics (benzene, toluene, ethylbenzene, xylenes) while inside the automobile can exceed ambient levels by a factor of six or so (SCAQMD 1989).

Particles

Three large-scale studies of airborne particles inside U.S. homes have been performed. In chronological order, these are:

- The Harvard Six-City study, carried out by the Harvard School of Public Health beginning in 1979 and continuing through 1988, with measurements taken in at least 1400 homes (Spengler *et al.* 1981)
- The New York State study, carried out by Research Triangle Institute in 433 homes in two New York State counties in 1986 (Sheldon *et al.* 1989), and
- The USEPA Particle TEAM (PTEAM) Study, carried out by Research Triangle Institute and Harvard University School of Public Health in 178 homes in Riverside, California in 1990 (Pellizzari *et al.* 1991, Ozkaynak *et al.* 1993, 1996a,b).

These are the most complete investigations of indoor and outdoor concentrations of particles.

The Harvard Six-City Study. The Harvard Six-City study is a prospective epidemiological study of the effects of particles and sulfur oxides on the health of grade school children. The six cities were chosen to represent low (Portage, Wisconsin and Topeka, Kansas), medium (Watertown, Massachusetts, and Kingston-Harriman, Tennessee), and high (St. Louis, Missouri and Steubenville, Ohio) outdoor particle and sulfate levels.

The study took place in two measurement phases. The first phase involved monitoring about 10 homes in each city for respirable particles ($PM_{3.5}$) (Spengler *et al.* 1980, 1985, Dockery and Spengler 1981a,b). The homes were measured every sixth day (24-h samples) for one to two years. In the second phase, a larger sample of 200 to 300 homes was selected from each city, with two consecutive week-long $PM_{2.5}$ samples collected both indoors and outdoors in summer and again in winter. More than 1400 homes were monitored.

Spengler *et al.* (1981) described the first five years of the Harvard Six-City study. During that time pulmonary function measurements were administered to 9000 adults and 11000 children in grades 1 to 6. Mean indoor concentrations exceeded the outdoor levels in all cities except Steubenville, where the outdoor levels of about 46 $\mu g/m^3$ slightly exceeded the indoor mean of about 43 $\mu g/m^3$. The authors noted that the major source of indoor particles is cigarette smoke, with levels of 24 $\mu g/m^3$ in homes without smokers, 36 in homes with one smoker, and 70 in homes with two smokers.

Neas *et al.* (1994) presented summary results for the entire second phase of the Six-City Study (1983–88). Homes with children 7 to 11 years of age, who had never smoked, were eligible for the study. A total of 1,273 homes with white children completed two weeks of summer and winter monitoring indoors and outdoors for $PM_{2.5}$ using the Harvard impactor. The annual (winter and summer) household $PM_{2.5}$ mean concentration for the 580 children living in consistently smoking households was 48.5 ± 1.4 (SE) $\mu g/m^3$ compared to 17.3 ± 0.5 $\mu g/m^3$ for the 470 children in consistently nonsmoking households. The distribution of exposures for children in three smoking categories is shown in Figure 4-2.

Spengler *et al.* (1987) reported on a new round of measurements in three communities—Watertown, St. Louis, and Kingston-Harriman,—within the Six-City study. In each community, about 300 children were selected to take part in a year-long diary and indoor air quality study. Measurements of $PM_{2.5}$ were taken indoors at home for two consecutive weeks in winter and again in summer. The sampler was the automated Harvard sampler (Marple *et al.* 1987), which collected an integrated sample for the week except for the five 8AM to 4PM weekday periods when the child was at school. During this 40-hour period, samples were taken in one classroom in each of the elementary schools involved. Results were presented for smoking and nonsmoking homes in each city by season (Figure 4-3). The authors noted that mean concentrations in homes with smokers were about 30 $\mu g/m^3$ greater than homes without smokers. The difference was greater in winter than in summer in all cities.

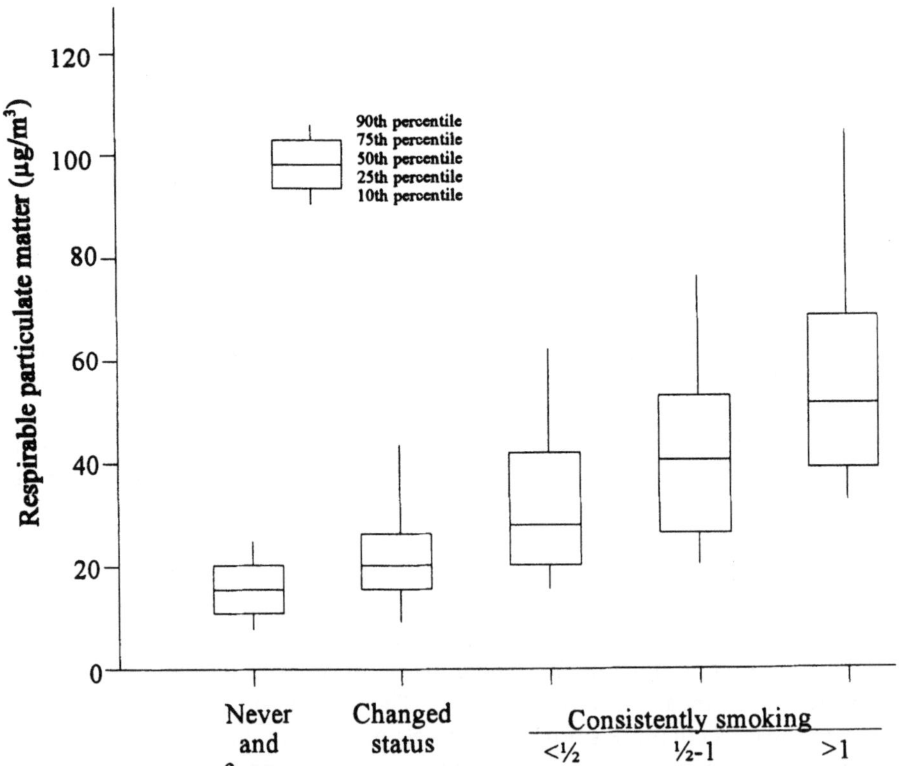

Figure 4-2. Annual average concentrations of indoor $PM_{2.5}$ by household smoking status and estimated number of cigarettes smoked in the home. Neas *et al.* 1994.

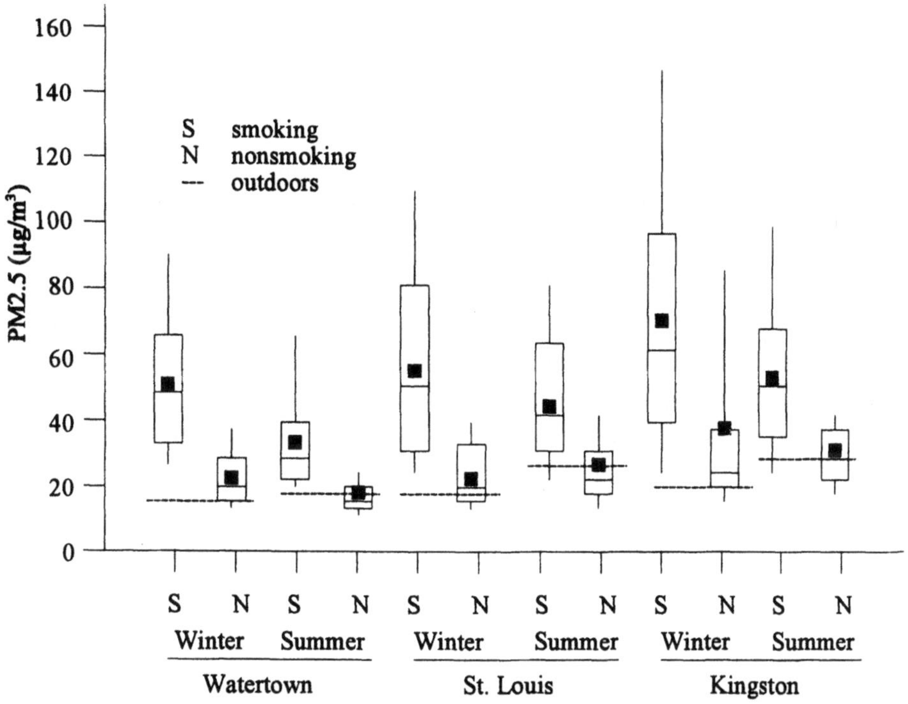

Figure 4-3. $PM_{2.5}$ ($\mu g/m^3$) in smoking (S) and nonsmoking (N) homes in three of the Harvard Six-City study sites.

Santanam *et al.* (1990) reported on a more recent and larger-scale monitoring effort in Steubenville and Portage as part of the Harvard Six-City study. One-hundred-and-forty homes in each city, split evenly between those with and without smoking, were monitored for one week in summer and in winter. The Harvard impactor sampler was used with an automatic time unit to collect $PM_{2.5}$ samples between 4 pm and 8 am on weekdays and all day on weekends, corresponding to likely times of occupancy for school-age children. Outdoor samples were collected from one site in each city. Elements were determined by x-ray fluorescence (XRF). Source apportionment using principal components analysis (PCA) and linear regressions on the elemental data was performed. Cigarette smoking was the single largest source, accounting for 20 to 27 $\mu g/m^3$ indoor $PM_{2.5}$ in Steubenville, and 10 to 25 $\mu g/m^3$ in Portage. Homes with smokers exceeded outdoor levels by 25 (summer) and 20 (winter) $\mu g/m^3$ in Steubenville, and 24 and 11 $\mu g/m^3$ in Portage.

The New York State Study. Sheldon *et al.* (1989) studied $PM_{2.5}$ and other agents in 433 homes in two New York State counties. One goal of the study was to determine the effect of kerosene heaters, gas stoves, wood stoves or fireplaces, and cigarette smoking on indoor concentrations of combustion products. A stratified design to include all 16 combinations of the four combustion sources was implemented, requiring about 22,000 telephone calls.

The sampler was a portable dual-nozzle impactor (Marple *et al.* 1987). Samples were collected in the main living area and in one other room (containing a combustion source if possible) during alternate 15-minute periods over a 7-day period. Outdoor samples were collected at a subset of 57 homes. All samples were collected during the winter (January to April) of 1986.

Mean indoor $PM_{2.5}$ concentrations were approximately double those outdoors in both counties (Table 4-3). Of the four combustion sources, only smoking created significantly higher indoor $PM_{2.5}$ concentrations in both counties. Use of kerosene heaters was associated with significantly higher concentrations in Suffolk (N = 22) but not in Onondaga (N = 13). Use of wood stoves/fireplaces and gas stoves did not elevate indoor concentrations in either county.

Leaderer and Hammond (1991) continued their analysis of the New York State data by selecting a subset of 96 homes for which both nicotine and $PM_{2.5}$ data were obtained. In the 47 homes in which nicotine was detected (detection limit = 0.1 µg/m³), the mean concentration of RSP was 44.1 (±25.9 SD) µg/m³ compared to 15.2 (±7.4) µg/m³ in the 49 homes where no nicotine was detected. Thus homes with smoking had an increased weekly geometric mean $PM_{2.5}$ concen-tration of about 29 µg/m³. Regressions of $PM_{2.5}$ on total number of cigarettes smoked during the week (N_{cig}) produced the following results:

$$PM_{2.5} = 17.7 + 0.322N_{cig} \quad (N = 96; R^2 = 0.55)$$
$$PM_{2.5} = 24.8 + 0.272N_{cig} \quad (N = 47; R^2 = 0.40)$$

where the first regression includes all homes but the second includes only homes with measurable nicotine levels. Thus each cigarette produces an increase of about 0.3 (±0.03) µg/m³ in the weekly mean $PM_{2.5}$ concentration, equivalent to an increase of 2.1 (±0.2) µg/m³ in the daily concentration.

Koutrakis *et al.* (1992) also analyzed the New York State data, using a mass-balance model to estimate $PM_{2.5}$ and elemental source strengths for cigarettes, wood burning stoves, and kerosene heaters. $PM_{2.5}$ source strength for smoking was estimated at 12.7 ± 0.8 (SE) mg/cigarette. For a final category of all other residual indoor sources, a source strength of 1.16 mg/h was calculated. For nonsource homes (N = 49) the authors estimated that 60 percent (9 µg/m³) of the total $PM_{2.5}$ mass was from outdoor sources, and 40 percent (6 µg/m³) from unidentified indoor sources. For smoking homes, they estimated that 54 percent (26 µg/m³) of the $PM_{2.5}$ mass was from smoking, 30 percent (15 µg/m³) from outdoor sources, and 16 percent (8 µg/m³) from unidentified sources. These authors also developed an elemental emissions profile for $PM_{2.5}$ particles from cigarettes, woodburning, and kerosene heaters. For cigarettes, the elemental profile included potassium (160 µg/cig), chlorine (69 µg/cig), and sulfur (65 µg/cig), as well as smaller amounts of bromine, cadmium, vanadium, and zinc. The woodburning profile included three elements: potassium (92 µg/h), silicon (44 µg/h) and calcium (38 µg/h). The kerosene heater profile included a major contribution from sulfur (1500 µg/h) and fairly large inputs of silicon (195 µg/h) and potassium (164 µg/h). A drawback of the mass-balance model was an inability to separately estimate the value of the penetration coefficient P and the decay rate k for particles and elements. Koutrakis *et al.* assumed a constant rate for k of $0.36\,h^{-1}$, and then solved for P.

The EPA Particle TEAM (PTEAM) Study. In 1986, the U.S. Congress mandated that USEPA's Office of Research and Development "carry out a TEAM Study of human exposure to particles." The USEPA's Atmospheric Research and Exposure Assessment Laboratory (AREAL) joined with the California Air Resources Board to sponsor a study in the Los Angeles Basin. The study was carried out primarily by the Research Triangle Institute and the Harvard School of Public Health, with additional support from Lawrence Berkeley Laboratory, Acurex, and AREAL. The main goal of the study was to estimate the frequency distribution of exposures to particles for Riverside residents. Another goal was to determine particle concentrations in the participants' homes and immediately outside the homes.

A pilot study was undertaken in nine homes in Azusa, California in March of 1989 to test the sampling equipment. Newly-designed personal exposure monitors (PEMs) were equipped with inhalable (PM_{10}) and fine ($PM_{2.5}$) particle inlets. The PEMs were impactors with 4 Lpm Casella pumps. Two persons in each household wore the PEMs for two consecutive 12-h periods (night and day). Each day they alternated inlet nozzles. The first five households were monitored concurrently for 7 consecutive days. The last four households were then monitored concurrently for 4 consecutive days. This resulted in approximately 100 PEM samples for each size fraction. Indoor and outdoor particle concentrations were monitored using microenvironmental exposure monitors (MEMs). These monitors were the Harvard "black boxes" (Marple *et al.* 1987) employing a 10 Lpm pump. Several indoor MEMs were placed in different rooms in each house to determine the magnitude of room-to-room variation. These monitors were capable of monitoring both fine and inhalable particles simultaneously.

Room-to-room variation of 12-h average particle levels was generally less than 10 percent. It was decided that this finding would justify using only one indoor monitor in the subsequent full-scale study.

The personal exposures were about twice as great as the indoor or outdoor concentrations for both PM_{10} and $PM_{2.5}$. Considerable effort was expended to determine whether this was a sampling artifact, due for example to the constant motion of the personal sampler. However, no evidence could be found for an artifactual effect. Nonetheless, to reduce chances for an artifactual finding in the main study, it was decided to use identical PEMs for both the personal and fixed (indoor-outdoor) samples in the main study.

Regressions of outdoor on indoor concentrations showed low R^2 values (1 to 30 percent) for both PM_{10} and $PM_{2.5}$ size fractions, as did regressions of daytime indoor on personal concentrations ($R^2 = 0$ to 18 percent). Overnight indoor concentrations had somewhat better ability to explain personal exposures ($R^2 = 14$ to 58 percent), as might be expected from the fact that the personal monitor was placed on the bedside table during the sleeping period. Personal exposures were essentially uncorrelated with outdoor concentrations ($R^2 = 0$ to 2 percent).

For the main study the following year, a three-stage probability sampling procedure was adopted (Pellizzari *et al.* 1991). Ultimately 178 residents of Riverside, California took part in the study in the fall of 1990. Respondents represented 139,000 ± 16,000 (S.E.) nonsmoking Riverside residents aged 10 and above. Their homes represented about 60,000 Riverside homes.

Each participant wore the PEM for two consecutive 12-hour periods. Concurrent PM_{10} and $PM_{2.5}$ samples were collected by the stationary indoor monitor (SIM) and stationary outdoor monitor (SAM) at each home. The SIM and SAM were essentially identical to the PEM. A total of 10 particle samples were collected for each household (day and night samples from the PEM_{10}, SIM_{10}, $SIM_{2.5}$, SAM_{10}, and $SAM_{2.5}$). Air exchange rates were also determined for each 12-hour period.

Up to four participants per day could be monitored, requiring 48 days in the field. A central outdoor site was maintained over the entire period (September 22 to

November 9, 1990). The site had two high-volume samplers (Wedding & Assoc.) with 10-μm inlets (actual cutpoint about 9.0 μm), two dichotomous PM_{10} and $PM_{2.5}$ samplers (Sierra-Andersen) (actual cutpoint about 9.5 μm), one PEM, one PM_{10} SAM, and one $PM_{2.5}$ SAM.

More than 2750 particle samples were collected, about 96 percent of those attempted. All filters were analyzed by x-ray fluorescence (XRF) for a suite of 40 metals. More than 1000 12-h average air exchange rate measurements were made.

A complete discussion of the quality of the data is found in Pellizzari *et al.* (1991), and in Thomas *et al.* (1993). Limits of detection (LODs), based on three times the standard deviation of the blanks, were on the order of 7 to 10 μg/m³. All field samples exceeded the LOD. Duplicate samples (N = 363) showed excellent precision for all types of particle samplers at all locations, with median relative standard deviations ranging from 2 to 4 percent.

As presented in Table 4-4, daytime mean personal PM_{10} concentrations (150 μg/m³) were more than 50 percent higher than either indoor or outdoor levels (95 μg/m³) (Clayton *et al.* 1993). Overnight mean personal PM_{10} concentrations (77 μg/m³) were similar to the indoor (63 μg/m³) and outdoor (86 μg/m³) levels. The higher personal concentrations do not appear to be due to skin flakes or clothing fibers. Many skin flakes were found on filters (up to an estimated 150,000 per filter) in subsequent scanning electron microscope analyses, but their mass does not appear to account for more than 10 percent of the excess personal exposure.

Mean $PM_{2.5}$ daytime concentrations were similar indoors (48 μg/m³) and outdoors (49 μg/m³), but indoor concentrations fell off during the sleeping period (36 μg/m³) compared to 50 μg/m³ outdoors. Thus, the fine particle contribution to PM_{10} concentrations averaged about 51 percent during the day and 58 percent at night both indoors and outdoors.

Table 4-4. Population weighted[a] PM concentrations and standard errors (μg/m³). Particle TEAM study in Riverside, California (Pellizzari *et al.* 1992)

Sample Type	N	Geometric Mean	GSD	Arithmetic Mean ± SE	Percentile	
					90 Percent ± SE	98 Percent
Daytime PM_{10}						
Personal	171	129	1.75	150 ± 9	260 ± 12	380
Indoor	169	78	1.88	95 ± 6	180 ± 11	240
Outdoor	165	83	1.68	94 ± 6	160 ± 7	240
Overnight PM_{10}						
Personal	168	68	1.64	77 ± 4	140 ± 10	190
Indoor	163	53	1.78	63 ± 3	120 ± 5	160
Outdoor	162	74	1.74	87 ± 4	170 ± 5	210
Daytime $PM_{2.5}$						
Indoor	173	35	2.25	48 ± 4	100 ± 7	170
Outdoor	167	38	2.07	49 ± 3	100 ± 6	170
Overnight $PM_{2.5}$						
Indoor	166	27	2.21	36 ± 6	83 ± 6	120
Outdoor	161	37	2.23	51 ± 4	120 ± 5	160

[a] Personal samples weighted to represent nonsmoking population of 139,000 Riverside residents aged 10 or above. Indoor-outdoor samples were weighted to represent 60,000 Riverside homes.

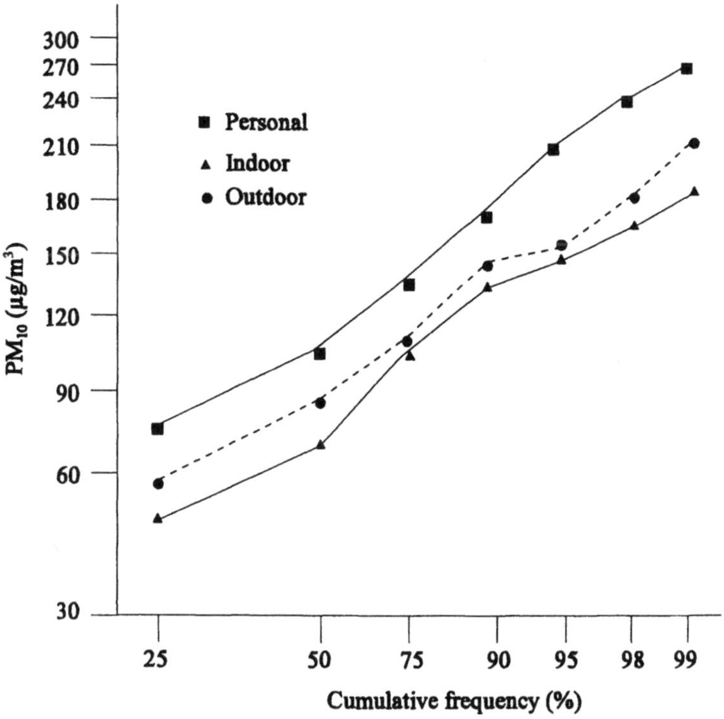

Figure 4-4. Cumulative frequency distribution of 24-h personal, indoor, and outdoor PM_{10} concentrations in Riverside, California.

Unweighted distributions are displayed in Figure 4-4 and Figure 4-5 for 24-h average PM_{10} and $PM_{2.5}$ personal, indoor, and outdoor concentrations. About 25 percent of the population of Riverside was estimated to have 24-h personal PM_{10} exposures exceeding the 150 μg/m³ 24-h NAAQS for ambient air.

Central-site $PM_{2.5}$ and PM_{10} concentrations agreed well with back yard concentrations. Overall, the data strongly suggest that a single central-site monitor can represent well the $PM_{2.5}$ and PM_{10} concentrations throughout a wider area such as a town or small city, at least in the Los Angeles basin.

Stepwise regressions resulted in smoking, cooking, and either air exchange rates or house volumes being added to outdoor concentrations as significant variables. Smoking added about 30 μg/m³ to the total $PM_{2.5}$ concentrations. Cooking added 13 μg/m³ to the daytime $PM_{2.5}$ concentration, but was not significant during the overnight period. At night, an increase in air exchange of one air change per hour resulted in a small increase of about 4.5 μg/m³ to the $PM_{2.5}$ concentration, but was not significant during the day. The house volume was not significant at night, but was significant during the day, with larger homes resulting in smaller $PM_{2.5}$ concentrations. Since air exchange and house volume were weakly correlated (negatively), they were not included together in the same regression.

A model developed in Koutrakis *et al.* (1992) was solved using nonlinear least squares to estimate penetration factors, decay rates, and source strengths for particles and elements from both size fractions. Penetration factors are very close to unity for nearly all particles and elements. The same conclusion was reached by Thatcher and Layton

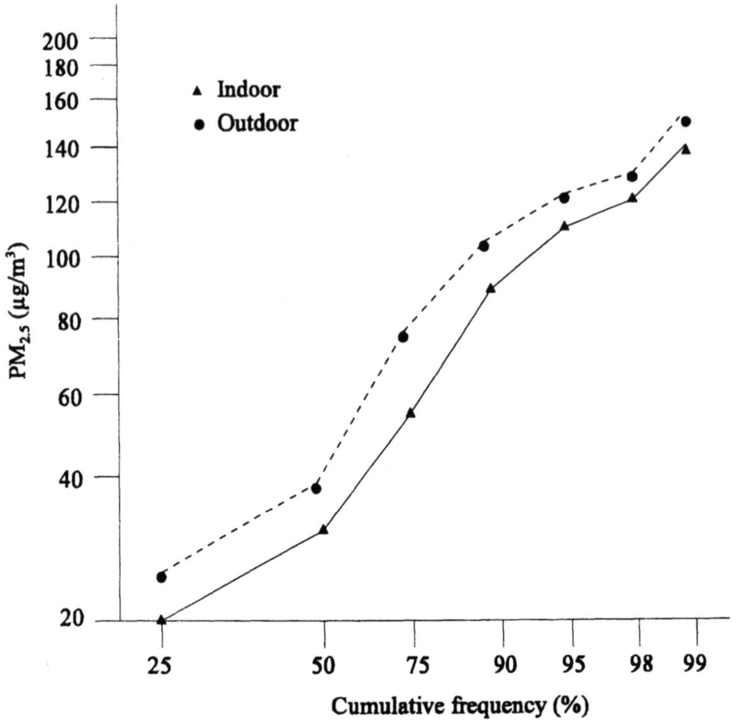

Figure 4-5. Cumulative frequency distribution of 24-h indoor and outdoor $PM_{2.5}$ concentrations in Riverside, California.

(1995). The calculated decay rate for fine particles is $0.39 \pm 0.16\,h^{-1}$, and for PM_{10} is $0.65 \pm 0.28\,h^{-1}$. Since PM_{10} contains the $PM_{2.5}$ fraction, a separate calculation was made for the coarse particles (PM_{10} to $PM_{2.5}$) with a resulting decay rate of $1.01 \pm 0.43\,h^{-1}$. Each cigarette emits 22 ± 8 mg of PM_{10} on average, about two-thirds of which (14 ± 4 mg) is in the fine fraction. Cooking emits 4.1 ± 1.6 mg/min of inhalable (PM_{10}) particles, of which about 40 percent (1.7 ± 0.6 mg/min) is in the fine fraction. All elements emitted by cooking were limited almost completely to the coarse fraction. Presumably carbon or other elements not measured by XRF were contained in the fine fraction. Sources other than cooking and smoking emit about 5.6 ± 3.1 mg/h of PM_{10}, of which only about 1.1 mg/h $\pm$ 1.0 (20 percent) is in the fine fraction.

Decay rates for elements associated with the fine fraction were generally lower than for elements associated with the coarse fraction, as would be expected, due to their lower settling velocities. For example, sulfur, which has the lowest mass median diameter of all the elements, had calculated decay rates of 0.16 ± 0.04 and $0.21 \pm 0.04\,h^{-1}$ for the $PM_{2.5}$ and PM_{10} fractions, respectively. The crustal elements (Ca, Al, Mn, Fe), on the other hand, had decay rates ranging from 0.6 to $0.8\,h^{-1}$.

Based on the mass-balance model, outdoor air was the major source of indoor particles in Riverside, providing about 75 percent of fine particles and 65 percent of inhalable particles in the typical home. It was also the major source for most elements, providing 70 to 100 percent of the observed indoor concentrations for 12 of the 15 elements. Only copper and chlorine were predominantly due to indoor sources in both the fine particle and inhalable particle fractions. However, these conclusions are strictly applicable only to Riverside. In five of the six cities studied by Harvard, and in both New York counties, outdoor air could not have provided as much as half of the indoor air particles, since the observed indoor-outdoor ratios of mean concentrations were ≥2.

Unidentified indoor sources accounted for most of the remaining particle and elemental mass collected on the indoor monitors. The nature of these sources is not yet understood. They do not include smoking, other combustion sources, cooking, dusting, vacuuming, spraying, or cleaning, since all these sources together account for less than the unidentified sources. For example, the unidentified sources accounted for 26 percent of the average indoor PM_{10} particles, whereas smoking accounted for 4 percent and cooking for 5 percent.

Of the identified indoor sources, the two most important were smoking and cooking. Smoking was estimated to increase 12-h average indoor concentrations of PM_{10} and $PM_{2.5}$ by 2 and 1.5 $\mu g/m^3$ per cigarette, respectively. Homes with smokers had mean PM_{10} concentrations about 30 $\mu g/m^3$ higher than homes without smokers. Most of this increase was in the fine fraction. Cooking increased indoor concentrations of PM_{10} by about 6 $\mu g/m^3$ per minute of cooking, with most of the increase in the coarse particles.

Emission profiles for elements were obtained for smoking and for cooking. Major elements emitted by cigarettes were potassium, chlorine, and calcium. Elements associated with cooking included aluminum, iron, calcium, and chlorine.

Other household activities such as vacuuming and dusting appeared to make smaller contributions to indoor particle levels. An interesting finding was that commuting and working outside the home resulted in lower daytime particle exposures than for persons staying at home.

The three studies are compared in Table 4-5. Multivariate calculations in all three studies result in rather similar estimates of the effect of smoking on indoor fine particle

Table 4-5. Indoor-outdoor mean concentrations ($\mu g/m^3$) of fine particles in three large-scale studies

Study Name	Homes	Out	In	In/Out
Harvard Six-City Study (Spengler *et al.* 1981)[a]				
Portage, Wisconsin	11	10	20	2.0
Topeka, Kansas	10	10	22	2.2
Kingston-Harriman, Tennessee	8	18	44	2.4
Watertown, Massachusetts	8	15	29	1.9
St. Louis, Missouri	10	18	42	2.3
Steubenville, Ohio	8	45	42	0.9
New York State ERDA Study (Sheldon *et al.* 1989)[b]				
Onondaga County	224	17	37	2.2
Suffolk County	209	22	46	2.1
EPA Particle TEAM Study (Pellizzari *et al.* 1993)[c]				
Riverside, California	178	50	43	0.9

[a] PM-3.5 measured using a cyclone sampler. Samples were collected every sixth day for one year (May 1986 to April 1987).

[b] PM-2.5 measured using an impactor developed at the Harvard School of Public Health. Samples were collected for one week at each household between January and April 1986.

[c] PM-2.5 measured using a Marple-Harvard-EPA sampler. Samples were collected for two 12-hour periods at each home between September and November 1990.

concentrations. Spengler *et al.* (1981) estimated an increase of about 20 μg/m³ per smoker, or 25 μg/m³ per smoking home, based on 55 monitored over a year in all six cities. Spengler *et al.* (1985) found a smoking effect of about 32 μg/m³ for smoking homes in multivariate models based on the Kingston-Harriman data. Sheldon *et al.* (1989) found an increase of 45 (Onondaga) and 47 (Suffolk) μg/m³ per smoking home in a multivariate model of the New York State data. Özkaynak *et al.* (1993) found an increase of about 30 to 35 μg/m³ in smoking homes in a multivariate regression model of the PTEAM data. Thus the estimates of the effect of smoking on indoor fine particle concentrations range from about 25 to 45 μg/m³, with the higher value occurring in more northerly and therefore more insulated homes, with lower air exchange rates. Other studies of indoor-outdoor concentrations of pesticides and elements include the Arnhem lead study (Diemel *et al.* 1981) and the Phillipsburg, New Jersey study (Lioy *et al.* 1990).

Carbon Monoxide

The largest personal monitoring study of CO exposures was carried out by the USEPA in Washington, D.C. and Denver, Colorado in the winter of 1982–83 (Akland *et al.* 1985; Hartwell *et al.* 1984, Johnson *et al.* 1983, 1986, Wallace *et al.* 1988). About 800 persons in Washington and 450 in Denver were monitored for 24 hours (48 hours in Denver) using electrochemical CO monitors with specially designed data loggers. The data loggers were capable of sampling the current from the CO monitor about 4 times per second. They were equipped with buttons that the subject could press when one activity ended and the next began. At that point, the logger would average all preceding values from the time the activity began. (There was also an automatic averaging every hour on the hour.) The result was an extraordinarily rich data base, with approximately 1200 persons averaging 40 activities per day, each with an associated CO average. At the end of the monitoring period, each subject provided a breath sample, which could be related to the previous exposure using the Coburn-Forster-Kane equation (Coburn *et al.* 1965).

Major findings of the study included the following:

- Commuters had the highest exposures to CO in general, averaging up to 13 ppm. Parking garages had the highest CO levels of any microenvironment, with churches and schools among the lowest.
- The main indoor sources of CO were gas stoves and cigarettes. Gas stoves increased levels by about 2.5 ppm when being used; homes with smokers had increases of about 1.5 ppm on average.
- Personal exposures were higher than would be predicted by the fixed stations. About 10 percent of Washington residents appeared to exceed the 8-hour outdoor standard of 9 ppm, as determined by their breath concentrations, although only one of the 11 fixed stations exceeded the standard during the monitoring period.

A study of California homes (Wilson *et al.* 1993a,b, 1995, Colome *et al.* 1994), each monitored for 48 hours, indicated that 13 of 277 homes (about 5 percent) had indoor 8-h averages exceeding 9 ppm (the outdoor standard). Since the outdoor standard is to be exceeded only once per year, it is clear that the fraction of homes with 8-h indoor averages exceeding 9 ppm more than once per year would be larger than the 5 percent observed in the single 48-h monitoring period. Homes with gas stoves and gas furnaces had indoor source terms for CO that were about 3 times higher than homes without such sources. Homes with wall furnaces had higher CO levels than homes with forced-air gas furnaces. Homes with smokers (N = 85) had CO levels about 0.5 ppm higher than homes without smokers (N = 190). Malfunctioning gas furnaces were a major cause of elevated CO concentrations. However, the homes with the highest CO levels also included some with electric stoves and electric heat, suggesting that other sources of CO were present in these homes. Such sources could include cars idling in attached garages or unvented gas or kerosene space heaters.

Pesticides

The main study of personal exposure and indoor concentrations of pesticides was sponsored by USEPA from 1984 to 86 (Immerman and Schaum 1990a,b, Immerman 1989). Two cities, one with expected high pesticide use (Jacksonville, Florida) and one with expected low use (Springfield, Massachusetts) were chosen. About 250 persons, selected to represent the populations of the two cities, were monitored for one day (24 hours). Personal, indoor, and concurrent outdoor concentrations were measured for about 32 target pesticides using a sampler containing polyurethane foam (PUF) as a sorbent.

Major findings included the following:

- Both personal exposures and indoor concentrations were much greater than outdoor concentrations for all target pesticides. In most cases, indoor sources contributed well over 90 percent of total exposure.
- The pesticides with the highest cancer risks were the chlorinated hydrocarbons such as chlordane, heptachlor, aldrin and dieldrin. Although all these pesticides had been banned for some years, they were still present in indoor air, presumably due to volatilization from termite-treated soil.

Nitrogen Dioxide

Nitrogen dioxide (NO_2), like CO and particles, is a USEPA criteria pollutant, with a National Ambient Air Quality Standard. NO_2 is emitted by industrial processes, but also by indoor combustion appliances such as gas stoves and furnaces. Several studies in the 1970s (Goldstein *et al.* 1979) suggested that children in homes with gas stoves suffered more infectious disease than children in homes with electric stoves. A possible connection with NO_2 (in lowering resistance) was postulated. Therefore several large-scale population-based studies of NO_2 in homes have been carried out in the past two decades. One of the first such studies was part of the Harvard six-city study (Quackenboss *et al.* 1986, Spengler *et al.* 1983).

An early study was sponsored by the Southern California Gas Corporation (Colome *et al.* 1987, Wilson *et al.* 1986). More than 500 homes served by the company were selected to participate in a study of week-long NO_2 concentrations in homes. The homes were visited during three seasons: spring and summer of 1984 and winter (January) of 1985. Major findings included

- Homes with gas stoves had higher concentrations than homes with electric stoves
- Homes with gas-fired wall furnaces had higher concentrations than homes with gas floor furnaces.

- Pilot lights accounted for about a third of total NO_2 emissions from gas stoves so equipped.
- In homes with gas stoves, kitchens had higher concentrations than living rooms, which in turn had higher concentrations than bedrooms.

Additional major studies of indoor concentrations of and personal exposures to NO_2 took place in Boston (Ryan *et al.* 1988a, 1988b, 1992) and Los Angeles (Spengler *et al.* 1992). Personal exposures were found to be well correlated with bedroom measurements ($R^2 = 0.48$ in Boston, 0.53 in Los Angeles), and also reasonably well correlated with outdoor concentrations in Los Angeles ($R^2 = 0.41$). However, in Boston, which has lower outdoor NO_2 than Los Angeles, personal exposures were poorly correlated with outdoor concentrations ($R^2 = 0.09$). The best model for personal exposure included season and cooking fuel in Boston, and season and range/furnace type in Los Angeles. In both cities, persons with gas ranges had about 10 ppb higher exposures than those without gas ranges.

Radon

Several early sets of measurements were undertaken with conventional aerosol sampling systems such as tube or screen diffusion batteries (Sinclair *et al.* 1977, George and Breslin 1980, Knutson *et al.* 1984, George *et al.* 1984), which had a lower size limit of about 5 nm. All studies found a bimodal distribution of the energy-weighted values around sizes of 5 to 10 nm and 100 to 150 nm, with the bulk of the energy (about 90 percent) in the larger size.

Improvements to the measurement systems allowed extension to smaller about 0.5 to 1 nm) sizes (Reineking *et al.* 1985, Reineking and Porstendörfer 1986, Reineking *et al.* 1988, Holub and Knutson 1987, Tu and Knutson 1988a, 1988b, Strong 1988, 1989) with the result that trimodal distributions can now be observed. Intercomparisons of the methods showed good agreement (Hopke *et al.* 1992).

A semi-continuous automated system has been developed by Ramamurthi and Hopke (1991). The sampled air is drawn through the slit in each sampler-detector unit between the detector and the Millipore filter. The slits of five units are covered with a combination of wire screens that has been carefully chosen in order to obtain the size distribution in the range of 0.5 to 500 nm. One unit is operated with an uncovered slit, thus providing the total ambient radon decay product concentrations. The six sampling heads with various combinations of screens with differing d_p(50 percent) values.

This system has now been used extensively to measure the exposure in six homes in the northeastern U.S. and two homes in southeastern Ontario, Canada (Hopke *et al.* 1995). Two homes included smokers. These studies have provide 565 measurements of the activity-weighted size distributions in these houses. The median value for the equilibrium fraction was 0.408 as compared with the previously employed value of 0.50. Using the recently adopted ICRP lung deposition and dosimetry model, the hourly equivalent lung dose rate per unit radon exposure was estimated for each measured size distribution. The mean equivalent dose rate per unit ^{222}Rn gas concentration was approximately 150 $nSv\,h^{-1}$ per $Bq\,m^{-3}$. It was found that the equivalent dose was strongly correlated with the fraction of airborne progeny activity carried by particles smaller than 15 nm in diameter with a correlation coefficient of 0.749. The correlation coefficient with the ≤2-nm size fraction (the "unattached" fraction) was 0.169, reflecting no significant relationship with the unattached fraction. Differences between houses with smokers present and absent were noted in the exposure conditions, but the resulting dose rate per unit radon gas concentration was essentially the same for the two groups. Thus, unless complex measurements of both decay product activity concentration and particle size are measured, it is better to measure radon gas as an acceptable indicator of the exposure such that the physical dose to the bronchial epithelium can be assessed. It is concluded that automated measurement systems are now available that can determine the concentration and size of radioactive particles in indoor air.

The results of 19 local surveys involving 552 homes were combined into a single estimated distribution of radon concentrations in all U.S. single-family homes (Nero *et al.* 1986). The distribution was log-normal, with a geometric mean of $36\,Bq\,m^{-3}$ ($0.96\,pCi\,L^{-1}$) and a geometric standard deviation of 2.84. This would imply that about 8 percent of U.S. single-family homes may exceed the USEPA guideline of $4\,pCi\,L^{-1}$ ($150\,Bq\,m^{-3}$) (USEPA 1992). (Because people move, the percent of the population with lifetime exposures at this level would be considerably smaller.) The arithmetic mean of the distribution was $55\,Bq\,m^{-3}$ ($1.5\,pCi\,L^{-1}$). Allowing for the fact that about a quarter of the population lives in multiple-unit dwellings, which are typically at lower levels of radon exposure, Nero (1988) estimated an average population exposure rate of about 0.25 WLM per year, to which he attributed a lifetime risk of lung cancer on the order of 0.4 percent. This is quite high compared to the lung cancer risk for nonsmokers of about 1 percent, implying that radon decay product exposure is responsible for a substantial fraction of nonsmokers' lung cancer.

A USEPA survey of 11,000 homes from 125 counties estimated an average annual radon concentration in U.S. homes of $1.25\,pCi\,L^{-1}$, with a median value of $0.67\,pCi\,L^{-1}$ (Marcinowski 1992). Since the measurements were only short-term (3–4 days), the estimate may be quite uncertain.

Ancillary Measurements

Air Exchange. Air exchange rates can have an important influence on pollution levels in the home. For example, if outdoor air is cleaner than indoor air, as is true for most VOCs and most pesticides (as we have seen above), then it will improve matters to open the windows, turn on the attic fan, or otherwise increase the air exchange rate. On the other hand, during times of high outdoor air pollution, closing the windows or otherwise reducing air exchange rates can have a protective effect. This is particularly true for particles, which deposit on walls and are thus removed from indoor air—the lower the air exchange rate, the longer the particles have to be removed from the air.

The largest study to measure air exchange rates in homes was the Southern California Gas Corporation study of NO_2 in southern California homes (Wilson *et al.* 1986). In this study, NO_2 was monitored indoors and outdoors for one week at 597 homes in March of 1984, 444 of those homes in July of the same year and 405 of the same homes in January of 1985.

Another large study was the New York State study of particles mentioned above; more than 400 homes in Suffolk and Onondaga County were monitored for particle levels over a period of one week. Week-long average air exchange rates were measured simultaneously.

The USEPA's PTEAM Study, discussed above, measured daytime and overnight air exchange rates in 178 homes in Riverside, Caifornia.

The USEPA's 1987 TEAM Studies in Baltimore, MD and Los Angeles CA also measured air exchange rates in 150 Baltimore homes and in 50 Los Angeles homes, the latter over two seasons.

The results of these large-scale studies are summarized in Table 4-6 and Figure 4-6. It can be seen that the southern California homes have higher air exchange rates than the Baltimore or New York homes, as might be expected from the balmier southern California climate.

MODELING

Basic Mass-Balance Model

Mass balance models have been used for more than a century in various branches of science. All such models depend on the law of the conservation of mass. They simply state that the change in mass of a substance in a given volume is equal to the amount of mass entering that volume minus the amount leaving the volume. Usually they are written in the form of first-order linear differential equations. That is, consider a volume V filled with a gas of mass m. The change in mass (Δm) over a small time (Δt) will simply be the difference between the mass entering the volume (m_{gain}) and the mass leaving the volume (m_{loss}):

$$\frac{\Delta m}{\Delta t} = \frac{(m_{gain} - m_{loss})}{\Delta t} \tag{2}$$

Taking the limit as (Δt) approaches zero, we have the differential equation for the rate of change of the mass:

$$\frac{dm}{dt} = \frac{d}{dt}(m_{gain} - m_{loss}) \tag{3}$$

If we require that the mass be uniformly distributed throughout the volume at all times, we have a condition that the physical chemists call "well-mixed." We assume that any mass gained or lost in the volume (V) is instantaneously distributed evenly throughout the volume. We may then replace the mass terms by the concentration ($C = m/V$):

$$\frac{VdC}{dt} = \frac{d}{dt}(m_{gain} - m_{loss}) \tag{4}$$

The above equation is the basis for all mass-balance models dealing with well-mixed compartments. It takes on many forms depending on the type of processes involved in transporting mass into or out of the volume being considered. A large class of models assume that the volume is a single compartment. More complex models assume multiple compartments. As an example of a single compartment model, we may consider a room of volume (V) that exchanges air with the outside at a constant flow rate (Q). We also assume that a mass of gas has been released in the room at time ($t = 0$), and that the outdoor concentration of this gas is 0. (This is the situation, for example, when a tracer gas such as SF_6 is released to determine the air exchange rate.) In this case, the gain in mass (m_{gain}) is zero and the loss in mass is equal to the flow rate (Q) out of the house times the concentration (C), so that Equation 4 becomes:

Table 4-6. Air exchange rates (l/hr) in different studies, ordered by length of monitoring period

				Arithmetic		Geometric	
Site/Time	Season	N	Period	Mean	SD	Mean	SD
Riverside, California	Fall	175	Day	1.37	1.37	0.91	2.45
Dundalk, Maryland	Spring	67	Day	0.77	0.66	0.59	1.92
Parkville-Overlea	Spring	62	Day	0.62	0.54	0.44	2.35
Los Angeles, California	Winter	46	Day	1.00	0.74	0.80	1.96
Los Angeles, California	Summer	41	Day	3.50	2.64	2.67	2.26
							2.19 ± 0.21
Riverside, California	Fall	174	Night	1.12	1.00	0.83	2.14
Dundalk, Maryland	Spring	68	Night	0.65	0.39	0.54	2.09
Parkville-Overlea	Spring	66	Night	0.40	0.32	0.32	2.14
Los Angeles, California	Winter	45	Night	0.80	0.85	0.64	1.81
Los Angeles, California	Summer	41	Night	2.05	2.06	1.45	2.12
							2.06 ± 0.13
Riverside, California	Fall	174	24-hr	1.24	1.11	0.92	2.16
Dundalk, Maryland	Spring	66	24-hr	0.71	0.47	0.59	1.82
Parkville-Overlea	Spring	62	24-hr	0.51	0.39	0.40	2.27
Los Angeles, California	Winter	45	24-hr	0.91	0.73	0.75	1.84
Los Angeles, California	Summer	41	24-hr	2.77	2.11	2.16	2.03
							2.02 ± 0.18
Suffolk County, New York	Winter	194	Week	0.59	0.42	0.50	1.68
Onondaga County, New York	Winter	201	Week	0.50	0.27	0.44	1.65
Ontario, Canada	Winter	98	Week	0.38	0.25	0.30	2.12
Southern California	Summer	426	Week	1.51	1.47	1.08	2.32
Southern California	Spring	571	Week	0.78	0.63	0.62	1.86
Southern California	Winter	371	Week	0.58	0.47	0.45	1.84
Northwest Lawrence Berkeley Lab	Winter-Spring	62	Week	0.65	0.37	0.54	1.93
New York (Fleming)	Winter-Spring	44	Week	0.20	0.15	0.14	2.64
Maryland (Geomet)	Winter-Spring	76	Week	0.76	0.58	0.57	2.50
							2.06 ± 0.34
Espoo, Finland	NA	NA	2 Weeks	0.52	0.27	0.45	NA
Canada	Winter	126	Month	0.32	0.26	0.25	2.14
New York State	Winter	57	6 Weeks	0.61	0.38	0.53	1.67
Northwest (Lambert)	Winter	284	Quarter	0.33	0.26	0.27	1.81
							1.87 ± 0.20
Total measurements	—	3683	—	0.93	0.76	0.72	2.06 ± 0.26

$$\frac{VdC}{dt} = -QC \quad (5)$$

Integrating this equation by separation of variables, we have

$$C = C_0 e^{-at} \quad (6)$$

where

$a = Q/V$ is the air exchange rate, and
C_0 is the concentration at time $t = 0$.

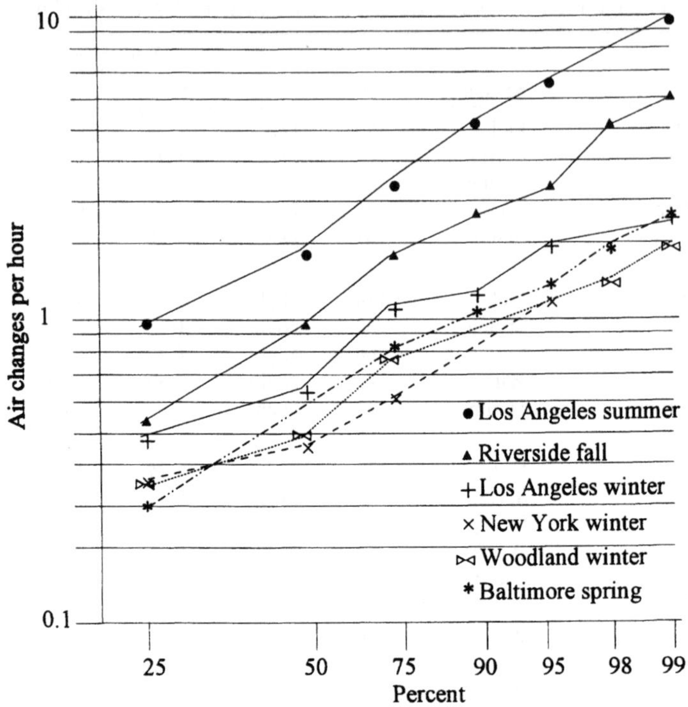

Figure 4-6. Air exchange measurements in five studies. Number of homes: Baltimore 130, Los Angeles 45, Riverside 175, New York 400, and Woodland 200.

Thus, we find that the original concentration of the tracer in the room decays with a time constant (a), the air exchange rate.

For a nonreactive gas with a nonzero outdoor concentration (*e.g.*, carbon monoxide), the mass balance equation takes the form:

$$\frac{dC_{in}}{dt} = a(C_{out} - C_{in}) \tag{7}$$

Depending on the variation with time of (C_{out}), this equation has a number of solutions. If (C_{out}) is constant, for example, and the initial indoor concentration is zero, then the indoor concentration rises at a rate determined by the air exchange rate to approach an asymptotic value equal to the outdoor concentration:

$$C_{in} = C_{out}(1 - e^{-at}) \tag{8}$$

An early effort at developing an indoor air quality model was made by Shair and Heitner (1974). This was a mass balance model in which the building was represented as a single well-mixed chamber. A single first-order linear differential equation represented the change in mass of a agent due to infiltration, exfiltration, recirculation, source generation, and removal due to filters in the circulation system or deposition on surfaces. Shair and Heitner solved the equation for certain simple inputs, such as a linearly increasing or decreasing outdoor concentration:

$$C_{out} = mt + b \tag{9}$$

Since the outdoor concentration normally is a slowly-varying function, Shair and Heitner's linear approximation is actually an excellent approximation for time intervals of moderate length.

If an indoor source (S(t)) exists, it enters the mass balance model in the following way:

$$\frac{dC_{in}}{dt} = a(C_{out} - C_{in}) + \frac{S(t)}{V} \tag{10}$$

where (S(t)) has the units of mass per unit time.

If the source has a constant generation rate (*e.g.*, CO_2 emissions from a person at rest), then (S(t)) is a constant value (S_0) and the equation becomes:

$$\frac{dC_{in}}{dt} = a(C_{out} - C_{in}) + \frac{S_0}{V} \tag{11}$$

If the substance of interest reacts or is adsorbed on surfaces while indoors, the equation becomes

$$\frac{dC_{in}}{dt} = aC_{out} - (a+k)C_{in} + \frac{S(t)}{V} \tag{12}$$

where (k) represents the loss of the substance due to chemical reaction, adsorption on surfaces, sedimentation, *etc.* The decay rate (k) has the same units as the air exchange rate a (1/time). Their sum (a + k) may be thought of as an effective air exchange rate. The decay rate (k) is often used to apply to particles, which disappear faster indoors than a nonreactive gas, such as CO. Since particles may experience more difficulty than molecules of a gas in penetrating the building envelope, a penetration factor (P) is often applied that multiplies the outdoor concentration in Equation 12 above.

As described above, Koutrakis *et al.* (1992) used least-squares analysis to solve a simplified form of the mass-balance model to determine source emission rates for particles and elements due to cigarettes, woodsmoke, and kerosene heater use. Koutrakis assumed a value for (k) in order to solve the equation for (P) and the source emission rates. Özkaynak *et al.* (1993) improved on Koutrakis' approach by using nonlinear least-squares analysis of the PTEAM results to solve the equation simultaneously for (k), (P), and source emission rates for $PM_{2.5}$ and PM_{10} particles and associated elements for smoking and for cooking.

Traynor *et al.* (1989) developed a "macromodel" based on Monte Carlo simulations using global input data such as house volumes, air exchange rates, and emissions from combustion sources to assess residential concentrations of combustion-source agents such as CO, NO_2, and respirable suspended particles. For a home with only one combustion source during winter in upstate New York, at an outdoor temperature of 0 °C, and an outdoor RSP geometric mean concentration of 19 µg/m^3, the model predicted geometric mean concentrations of about 80 µg/m^3 in a home with smoking, 75 µg/m^3 for a radiant kerosene heater, about 60 µg/m^3 for a convective unvented gas space heater and a non-airtight wood stove, and about 40 µg/m^3 for an infrared unvented gas space heater. An airtight wood stove was predicted to produce a geometric mean about 30 μg/m^3. Gas ovens, dryers, hot water heaters, boilers, and forced-air furnaces were predicted to result in low indoor concentrations of 10 to 15 μg/m^3, unless the gas oven was used for heating, in which case the predicted geometric mean was about 20 μg/m^3.

At present, the most complete form of the mass-balance indoor air quality model has been presented by Nazaroff and Cass (1989). These authors developed the model to allow for changes in particle size and chemical composition, including terms for homogenous turbulence, natural convection, thermophoresis, advection, and Brownian motion. Coagulation of particles is also included. The computer form of the model required 40 to 60 minutes of CPU time to determine an 11-hour evolution of an aerosol mixture of 16 different sizes. The model was partially validated by checking it against the results of a chamber study using cigarette-generated aerosol to determine the effectiveness of air cleaners (Offermann *et al.* 1985).

A crucial unknown parameter in the mass-balance model for particles is the rate of decay to surfaces. This rate of decay (k) may be expressed as the product of a deposition velocity (k_d) with the surface to volume ratio in the room or building:

$$k = k_d \frac{S}{V} \tag{13}$$

The deposition velocity will vary with particle size.

Both the Nazaroff study above and the series of studies by Weschler and colleagues below have provided useful data on deposition velocities for important anions such as sulfates.

A series of studies, also concerned with the effects of indoor particles on materials, were carried out by Weschler and colleagues at AT&T Bell Laboratories (Weschler *et al.* 1989, Sinclair *et al.* 1988, 1990, 1992). Studies of buildings with low occupancy, large amounts of electronic equipment, and high-quality filtering and HVAC systems succeeded in determining deposition velocities for coarse particles and various fine particle ions. For coarse particles, these velocities were about equal to velocities predicted for gravitational settling, similar to the results of Nazaroff *et al.* (1990a) described above. For fine particles, however, the deposition velocity was greater than that predicted for gravitational settling alone. For sulfates, the dominant anion in fine particles, deposition velocities at four buildings in Wichita, Lubbock, Newark, and Neenah were 0.004, 0.005, 0.005, and 0.004 cm/s, respectively (Sinclair *et al.* 1992).

Nazaroff *et al.* (1993) reviewed these and other studies of deposition velocity. The authors pointed out that the studies by Weschler and colleagues and also one study in Helsinki (Raunemaa *et al.* (1989) had produced values ranging from 0.003 to 0.005 cm/s for fine-mode sulfate, but that studies by Nazaroff and colleagues (Nazaroff *et al.* 1990a,b) resulted in much smaller values ranging from 0.00002 to 0.001 cm/s. It is not clear whether the differences are due to the many differences in surface materials and filtration systems in the different types of buildings (museums *versus* telephone equipment buildings) or to the different methods of determining deposition velocities. However, the discrepancy is clear evidence that further work is needed.

Because all large-scale studies of indoor air quality agree that the most important indoor source of fine particles is cigarette smoke, a brief review of models and chamber studies dealing with ETS is provided here.

Using the mass balance approach, a number of efforts have been undertaken to model mathematically the agent concentrations from tobacco smoke in indoor locations.

For example, Brief (1960) proposed a simple graph to determine transient concentrations for agents in indoor settings that is based on an exponential decay as a function of time. Turk (1963) proposed a general equation for calculating the concentrations in a chamber that includes both exterior and interior sources, as well as the removal effect of agents by air treatment systems. Bridge and Corn (1972) reported that a solution to the equations proposed by Turk (1963) adequately predicts tobacco smoke in occupied spaces. Jones and Fagan (1974) used Turk's equation to calculate carbon monoxide (CO) concentrations versus time from cigarette smoke in an office building and a single-family dwelling. Ishizu (1980) examined experimentally the inclusion of a mixing factor in these models. Nazaroff and Cass (1989) developed a rigorous and detailed model incorporating particle growth with aging.

Repace and Lowrey (1980) developed a modification of the Turk equation incorporating a mixing factor. They modeled the concentration as a function of time (assuming an initial concentration of zero) as:

$$C = C_{\infty}(1 - \exp(-(a + k)mt)) \tag{14}$$

where

C_{∞} = $G/Vm(a + k)$ is the equilibrium concentration of ETS particles
m = a "mixing factor" to account for imperfect mixing, and
G = the generation rate, a function of the number of cigarettes being smoked and the total particle mass emitted from sidestream and exhaled mainstream smoke.

Using an estimated value for (G) of 24 mg/cigarette, Repace (1987a) arrived at an estimate for the steady-state concentration of $PM_{3.5}$ (in $\mu g/m^3$) due to smoking:

$$PM_{3.5} = 217\frac{D_{hs}}{a} \tag{15}$$

where (D_{hs}) is the density of "habitual smokers" (*i.e.*, those who smoke two cigarettes per hour) per $100\,m^3$, and the value of 217 incorporates a mixing factor and a particle decay rate based on empirical observations in a number of locations.

As reported in Repace and Lowrey (1982), this equation was partially validated by a chamber experiment of Leaderer *et al.* (1984), in which the measured equilibrium concentration was $620\,\mu g/m^3$ compared to $478\,\mu g/m^3$ predicted by the equation, a difference of about 23 percent. Since the Repace-Lowrey equation assumes a nation-wide sales-weighted average tar content, a small experiment using one or two types of cigarettes would be expected to give somewhat different values.

In recent research, the basic mass-balance model was adapted to the case of a sequence of cigarettes smoked one after another, and its effectiveness in predicting the agent concentrations as a function of time (the "time series" of concentrations) was tested using real-time monitoring instruments (Ott *et al.* 1992). This work has developed a model for computing the time series of agents generated by sequential cigarette smoking activity patterns, the Sequential Cigarette Exposure Model (SCEM). If (t_i) is the time at which cigarette (i) begins, and (s_i) is the duration of cigarette (i), then during the time the cigarette is being smoked, the particle concentration rises according to:

$$C = \frac{S}{V(a+k)}(1-(\exp(-(a+k)(t-t_i)))) + C_{2i-2}\exp(-(a+k)(t-t_i)) \quad (16)$$

whereas between cigarettes the concentration declines:

$$C = C_{2i-1}\exp(-(a+k)(t-s_i-t_i)) \quad (17)$$

In these equations, $C_{2i-2} = C(t_i)$ and $C_{2i-1} = C(t_i + s_i)$.

A computer program for the SCEM has been written in Microsoft QuickBASIC Version 4.5 programming language to apply to the two equations above. This research also has derived theoretical equations for the minimum, maximum, and mean agent concentration in a well-mixed microenvironment for any cigarette smoking activity pattern. General expressions also have been derived for the case of the habitual smoker (uniform cigarette duration and same time between cigarettes) and for the case of multiple habitual smokers.

The equations used to derive the SCEM are general and are consistent with earlier ETS indoor air quality models that were derived for special cases (for example, the Repace and Lowrey equation described above). The SCEM considers the general case in which each habitual smoker can have any smoking rate, and concentration is measured on a "real-time" (*i.e.*, continuous) basis.

Solutions to the mass balance equation provide a theoretical basis for calculating all parameters of the model—air exchange rate, source strength, and sink removal terms—in a single experiment. Because of the SCEM's fine time resolution, experiments to validate the model require monitoring instruments that operate with fine time resolution (minutes or seconds). The air exchange rate is determined from the exponential decay of concentrations in the microenvironment (Ott *et al.* 1992). The source strength is determined from the equilibrium concentration with continuous smoking. The sink removal term for agents that adhere to surfaces, such as particles, is determined by subtracting the particle decay rate from the decay rate for a agent that has no surface sinks, such as CO.

Klepeis *et al.* (1995) applied an extension of the SCEM model (now called the Multiple Cigarette Exposure Model, or MCEM) to smoking lounges at the San Francisco and San Jose airports. Three Piezobalances were placed at each end and the center of the lounges. One investigator counted the number of lit cigarettes every minute while the other took readings from the Piezobalances. A Langan CO monitor recorded carbon monoxide readings continuously. Five visits to each airport were made. During the fifth visit, an experiment was performed to calculate the air exchange rate when there were few or no smokers present: several cigars were smoked and the decay of the CO level was measured, giving air exchange rates of 10.7 and 13.0 air changes per hour (ach) at the two airports. The decay rate of the ETS-related $PM_{3.5}$ was estimated to be 0.048 and 0.034 min^{-1}, or 2.88 and 2.04 h^{-1}. The calculated $PM_{3.5}$ source strengths for cigarettes during these two visits were identical at 1,340 μg/min. Air exchange rates were not determined at the other visits, but assuming the same rates resulted in an average $PM_{3.5}$ emission rate of 1,450 μg/min.

Ott *et al.* (1996) tested the model in a tavern before and after smoking was prohibited. During 26 visits over a period of two years while smoking was allowed, indoor concentrations averaged 56.9 μg/m^3 above outdoor levels, compared to 5.9 μg/m^3 above outdoor levels on 24 visits in the first six weeks after smoking was prohibited. A second set of follow-up visits (matched by time of day, day of week, and season to the earlier visits) yielded an average concentration 13.1 μg/m^3 above outdoor levels. Using cigarette emission rates from the literature, the measured tavern volume of $521\,m^3$, and a measured air exchange rate under "typical" conditions of 7.5 ach, the mass-balance model predicted $42.5\,\mu g/m^3$ for an average "continuous smoking" count of 1.17 cigarettes, comparing favorably with the observed average of $43.8\,\mu g/m^3$.

Ott *et al.* (1995) used Laplace transforms to apply the model to an experiment in which three Kentucky reference 2R1 cigarettes were smoked one after another in a $25.7\,m^3$ bedroom. Resulting $PM_{3.5}$ concentrations were measured in the bedroom and living room with a Piezobalance. Peak values reached $5{,}500\,\mu g/m^3$, and about $2{,}000\,\mu g/m^3$ in the living room. The living room window was opened two hours later, but concentrations were still in excess of $100\,\mu g/m^3$ after four hours.

Stochastic (Microenvironment) Exposure Models

Whereas indoor air quality is typically modeled by mass balance models like those described above, human exposure to airborne agents is often modeled using a completely different approach: time-weighted summations of exposures in the places we spend time in each day—home, work, in vehicles, and outdoors. The basic equation for human exposure is usually written as:

$$E_i = \sum c_{ij} t_j \qquad (18)$$

where

E_i = the exposure of person (i)
c_{ij} = the concentration person (i) encounters in microenvironment (j)
t_j = the time spent in microenvironment (j)

Here a microenvironment is defined as a place in which the concentration of the agent is relatively homogeneous (Duan 1982, 1985, 1991). For example, a home might be a microenvironment if it is fairly well-mixed, the back yard might be another, the vehicle a third microenvironment. However, if a strong source exists in say, the kitchen, then when the source is on the kitchen might be one microenvironment, and when it is off the kitchen is a different microenvironment.

The richest data base for testing models of this sort was the Washington-Denver CO study described above, which included about 40 microenvironments per person per day for about 1,200 participants. Ott *et al.* (1988) developed the Simulated Human Air Pollution Exposure (SHAPE) model to model the CO exposures of the participants.

Models designed to predict indoor concentrations of bioaerosols have not been developed. However, the several efforts that have been made to predict outdoor pollen levels are relevant since outdoor air is often the primary source for indoor pollen aerosols. Arizmendi *et al.* (1993) used time series analysis to predict pollen concentrations, but did not allow for the effects of meteorological variables. Stephen *et al.* (1990) estimated diurnal rhythm with a one-parameter time series model that provided good short-term forecasts. The one-step prediction error variance was reduced by 88 percent for *Cladosporium* and by 98 percent for basidiospores. Stark *et al.* (1997) have developed a model to predict ragweed pollen concentrations using data from one of the American Academy of Allergy, Asthma and Immunology Aeroallergen Network counting stations (Kalamazoo, Michigan). The model is designed to forecast ragweed pollen levels one day in advance, and uses Poisson Regression and the weather forecast for the next day (rain, wind speed, average temperature), and the point within the ragweed season. This approach differs from previous attempts in that the model is fit iteratively each day. By updating the coefficients of the model based on the information to date, the method does not rely on the hope that ragweed pollen levels will follow a trend similar to the previous year. In addition, the model is responsive to weather anomalies.

FUTURE DIRECTIONS

A promising new development is our improving ability to measure environmental agents during real time. A useful monitor has been developed for total particle-bound PAHs that operates on a photoelectric principle and is capable of measuring down to the $10 ng/m^3$ level (Wilson *et al.* 1994). Real-time particle monitors using laser technology promise to provide a wealth of data on the sources and deposition behavior of different size fractions of particles. A continuous breath sampling method employing tandem mass spectrometers has been developed (Gordon *et al.* 1992), which can provide abundant data on uptake, distribution and excretion of VOCs.

REFERENCES

AAAAI (American Academy of Allergy, Asthma, and Immunology). 1994. National Allergy Bureau Pollen and Mold Report. Milwaukee, Wisconsin.

AAAAI (American Academy of Allergy, Asthma, and Immunology). 1995. National Allergy Bureau Pollen and Mold Report. Milwaukee, Wisconsin.

Akland, G., T.D. Hartwell, T.R. Johnson, and R. Whitmore. 1985. Measuring human exposure to carbon monoxide in Washington, DC and Denver, CO during the winter of 1982–83. Environ. Sci. Technol. 19:911–918.

Albrecht, E. and A. Kaul. 1967. Continuous registration of ^{222}Rn in air varying with time. Pages 643–650 in: Assessment of airborne radioactivity in nuclear operations. International Atomic Energy Agency. Vienna, Austria.

Andelman, J.B. 1985a. Human exposures to volatile halogenated organic chemicals in indoor and outdoor air. Environ. Health Perspect. 62:313–318.

Andelman, J.B. 1985b. Inhalation exposure in the home to volatile organic contaminants of drinking water. Sci. Total Environ. 47:443–460.

Andelman, J.B. 1990. Total exposure to volatile organic compounds in potable water. Chapter 20 in: Ram, N.M., R.F. Christman, and K.P. Cantor (eds.). Significance and treatment of volatile organic compounds in water supplies. Lewis Publishers. Chelsea, Michigan.

Andersen, A.A. 1958. New sampler for the collection, sizing, and enumeration of viable airborne particles. J. Bacteriol. 76:471–484.

Arizmendi, C.M., J.R. Sanchez, N.E. Ramos, and G.I. Ramos. 1993. Time series predictions with neural nets: application to airborne pollen forecasting. Internat. J. Biometeorol. 37:139–144.

Arnow, P.M., D. Weil, and M.F. Para. 1985. Prevalence and significance of Legionella pneumophila contamination of residential hot-tap water systems. J. Infect. Dis. 152:145.

Ashley, D.L., M.A. Bonin, L. Carkinali, J.M. McGraw, and J.V. Wooten. 1994. Blood concentrations of volatile organic compounds in a nonoccupationally exposed US population and in groups with suspected exposure. Clin. Chem. 40:1401–1404.

Barkley, J., J. Bunch, J.T. Bursey, N. Castillo, S.D. Cooper, J.M. Davis, M.D. Erickson, B.S.H. Harris III, M. Kirkpatrick, L. Michael, S.P. Parks, E. Pellizri, M. Ray, D. Smith, K. Tomer, R. Wagner, and R.A. Zweidenger. 1980. Gas chromatography mass spectrometry computer analysis of volatile halogenated hydrocarbons in man and his environment—a multimedia environmental study. Biomed. Mass Spectrometry. 7(4):139–147.

Berglund, B., I. Johansson, and T. Lindvall. 1982a. A longitudinal study of air contaminants in a newly built preschool. Environ. Internat. 8:111–115.

Berglund, B., I. Johansson, and T. Lindvall. 1982b. The influence of ventilation on indoor/outdoor air contaminants in an office building. Environ. Internat. 8:395–399.

Berglund, B., I. Johansson, and T. Lindvall. 1987. Volatile organic compounds from building materials in a simulated chamber study. Pages 16–21 in: Seifert, B., H. Esdorn, M. Fischer, H. Rüden, and J. Wegner (eds.). Indoor Air '87. Proceedings of the 4th International Conference on Indoor Air Quality and Climate. August 17–21. Volume 1. Institute for Soil, Water, and Air Hygiene. Berlin (West).

Bourdillon, R.B., O.M. Lidwell, and J.C. Thomas. 1941. A slit sampler for collecting and counting airborne bacteria. J. Hyg. (Camb). 14:197–224.

Breysse, P.A. 1984. Formaldehyde levels and accompanying symptoms associated with individuals residing in over 1,000 conventional and mobile homes in the State of Washington. Pages 403–408 in: Berglund, B. T. Lindvall, and J. Sundell (eds.). Indoor Air Sensory and Hyperactivity Reactions to Sick Buildings. Volume 3. Swedish Council for Building Research. Stockholm, Sweden.

Bridge, D.P. and M. Corn. 1972. Contribution to the assessment of exposure of nonsmokers to air pollution from cigarette and cigar smoke in occupied spaces. Environ. Res. 5:192–209.

Brief, R.S. 1960. A simple way to determine air contaminants. Air Engineering. 2:39–51.

Brown, V.M. and D.R. Crump. 1996. Volatile organic compounds. In: Berry, R.W., V.M. Brown, S.K.D. Coward, D.R. Crump, M. Gavin, C.P. Grimes, D.F. Higham, A.V. Hull, C.A. Hunter, I.G. Jeffery, R.G. Lea, J.W. Llewellyn, and G.J. Raw (eds.). Indoor air quality in homes: Part I. The building research establishment indoor environment study. Construction Research Communications. London, United Kingdom.

Busigin, A., A.W. Van der Vooren, J.C. Babcock, and C.R. Phillips. 1981. The nature of unattached ^{218}Po (RaA) particles. Health Phys. 40:333–343.

Buttner, M.P. and L.D. Stetzenbach. 1991. Evaluation of four aerobiological sampling methods for the retrieval of aerosolized *Pseudomonas syringae*. Appl. Environ. Microbiol. 57:1268–1270.

Cartwright, B.G. and E.K. Shirk. 1978. A nuclear-track-recording polymer of unique sensitivity and resolution. Nucl. Instrum. Meth. 153:457–460.

Cheng, Y.S. and H.C. Yeh. 1980. Theory of a screen-type diffusion battery. J. Aerosol Sci. 11:313–320.

Cheng, Y.S., J.A. Keating, and G.M. Kanapilly. 1980. Theory and calibration of a screen-type diffusion battery. J. Aerosol Sci. 11:549–556.

Chew, G.L., H.A. Burge, D.W. Dockery, M.L. Muilenberg, S.T. Weiss, and D.R. Gold. 1998. Limitations of a home characteristics questionnaire as a predictor of indoor allergen levels. Amer. J. Resp. Crit. Care Med. In press.

Chu, K.D. and P.K. Hopke. 1985. Continuous monitoring method of the neurtralizing phenomena of Polonium-218. Presented at the 78th Annual Meeting of the Air Pollution Control Association. June. Detroit, Michigan. Paper Number 85-85.5. Air Pollution Control Association. Pittsburgh, Pennsylvania.

Chu, K.D. and P.K. Hopke. 1988. Neutralization kinetics for polonium-218. Environ. Sci. Technol. 22:711–717.

Clayton, C.A., R.L. Perritt, E.D. Pellizzari, K.W. Thomas, R.W. Whitmore, H. Özkaynak, J.D. Spengler, and L.A. Wallace. 1993. Particle total exposure assessment methodology (PTEAM) study: Distributions of aerosol and elemental concentrations in personal, indoor, and outdoor air samples in a Southern California community. J. Exposure Anal. Environ. Epidemiol. 3:227–250.

Coburn, R.F., R.E. Forster, and P.B. Kane. 1965. Considerations of the physiological variables that determine the blood carboxyhemoglobin concentration in man. J. Clin. Invest. 43:1098–1103.

Cohen, B.L. and R. Nason. 1986. A diffusion barrier charcoal adsorption collector for measuring Rn concentrations in indoor air. Health Phys. 50:457–463.

Colome, S.D., A.L. Wilson, and Y. Tian. 1994. California residential indoor air quality study. Volume 2: Carbon monoxide and air exchange rate: a univariate and multivariate analysis. Integrated Environmental Services. Irvine, California. 200 pages.

Colome, S.D., A.L. Wilson, E.W. Becker, S.J. Cunningham, and P.E. Baker. 1987. Analysis of factors associated with indoor residential nitrogen dioxide: multivariate regression results. Pages 405–409 in: Seifert, B., H. Esdorn, M. Fischer, H. Rüden, and J. Wegner (eds.). Indoor Air '87. Proceedings of the 4th International Conference on Indoor Air. August 17–21. Volume 1. Institute for Water, Soil, and Air Hygiene. Berlin (West).

Countess, R.J. 1970. ^{222}Rn flux measurement with a charcoal canister. Health Phys. 31:455–456.

Coutant, R.W., R.G. Lewis, and J. Mulik. 1985. Passive sampling devices with reversible adsorption. Anal. Chem. 57:219–223.

Coutant, R.W., R.G. Lewis, and J. Mulik. 1986. Modification and evaluation of a thermally desorbable passive sampler for volatile organic compounds in air. Anal. Chem. 58:445–448.

Crook, B. 1997. Non-inertial samplers: Biological perspectives. Pages 269–284 in: Cox, C.S. and C.M. Wathes. Bioaerosols Handbook. CRC/Lewis Press. Boca Raton, Florida.

De Bortoli, M., H. Knöppel, E. Pecchio, A. Peil, L. Rogora, H. Schauenburg, H. Schlitt, and H. Vissers. 1986. Concentrations of selected organic pollutants in indoor and outdoor air in northern Italy. Environ. Internat. 12(1–4):343–350.

Delfino, R.J., B.D. Coate, R.S. Zeiger, J.M. Seltzer, D.H. Street, and P. Koutrakis. 1996. Daily asthma severity in relation to personal ozone exposure and outdoor fungal spores. Am. J. Respir. Crit. Care Med. 154:633–641.

Diemel, J.A.L., B. Brunekreef, J.S.M. Boleij, K. Biersteker, and S.J. Veenstra. 1981. The Arnhem lead study: II. Indoor pollution and indoor-outdoor relationships. Environ. Res. 25:449–456.

Dietz, R.N. and E.A. Cote. 1982. Air infiltration measurements in a home using a convenient perfluorocarbon tracer technique. Environ. Int. 8:419–433.

Dockery, D. and J. Spengler. 1981a. Indoor-outdoor relationships of respirable sulfates and particulates. Atmos. Environ. 15:335–343.

Dockery, D.W. and J.D. Spengler. 1981b. Personal exposure to respirable particulates and sulfates. JAPCA 31(2):153–159.

Dowty, B., D. Carlisle, J.L. Laseter, and J. Storer. 1975. Halogenated hydrocarbons in New Orleans drinking water and blood plasma. Science 187:75–77.

Dua, S.K., P. Kotrappa, and P.C. Gupta. 1983. Influence of relative humidity on the charged fraction of decay products of radon and thoron. Health Phys. 45:152–157.

Duan, N. 1982. Models for human exposure to air pollution. Environ. Internat. 8:305–309.

Duan, N. 1985. Application of the microenvironment monitoring approach to assess human exposure to carbon monoxide. Rand Corporation. Report R-322-EPA. Santa Monica, California. 50 pages.

Duan, N. 1991. Stochastic microenvironment models for air pollution exposure. J. Exposure Anal. Environ. Epidemiol. 2:235–257.

EML (Environmental Measurements Laboratory). 1990. Procedures Manual. 27th Edition. Report Number HASL-300. U.S. Department of Energy. November. New York, New York.

Fantini, M. and G. Renard. 1981. Influence of chemical and physical parameters on the quality and reproducibility of tracks in cellulose nitrate films. Pages 193–196 in: Fowler, P.H. and V.M. Clapham (eds.). Solid state nuclear track detectors. Pergamon Press. Oxford, United Kingdom.

Fellin, P. and R. Otson. 1993. Seasonal trends of volatile organic compounds (VOCs) in Canadian homes. Pages 339–343 in: Jaakola, J.J.K., R. Ilmarinen, and O. Seppänen (eds.). Indoor Air '93: Proceedings of the 6th International Conference on Indoor Air Quality and Climate. Volume 1. Helsinki University of Technology. Espoo, Finland.

Frey, G.J., P.K. Hopke, and J.J. Stukel. 1981. The effects of trace gases and water vapor on the diffusion coefficient of Po-218. Science 211:480–481.

George, A.C. 1972. Measurement of the uncombined fraction of radon daughters with wire screens. Health Phys. 23:390–392.

George, A.C. 1976. Scintillation Flasks for the Determination of Low Level Concentrations of Radon. Presentation by the Health and Safety Laboratory, U.S. Department of Energy, at the Ninth Midyear Health Physics Symposium. February. Denver, Colorado.

George, A.C. 1977. A passive environmental radon monitor. Pages 25–30 in: Breslin, A.J. (ed.). Radon Workshop. February. Health and Safety Laboratory, U.S. Department of Energy. HASL-325. New York, New York.

George, A.C. 1984. Passive integrated measurement of indoor radon using activated carbon. Health Phys. 46:867–872.

George, A.C. and A.J. Breslin. 1980. The distribution of ambient radon and radon daughters in residential buildings in the New Jersey-New York area. Page 1272 in: National Radiation Environment III. Volume 2. Technical Information Center, U.S. Department of Energy. CONF-780422. Washington, D.C.

George, A.C. and T. Weber. 1990. An improved passive activated C collector for measuring environmental ^{222}Rn in indoor air. Health Phys. 58:583–589.

George, A.C., M.H. Wilkening, E.O. Knutson, D. Sinclair, and L. Andrews. 1984. Measurements of radon and radon daughter aerosols in Socorro, New Mexico. Aerosol Sci. Technol. 3:277–281.

Giardino, N.J., E. Gumerman, N.A. Esmen, J.B. Andelman, C.R. Wilkes, and M.J. Small. 1990. Real-time air measurements of trichloroethylene in domestic bathrooms using contaminated water. Pages 707–712 in: Walkinshaw, D. (ed.). Indoor Air '90: Proceedings of the 5th international conference on indoor air quality and climate. July 29–August 3. Volume 2. Toronto, Canada. Canada Mortgage and Housing Association. Ottawa, Ontario.

Girman, A.T., Hodgson, A.W. Newton, and A.W. Winkes. 1986. Volatile organic emissions from adhesives with indoor applications. Environ. Internat. 12(1–4):317–321.

Girman, J.R., L. Alivantis, G. Kullasingam, M. Petreas, and L. Webber. 1987a. Bake-out of an office building. Pages 22–26 in: Seifert, B., H. Esdorn, M. Fischer, H. Rüden, and J. Wegner (eds.). Indoor Air '87. Proceedings of the 4th International Conference on Indoor Air Quality and Climate. August 17–21. Volume 1. Institute for Soil, Water, and Air Hygiene. Berlin (West).

Girman, J.R., A.T. Hodgson, and M.L. Wind. 1987b. Considerations in evaluating emissions from consumer products. Atmos. Environ. 21:315–320.

Girman, J.R. and A.T. Hodgson. 1987. Exposure to methylene chloride from controlled use of a paint remover in a residence. Presented at the 80th Annual Meeting of the Air Pollution Control Association. June 21–26. New York. Report Number LBL 23078. Lawrence Berkeley Laboratory. Berkeley, California.

Goldstein, B.D., R.J. Melia, S. Chinn, C.V. Florey, D. Clark, and H.H. John. 1979. The relationship between respiratory illness in primary school children and the use of gas for cooking: II. Factors affecting nitrogen dioxide levels in the home. Internat. J. Epidemiol. 8(4):339–345.

Goldstein, B.D., R.G. Tardiff, S.R. Baker, G.F. Hoffnagle, D.R. Murray, P.A. Catizone, R.A. Kester, and D.G. Caniparoli. 1992. Valdez Air Health Study. Alyeska Pipeline Service Company. Anchorage, Alaska.

Goldstein, S.D. and P.K. Hopke. 1985. Environmental neutralization of polonium-218. Environ. Sci. Technol. 19:146–150.

Gordon, S.M., D.V. Kenny, and T.J. Kelly. 1992. Continuous real-time breath analysis for the measurement of half-lives of expired volatile organic compounds. J. Exposure Anal. Environ. Epidemiol. Suppl. 1:41–54.

Gray, D.J. and S.T. Windham. 1987. EERF standard operating procedures for radon-222 measurement using charcoal canisters. U.S. Environmental Protection Agency. Report Number EPA 520/5-87-005. Washington, D.C. 30 pages.

Guerin, M.R., C.E. Higgins, and R.A. Jenkins. 1987. Measuring environmental emissions from tobacco combustion: sidestream cigarette smoke literature review. Atmos. Environ. 21:291–297.

Gundel, L.A., V.C. Lee, K.R.R. Mahanama, R.K. Stevens, and J.M. Daisey. 1995. Direct determination of the phase distributions of semi-volatile polycyclic aromatic hydrocarbons using annular denuders. Atmos. Environ. 29(14):1719–1733.

Hartwell, T.D., C.A. Clayton, R.W. Michie, R.W. Whitmore, H.S. Zelon, S.M. Jones, and D.A. Whitehurst. 1984. Study of carbon monoxide exposure of residents of Washington, DC and Denver, CO. Report Number EPA-600/S4-84-031, PB84-183516. Environmental Monitoring Systems Laboratory, U.S. Environmental Protection Agency. Research Triangle Park, North Carolina.

Hartwell, T.D., C.A. Clayton, R.W. Michie, R.W. Whitmore, H.S. Zelon, S.M. Jones, and D.A. Whitehurst. 1986. Study of carbon monoxide exposure of residents of Washington, DC and Denver, CO. EPA-600/S4-84-031, PB84-183516. Environmental Monitoring Systems Laboratory, U.S. Environmental Protection Agency. Research Triangle Park, North Carolina.

Higgins, C.E., R.R. Jenkins, and M.R. Guerin. 1983. Application of Tenax trapping to cigarette smoking. JAOAC 66:1074–1083.

Higgins, C.E. 1987. Organic vapor phase composition of sidestream and environmental tobacco smoke from cigarettes. Pages 140–151 in: Proceedings of the 1987 EPA/APCA Symposium on Measurement of Toxic and Related Air Pollutants. EPA Report Number 600/9-87-010. Research Triangle Park, North Carolina. May.

Highsmith, V.R., R.B. Zweidenger, and R.G. Merrill. 1988. Characterization of indoor and outdoor air associated with residences using woodstoves: a pilot study. Environ. Internat. 14:213–219.

Hirst, J.M. 1952. An automatic volumetric spore trap. Ann. Appl. Biol. 39:257–265.

Hocking, A.D. and J.I. Pitt. 1980. Dichloran-glycerol medium for enumeration of xerophilic fungi from low-moisture foods. Appl. Environ. Microbiol. 39:488–492.

Hodgson, A.T., J.R. Girman, and J. Binenboym. 1986. A multi-sorbent sampler for volatile organic compounds in indoor air. Presented at the 79th Annual Meeting of the Air Pollution Control Association. June. Minneapolis, Minnesota. APCA Paper Number 86-37.1. Pittsburgh, Pennsylvania.

Hodgson, A.T.K., Garbesi, R.G. Sextro, and J.M. Daisey. 1988. Evaluation of soil-gas transport of organic chemicals into residential buildings: Final report. Lawrence Berkeley Laboratory. Report Number LBL-25465. Berkeley, California. 100 pages.

Hollowell, C.D. and R.R. Miksch. 1981. Sources and concentrations of organic compounds in indoor environments. Bull. N.Y. Acad. Med. 57(10):962–977.

Holub, R.F. and E.O. Knutson. 1987. Measuring polonium-218 diffusion-coefficient spectra using multiple wire screens. Pages 340–356 in: Hopke, P.K. (ed.). Radon and its decay products: Occurrence, properties and health effects. Symposium Series 331. American Chemical Society. Washington, D.C.

Hopke, P.K. 1989. Use of electrostatic collection of ^{218}Po for measuring Rn. Health Phys. 57:39–42.
Hopke, P.K. 1992. Some thoughts on the "unattached" fraction of radon decay products. Health Phys. 63(2):209–212.
Hopke, P.K., B. Jensen, C.S. Li, N. Montassier, and P. Wasiolek. 1995. Assessment of the exposure to and dose from radon decay products in normally occupied homes. Environ. Sci. Technol. 29:1359–1364.
Hopke, P.K., M. Ramamurthi, E.O. Knutson, K.W. Tu, P. Scofield, R.F. Holub, Y.S. Cheng, Y.F. Su, W. Winklmayr, J.C. Strong, S. Solomon, and A. Reineking. 1992. The measurement of activity-weighted size distribution of radon progeny: methods and laboratory intercomparison studies. Health Phys. 63(5):560–570.
Immerman, F.W. 1989. Nonoccupational pesticide exposure study (NOPES). Final Summary Report. U.S. Environmental Protection Agency. Contract Number. 68-02-4544. Research Triangle Park, North Carolina.
Immerman, F.W. and J.L. Schaum. 1990a. Nonoccupational pesticide exposure study: Final report. EPA/600-3-90-003. U.S. Environmental Protection Agency. Washington, D.C.
Immerman, F.W. and J.L. Schaum. 1990b. Nonoccupational pesticide exposure study (NOPES). Project summary. U.S. Environmental Protection Agency. EPA 600/S3-90/003. Research Triangle Park, North Carolina.
Ingram, J.M., R. Sporik, G. Rose, R. Honsigner, M. Chapman, and T.A.E. Platts-Mills. 1995. Quantitative assessment of exposure to dog (Can f 1) and cat (Fel d 1) allergens: Relation to sensitization and asthma among children living in Los Alamos, New Mexico. J. Allergy Clin. Immunol. 96(4):449–456.
Ishizu, Y. 1980. General equation for the estimation of indoor pollution. Environ. Sci. Technol. 14(10):1254–1257.
James, A.C., G.F. Bradford, and D.M. Howell. 1972. Collection of unattached RaA atoms using wire gauge. J. Aerosol Sci. 3:243–250.
Jarke, F.H. and S.M. Gordon. 1981. Recent investigations of volatile organics in indoor air at sub-ppb levels. Presented at the 74th Annual Meeting of the Air Pollution Control Association. June. Paper Number 81-57.2. Pittsburgh, Pennsylvania.
Jermini, C., A. Weber, and E. Grandjean. 1976. Quantitative determination of various gas-phase components of the sidestream smoke of cigarettes in room air. Internat. Arch. Occup. Environ. Health (German). 36:169–181.
Jo, W.K., C.P. Weisel, and P.J. Lioy. 1990a. Routes of chloroform exposure and body burden from showering with contaminated tap water. Risk Anal. 10:575–580.
Jo, W.K., C.P. Weisel, and P.J. Lioy. 1990b. Chloroform exposure and the health risk associated with multiple uses of chlorinated tap water. Risk Anal. 10:581–585.
Johansson, I. 1978. Determination of organic compounds in indoor air with potential reference to air quality. Atmos. Environ. 12:1371–1377.
Johnson, T. 1983. A study of personal exposures to carbon monoxide in Denver, Colorado. Final report. Contract Number 68-02-3755. U.S. Environmental Protection Agency. Research Triangle Park, North Carolina.
Johnson, T., J. Capel, and L. Wijnberg. 1986. Selected data analyses relating to studies of personal carbon monoxide exposure in Denver and Washington, DC. Final report. Contract Number 68-02-3496. U.S. Environmental Protection Agency. Research Triangle Park, North Carolina.
Joly, J.R., E. Dewaily, L. Bernard, D. Ramsey, and J. Brisson. 1986. Legionella and domestic water heaters in Quebec City area. Can. Med. Assoc. J. 132:160.
Jones, R.M. and R. Fagan. 1974. Application of mathematical model for the buildup of carbon monoxide from cigarette smoking in rooms and houses. ASHRE Journal 16:49–53.
Jones, W., K. Morring, P. Morey, and W. Sorenson. 1985. Evaluation of the Andersen viable impactor for single stage sampling. AIHA Journal 46:294–298.
Kleipis, N.E., W.R. Ott, and P. Switzer. 1995. Modeling the time series of respirable suspended particles and carbon monoxide from multiple smokers: Validation in two public smoking lounges. Presented at the 88th Annual Meeting of the Air and Waste Management Association. June. Paper Number A-1233. Air and Waste Management Association. Pittsburgh, Pennsylvania.
Knöppel, H. and H. Schauenburg. 1987. Screening of household products for the emission of volatile organic compounds. Pages 27–31 in: Seifert, B., H. Esdorn, M. Fischer, H. Rüden, and J. Wegner (eds.). Indoor Air '87. Proceedings of the 4th International Conference on Indoor Air Quality and Climate. August 17–21. Volume 1. Institute for Water, Soil, and Air Hygiene. Berlin (West).
Knutson, E.O. 1989. Personal computer programs for use in radon and thoron progeny measurements. Report Number EML-517. Environmental Measurements Laboratory, U.S. Department of Energy. New York, New York.

Knutson, E.O., A.C. George, R.H. Knuth, and B.R. Koh. 1984. Measurements of radon daughter particle size. Radiat. Prot. Dosim. 7:121–125.

Kotrappa, P., J.C. Dempsey, J.R. Hickey, and L.R. Steiff. 1988. A practical E-PERMÔ (electret passive environmental radon monitor) system for indoor ^{222}Rn measurements. Health Phys. 54:47–56.

Koutrakis, P., S.L.K. Briggs, and B.P. Leaderer. 1992. Source apportionment of indoor aerosols in Suffolk and Onondaga counties, New York. Environ. Sci. Technol. 26:521–527.

Krause, C., W. Mailahn, R. Nagel, C. Schulz, B. Seifert, and D. Ullrich. 1987. Occurrence of volatile organic compounds in the air of 500 homes in the Federal Republic of Germany. Pages 102–106 in: Seifert, B., H. Esdorn, M. Fischer, H. Rüden, and J. Wegner (eds.). Indoor Air '87. Proceedings of the 4th International Conference on Indoor Air Quality and Climate. August 17–21. Volume 1. Institute for Soil, Water, and Air Hygiene. Berlin (West).

Krost, K.J., E.D. Pellizari, S.G. Walburn, and S.A. Hubbard. 1982. Collection and analysis of hazardous organic emissions. Anal. Chem. 54:810–817.

Leaderer, B.P. and S.K. Hammond. 1991. Evaluation of vapor-phase nicotine and respirable suspended particle mass as markers for environmental tobacco smoke. Environ. Sci. Technol. 25:770–777.

Leaderer, B.P., W.S. Cain, R. Isserof, and L.G. Berglund. 1984. Ventilation requirements in buildings—II. Particulate matter and carbon monoxide from cigarette smoking. Atmos. Environ. 18:99–106.

Lebret, E., H.J. Van de Weil, D. Noij, and J.S.M. Boleij. 1986. Volatile hydrocarbons in Dutch homes. Environ. Internat. 12(1–4):323–332.

Lewis, R.G. and A.E. Bond. 1987. Nonoccupational exposure to household pesticides. Pages 195–199 in: Seifert, B., H. Esdorn, M. Fischer, H. Rüden, and J. Wegner (eds.). Indoor Air '87. Proceedings of the 4th International Conference on Indoor Air Quality and Climate. August 17–21. Volume 1. Institute for Soil, Water, and Air Hygiene. Berlin (West).

Lewis, R.G. and K.E. MacLeod. 1982. A portable sampler for pesticides and semivolatile industrial organic chemicals in air. Anal. Chem. 54:310–315.

Lewis, R.G., J.D. Mulik, R.W. Coutant, G.W. Wooten, and C.R. McMillan. 1985. Thermally desorbable passive sampling device for volatile organic chemicals in ambient air. Anal. Chem. 57:214–219.

Lioy, P.J., J.M. Waldman, T. Buckley, J. Butler, and C. Pietarinen. 1990. The personal, indoor and outdoor concentrations of PM-10 measured in an industrial community during the winter. Atmos. Environ. 24:57–66.

Löfroth, G., B. Burton, L. Forehand, S.K. Hammond, R. Seila, R. Zweidenger, and J. Lewtas. 1989. Characterization of environmental tobacco smoke. Environ. Sci. Technol. 23:610–614.

Lucas, H.F. 1957. Improved low-level alpha scintillation counter for radon. Rev. Sci. Instr. 28:258–259.

Madelin, T.M. and M.F. Madelin. 1997. Biological analysis of fungi and associated molds. Pages 361–386 in: Cox, C.S. and C.M. Wathes. Bioaerosols Handbook. CRC/Lewis Press. Boca Raton, Florida.

Mailahn, W., B. Siefert, and D. Ullrich. 1987. The use of a passive sampler for the simultaneous determination of long-term ventilation rates and VOC concentrations. Pages 149–153 in: Seifert, B., H. Esdorn, M. Fischer, H. Rüden, and J. Wegner (eds.). Indoor Air '87. Proceedings of the 4th International Conference on Indoor Air Quality and Climate. August 17–21. Volume 1. Institute for Soil, Water, and Air Hygiene. Berlin (West).

Marcinowski, F. 1992. Nationwide survey of residential radon levels. Health Phys. 62(6):S13 (Supplement).

Marple, V.A., K.L. Rubow, W. Turner, and J.D. Spengler. 1987. Low flow rate sharp cut impactors for indoor air sampling: Design and calibration. JAPCA 37:1303–1307.

McKone, T.E. 1987. Human exposure to volatile organic compounds in household tap water: The indoor inhalation pathway. Environ. Sci. Technol. 21:1194–1201.

Milton, D., R. Gere, H. Feldman, and I. Greaves. 1990. Endotoxin measurement aerosol sampling and application of a new Limulus method. AIHA Journal 51:331–337.

Milton, D.K., H.A. Feldman, D.S. Neuberg, R.J. Bruckner, and I.A. Greaves. 1992. Environmental endotoxin measurement: The kinetic *Limulus* assay with resistant-parallel line estimation. Environ. Res. 57:212–230.

Mølhave, L. 1982. Indoor air pollution due to organic gases and vapours of solvents in building materials. Environ. Internat. 5(1–6):117–127.

Mølhave, L. and J. Møller. 1979. The atmospheric environment in modern Danish dwellings: Measurements in 39 flats. Pages 171–186 in: Fänger, O. and O. Valbjõrn (eds.). Indoor Climate. SBI. Hõrsholm, Denmark.

Nazaroff, W.W. 1984. Optimizing the total-alpha three-count technique for measuring concentrations of radon progeny in residences. Health Phys. 24:387–395.

Nazaroff, W.W. and G.R. Cass. 1989. Mathematical modeling of indoor aerosol dynamics. Environ. Sci. Technol. 23:157–166.

Nazaroff, W.W., L.G. Salmon, and G.R. Cass. 1990a. Concentration and fate of airborne particles in museums. Environ. Sci. Technol. 24:66–77.

Nazaroff, W.W., M.P. Ligocki, T. Ma, and G.R. Cass. 1990b. Particle deposition in museums: Comparison of modeling and measurement results. Aerosol Sci. Technol. 13:332–348.

Nazaroff, W.W., A.J. Gadgil, C.J. Weschler. 1993. Critique of the use of deposition velocity in modeling indoor air quality. Pages 81–104 in: Nagda, N.L. (ed.). Modeling of indoor air quality and exposure. ASTM STP 1205. American Society for Testing and Materials. Philadelphia, Pennsylvania.

Neas, L.M., D.W. Dockery, J.H. Ware, J.D. Spengler, B.G. Ferris, Jr., and F.E. Speizer. 1994. Concentration of indoor particulate matter as a determinant of respiratory health in children. Am. J. Epidemiol. 139:1088–1099.

Neas, L.M., D.W. Dockery, H. Burge, P. Koutrakis, and F.E. Speizer. 1996. Fungus spores, air pollutants, and other determinants of peak expiratory flow rate in children. Am. J. Epidemiol. 143(8):797–807.

Nero, A.V. 1988. Estimated risk of lung cancer from exposure to radon decay products in U.S. homes: A brief review. Atmos. Environ. 22:2205–2211.

Nero, A.V., Schwehr, M.B., Nazaroff, W.W. and Rezvan, K.L. (1986). Distribution of airborne radon 222 concentrations in U.S. homes. Science 234:992–997.

NTP (National Toxicology Program). 1986. Technical report on the toxicity and carcinogenesis of 1,4-dichlorobenzene (CAS #106-46-7) in F344/n rats and B6C3F1 mice (gavage study). National Toxicology Program Technical Report Number 319. Board Draft. March 1986.

NTP (National Toxicology Program). 1988. Technical report on the toxicity and carcinogenesis of d-limonine (CAS #106-46-7) in F344/n rats and B6C3F1 mice (gavage study). National Toxicology Program Technical Report Number 347. NIH Publication Number 88–2802.

Nuchia, E. 1986. MDAC—Houston materials testing database users' guide. National Aeronautics and Space Administration. Washington, D.C.

Offerman, F.J., R.G. Sextro, W.J. Fisk, D.T. Grimsrud, W.W. Nazaroff, A.V. Nero, K.L. Revzan, and J. Yater. 1985. Control of respirable particles in indoor air with portable air cleaners. Atmos. Environ. 19:1761–1771.

Oliver, K.D., J.D. Pleil, and W.A. McClenny. 1986. Sample integrity of trace level volatile organic compounds in ambient air stored in summa polished canisters. Atmos. Environ. 20:1403.

Ott, W.R. 1985. Total human exposure: an emerging science focuses on humans as receptors of environmental pollution. Environ. Sci. Technol. 19:880.

Ott, W.R., J. Thomas, D. Mage, and L.Wallace. 1988. Validation of the simulation of human activity and agent exposure (SHAPE) model using paired days from the Denver, CO carbon monoxide study. Atmos. Environ. 22:2101–2113.

Ott, W.R., L. Langan, and P. Switzer. 1992. A time series model for cigarette smoking activity patterns: model validation for carbon monoxide and respirable particles in a chamber and an automobile. J. Exposure Anal. Environ. Epidemiol. 2(Suppl. 2):175–200.

Ott, W.R., P. Switzer, and J. Robinson. 1995a. Particle concentration inside a tavern before and after prohibition of smoking: evaluating the performance of an indoor air quality model. JAWMA 46:1120–1134.

Ott, W.R., N.E. Kleipis, and P. Switzer. 1995b. Modeling environmental tobacco smoke in the home using transfer functions. Presented at the 88th Annual Meeting of the Air and Waste Management Association. June. Paper Number A-1043. Air and Waste Management Association. Pittsburgh, Pennsylvania.

Özkaynak, H., P.B. Ryan, L.A. Wallace, W.C. Nelson, and J.V. Behar. 1987. Sources and emission rates of organic chemical vapors in homes and buildings. Pages 3–7 in: Seifert, B., H. Esdorn, M. Fischer, H. Rüden, and J. Wegner (eds.). Indoor Air '87. Proceedings of the 4th International Conference on Indoor Air Quality and Climate. August 17–21. Volume 1. Institute for Soil, Water, and Air Hygiene. Berlin (West).

Özkaynak, H., J.D. Spengler, J. Xue, P. Koutrakis, E.D. Pellizari, and L.A. Wallace. 1993. Sources and factors influencing personal and indoor exposures to particles: Findings from the particle TEAM pilot study. Pages 457–462 in: Indoor Air '93: Proceedings of the 6th International Conference on Indoor Air Quality and Climate. Volume 3. Helsinki, Finland.

Özkaynak, H., J. Xue, R. Weker, D. Butler, P. Koutrakis, and J.D. Spengler. 1996a. The particle TEAM (PTEAM) study: Analysis of the data. Volume III. Final report. U.S. Environmental Protection Agency. Contract Number 68-0204544. Research Triangle Park, North Carolina. 500 pages.

Özkaynak, H., J. Xue, J.D. Spengler, L.A. Wallace, E.D. Pellizari, and P. Jenkins. 1996b. Personal exposure to airborne particles and metals: Results from the particle TEAM study in Riverside, California. J. Exposure Anal. Environ. Epidemiol. 6:57–78.

Palmes, E.D., A.F. Gunnison, J. DiMattio, and C. Tomczyk. 1986. Personal sampler for NO_2. AIHA Journal. 37:570–577.

Pellizari, E.D. 1979. Analysis of organic air pollutants by gas chromatography and mass spectrometry. U.S. Environmental Protection Agency. Office of Research and Development. Research Triangle Park, North Carolina. 300 pages.

Pellizzari, E.D., K. Perritt, T.D. Hartwell, L.C. Michael, R. Whitmore, R.W. Handy, D. Smith, and H. Zelon. 1987a. Total exposure assessment methodology (TEAM) study: Elizabeth and Bayonne, New Jersey; Devils Lake, North Dakota; and Greensboro, North Carolina. Volume II. U.S. Environmental Protection Agency. Washington, D.C. 800 pages.

Pellizzari, E.D., K. Perritt, T.D. Hartwell, L.C. Michael, R. Whitmore, R.W. Handy, D. Smith, and Zelon, H. 1987b. Total exposure assessment methodology (TEAM) study: Selected communities in northern and southern California. Volume III. U.S. Environmental Protection Agency. Washington, D.C. 500 pages.

Pellizzari, E.D., L.C. Michael, R. Perritt, D.J. Smith, T.D. Hartwell, and J. Sebestik. 1988. Comparison of indoor and outdoor toxic air agent levels in several southern California communities. Final report. Contract Number 68-02-4544. U.S. Environmental Protection Agency. Research Triangle Park, North Carolina. 500 pages.

Pellizzari, E.D., L.C. Michael, R. Perritt, D.J. Smith, T.D. Hartwell, and J. Sebestik. 1989. Comparison of Indoor and Outdoor Toxic Air Pollutant Levels in Several Southern California communities. Final Report, Contract #68-02-4544. U.S. Environmental Protection Agency. Research Triangle Park, North Carolina.

Pellizzari, E.D., K.W. Thomas, C.A. Clayton, R.W. Whitmore, R.C. Shores, H.S. Zelon, and R.L. Perritt. 1991. Particle total exposure assessment methodology (PTEAM): Riverside, California pilot study. Final report. Volume 1. U.S. Environmental Protection Agency. Contract Number 68-02-4544. Research Triangle Park, North Carolina. 500 pages.

Pellizzari, E.D., K.W. Thomas, C.A. Clayton, R.W. Whitmore, R.C. Shores, H.S. Zelon, and R.L. Perritt. 1992. Particle Total Exposure Assessment Methodology (PTEAM): Riverside, California Pilot Study. Final Report. Vol. 1. EPA Contract #68-02-4544. Research Triangle Park, North Carolina.

Platts-Mills, T.A.E., M.L. Hayden, M.D. Chapman, and S.R. Wilkins. 1986. Seasonal variation in the dust mite and grass pollen allergens in dust from the houses of patients with asthma. J. Allergy Clin. Immunol. 79:781.

Porstendörfer, J. and T.T. Mercer. 1979. Influence of electric charge and humidity upon the diffusion coefficient of radon decay products. Health Phys. 15:191–199.

Prichard, H.M. and K.A. Marien. 1985. A passive diffusion Rn-222 sampler based on activated carbon absorption. Health Phys. 48:797–803.

Quackenboss, J.J., J.D. Spengler, M.S. Kanarek, R. Letz, and C.P. Duffy. 1986. Personal exposure to nitrogen dioxide: Relationship to indoor/outdoor air quality and activity patterns. Environ. Sci. Technol. 20:775–783.

Raabe, O.G. and M.E. Wrenn. 1969. Analysis of the activity of radon daughter samples by weighted least squares. Health Phys. 17:593–605.

Ramamurthi, M. and P.K. Hopke. 1989. On improving the validity of wire screen unattached fraction daughter measurements. Health Phys. 56:189–194.

Ramamurthi, M. and P.K. Hopke. 1991. An automated, semi-continuous system for measuring indoor radon progeny activity-weighted size distributions, d_p: 0.5–500 nm. Aerosol Sci. Technol. 14:82–92.

Ramamurthi, M., R. Strydom, and P.K. Hopke. 1990. Assessment of wire and tube penetration theories using a $^{218}PoO_x$ cluster. J. Aerosol Sci. 21:203–211.

Raunemaa, T., M. Kulmala, J.H. Saari, M. Olin, and M.H. Kulmala. 1989. Indoor air aerosol model: Transport indoors and deposition of fine and coarse particles. Aerosol. Sci. Technol. 11:11–25.

Reineking, A. and J. Porstendörfer. 1986. High-volume screen diffusion batteries and a-spectroscopy for measurement of the radon daughter activity size distributions in the environment. J. Aerosol Sci. 17:873–879.

Reineking, A., K.H. Becker, and J. Porstendörfer. 1985. Measurements of the unattached fractions of radon daughters in houses. Sci. Total Environ. 45:261–270.

Reineking, A., K.H. Becker, and J. Porstendörfer. 1988. Measurement of activity size distributions of the short-lived radon daughters in the indoor and outdoor environment. Radiat. Prot. Dosim. 24:245–250.

Repace, J.L. 1987a. Indoor concentrations of environmental tobacco smoke: Models dealing with effects of ventilation and room size. Pages 25–41 in: O'Neill, I.K., K.D. Brunnemann, B. Dodet, and D. Hoffmann (eds.). Environmental carcinogens methods of analysis and exposure measurement. Volume 9: Passive Smoking. International Agency for Research on Cancer. Lyon, France.

Repace, J.L. 1987b. Indoor concentrations of environmental tobacco smoke: field surveys. Pages 141–162 in: O'Neill, I.K., K.D. Brunnemann, B. Dodet, and D. Hoffmann (eds.). Environmental carcinogens methods of analysis and exposure measurement. Volume 9: Passive Smoking. International Agency for Research on Cancer. Lyon, France.

Repace, J.L. and A.H. Lowry. 1980. Indoor air pollution, tobacco smoke, and public health. Science. 208:464–472.

Repace, J.L. and A.H. Lowry. 1982. Tobacco smoke, ventilation, and indoor air quality. ASHRE Transactions. 88(Pt.1):895–914.

Repace, J.L. and A.H. Lowry. 1985. A quantitative estimate of non-smokers' lung cancer risk from passive smoking. Environ. Internat. 11:3–22.

Reponen, T., A. Nevalainen, M. Jantunen, M. Pellikka, and P. Kalliokosk. 1992. Normal range criteria for indoor air bacteria and fungal spores in a subarctic climate. Indoor Air 2:26–31.

Roberts, J.W., W.T. Budd, M.G. Ruby, D.E. Camann, R.C. Fortmann, R.G. Lewis, L.A. Wallace, and T.M. Spittler. 1992. Human exposure to pollutants in the floor dust of homes and offices. J. Exposure Anal. Environ. Epidemiol. 1(Suppl. 1):127–146.

Rodriguez, G.G., D. Phipps, K. Ishiguro, and H.F. Ridgeay. 1992. Use of fluorescent redox probe for direct visualization of actively respiring bacteria. Appl. Environ. Microbiol. 58:1801.

Ryan, P. B., M.L. Soczek, J.D. Spengler, and I.H. Billick, 1988a. The Boston residential NO_2 characterization study: I. A preliminary evaluation of the survey methodology. JAPCA 38:22–27.

Ryan, P. B., M.L. Soczek, R.D. Treitman, J.D. Spengler, and I.H. Billick. 1988. The Boston residential NO_2 characterization study: II. Survey methodology and population concentration estimates. Atmos. Environ. 22:2115–2125.

Ryan, P.B., M. Schwab, and J.D. Spengler. 1992. Nitrogen dioxide exposure studies—Volume 3: Personal exposure to nitrogen dioxide in Boston: A microenvironmental approach. Gas Research Institute. Chicago, Illinois. 250 pages.

Sakaguchi, M., S. Inouye, R. Sasaki, M. Hashimoto, C. Kobayashi, and H. Yasueda. 1996. Measurement of airborne mite allergen exposure in individual subjects. J Ref for filter cassettes and immunoassays. J. Allergy Clin. Immunol. 97(5):1040–1044.

Santanam, S., J.D. Spengler, and P.B. Ryan. 1990. Particulate matter exposures estimated from an indoor-outdoor source apportionment study. Pages 583–588 in: Walkinshaw, D. (ed.). Indoor Air '90: Proceedings of the 5th International Conference on Indoor Air Quality and Climate. Toronto, Canada. Volume 2. Canada Mortgage and Housing Corporation. Ottawa, Ontario.

Saraf, A., L. Larsson, H. Burge, and D. Milton. 1997. Quantification of ergosterol and 3-hydroxy fatty acids in settled house dust by gas chromatography-mass spectrometry: comparison with fungal culture and determination of endotoxin by Limulus assay. Appl. Environ. Microbiol. 63(7):2554–2559.

Schappi, G.F., C. Suphioglu, P.E. Taylor, and R.B. Knox. 1997. Concentrations of the major birch tree allergen Bet v 1 in pollen and respirable fine particles in the atmosphere. J. Allergy Clin. Immunol. 100(5):656–661.

Scheibel, H.G., and J. Porstendorfer. 1984. Penetration measurements for tube and screen-type diffusion batteries in the ultrafine size range. J. Aerosol Sci. 15:673–682.

SCAQMD (South Coast Air Quality Management District). 1989. In-vehicle characterization study in the South Coast air basin. Los Angeles. 100 pages.

Schwab, M., J.D. Spengler, P.B. Ryan, S. Colome, and A.L. Wilson. 1990. Nitrogen dioxide exposure studies. Volume IV: Human exposure to nitrogen dioxide in the Los Angeles basin. Report prepared for Gas Research Institute (Chicago, Illinois) and Southern California Gas Company (Los Angeles, California).

Seifert, B. and H.J. Abraham. 1982. Indoor air concentrations of benzene and some other aromatic hydrocarbons. Ecotoxicol. Environ. Safety. 6:190–192.

Seifert, B. and H.J. Abraham. 1983. Use of passive samplers for the determination of gaseous organic substances in indoor air at low concentration levels. Internat. J. Environ. Anal. Chem. 13:237–253.

Shair, F.H. and K.L. Heitner. 1974. Theoretical model for relating indoor pollutant concentrations to those outside. Environ. Sci Technol. 8:444–451.

Sheldon, L.S., R.W. Handy, T.D. Hartwell, R.W. Whitmore, H.S. Zelon, and E.D. Pellizari. 1988a. Indoor air quality in public buildings. Volume 1. U.S. Environmental Protection Agency. Washington, D.C. EPA 600/6-88/009a. National Technical Information Service PB 89-102511/AS. Springfield, Virginia.

Sheldon, L.S., H.S. Zelon, J. Sickles, C. Eaton, and T. Hartwell. 1988b. Indoor air quality in public buildings. Volume 2. U.S. Environmental Protection Agency. Research Triangle Park, North Carolina. EPA 600/6-88/009b. National Technical Information Service PB 89-102511/AS. Springfield, Virginia.

Sheldon, L.S., T.D. Hartwell, B.G. Cox, J.E. Sickles II, E.D. Pellizzari, M.L. Smith, R.L. Perritt, R.L., and S.M. Jones. 1989. An investigation of infiltration and indoor air quality. Final report. Contract No. 736-CON-BCS-85. New York State Energy Research and Development Authority. Albany, New York. 500 pages.

Sheldon L.S., A. Clayton, B. Jones, J. Keever, R. Perritt, D. Smith, D. Whitaker, and R. Whitmore. 1992. Indoor pollutant concentrations and exposures. Contract #A833-156. Final report. California Air Resources Board, Sacramento, California.

Sigsby, J.E., S.B. Tejada, W.D. Ray, J.M. Lang, and J.D. Duncan. 1987. Volatile organic compound emissions from 46 in-use passenger cars. Environ. Sci. Technol. 21-466–475.

Sinclair, J.D., A.C. George, and E.O. Knutson. 1977. Application of diffusion batteries to measurement of submicron radioactive aerosols. Pages 103–114 in: Airborne Radioactivity. American Nuclear Society. Lagrange Park, Illinois.

Sinclair, J.D., L.A. Psota-Kelty, and C.J. Weschler. 1988. Indoor/outdoor ratios and indoor surface accumulations of ionic substances at Newark, New Jersey. Atmos. Environ. 22:461–469.

Sinclair, J.D., L.A. Psota-Kelty, C.J. Weschler, and H.C. Shields. 1990. Measurement and modeling of airborne concentrations and indoor surface accumulation rates of ionic substances at Neenah, Wisconsin. Atmos. Environ. 24A:627–638.

Sinclair, J.D., L.A. Psota-Kelty, G.A. Peins, and A.O. Ibidunni. 1992. Indoor-outdoor relationships of airborne ionic substances: Comparison of electronic equipment room and factory environments. Atmos. Environ. 26A:871–882.

Skov, P., O. Valbjõrn, B.V. Pedersen, and Danish Indoor Climate Study Group. 1990. Influence of indoor climate on the sick building syndrome in an office environment. Scand. J. Work Environ. Health. 15:286–295.

Solomon, S. and T. Ren. 1992. Counting efficiencies for alpha particles emitted from wire screens. Aerosol Sci. Technol. 17:69–83.

Solomon, W.R. 1976. Volumetric studies of aeroallergen prevalence. I. Pollens of weedy spores at a midwestern station. J. Allergy Clin. Immunol. 57(4):318–327.

Solomon, W.R., H.A. Burge, and M.L. Muilenberg. 1983. Allergen carriage by atmospheric aerosol. I. Ragweed pollen determinants in smaller micronic fractions. J. Allergy Clin. Immunol. 72(5):443–447.

Spengler, J.D., D.W. Dockery, W.A. Turner, J.M. Wolfson, and B.G. Ferris, Jr. 1981. Long-term measurements of respirable sulfates and particles inside and outside homes. Atmos. Environ. 15:23–30.

Spengler, J.D., C.P. Duffy, R. Letz, T.W. Tibbits, and B.G. Ferris, Jr. 1983. Nitrogen dioxide concentrations inside and outside 137 homes and implications for ambient air quality standards and health effects research. Environ. Sci. Techol. 17:164–168.

Spengler, J.D., R.D. Treitman, T.D. Tosteson, D.T. Mage, and M.L. Soczek. 1985. Personal exposures to respirable particulates and implications for air pollution epidemiology. Environ. Sci. Technol. 19:700–707.

Spengler, J.D., J. Ware, F. Speizer, B. Ferris, D. Dockery, E. Lebret, and B. Brunnekreef. 1987. Harvard's indoor air quality respiratory health study. Pages 742–746 in: Seifert, B., H. Esdorn, M. Fischer, H. Rüden, and J. Wegner (eds.). Indoor Air '87. Proceedings of the 4th International Conference on Indoor Air Quality and Climate. August 17–21. Volume 2. Institute for Water, Soil, and Air Hygiene. Berlin (West).

Spengler, J.D., P.B. Ryan, and M. Schwab. 1992. Nitrogen dioxide exposure studies—Volume 4. Personal exposure to nitrogen dioxide in the Los Angeles basin. Gas Research Institute. Chicago, Illinois. 200 pages.

Spicer, C.W., M.W. Holdren, L.E. Slivon, R.W. Contant, M.E. Graves, D.S. Shadwick, W.A. McClenny, J.D. Mulik, and T.R. Fitz-Simmons. 1986. Intercomparison of sampling techniques for toxic organic compounds in indoor air. Pages 45–60 in: Hochheiser, S. and R.K.M. Jayanti (eds.). Proceedings of the 1986 EPA/APCA Symposium on the Measurement of Toxic Air Pollutants. Air Pollution Control Association. Pittsburgh, Pennsylvania.

Spitz, H. and M.E. Wrenn. 1977. Design and application of a continuous, digital-output, environmental radon measuring instrument. Pages 48–64 in: Breslin, A.J. (ed.). Radon workshop. February. HASL-325. Health and Safety Laboratory. U.S. Department of Energy. New York, New York.

Stark, P.C., L.M. Ryan, J. MacDonald, and H.B. Burge. 1997. Using meteorologic data to model and predict ragweed pollen levels. Aerobiologia 13:177–184.

Stephen, E., A.E. Raftery, and P. Dowding. 1990. Forecasting spore concentrations: A time series approach. Internat. J. Biometeorol. 34(2):87–89.

Strong, J.C. 1988. The size of attached and unattached radon daughters in room air. J. Aerosol Sci. 19:1327–1330.

Strong, J.C. 1989. Design of the NRPB activity size measurement system and results. Presented at the Workshop on "Unattached" Fraction Measurements. April. University of Illinois. Urbana, Illinois. 32 pages.

Su, H.J., A. Rotnitzky, H.A. Burge, and J.D. Spengler. 1992. Examination of fungi in domestic interiors by using factor analysis. Correlations and associations with home factors. Appl. Environ. Microbiol. 58:181–186.

Su, H.J., J.D. Spengler, and H.A. Burge. 1998. Influence of residential airborne fungal concentrations on childhood respiratory symptoms. Appl. Environ. Microbiol. Submitted for publication.

Swanson, M.C., M.K. Agarwal, and C.E. Reed. 1985. An immunochemical approach to indoor aeroallergen quantitation with a new volumetric air sampler: studies with mite, roach cat, mouse, and guinea pig antigens. J. Allergy Clin. Immunol. 76:721.

Swanson, M.C., A.R. Campbell, M.J. Klauck, and C.E. Reed. 1989. Correlations between levels of mite and cat allergens in settled and airborne dust. J. Allergy Clin. Immunol. 83:776.

Thatcher, T.L. and D.W. Layton. 1995. Deposition, resuspension, and penetration of particles within a residence. Atmos. Environ. 29:1487–1497.

Thomas, J.W. 1970. Modification of the Tsivoglou method for radon daughters in air. Health Phys. 19:691–693.

Thomas, J.W. 1972. Measurement of radon daughters in air. Health Phys. 23:783–789.

Thomas, K.W., E.D. Pellizzari, R.L. Perritt, and W.C. Nelson. 1991. Effect of dry-cleaned clothes on tetrachloroethylene levels in indoor air, personal air, and breath for residents of several New Jersey homes. J. Exposure Anal. Environ. Epidemiol. 1:475–490.

Thomas, K.W., E.D. Pellizzari, C.A. Clayton, R.L. Perritt, R.N. Dietz, R.W. Goodrich, W.C. Nelson, P.R. and L.A. Wallace. 1993. Temporal variability of benzene exposure for residents in several New Jersey homes with attached garages or tobacco smoke. J. Exposure Anal. Environ. Epidemiol. 3:49–73.

Thomas, K.W., E.D. Pellizzari, C.A. Clayton, D.A. Whitaker, R.C. Shores, J.D. Spengler, H. Özkaynak, and L.A. Wallace. 1993. Particle total exposure assessment methodology (PTEAM) study: Method performance and data quality for personal, indoor, and outdoor aerosol monitoring at 178 homes in southern California. J. Exposure Anal. Environ. Epidemiol 3:203–226.

Tichenor, B.A. and M.A. Mason. 1987. Organic emissions from consumer products and building materials to the indoor environment. JAPCA 38:264–268.

Tichenor, B.A., L.E. Sparks, J.B. White, and M.D. Jackson. 1990. Evaluating sources of indoor air pollution. JAWMA 41:487–492.

Traynor, G.W., J.C. Aceti, M.G. Apte, B.V. Smith, L.L. Green, A. Smith Reisner, K.M. Novak, and D.O. Moses. 1989. Macromodel for assessing residential concentrations of combustion-generated pollutants: model development and preliminary predictions for CO, NO_2, and rerspirable suspended particulates. LBL-25211. Lawrence Berkeley Laboratory. Berkeley, California. 200 pages.

Traynor, G.W., B.V. Apte, H.A. Sokol, J.C. Chuang, W.G. Tucker, and J.L. Mumford. 1990. Selected organic pollutant emissions from unvented kerosene space heaters. Environ. Sci. Technol. 24:1265–1270.

Tsivoglou, E.C., H.E. Ayers, and D.A. Holiday. 1953. Occurrence of non-equilibrium atmospheric mixtures of radon and its daughters. Nucleonics 11(9):40.

Tu, K.W., and E.O. Knutson. 1988a. Indoor radon progeny particle size distribution measurements made with two different methods. Radiat. Protect. Dosim. 24:251–255.

Tu, K.W., and E.O. Knutson. 1988b. Indoor outdoor aerosol measurements for two residential buildings in New Jersey. Aerosol Sci. Technol. 9:71–82.

Turk, A. 1963. Measurement of odorous vapors in test chambers: Theoretical. ASHRE Journal. 5(10):55–58.

USEPA (U.S. Environmental Protection Agency). 1986. Broad Scan Analysis of the FY82 National Human Adipose Tissue Survey Specimens. Volumes I–V. Office of Research and Development. Washington, D.C. 1200 pages.

USEPA (U.S. Environmental Protection Agency). 1989. Indoor air quality and work environment survey: EPA headquarters' buildings. Volume I: Employee surveys. Final Report. Office of Research and Development. EPA Contract Number 68-01-7359. Washington, D.C. 250 pages.

USEPA (U.S. Environmental Protection Agency). 1990a. Indoor air quality and work environment survey: EPA headquarters' building. Volume II: Results of indoor air environmental monitoring study. Office of Research and Development. Washington, D.C. 200 pages.

USEPA (U.S. Environmental Protection Agency). 1990b. Indoor air quality and work environment survey: EPA headquarters' buildings. Volume III: Relating employee responses to the follow-up questionnaire with environmental measurements of indoor air quality. Office of Research and Development. May 1990. Washington, D.C. 150 pages.

USEPA (U.S. Environmental Protection Agency). 1991. Indoor air quality and work environment survey: EPA headquarters' buildings. Volume IV: Multivariate statistical analysis of health comfort and odor perception as related to personal and workspace characteristics. Office of Research and Development. Washington, D.C. 250 pages.

USEPA (U.S. Environmental Protection Agency). 1992. A citizen's guide to radon (2nd edition). ANR-464. Office of Air and Radiation. Washington, D.C.

USEPA (U.S. Environmental Protection Agency). 1996. Exposure factors handbook. Volume III: Activity factors. EPA/600/P-95/002Bc. Office of Research and Development. Washington, D.C. 400 pages.

Verhoeff, A. and H.A. Burge. 1997. Health risk assessment of fungi in home environments. Ann. Allergy 78(6):544–559.

Verhoeff, A.P., J.H. Van Wijnen, E.S. Van Reenen-Hoekstra, R.A. Samson, R.T. Van Storen, and B. Brunekreef. 1994. Fungal propagules in house dust. II. Relation with residential characteristics and respiratory symptoms. Allergy 79:540–547.

Wallace, L.A. 1987. The TEAM study: summary and analysis. Volume I. U.S. Environmental Protection Agency. Office of Research and Development. EPA 600/6-87/002a. National Technical Information Service. PB 88-100060. Springfield, Virginia.

Wallace, L.A. and Clayton, C. A. 1987. Volatile Organic Compounds in 600 U.S. Homes: Major Sources of Personal Exposure. Pages 183–187 in: B. Seifert, H., Esdorn, M., Fischer, H., Rüden, and J. Wegner (eds.). Indoor Air '87: Proceedings of the 4th International Conference on Indoor Air Quality and Climate, Berlin, 17–21 August 1987: Vol. 1. Institute for Water, Soil, and Air Hygiene. Berlin.

Wallace, L.A. 1989. The exposure of the general population to benzene. Cell Biol. Toxicity 5:297–314.

Wallace, L.A. 1990. Major sources of exposure to benzene and other volatile organic compounds. Risk Anal. 10:59–64.

Wallace, L.A. 1991. Comparison of risks from outdoor and indoor exposure to toxic chemicals. Environ. Health Perspect. 95:7–13.

Wallace, L.A. 1997. Unpublished data. U.S. Environmental Protection Agency. 12201 Sunrise Valley Drive, 555 National Center, Reston, Virginia 20192.

Wallace, L.A., R. Zweidinger, M. Erickson, S. Cooper, D. Whitaker, and E. Pellizzari. 1982. Monitoring individual exposure: Measurement of volatile organic compounds in breathing-zone air, drinking water, and exhaled breath. Environ Internat. 8:269–282.

Wallace, L.A., E. Pellizzari, T. Hartwell, R. Rosenzweig, M. Erickson, C. Sparacino, and H. Zelon. 1984. Personal exposure to volatile organic compounds: I. Direct measurement in breathing-zone air, drinking water, food, and exhaled breath. Environ. Res. 35:293–319.

Wallace, L.A., E. Pellizzari, T. Hartwell, C. Sparacino, L. Sheldon, and H. Zelon. 1985. Personal exposures, indoor-outdoor relationships and breath levels of toxic air agents measured for 355 persons in New Jersey. Atmos. Environ. 19:1651–1661.

Wallace, L.A., E. Pellizzari, T. Hartwell, R. Whitmore, C. Sparacino, and H. Zelon. 1986. Total Exposure Assessment Methodology (TEAM) study: Personal exposures, indoor-outdoor relationships, and breath levels of volatile organic compounds in New Jersey. Environ. Internat. 12:369–387.

Wallace, L.A., E.D. Pellizari, T.D. Hartwell, C. Sparacino, R. Whitmore, L. Sheldon, H. Zelon, and R. Perrit. 1987a. The TEAM study: Personal exposures to toxic substances in air, drinking water, and breath of 400 residents of New Jersey, North Carolina, and North Dakota. Environ. Res. 43:290–307.

Wallace, L.A., E. Pellizari, T. Hartwell, K. Perritt, and R. Ziegenfus. 1987b. Exposures to benzene and other volatile organic compounds from active and passive smoking. Arch. Environ. Health. 42:272–279.

Wallace, L.A., E. Pellizari, B. Leaderer, T. Hartwell, R. Perritt, H. Zelon, and L. Sheldon. 1987c. Emissions of volatile organic compounds from building materials and consumer products. Atmos. Environ. 21:385–393.

Wallace, L.A., E.D. Pellizzari, T.D. Hartwell, R. Whitmore, R. Perritt, and L. Sheldon. 1988. The California TEAM study: Breath concentrations and personal exposures to 26 volatile compounds in air and drinking water of 188 residents of Los Angeles, Antioch, and Pittsburgh, CA. Atmos. Environ. 22:2141–2163.

Wallace, L.A., E.D. Pellizzari, T.D. Hartwell, V. Davis, L.C. Michael, and R.W. Whitmore. 1989. The influence of personal activities on exposure to volatile organic compounds. Environ. Res. 50:37–55.

Wallace, L.A., W.C. Nelson, R. Ziegenfus, and E. Pellizzari. 1991a. The Los Angeles TEAM study: Personal exposures, indoor-outdoor air concentrations, and breath concentrations of 25 volatile organic compounds. J. Exposure Anal. Environ. Epidemiol. 1(2):37–72.

Wallace, L.A., E. Pellizzari, and C. Wendel. 1991b. Total volatile organic concentrations in 2700 personal, indoor, and outdoor air samples collected in the USEPA TEAM studies. Indoor Air. 4:465–477.

Wellish, E.M. 1913. The distribution of the active deposit of radium in an electric field. Phil. Mag. 6(xxvi):623–635.

Weschler, C.J., H.C. Shields, S.P. Kelty, L.A. Psota-Kelty, and J.D. Sinclair. 1989. Comparison of effects of ventilation, filtration and outdoor air at telephone office buildings. Pages 9–34 in: Nagda, N.L. and J.P. Harper (eds.). Design and protocol for monitoring indoor air quality. ASTM STP 1002. American Society for Testing and Materials. Philadelphia, Pennsylvania.

Wilson, A.L., S.D. Colome, P.E. Baker, and E.W. Becker. 1986. Residential indoor air quality characterization study of nitrogen dioxide. Phase I Final Report. October. Southern California Gas Company. Los Angeles, California.

Wilson A.L., S.D. Colome, and Y. Tian. 1993a. California residential indoor air quality study. Volume I: Methodology and descriptive statistics. Integrated Environmental Services. Irvine, California.

Wilson A.L., S.D. Colome, and Y. Tian. 1993b. California residential indoor air quality study. Volume I: Appendix. Integrated Environmental Services. Irvine, California.

Wilson, N.K., R.K. Barbour, J.C. Chuang, and R. Mukund. 1994. Evaluation of a real-time monitor for fine particle-bound PAH in air. Pages 167–174 in: Polycyclic aromatic compounds. Volume 5. Gordon and Breach Science Publishers. New York, New York.

Wilson A.L., S.D. Colome, and Y. Tian. 1995. California residential indoor air quality study. Volume 3: Ancillary and exploratory analyses. Integrated Environmental Services. Irvine, California.

Wrenn, M.E., H. Spitz, and N. Cohen. 1975. Design of a continuous digital output environmental radon monitor. Institute of Electrical and Electronics Engineers. Trans. Nuclear Sci. NS-22:645–648.

Yanagisawa, Y. and H. Nishimura. 1982. A badge-type personal sampler for measurement of personal exposure to NO_2 and NO in ambient air. Environ. Internat. 8:235–239.

Yeh, H.C., Y.S. Cheng, and M.M. Orman. 1982. Evaluation of various types of wire screens as diffusion battery cells. J. Colloid Interface Sci. 86:12–16.

Zweidinger, R.B., J.E. Sigsby, S.B. Tejada, F.D. Stump, D.L. Dropkins, and W.D. Ray. 1988. Detailed hydrocarbon and aldehyde mobile source emissions from roadway studies. Environ. Sci. Technol. 22:956–962.

5

DERMAL EXPOSURES IN RESIDENCES

Dermal exposure is clearly an important route of entry for environmental materials not just into the skin itself (where undesirable endpoints may include dermatitis, skin irritation, sensitization or even skin cancer), but also into the body as a whole. Once an agent has passed this first line of defense and has entered the systemic circulation, it has the opportunity to produce effects in the rest of the body, such as liver toxicity, kidney damage, and cancer.

The basic issue of dermal exposure in the residential environment can be broken down into two steps:

- What is the skin exposed to (external exposure)? How can this exposure be quantified (concentration, time, and area—see *Pathways and Exposure Scenarios* and *Deposition onto the Skin*)? This may seem on the surface to be a simple question, but in reality is often the least accurately characterized aspect of the whole process, often leading to much uncertainty in the overall risk assessment process.
- How much actually penetrates into the skin itself and finally into the systemic circulation (internal exposure)? How can internal exposure be best quantified and predicted from what is known about the external exposure, and about the properties of the agent and the medium in which it is applied to the skin (see *Availability for Absorption*)?

Models of external exposure require information on *concentration, time, and area.* These in turn depend on such things as: (1) use frequency for consumer products, habits and practices data, ambient levels, deposition onto skin and clothing, or (2) models of distribution of chemicals in the residential environment for such agents as building material components, paint, chemicals in shower water (see *Deposition onto the Skin*). External exposure calculations also require knowledge of various physiological parameters, particularly regional body surface areas (see *Pathways and Exposure Scenarios*). Many of these factors have been put together in consistent systematic form in various computer programs (*e.g.*, CONSEXPO, THERdbASE). The last section of this chapter, *Software*, describes these programs.

Of particular interest in this chapter is the role of absorption through the skin as a route of exposure to the body cells and tissues in the complex chemical environment of the home. Once the skin is exposed to an agent, the stratum corneum is first line of defense. Functioning essentially as a lipid membrane, it is particularly effective for preventing the entry of hydrophilic agents. In addition to protecting the body as a whole by preventing entry into the systemic circulation, the stratum corneum also protects the lower structures of the skin itself. This latter function is particularly important for end-points such as contact sensitization or irritation, as well as skin cancer.

The other layers of the skin (lower epidermis and upper dermis) also perform barrier functions. For example, although more lipophilic agents may be able to penetrate the stratum corneum, in order to reach the underlying structures, and in particular the capillary bed of the skin (their entry to the rest of the body), they must successfully penetrate more aqueous environments in the lower parts of the epidermis and the dermis. Hair shafts and sweat glands may also modulate the overall permeability properties of the skin by providing localized regions and pathways with potentially different permeability properties (see Figure 5-1).

Once an agent penetrates the stratum corneum and diffuses through the underlying skin layers under its chemical potential gradients, it may move in a number of different pathways depending on its chemical and biological properties (see Figure 5-2 and Figure 5-3). Apart from losses due to evaporation, skin turnover, and sloughing, some agents undergo significant metabolism and/or irreversible binding in the viable skin layers. Each agent will also partition into the various lipid and aqueous components of the skin according to its specific affinities. Finally, it will be cleared from the skin by the exten-

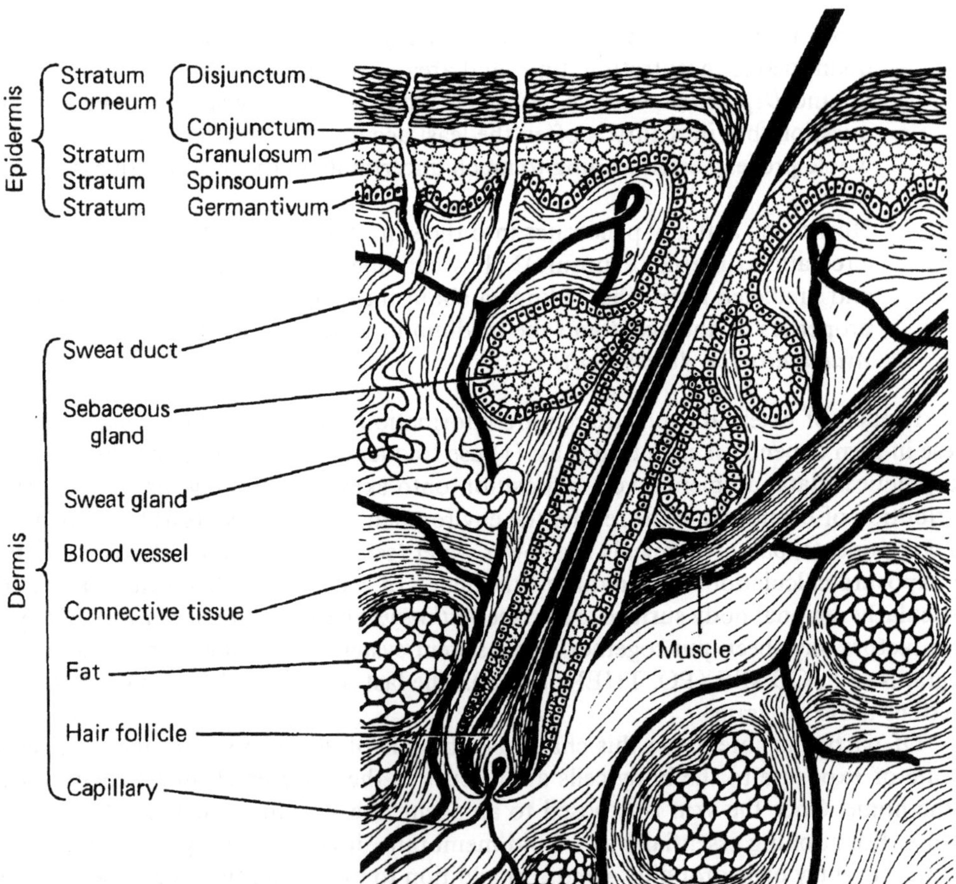

Figure 5-1. Cross-section through the skin (adapted from Doull *et al.* 1991).

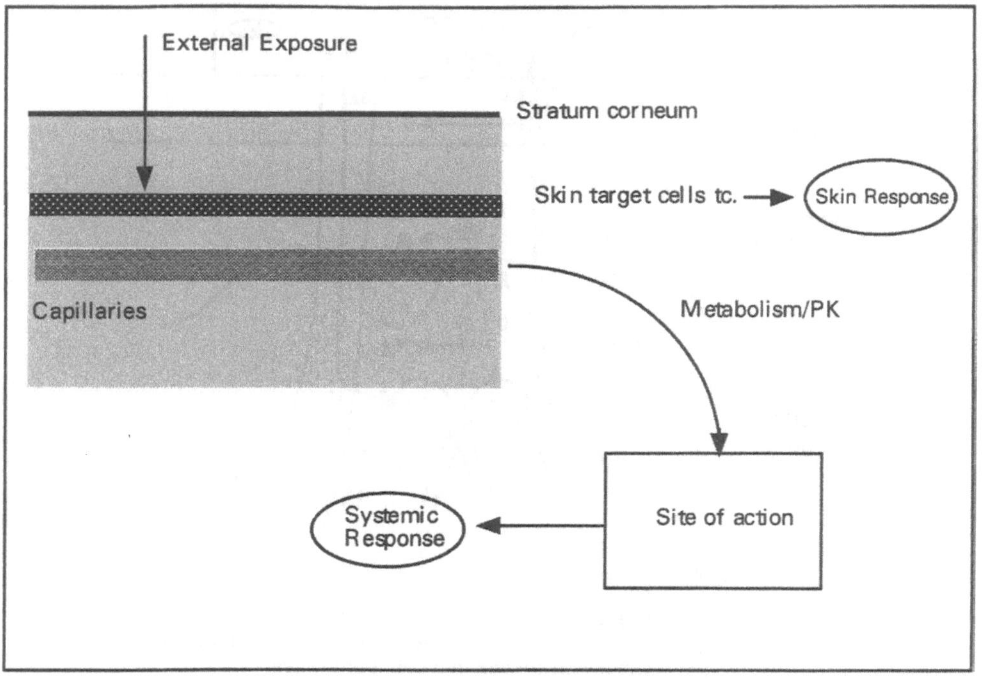

Figure 5-2. Schematic representation of dermal penetration pathways.

sive network of capillaries that tend to reside primarily near the juncture of the epidermis and dermis.

For agents that exert systemic effects, it is the movement from the skin into the circulation (clearance) that is of primary interest. Systemic absorption needs to be compared with some kind of independently established "safe" dose, typically from an animal bioassay. This raises the issues of interspecies, low-dose, and route-to-route extrapolations. The latter is crucial when dermal exposure is to be compared with an oral cancer bioassay, for example. Virtual Safe Doses (VSDs), No Significant Risk Limits (NSRLs), or other estimates of "safe" doses derived from such (oral) bioassays are typically expressed in terms of average daily exposure in mg/kg/d. Refining a dermal exposure estimate to determine the *absorbed* dose of an agent and comparing it with an (external, oral) VSD or NSRL implies the assumption that the oral dose is 100 percent absorbed. This introduces a certain lack of conservatism into the calculation. Ideally, comparisons of exposures with VSDs or NSRLs should be made with all measures expressed as absorbed doses. Examples of agents where systemic effects may be important include pesticides and trace contaminants in consumer products.

For contact sensitizers or irritants, or agents that have other local effects in the skin, it is their concentration in specific areas of the skin itself that will likely be the primary focus. For example, it is likely that the local concentration of potential sensitizers at the Langerhans cells of the skin mediates the severity of the sensitization response. Examples of such agents may include some cosmetic ingredients or fragrance materials in household products.

It should also be noted that some agents (or mixtures) may affect their own absorption, either by altering the permeability characteristics of the skin itself, or by modulat-

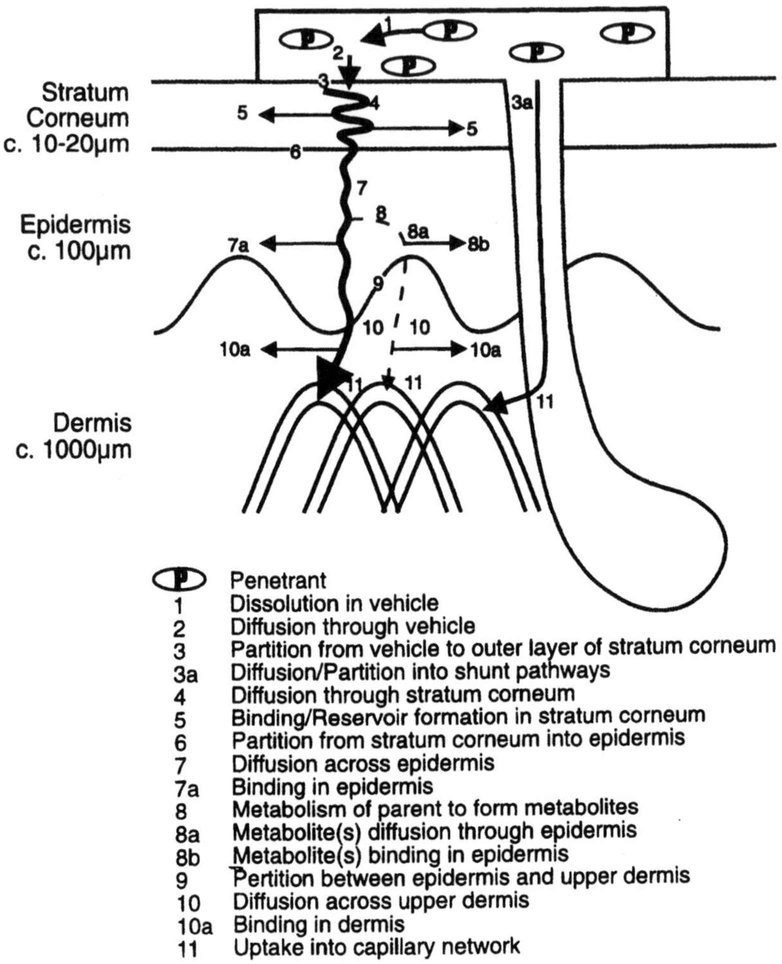

Figure 5-3. Transport and loss processes occurring in the skin (ECETOC 1993).

ing other chemical or physiological processes, such as skin blood flow. These generally uncommon effects require consideration on a case-by-case basis. Detailed exploration of the skin as a self-regulating barrier, and of the topically applied dose as a complex mixture of interacting ingredients is still in its infancy (Williams *et al.* 1996, Qiao *et al.* 1996, Baynes *et al.* 1996).

Internal exposure hinges on the chemical characteristics of the agent and its vehicle, as well as the condition and type of skin itself. When little experimental information is available on, for example, the skin penetration characteristics for a particular agent, it may still be possible to predict something about its absorption from its physicochemical characteristics and/or structure. This may be useful as an initial screen or as a basis for further investigation. Such further investigation may take the form of *in vitro* studies using human or animal skin or skin surrogates, or *in vivo* animal or clinical studies. Results from such studies may, in turn, be used to refine and improve the predictive capacity of the general mathematical models (see *Availability for Absorption*).

IDENTIFICATION OF PATHWAYS AND EXPOSURE SCENARIOS

It is quite challenging to define all of the variables which impact dermal exposure over time. For example, the characteristics of P, the contact function, can become complex, particularly as both primary and secondary exposures are considered. Consider use of a household product applied with a cloth to clean a surface. It is reasonable to assume that the individual applying the product with a cloth may contact the household product. If the product is not completely removed from the surface (via absorption onto the cloth, volatilization, *etc.*), there is potential for other individuals to contact the product post-application. There may be potential for further contact to the product during laundering of the cloth which was used to apply the product to the surface. Finally, if the hands of the individual who applied the product are not thoroughly cleaned and other surfaces in the house are contacted, the individual may transfer the household product to other surfaces (*e.g.*, utensils, books, magazines, television remote control, telephone) which serve as secondary sources of exposure.

Another common example of secondary sources of exposure are dislodgeable residues from lawns treated with pesticides, which are then picked up by clothing or shoes and deposited on carpets, floors, and other surfaces in the household. Frequently, more exposure results from the secondary sources than from the primary.

The exposure, E, is represented by the time course of the external exposure as a function of the path (*i.e.*, position of the individual with respect to the source) and time. Consider a swimming exposure scenario where an individual is in the vicinity of the pool for two hours, but only swims in the pool during three swims lasting fifteen minutes each. Other examples of potential exposures as a result of swimming include secondary sources where water from the pool is transferred to other surfaces outside or inside the household, and exposures during laundering of clothing or towels used during the swim. Characterization of the specific exposure activities and their accompanying duration and frequency of exposure are critical to adequately assess residential dermal exposure. Detailed analyses of other behaviors and tasks throughout the home that affect dermal exposure are considered throughout this book.

Information on skin surface areas in contact with an agent are also critical to any residential dermal exposure assessment. Depending on the exposure scenario, whole body surface area or body part surface area measurements or estimations must be developed for the assessment of dermal exposure. Statistical distributions of skin surface area have been developed for both adults and children, and are readily available. Statistical distributions of skin surface area are particularly useful for Monte Carlo analyses.

Skin Surface Areas

Table 5-1 gives distributions of skin surface areas of various parts of the body of adults, together with some indications of their statistical distributions. The distributions are particularly useful for "worst case" estimates or for Monte Carlo analyses. Another useful source of such numbers (as well as their changes with the maturation of the individual) can be found in AIHC (1994).

Table 5-1. Approximate surface areas of selected body parts (USEPA 1985, 1989)

	Men				
Body Part	Mean	Standard Deviation	Minimum	Maximum	N[a]
Head	0.118	0.0160	0.090	0.161	32
Trunk	0.569	0.104	0.306	0.893	32
Upper extremities	0.319	0.0461	0.169	0.429	48
Arms	0.228	0.0374	0.109	0.292	32
Upper arms	0.143	0.0143	0.122	0.156	6
Forearms	0.114	0.0127	0.0945	0.136	6
Hands	0.084	0.0127	0.0596	0.113	32
Lower extremities	0.636	0.0994	0.283	0.868	48
Legs	0.505	0.0885	0.221	0.656	32
Thighs	0.198	0.1470	0.128	0.403	32
Lower legs	0.207	0.0379	0.093	0.296	32
Feet	0.112	0.0177	0.0611	0.156	32
Total	1.94	0.00374[b]	1.66	2.28[c]	—

[a] number of observations.
[b] median (standard error).
[c] percentiles (5th–95th).

DEPOSITION ON THE SKIN

In order to begin to assess dermal exposure of an individual to a particular agent, the concentrations to which the skin is actually exposed must be determined. For solids or liquids, often the most convenient way of expressing this is in terms of deposition (*e.g.*, mg of material per cm^2 of skin). (For vapors, the concentration of the component of interest in the vapor phase, in ppm or mg/ml, is perhaps most relevant). Estimates of deposition do not provide estimates of dermal exposure. They need to be combined with (1) considerations of exposed surface area and time of exposure to give an estimate of the overall exposure of the individual to the material, together with (2) estimates of the availability of the agent from the material or matrix in which it is embedded (such as soil or a complex consumer product).

A number of methods have been devised to measure the deposition of materials of various kinds onto the skin.

Measurement Methods

The amount of an agent deposited on the skin has been shown to vary by activity and, for solids, parameters such as particle size may influence the deposition rate (Kissel *et al.* 1996). There are a variety of methods used for quantifying dermal exposure, and numerous summaries have been prepared (McArthur 1992, Fenske 1993, Ness 1994). There are numerous methods used for measuring dermal exposure or the source of the exposure: absorbent pad and clothing sampling, glove and hand wash sampling, surface wipe sampling, fluorescent tracer, and other techniques.

Absorbent Pads and Clothing. Absorbent pad and clothing sampling are probably the most frequently used method overall. This technique involves attaching gauze pads, treated cloth, or alpha-cellulose pads to various sites on a worker's outer clothing or skin to capture the contaminant that would have been deposited on the skin. An estimate of the potential dermal exposure can then be developed by multiplying the amount of contaminant deposited on a unit area of the absorbent pad by the surface area of the body area for which the pad is positioned to represent. The assumptions made are that (1) the contaminant deposited on the pad is uniformly distributed, and (2) the deposit of contaminant on a given skin surface area is represented by that deposited on the absorbent pad. This technique is generally used for sampling nonvolatile contaminants or those with very low vapor pressure, although charcoal-impregnated cloth has been evaluated for sampling of volatile agents. There are advantages and limitations associated with the use of absorbent pad sampling, and interpretation of the data can be quite challenging. One of the most prominent uncertainties is extrapolating the amount of contaminant captured on a small area covered by the pad to the particular section of the skin surface area. For this reason, "whole body" exposure assessment methods (*e.g.*, using an entire garment as the absorbent pad) is considered to be superior to sampling using pads. U.S. Environmental Protection Agency (USEPA) guidance developed for measuring pesticide dermal exposures using absorbent pads and hand rinse techniques may be helpful in developing sampling protocols (USEPA 1987). Removal of the pads and handling of pads is a critical aspect of the procedure and care should be taken not to contaminate the pads or to inadvertently cause loss of contaminant from the pad.

Glove and Hand Wash. Glove and hand-wash sampling are also fairly common methods for evaluating dermal hand exposure, although interpretation of the results can be quite difficult. Sections of lightweight absorbent gloves may be analyzed. The hands may be swabbed or rinsed with appropriate liquids, using various techniques. Fenske and Lu (1994) have demonstrated that use of standard hand washing techniques for removal of pesticides is not likely to be representative of actual exposures because of uncertainty associated with the removal efficiency. Other authors note that bag rinsing—particularly with solvents—is of questionable value and can be quite dangerous (Fenske 1993). Standardization of the hand rinse procedure is important, as the removal efficiency can vary with the chemical loading on the skin, the time between exposure and washing, and the type of wash solution used (Fenske and Lu 1994).

Wipe Samples. Wipe sampling is also a fairly common technique used to estimate the amount of dermal exposure occurring from contact with contaminated surfaces. The interpretation of wipe sampling data can be difficult, and standardization of the sampling technique is important. Often, moistened cotton gauze pads are used to sample a standardized area of contamination. The number of times a surface is wiped has not been standardized, and a weight may be added to the sampling pad to exert a uniform pressure on the sampling device. Fenske (1993) recommends that the goal of wipe sampling should be to estimate the potential transfer of surface residue to the skin, with an estimated Dermal Transfer Coefficient estimated for specific activities. Lioy *et al.* (1993) developed and validated a standardized wipe sampling technique to measure dust on flat surfaces.

Hand Press. A variation of the wipe sampling technique uses the hand as the sampling device. The palm of the hand is pressed onto a contaminated surface of a specified area. This approach has been used to validate the polyurethane foam (PUF) roller sampling technique discussed below. Standardization of the hand press method is important but may be difficult to control due to individual variability and the nature of the technique.

Fluorescent Tracers. Fluorescent tracers have been used to identify contaminated areas on the skin or cloth and to quantify dermal exposure. Quantification of exposure requires calibration of the fluorescence for a known concentration and comparison with the sampled surface area. Fluorescent tracers have been combined with video imaging measurements to produce estimates of exposure for virtually the entire body, during which pre- and post-exposure images were taken. A standard curve relating dermal fluorescence to skin-deposited tracer was developed, and chemical residue sampling was used to estimate the quantity of chemical deposited on the skin (Fenske *et al.* 1985, 1986).

Several other techniques have been developed for measuring residues from contaminated indoor surfaces that can become dislodged and available for contact. Children are often the population of concern, and many of the techniques have been specifically designed with children's activities, such as crawling, in mind. These sampling techniques have been used to measure exposure to pesticides used indoors and around residences, but have wide applicability to other exposure scenarios as well.

Polyurethane Foam (PUF) Roller. This device was originally designed for use on indoor surfaces to measure dislodgeable residues from contaminated surfaces that may be contacted by a child during activities such as crawling. The device consists of a polyurethane foam (PUF) ring fitted around an aluminum or stainless steel roller and attached to a handle. To simulate the moistness of human skin, the PUF can be moistened with water. Modifications have been made to facilitate taking samples on turf. The roller is attached to a wheeled handle and is weighted to approximate the pressure of a crawling or standing child. After two rolls the PUF ring is slit and carefully removed from the roller for analysis.

California Cloth Roller. This technique was also originally designed to measure dislodgeable residues on indoor surfaces which may be contacted by a child. A sheet of percale cotton/polyester cloth is placed over the contaminated surface and covered with a sheet of protective plastic. Then a weighted foam-covered roller is rolled backward and forward over the entire area ten times. The cloth is then carefully collected and analyzed.

Drag Sleds. The Dow drag sled technique consists of a weighted plywood block which is attached to a removable denim patch. The sled is dragged over a 3 in × 4 ft carpet strip once, and the denim patch is carefully removed after each drag for analysis. The technique was developed to estimate the transfer of pesticide from a contaminated carpet surface to the skin.

Vacuum Cleaners. Standard home vacuum cleaners have also been used to measure contaminants in house dust. In addition, specially designed vacuum devices have been used. The high-volume surface samplers (HSV3, HSV2) have been specifically designed with a stainless steel sampling train to collect dust from carpets. A cyclone has also been incorporated to separate larger particles for collection.

Coupons. Sampling coupons have also been used to measure the residue deposited on indoor surfaces following pesticide application. The "Gunther/Iwata" coupon approach uses surrogate samples of carpet, tile, glass, *etc.*, placed near or on the surface that they represent. Replicate sampling coupons can be used, and coupons are collected before and after pesticide application at specified intervals. Sampling coupons are then extracted and analyzed.

Key Monitoring Studies

Pesticide Studies (PHED). The Pesticide Handlers Exposure Database (PHED) is a generic database of measured exposure data collected from workers handling or applying pesticides in the field. It was designed by a task force of representatives from Health Canada, the USEPA, and the American Crop Protection Association (Health Canada *et al.* 1995). PHED is comprised of data generated by the USEPA or the pesticide registrant, in accordance with USEPA guidelines and using a standard data reporting format. The database possesses uniformity in data definition, and quality assurance and quality control objectives. Thus, it represents a large set of data from which one can make more statistically valid estimates of pesticide exposures than from a single exposure study. The current version of the database (PHED V1.1, 1995) contains over 1700 records, each record including one replicate of data representing a single worker performing a given activity for one day or less, and can be used as surrogate data for estimating exposures during pesticide handling and application. The use of PHED's surrogate data in estimating pesticide exposures is based on the assumption that exposure to pesticide users is largely a function of the physical parameters of handling and application rather than of the chemical properties of the pesticide active ingredient.

PHED is separated into four files: Mixer/Loader, Applicator, Flagger, and Mixer/Loader/Applicator. The database currently contains data collected during handling and application of pesticides in the field, and thus may not be applicable to estimates of residential exposure to pesticides. As additional data are submitted to the USEPA, the database will be updated and modified, and it is anticipated that residential exposure data will be included within PHED in the future.

Film Thickness Data. A study was conducted for the USEPA to measure the film thickness of liquids deposited on the skin following certain activities (USEPA 1992b). Initially, three aqueous and three nonaqueous liquids were applied to the hands of test subjects via handling of a contaminated rag, application of liquid to the hands from a saturated cloth, immersion of the hands in the liquid, or during spill cleanup.

Due to difficulties in accounting for volatilization and evaporation, only the data from the nonaqueous liquids (*i.e.*, mineral oil, cooking oil, and bath oil) were retained. The study also found that retention varied with the individual, and with the technique used to apply and remove the liquid from the hands. The data were reported on a per-event basis.

Estimation of Deposition

Assumptions. Many screening-level methods are based on a deposition approach, which assumes that a finite mass of the contaminant or thin film of liquid is deposited on the skin, and subsequently absorbed. This approach is commonly used to estimate skin uptake of chemicals in soil (USEPA 1992a). It assumes uniform deposition over the exposed skin surface area, and that the process of absorption is sufficiently slow that one can separate the deposition from the absorption process. This method can also be used to preliminarily interpret wipe sampling data (Leung and Paustenbach 1994).

A considerable amount of deposition data has been collected to assess exposure to pesticide handlers and operators (see *Deposition on the Skin: Key monitoring studies*), although available data for residential exposures are limited. Input values for the quantity deposited on the skin and the absorption fraction will undoubtedly contribute the most to the uncertainty associated with modeled estimates. The quantity deposited on the skin is estimated either based on a mass basis, or using a thin film approach.

The approach for estimating deposition based on mass is:

$$DA_{event} = A \times Q \times WF \times \% \times 100 \quad (1)$$

where

DA_{event} = Absorbed dose per event, mg/event
A = Surface area of skin available for contact, cm^2
Q = Quantity of mixture deposited on the skin, mg/cm^2 per event
WF = Weight fraction of the substance in the mixture, unitless
ABS = Percent absorbed through the skin, percentage

Methods for estimating and predicting the percent absorbed through the skin are described in *Estimation of Absorbed Dose.*

Data Availability and Data Gaps. There are several studies in which soil adherence rate for adults and children during environmental exposure to soil have been measured (Kissel *et al.* 1996, Kissel *et al.* 1996, Que Hee *et al.* 1985, Driver *et al.* 1989). These studies confirm that the amount of soil retained on the skin is greatly influenced by the individual tasks. The properties of the soil (*e.g.*, particle size) may also influence soil adherence rates. Many of the soil adherence rates reported for adults ranged from approximately 0.04 mg/cm^2 to 1.5 mg/cm^2, depending on the activity (Kissel *et al.* 1996, Que Hee *et al.* 1985, Driver *et al.* 1989). High soil adherence rates were found for children playing in mud. Table 5-2 provides central loading estimates for soil adherence, based on the limited measurements available.

Validation and Uncertainty. While the hands are often identified as a site of higher exposure, other parts of the body may also receive substantial exposures. For example, during mixing/loading of pesticides in one study, high exposure to the lower leg occurred because most of the operations involved pouring liquid from one container to another below the waist level (Knaak *et al.* 1992). Another researcher found unusually high exposure at the legs due to frequent contact with spray nozzles during loading of pesticides (Knarr *et al.* 1985). In yet another study, high exposure to the thighs was reportedly due to workers rubbing their hands against their pants in the thigh area during pesticide operations (Lavy *et al.* 1980). A study conducted by Kissel *et al.* (1996) contains soil adherence values for several body parts, including the hands. The soil adherence studies are limited in their application due to the small number of measurements, and must be used with caution. Individual variability, and differences in soil adherence due to varying clothing, seasonal changes, temporal, and other environmental conditions have not been fully examined.

Use of Surrogate Data. Due to a lack of measurement data for many residential exposure scenarios, use of surrogate data for predicting exposure may be considered. In eval-

Table 5-2. Mean and range of soil adherence by activity and body region

Activity	N[a]	Hands	Arms	Face	Legs	Feet
Outdoor						
Soccer No. 1	8	0.11	0.011	0.031	0.012	—
		0.066–0.18	0.0058–0.019	0.010–0.093	0.083–0.016	
Soccer No. 2	8	0.035	0.0043	0.014	0.016	—
		0.011–0.11	0.0022–0.0083	0.0034–0.055	0.011–0.022	
Soccer No. 3	7	0.019	0.029	0.081	0.012	—
		0.013–0.028	0.0014–0.0060	0.0052–0.013	0.0078–0.018	
Grounds Keeper No. 1	2	0.15	0.0050	—	0.0021	0.018
Grounds Keeper No. 2	5	0.098	0.0021	0.0012	0.010	—
		0.04–0.24	0.00065–0.0067	0.00063–0.0021	0.0045–0.023	
Grounds Keeper No. 3	7	0.030	0.0023	0.0009	0.0047	0.0041
		0.014–0.065	0.0012–0.0043	0.00044–0.0019	0.0021–0.010	
Grounds Keeper No. 4	7	0.046	0.014	0.0008	0.0029	0.0018
		0.025–0.082	0.0079–0.023	0.00035–0.0018	0.0018–0.0044	
Grounds Keeper No. 5	8	0.032	0.023	0.0010	0.0037	—
		0.021–0.49	0.0098–0.052	0.0008–0.0014	0.0019–0.0073	
Irrigation Installers	6	0.019	0.018	0.0054	0.0063	—
		0.12–0.31	0.0053–0.062	0.0029–0.010	0.0047–0.0086	
Rugby Players	8	0.04	0.27	0.36	0.59	—
		0.26–0.62	0.18–0.40	0.23–0.55	0.026–0.13	
Farmers No. 1	4	0.41	0.59	0.0059	0.018	—
		0.20–0.84	0.0094–0.37	0.0012–0.028	0.011–0.030	
Farmers No. 2	6	0.47	0.13	0.037	0.041	—
		0.33–0.69	0.056–0.29	0.0088–0.16	0.013–0.13	
Reed Gatherers	4	0.66	0.036	0.16	—	0.63
		0.25–1.7	0.011–0.12	0.0047–5.4		0.028–14
Kids-in-mud No. 1	6	35	11	36	—	24
		15–84	1.7–73	18–75		6.2–9.3
Kids-in-mud No. 2	6	58	11	9.5	—	6.7
		24–140	2.6–44	4.0–23		0.47–97
Indoors						
Tae Kwon Do	7	0.0062	0.0019	0.0020	—	0.0024
		0.0036–0.011	0.0006–0.0062	0.0011–0.0034		0.0012–0.0049
Greenhouse Workers	2	0.043	0.0064	0.0015	0.0051	—

[a] N = number of subjects.
Source: Kissel *et al.* 1996.

uating the appropriateness of using surrogate data collected for other residential activities, a number of important factors should be carefully considered. One important factor is the activity being performed by the individual(s) in the environmental conditions for which monitoring data are available. If the activity is similar enough to the activity and the environmental conditions during the exposure to be predicted, it may be appropriate to use the surrogate data. The exposure received by an individual is greatly impacted by the specific tasks or activities performed by the individual. Another important consideration is the population being evaluated, and the likelihood that they would perform the activity in a similar manner to the population for which surrogate data is available. Finally, the clothing worn by the individual(s) monitored should be compared with the likely clothing that would be worn by the population under study.

AVAILABILITY FOR ABSORPTION

Once an agent has become deposited onto the skin, it does not necessarily become available for absorption. Two additional things have to happen. First, the compound must be released from its particular vehicle or matrix. Sometimes this process is no more involved than partitioning out from a carrier vehicle such as water into the stratum corneum of the skin. This process is usually taken to be included in the dermal penetration coefficient of an agent (see *Estimation of absorbed dose*). On the other hand, in some cases the process by which a material is released from its carrier matrix and becomes available for absorption into the skin is in itself a complex and ill-understood process. This is the case, for example, for materials that are exposed to the skin via the soil or other complex matrices.

Second, once made available for absorption, the compound or material must penetrate the stratum corneum of the skin, and likely other skin layers, to reach its site(s) of action in the body. This penetration process is discussed further in *Estimation of absorbed dose*.

Effect of Matrix or Vehicle on Absorption

One of the major factors modifying the dermal penetration potential of agents is the matrix in which they are applied to the skin. The matrix can influence absorption in a number of interrelated ways. First, it can affect bioavailability by altering the partitioning of the agent of interest into the stratum corneum lipids. Second, it affects the solubility and hence potentially the driving concentration of the agent into the skin. Third, the vehicle itself may alter the barrier properties of the skin itself over time. It may do this in a number of ways, for example by interacting directly with the membrane lipids by inducing changes in the physiology of the skin (irritation, sensitization, altered skin blood flow).

Artificial Membranes

A number of stratified human skin cell culture systems have been developed that exhibit some of the properties of excised skin, including a multi-layered structure and nascent barrier properties. However, they do not as yet seem to reproduce *quantitatively* the barrier properties of excised skin.

Barrier Effect of Clothing and Gloves

It is commonly assumed that clothing will prevent dermal contact and absorption of contaminants through the skin. However, it has been demonstrated that contaminants can be deposited on the inside of clothing worn during application and use of pesticides. In addition, a "pumping effect" can occur while wearing clothing, which may act to cause contaminants to move under loose clothing (USEPA 1992a). In addition, studies have demonstrated that hands cannot be considered to be protected even when waterproof gloves are worn (USEPA 1992a).

There are many factors affecting the protection provided by clothing or personal protective equipment, including permeation, degradation, and penetration. Consideration of the potential for permeation and degradation of the protective clothing requires such information as the characteristics of the chemical (or formulation) which is in contact with the clothing, the clothing construction and finish, the garment design, and the characteristics of the clothing material. Laboratory glove permeation testing is commonly conducted to evaluate the permeation characteristics of a given contaminant/glove matrix. Other factors, such as elevated temperature, stressing, and pressure applied to the glove during use, have been found to significantly reduce the protection provided during actual use when compared with laboratory glove permeation data (Zellers and Sulewski 1993, Keeble *et al.* 1993).

ESTIMATION OF ABSORBED DOSE

Mechanisms of Absorption

Figure 5-3 illustrates the progress of a dermally applied or exposed agent through the skin to the systemic circulation. As can readily be seen, there are multiple pathways into and through the skin, and many points along the way at which an agent can be sequestered or diverted. The relative importance of each of these (and no doubt other) processes depends on the agent itself, as well as the endpoint of interest.

Simplifying this process even further, movement through the skin can be thought of as taking place via three potential "pathways", intracellular, transcellular, and transappendageal (shunt pathway). The relative importance of each of these notional pathways depends on the nature of the agent (particularly its lipophilicity) and on the nature and condition of the skin itself.

The absorption process itself takes place in two phases (provided the exposure is long enough). In the first phase, soon after the initial contact, the compound partitions and penetrates into the outer layers of the skin (the stratum corneum), which initially act as an empty reservoir. Over time, this reservoir is filled (at a rate dependent on the lipophilicity of the agent) and the agent diffuses to deeper layers of the skin, finally reaching the level of the skin capillary bed. As the available compartments in the skin are progressively filled, the uptake rate settles down to a steady state value, dependent not on the nature and size of these compartments, but on the exposure conditions, the barrier properties of the skin itself, and the ability of the network of capillaries to remove the agent from the skin. Also, certain skin structures act as shunt pathways which may facilitate transdermal migration of the agent.

This time-dependent process is illustrated in Figure 5-4. The upper curve (a) represents the cumulative amount of an agent that has penetrated *into* the skin, while its

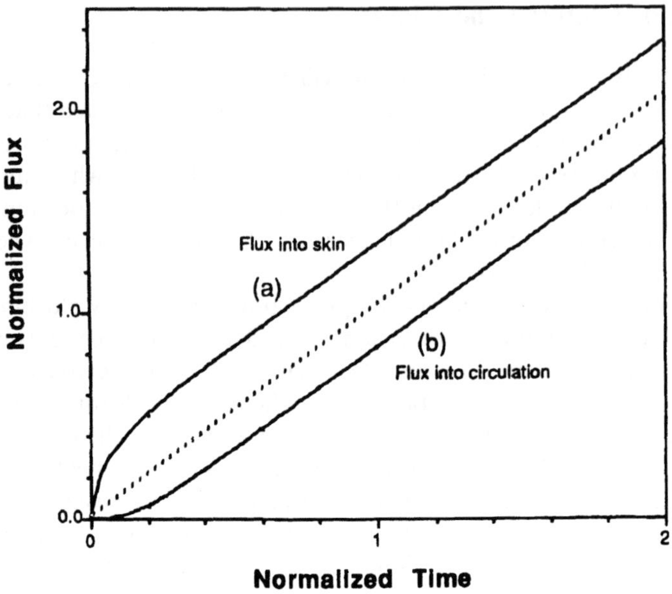

Figure 5-4. Fluxes into (a) and through the skin (b). The broken line gives the steady-state approximation of flux through the skin (see text).

slope gives the influx. As the skin reservoirs fill up, the initially-high influx decreases until a constant, steady-state concentration is reached. The lower curve, on the other hand, represents the amount of agent that has actually reached the systemic circulation at the same time (while its slope represents the *throughflux*). The throughflux is initially zero (when the agent is still entering the upper skin layers), and gradually increases to a steady-state concentration (equal to the influx).

Whether we are interested primarily in the initial phase of skin uptake, or in its steady-state value, depends of course on the exposure scenario. For long-term exposures (hours, days), the steady-state flux is of primary importance. For any particular agent in a particular vehicle, this steady-state flux is best characterized by a *dermal penetration coefficient*, K_p, (with units of cm/h). This parameter represents the normalized flux of an agent through the skin in the steady-state. If we know the K_p of an agent, we can determine the amount, Q, of that agent that would be absorbed over a time interval, T, from Equation 2:

$$Q = K_p \times c \times T \times A \tag{2}$$

where

c = applied concentration on the surface of the skin, and
A = exposed surface area.

A major advantage of using a dermal penetration coefficient to characterize dermal absorption, is that it is essentially independent of the specifics of the exposure scenario. It is a "normalized" flux, in the sense that it is defined as the amount of an agent absorbed per unit area of exposed skin, per unit time, and per unit of exposed concentration. The amount absorbed in any particular case can be determined by multiplying K_p by the appropriate exposure area, time, and concentration. This is to be contrasted with such measures of dermal penetration as the percent or fraction absorbed. These quantities are very much tied to the specifics of the exposure situation in which they were originally measured, and cannot be directly applied to other exposure scenarios. For example, absorption to an agent may be measured experimentally to be a certain percent of the original dose over 24 h. This "percent absorbed" value should not be applied uncritically to other exposure times, or when the skin is exposed markedly different doses of the material.

Values for K_p can be estimated in a number of ways, both experimentally and theoretically (from the chemical properties of the agent), as the following paragraphs illustrate.

In Vivo/Clinical Methods

These have perhaps been the most traditional and in some ways the most direct methods to measure dermal penetration. Numerous methods have been developed, and all involve exposing the skin of an intact animal or human to the agent of interest. Some methods then measure the depletion of the agent from the exposed dose by the animal while, more commonly, others measure the uptake of the agent into the animal, either by measuring the amounts taken up into the skin, plasma, tissues, urine, feces, or some combination of all of these. Clearly, the results of such experiments require careful evaluation, particularly when the results are used to extrapolate to other exposure scenarios (*e.g.*, from high to low doses, or from one species to another). Not the least of the problems associated with these methods are the confounding effects of distribution and metabolism in the rest of the body of the animal. Analytical considerations are also particularly crucial because of the effects of metabolism and the potential dilution of the absorbed dose in the body as a whole.

In Vitro Methods

In order to avoid some of the difficulties associated with *in vivo* and clinical methods to estimate permeability, as well as to address current issues related to whole animal usage, an number of in vitro methods have evolved. An exhaustive review of these methods is beyond the scope of the present chapter. However, some general points can be made.

Skin/Membrane Types. Although living human skin is in principle the closest approximation available in vitro to the ultimate exposure scenario of interest, there are a number of reasons to remain skeptical of the potential for using living human skin tissue for routine studies. These include:

- *Availability.* Access to living human skin usually requires a close working relationship with a hospital. This is difficult, particularly in Europe. Furthermore, the supply of skin is very limited, usually obtained from plastic surgery, breast reduction, *etc.* Such tissue is impractical for routine safety studies.
- *Reproducibility.* Skin obtained in this way is subject to wide and uncontrollable variations in handling, treatment, *etc.*, which may markedly affect the permeability properties of the skin.

- *Integrity*. Skin barrier integrity can be much better controlled using animal skin.
- *Metabolism*. In principle, the facility for relevant metabolic processing of agents is a major advantage of living human skin in those cases where metabolism is known to take place. The capacity of the skin for metabolism is, however, particularly subject to the reproducibility limitation above. For most agents, metabolism is simply not relevant.
- *Stratum corneum*. This is the major barrier for most agents. This barrier is essentially "dead" tissue, and its barrier properties are unaffected by whether or not the rest of the skin remains viable.

The choice of most appropriate skin type may depend on the particular agent of interest, or question asked. There may be cases where living human skin is appropriate. However, there are also cases where animal skin, cadaver skin, artificial skin systems or synthetic membranes may be sufficient and appropriate to provide the information needed. It is generally regarded as unnecessarily restrictive (and unrealistic) to require the use of living human skin for routine in vitro dermal penetration studies in support of safety assessments.

Alternative skin or membrane types include the following:
Pig skin
Hairless guinea pig skin
Rat skin
Mouse skin
Stratified human skin cell culture systems
Artificial membranes such as silastic sheeting

Many of these skin types may be useful for some purposes, but not for others. For example, silastic sheeting is in general not useful for predicting an agent's penetration through the skin, but it is useful for assessing the effects of different vehicles on the availability of an agent for absorption.

Figure 5-5 illustrates one view of the hierarchy of skin types for relevance of assessing dermal absorption in humans (as well as other factors). Although this is a useful guide, it is clear that the choice of skin type, as well as the choice of experimental or theoretical analysis of the data, depends on the particulars of the issue at hand.

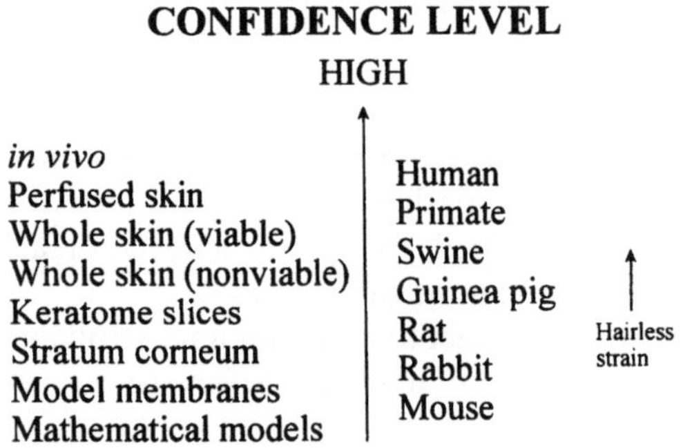

Figure 5-5. Hierarchy of skin types for measuring percutaneous absorption, as recommended by Howes *et al.* 1996.

Note that the placement of mathematical models at the bottom of the hierarchy in the above figure is misleading, since in many cases mathematical models incorporate, in a systematic way, experimental data from one or more of the skin types. Although we would not go so far as to say that a theoretical model prediction for organic molecules should supersede agent-specific experimentally determined values (USEPA 1992a), a well validated mathematical model, based on a sufficiently extensive database from an appropriate experimental system, would likely be a more reliable predictive tool than an inappropriate or ill-considered experiment on an agent of interest.

Diffusion Cells. Once a skin type has been chosen, it remains to decide how to isolate and mount that skin sample so as to be able to most appropriately measure its permeability properties to the material of interest. For example, it must be decided whether to use full-thickness or split-thickness skin. In addition, the apparatus into which the skin samples are mounted (diffusion cells), and the manner in which the skin is exposed, together with the methods for collecting the material that actually penetrates the skin, is crucial.

In general, diffusion cells can be classified into two types: static and flow-through diffusion cells. As their name implies, static diffusion cells allow the material that passes through the skin to diffuse into a reservoir solution (often stirred), than may be periodically sampled. Flow-through systems, on the other hand, allow the receptor solution to flow along the undersurface of the skin, removing any material that has penetrated the skin. It is thought that this set-up more closely approximates the clearance of agents from the skin by the capillaries in vivo. However, this is only partially true, since the capillaries have a much greater effective surface area than the undersurface of the skin. In addition, the composition of the receptor fluid will influence the efficiency of removal. This is particularly important for more lipophilic agents, which may not be at all inclined to partition into an aqueous receptor phase. In some cases, this can be compensated for by adding proteins or mild surfactants into the receptor solution, though it is a delicate balance to attempt to match the properties of blood within a network of capillaries.

There is also an issue around whether the amount of an agent measured in the skin itself should be considered part of the absorbed dose. Again, this would presumably depend on the particulars of the case in question.

Tape Stripping

An alternative approach to predicting dermal absorption has been developed principally by Rougier *et al.* (1983, 1985, 1987). This method involves allowing an agent to penetrate the skin of a human volunteer (or animal) for a fixed period of time (usually 30 min), washing the agent from the skin, and then measuring the amount remaining in the stratum corneum by tape stripping.

In the validation of the method, the amount remaining in the stratum corneum is correlated with the total amount absorbed over an extended period of time (usually 4 days) (total amount in tissues, excreta, *etc.*) under identical exposure conditions (without stripping). Relationships such as Equation 3, based on the absorption of four agents (benzoic acid, sodium benzoate, caffeine, and acetyl salicylic acid), are typically obtained (Rougier *et al.* 1987):

$$Q(\text{abs},\ 4\ \text{days}) = 1.83Q(\text{sc},\ 30\ \text{min}) - (0.52\ \text{nmol/cm}^2) \quad (3)$$

where

Q(abs, 4 days) = total amount penetrated within 4 days, and
Q(sc, 30 min) = amount measured in the stratum corneum (by tape stripping) 30 min after application.

The coefficients 1.83 and 0.52 are determined from the correlation curve, which includes data from multiple application sites (upper, outer arm, abdomen, post auricular, and forehead) (r = 0.97) (Rougier *et al.* 1987).

A more complete analysis extends the range of applicability and usefulness of this noninvasive method by taking into account molecular weight dependence of diffusion through the stratum corneum. Operational equations can be derived in which the parameters and coefficients have a direct mechanistic interpretation. These equations can be used to more accurately predict dermal penetration rates from the amount of material measured in the stratum corneum after a short period of time, such as 30 min.

Predictive Models and QSAR

In many cases it may be desirable to have an approximate estimate of the likely dermal penetration of an agent without the necessity of performing an experiment. This may be useful, for example, for screening purposes or for initial estimates in a risk assessment. In many cases, the chemical properties of the agent can be used to provide such an estimate.

Lipophilicity and Stratum Corneum Permeability

For many agents, the stratum corneum presents the major barrier for penetration into intact, healthy skin. Entry into and diffusion through the stratum corneum and skin tissue can be estimated by an analysis of the physicochemical characteristics of an agent, such as its octanol/water partition coefficient and molecular weight. For some agents, and under some conditions, it may be important to recognize other components of the skin as contributing to overall skin permeability, in particular a polar pathway through the stratum corneum and an aqueous diffusion barrier.

The stratum corneum is the outermost section of the skin consisting of multiple heterogeneous membranes comprised of morphometrically distinct lipid and protein domains. However, to a good approximation, for moderately lipid-soluble agents, it behaves very much like a homogeneous lipid membrane. Thus, penetration of an agent through the stratum corneum is expected to be dependent on its octanol-water partition coefficient and its molecular weight. A number of investigators have developed semi-empirical equations that predict K_{sc} values for the stratum corneum based on these defining chemical parameters. Of these equations, the most commonly used seems to be that of Potts and Guy (1992a), which is based, following the method of Kasting *et al.* (1987), on a multiple linear regression of measured log K_{sc} values on measured or calculated log K_{ow} values and molecular weights MW:

$$\log K_{sc} = -2.74 + 0.71 \log K_{ow} - 0.0061 MW \left(\frac{cm}{h}\right)$$

$$(r^2 = 0.67) \quad (4)$$

Equation 4 is based on the analysis of an extensive set of in vitro (usually cadaver skin) dermal penetration data collected by Flynn (1990). The subset of the Flynn data on which this equation is based includes all 93 agents for which K_{ow} values were reported. Unfortunately, this includes a number of agents for which this analysis is inappropriate because the assumption of steady-state for the determination of K_{sc} was not met. The agents in question are the very lipid soluble agents styrene, toluene and ethyl benzene, for which the skin permeability coefficient was determined in vivo by measuring the depletion of these agents from a reservoir over a period of time too short to meet the steady-state criteria (Kasting and Robinson 1993). Further discussion is presented below on the effects of failure to achieve steady-state on both the determination of dermal penetration coefficients from experimental data, and on the use of such coefficients to predict nonsteady-state dermal absorption from exposure data *in vivo*.

If these three questionable data points are removed from the analysis (and if a more appropriate value for K_{ow} for sucrose of −3.7 is taken), the remaining 90 points give a multiple linear regression somewhat different from Potts and Guy (1992):

$$\log K_{sc} = -2.74 + 0.62 \log K_{ow} - 0.0054\,\mathrm{MW}\left(\frac{\mathrm{cm}}{\mathrm{h}}\right)$$

$$(r^2 = 0.65) \qquad (4')$$

As can be seen, the coefficients of this regression analysis are seen to be highly dependent on the data used in the analysis. Care should be taken to ensure that the data meets the relevant criteria, and as more relevant data becomes available, the predictive equation should be modified accordingly.

The above analysis has a number of inherent limitations. Firstly, it assumes that the relation between log K_{sc}, log K_{ow}, and MW is linear. Although this appears to be a reasonable approximation for moderately lipophilic agents, there is reason to believe that nonlinearities may become progressively more apparent both at the very lipophilic and very hydrophilic ends of the range of K_{ow}. Some of these nonlinearities can be taken into account, as in the present composite model, by incorporation of additional pathways through the skin such as the polar pathway and the aqueous barrier of the skin. (Capillary clearance also can be a highly nonlinear process, but this is not relevant in the analysis of in vitro data.) Methods of analysis that have the flexibility to take nonlinearities into account clearly are also at an advantage. An example of such a method is neural network analysis (Hussain *et al.* 1993). There is also some evidence that proportionality between log K_{sc} and the square root of the molecular weight more accurately describes currently available literature data on penetration of agents through human skin.

Secondly, it is far from clear that all physicochemical parameters relevant to the prediction of an agent's penetration of the stratum corneum can be encapsulated in just the octanol-water partition coefficient and the molecular weight of an agent. Other parameters, such as charge and charge distribution may clearly be relevant. Such additional factors can be taken into account both in the "traditional" approach, and using neural network analysis (Hussain *et al.* 1993). However, despite these limitations, it appears that Equation 3 or Equation 4 present a reasonable approximation to stratum corneum permeability for moderately lipophilic agents, especially when combined explicitly in a composite model with other relevant pathways to give overall dermal penetration coefficients.

A truly comprehensive comparison of some 16 mathematical models based on empirical correlations between octanol/water partition coefficients and molecular weights with permeability coefficients has been conducted. However, the "goodness of fit" of these models is not evaluated, as was done with a smaller subset of models by Wilschut *et al.* (1995).

Composite Model

For rapidly penetrating agents, or for many agents under some conditions (*e.g.*, when the stratum corneum is compromised), it may be important to take into account clearance of an agent into the capillary bed of the skin. Such a model would take into account changes in skin capillary blood flow and permeability, and predict changes in penetration rate under altered physiological conditions.

As shown in Figure 5-6, such a composite model has recently been developed (Kasting and Robinson 1993, Wilschut *et al.* 1995).

The model may be summarized by the following equation:

$$K_p = \frac{1}{\frac{1}{K_{sc} + K_{pol}} + \frac{1}{K_{aq}} + \frac{1}{K_{cap}}} \qquad (5)$$

where

K_p (cm/h) = overall permeability coefficient for the skin
K_{sc} = permeability coefficient for the stratum corneum (usually the major barrier for absorption)
K_{aq} = permeability coefficient for the aqueous layers (epidermis and dermis down to the level of the capillaries), and
K_{cap} = clearance into the skin capillaries.

A number of empirical models for K_{sc} from aqueous vehicle have been developed, based on available experimental data (usually using cadaver skin *in vitro*). These models usually relate K_{sc} to the molecular weight and octanol-water partition coefficient of an agent. K_p values for human skin range from about 10^{-6} cm/h for very hydrophilic agents, up to about 10^{-1} cm/h for very lipophilic agents. The overall K_p value predicted by this model, and its dependence on the octanol-water partition coefficient, is illustrated in Figure 5-7.

Overall penetration of agents Q (mg) through the skin is given by the product result of Equation 2. This relationship is strictly applicable only in the steady-state

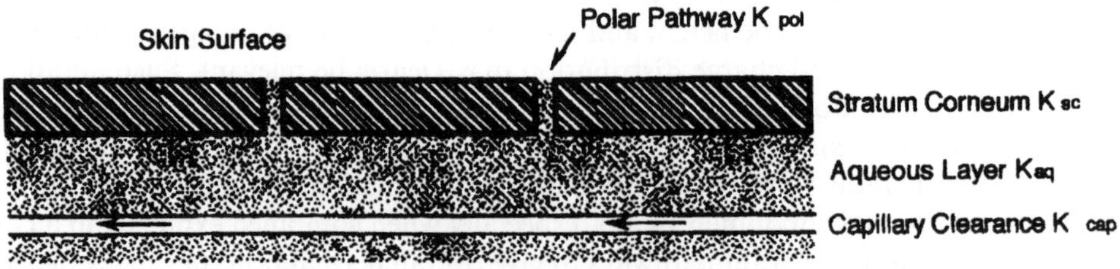

Figure 5-6. Schematic representation of the skin, showing 4 major components of composite model for dermal penetration, each characterized by a separate penetration coefficient K (cm/h).

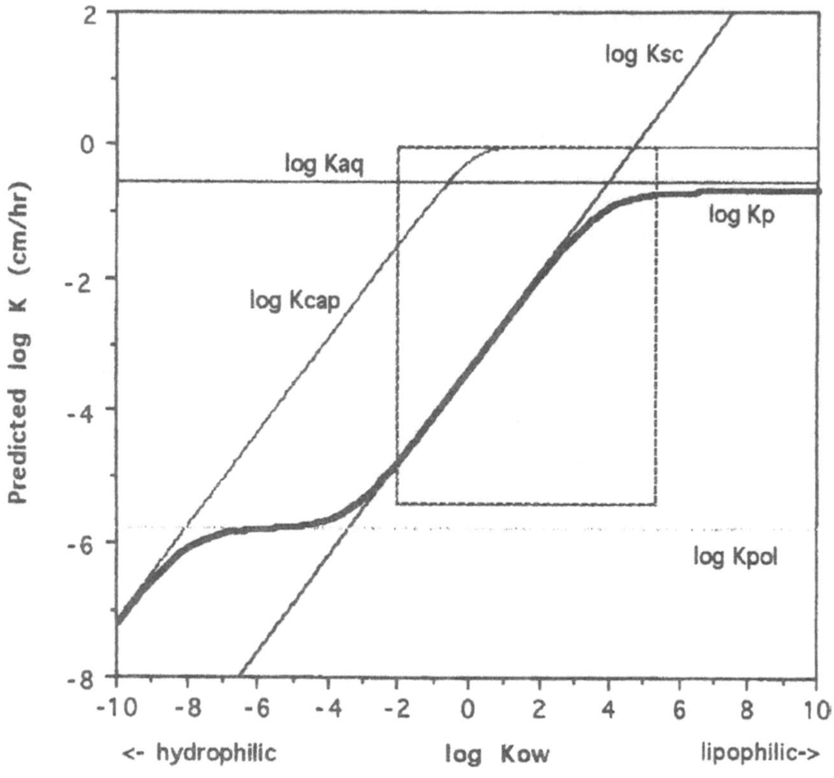

Figure 5-7. Overall skin penetration coefficient K_p against octanol/water partition coefficient K_{ow} (thicker curve) for agents of MW = 100 predicted by the composite model. Narrow lines represent penetration coefficients of each of the 4 major components that contribute to overall permeability: the aqueous layer of the skin K_{aq}, the polar pathway K_{pol}, the stratum corneum lipid pathway K_{sc}, and the capillary clearance term K_{cap}. The broken rectangle shows the range of K_{ow} and K_p values included in the database compiled by Flynn (1990).

although, with care, estimates of dermal penetration under nonsteady-state conditions can be made.

The maximum flux through the skin is limited by an agent's aqueous solubility. For very lipophilic agents this is very low, so that the maximum flux is low despite high K_p values.

Although the current model is principally generic and intended to give an approximate indication of an agent's potential to penetrate through the skin into the vasculature, its basis in the underlying physiology of the skin allows it to be readily extended to apply to such additional agent-specific processes that affect absorption as metabolism and binding in the skin, and thus form the basis of agent-specific mechanistic models for dermal penetration.

A recent published comparison (Wilschut *et al.* 1995) of five theoretical dermal penetration models, in relation to a critically compiled set of dermal penetration data, has shown that the present composite model performs best in describing the data, although it should be acknowledged that each model does a reasonable predictive job in the light of inherent data scatter.

Solubility and Maximum Flux

As the octanol water partition coefficients K_{ow} of agents continues to increase, dermal penetration does not increase indefinitely. Rather, it is limited by a number of factors. As we have seen in the preceding discussion, penetration into the systemic circulation for lipophilic agents is limited by their very modest capacities to penetrate the aqueous layers of the skin (dermis and viable epidermis), as well as their relative reluctance to enter the bloodstream.

These factors set a natural upper limit to the dermal penetration coefficients for very lipophilic agents of about 10^{-1} cm/h.

However, in many cases it is not the penetration coefficient itself which is of interest, but rather the flux of an agent through the skin. This flux is the product of a penetration coefficient and a driving concentration. Since the aqueous layers of the skin are the major impediment for lipophilic agents, solubility in water plays a major role in determining the maximum flux of such agents through the skin.

Like other parameters we have been considering, water solubility (WS) depends on the physico-chemical properties of agents, particularly the octanol-water partition coefficient and molecular weight. For example, MacKay (1982) has described this relationship as follows:

$$\log \mathrm{WS(mg/ml)} = 0.25 - \log \mathrm{K_{ow}} + \log \mathrm{MW} \tag{6}$$

The maximum flux Jmax is then given by the product:

$$J_{max} = K_p \times W_s \tag{7}$$

Jmax decreases as a function of K_{ow}, as shown in Figure 5-8.

"Steady-State" and Short Exposure Times

Figure 5-4 shows typical time courses of agents into and through the skin. Following the onset of dermal exposure, the cumulative influx *into* the skin typically follows the time-course shown in curve *a*, whereas the flux of the same agent *through* the skin into the vasculature displays a time-course more like curve *b*. The influx rate starts off at a higher rate because there is initially no agent in the skin itself, and the driving concentration gradient is at a maximum. At this point, the flux from the skin into the vasculature is negligible. After a while, material builds up in the skin and outflux into the blood increases. Eventually, sufficient material builds up in the skin itself that a steady-state is reached in which influx into the skin equals outflux from the skin into the vasculature. After such a period, which depends on the agent and may be an hour or more, the curves *a* and *b* have the same slope [given by the product of the dermal penetration coefficient K_p (cm/hr) and the applied concentration (mg/cm^2)]. However, at any given time (even when steady-state has been reached), the total amount of material that has entered the skin is always greater than the amount of material that has entered the vasculature: curve *a* is always higher than curve *b*. If the assumption is made right from the outset that the steady-state equation (Fick's Law) describes both influx into the skin and into the blood, then both fluxes are described by the broken straight line in Figure 5-4.

The assumption that one component (influx into the skin or into the blood) is at the steady-state when the other is not has different implications for different types of

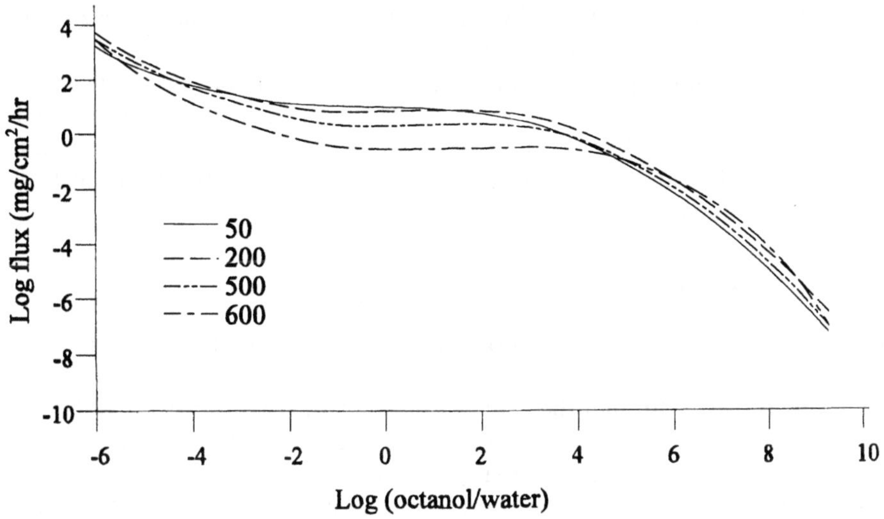

Figure 5-8. The maximum flux as a function of octanol/water partition coefficient for molecular weights of 50, 200, 500 and 800 (from Wilschut *et al.* 1995).

experiment that attempt to determine K_p. In experiments that measure influx into the skin (*e.g.*, by the depletion of an agent from an applied dose), failure to reach steady-state leads to an overestimate of K_p if the steady-state equation is applied indiscriminately. This is the case in the estimates of Dutkiewicz and Tyras (1967, 1968) for the very lipophilic agents toluene, styrene, and ethyl benzene, which form part of the data set collected by Flynn (1990). The USEPA is currently using this data set as the basis for some of its predictions of dermal penetration. On the other hand, if accumulation in the blood or receptor medium is measured, erroneous application of the steady-state equation leads to an underestimate of K_p at these early times.

Errors can also be made when applying even correctly measured K_p values and the steady-state flux equation to human exposure situations in which steady-state has not yet been achieved. In this case, influx into the skin is underestimated, whereas "throughflux" into the blood is overestimated. However, such "erroneous" values may still be useful, depending on what is being estimated in a dermal exposure situation. At the very least, the steady-state calculation directly gives a lower bound on influx into the skin and an upper bound on influx into blood, which may themselves both be useful. Furthermore, it is not always clear whether "influx" or "throughflux" is the most appropriate measure for dermal exposure. For example, if all of an agent that enters the stratum corneum eventually reaches the vasculature even when the skin surface is no longer exposed, and the total amount of the agent entering the blood is the most appropriate indicator of potential toxicity, then it would be prudent to try to estimate total influx during the exposure period. However, if a significant proportion of an agent that enters the skin leaves the skin by other means (*e.g.*, metabolism, sloughing of the stratum corneum containing the agent), or if peak or steady-state concentrations of the agent in the blood are more appropriate indicators of potential systemic toxic effects, then perhaps the "throughflux" into the vasculature may be the more important flux estimate. If it is not clear what happens to an agent when it enters the skin, or what its mode of possible toxic action is, then it may be best to provide a cumulative flux estimate that lies somewhere between these extremes. This can be provided by the steady-state influx equation, even when it is clear that steady-state is not yet achieved.

Such a rationale greatly extends the range of applicability of the simple steady-state equation beyond where it is strictly appropriate. This is particularly important in light of the fact that attempts to describe nonsteady-state dermal absorption kinetics often require much more data and knowledge of the system than are currently available for any particular agent of interest.

"Thin-Film" Calculations

In many dermal exposure situations a certain amount of product is applied or remains as a residue on the skin surface, and the question is asked: How much of a certain agent is likely absorbed through the skin? In place of the often unreasonably conservative assumption of 100 percent absorption of the applied dose, two calculations can be made using a known or estimated dermal penetration coefficient for the agent.

Assuming that the dose is applied over a certain surface area, A, and that the driving concentration, c, is the concentration in the formulation ("surface area" method), the percent absorbed can be approximated by:

$$\text{Percent} = K_p \times T \times A \times D \times \frac{100}{V} \tag{8}$$

where

K_p = dermal penetration coefficient (cm/h)
T = exposure time (often 24 h for a "leave-on" product)
D = deposition fraction, and
V = applied volume or dose (ml ~ g).

Note that the percent absorbed is independent of the concentration of the agent of interest in the formulation.

Assuming that the dose is diluted or spread in a thin film over the same surface area to a depth of h = 0.0025 cm ("thin film" method). In this case, the percent absorbed is given by:

$$\text{Percent} = K_p \times T \times D \times \frac{100}{1} \tag{9}$$

Note that again the percent absorbed is independent of the concentration of the agent of interest in the formulation. In addition, this value is independent of the surface area over which the formulation is applied.

The former calculation may be more appropriate for oily creams, while the latter assumption is clearly more appropriate for soap residues left behind on the skin following washing, bathing, *etc*. The diagram in Figure 5-9 shows the relationship between these two calculations. The shaded area below the diagonal line corresponds to situations where the thin film method predicts a lower percent absorbed than the surface area method. For example, if a dose of less than 1 g of a product is deposited on the face and neck (area 400 cm^2), then the thin-film method predicts a lower percent absorbed than the surface-area method.

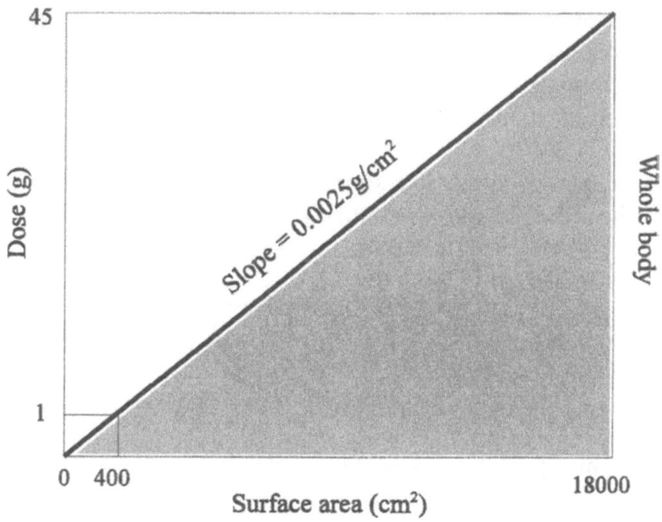

Figure 5-9. Relationship between percent absorbed using surface area (penetration coefficient) calculation versus thin film calculation method (see text).

Clearly, care must be taken in both of these calculations when a significant proportion of the dose is absorbed. In such cases, the dose can be approximated as a single compartment with a (slowly time-dependent) driving concentration. It should also be remembered these calculations are approximations, and that all the usual assumptions associated with the calculation of dermal absorption using dermal penetration coefficients (quasi steady-state, relative permanence of vehicle with respect to absorption, evaporation, *etc.*) also apply in these cases.

Application in Risk Assessment

Often, in a risk assessment context, animal bioassay data are available for one route of exposure to a chemical, whereas anticipated human exposure is via a different route. Apart from issues of animal-to-human extrapolation, there is an important question of what is the most appropriate dose comparison in such cases. Scientists are attempting to set up a set of reasonable but conservative assumptions to deal with such cases where there is a paucity of data.

Animal cancer bioassays are often performed via the oral route of exposure (either in feed, drinking water, or gavage), and a virtual safe dose (VSD) is calculated as mg/kg/day dosed (usually for one-in-one-million total excess lifetime risk). If the human exposure of concern is via the same route, then an appropriate "human" VSD can be calculated. The value of this human VSD depends on the method of animal-to-human extrapolation used, and is often determined by specific regulatory requirements. For example, the FDA assumes equivalence of mg/kg/day, whereas the procedures of the USEPA and California Proposition 65 require a "surface area" correction or assumption of the equivalence of $mg/kg^{(2/3)}/day$.

On the other hand, if the anticipated human exposure is by a route other than the oral, then use of mg/kg/day for dose may be inappropriate because of the potentially quite different absorption characteristics of different routes of exposure. Issues of absorbed dose or even tissue dose need to be explicitly addressed. For example, in calculating a VSD for human dermal exposure from an animal oral VSD, comparison of expected plasma levels in the two cases is the minimum requirement. Thus, for a dermally applied dose, Dd (= D*.BW/A, where D* is in units of mg/cm^2/d), to be equivalent to an orally absorbed dose, Do, in terms of systemic load (and for the systemic concentrations to be the same in the steady-state), the following relation must hold:

$$\frac{D_d}{D_o} = r = 1 \times V_{app} \times \frac{\#}{K_p \times A \times T} \tag{10}$$

where

K_p = dermal penetration coefficient (cm/h)
A = surface area (cm^2)
T = time of exosure (h)
\# = number of times per day dosed (d^{-1})
Vaap = applied volume (ml)
f = fraction absorbed orally.

Note that r is usually greater than 1 (*i.e.*, there is greater oral than dermal absorption). Thus, assuming that r = 1 assures a certain conservatism or "safety factor" for oral-to-dermal extrapolation.

Extrapolation can rationally be performed in this way only when the target site(s) are other than at the site of application. Thus, Equation 1 can be used to compute an equivalent VSD from the dermal route, given an oral VSD for tumors other than in the skin. (With the liver and gastrointestinal tract being likely exposed to a greater extent via the oral route, tumors in these tissues would lead to a conservative calculated dermal VSD.) When responses in the skin itself may be an issue—either because the skin may be uniquely sensitive or because exposure via the dermal route may expose the skin to higher local concentrations than via the systemic circulation alone—agent-specific experimental information is required to compare routes. The local skin concentration resulting from topical application may be higher than from oral dosing, even though the total absorbed dose may be less by this route. The present steady-state model is independent of such transient processes of accumulation (which affect the approach to steady-state but not the steady state fluxes themselves). The "safety factor," r, may be more than compensated for in such cases. Factors that may need to be taken into account in trying to compare local skin concentrations following these two routes of exposure include:

- Differential accumulation or pooling in skin lipids resulting from dermal exposure (unlikely except for very lipophilic agents). Note that the model does take into account distribution in lipids (including skin lipids) and other compartments that are the same via both routes.
- Local binding in skin
- Local metabolism in skin
- Site of action within the skin (*i.e.*, near capillaries versus near surface of skin), and

- Metabolism in the rest of the body, particularly first pass detoxification by the liver, that would reduce the effectiveness of the oral route for delivery of toxic moiety.

FUTURE RESEARCH NEEDS

There is a great need for additional research on residential dermal exposure. Available data with which to predict the amount of solids deposited on the skin are lacking for many residential exposure scenarios. Use of the available monitoring data is thus limited. Additional research is needed to adequately characterize the activity patterns associated with residential exposures (*i.e.*, "habits and practices" data). There is a need for additional standardization of approaches for conducting residential dermal monitoring studies for dermal exposure assessment, including data reporting and interpretation. Secondary exposure, in which material is transferred to clothing or other surfaces, and from there to the skin, is poorly explored. Availability of materials from complex vehicles/matrices (such as soil) is also poorly understood. Finally, our understanding of the impact of clothing and other protective equipment on mitigating exposure is limited at present, but the available information suggests that great care must be taken in collecting and interpreting these data.

On the more theoretical front, as predictive models of dermal penetration are established and validated, their range of application needs to be clearly defined. There are, at present, some areas where the existing simple predictive models cannot be applied. For example, because of the limited data available, models of stratum corneum permeability (such as the Potts-Guy model) are limited to predicting penetration from aqueous vehicle. Penetration from other vehicles at present needs to be addressed experimentally.

There are a number research efforts being undertaken to address this situation. Some of these are chemical-specific, where the mechanism of absorption for a particular chemical is explored in detail. Although undoubtedly valuable, such methods are necessarily restricted to a few "important" agents. However, other approaches are more generic in scope, and two of these may be mentioned here. Firstly, it is now possible to directly model intermolecular interactions as materials (including complex mixtures) pass through the various layers of the skin. An example of this approach is the work of Jin and Hopfinger (1996). Once key chemical properties have been identified in this way (beyond the familiar octanol-water partition coefficients and molecular weights), more accurate and more widely applicable predictive tools will no doubt emerge.

A more empirical approach is being explored by Reviere and his coworkers, Qiao *et al.* (1996) and Williams *et al.* (1996). This approach experimentally attempts (using an isolated perfused skin preparation) to tease out the matrix of interactions between the chemicals in a complex mixture with each other and with relevant components of the skin. Specific components of a complex vehicle (such as surfactants and vaso-active agents) are studied with respect to their effect on the skin and on the dermal penetration of other agents. It is envisaged that, eventually, this approach may lead to an understanding of phenomena such as irritation, that may emerge from such complex nonlinear systems of interactions and feedback loops.

Other, perhaps more mundane, issues also need clarification, such as more insight into the uncertainties involved in using such model predictions. Uncertainties can arise both from methodological (how good is the model?) and biological (variability) reasons. Use of such techniques as Monte Carlo analysis are very useful here.

REFERENCES

AIHC (American Industrial Health Council). 1994. Exposure factors sourcebook. Washington, D.C. 118 pages.

Baynes, R.E., C. Brownie, H. Freeman, and J.E. Reviere. 1996. In vitro percutaneous absorption of benzidine in complex mechanistically defined chemical mixtures. Toxicol. Appl. Pharmacol. 141:497–506.

Doull, J., C.D. Klaassen, and M.O. Amdur. 1991. Casarett and Doull's Toxicology. Macmillan. New York New York.

Driver, J.H., J.J. Konz, and G.K. Whitmyre. 1989. Soil adherence to human skin. Bull. Env. Contam. Toxicol. 43:814–820.

Dutkiewicz, T. and H. Tyras. 1967. A study of the skin absorption of ethylbenzene in man. Brit. Med. J. 24:330–332.

Dutkiewicz, T. and H. Tyras. 1968. Skin absorption of toluene, styrene, and xylene in man. Brit. J. Indust. Med. 25:243–246.

ECETOC (European Centre for Ecotoxicology and Toxicology of Chemicals). 1993. Technical Report No. 20: Percutaneous Absorption. August 1993. Brussels.

Fenske, R.A., J.T. Leffingwell, and R.C. Spear. 1985. Evaluation of fluorescent tracer methodology for dermal exposure assessment. In: Honeycutt, R.C., Zweig, G., and Ragsdale (eds.). Dermal exposure related to pesticide use. ACS Symposium Series Number 273. American Chemical Society. Washington, D.C.

Fenske, R.A., J.T. Leffingwell, and R.C. Spear. 1986. A video imaging technique for assessing dermal exposure. II. Fluorescent tracer testing. AIHA Journal 47:771–775.

Fenske, R.A. 1993. Dermal exposure assessment techniques. Ann. Occup. Hyg. 37(6):687–706.

Fenske, R.A., and C. Lu. 1994. Determination of handwash removal efficiency: Incomplete removal of the pesticide chlorpyrifos from skin by standard handwash techniques. AIHA Journal 55(5):425–432.

Flynn, G.L. 1990. Physico-chemical determinants of skin absorption. Pages 93–127. In: Gerity, T.R. and Henry, C.J. (eds.). Principles of route to route extrapolation for risk assessment. Elsevier Science Publishing Company, Inc. Amsterdam, The Netherlands.

Health Canada, U.S. Environmental Protection Agency, and American Crop Protection Association. 1995. Pesticide handler's exposure database (PHED). 1992. Version 1.1 update. March 1995.

Howes, D., R. Guy, J. Hadgraft *et al.* 1996. Methods for assessing percutaneous absorption. The report and recommendations of ECVAM workshop 13. ATLA 24:81–106.

Hussain, A.S., A.K. Kane, X. Yu, and R. Bronaugh. 1992. Chemical structure—skin permeability relationships: A neural network analysis. Abstract. Pharm. Res. 9:S191.

Hussain, A.S., R.D. Johnson, N.N. Vachharajani, and W.A. Ritschel. 1993. Feasibility of developing a neural network for prediction of human pharmacokinetic parameters from animal data. Pharm. Res. 10(3):466–469.

Jin, B. and A.J. Hopfinger. 1996. Characterization of lipid membrane dynamics by simulation: 3. Probing molecular transport across the phospholipid bilayer. Pharm Res 13:1786–1794.

Kasting, G.B., R.L. Smith, and E.R. Cooper. 1987. Effect of lipid solubility and molecular size on percutaneous absorption. Pharmacol. Skin 1:138–153.

Kasting, G.B. and P.J. Robinson. 1993. Skin permeability is limited by viable tissues. Pharm. Res. 10:930–931.

Kissel, J.C., K.Y. Richter, and R.A. Fenske. 1996. Factors affecting soil adherence to skin in hand-press trials. Bull. Environ. Contam. Toxicol. 56(5):722–728.

Keeble, V.B., L. Correll, and M. Ehrich. 1993. Evaluation of knit glove fabrics as barriers to dermal absorption of organophosphorous insecticide using an in vitro test system. Toxicol. 81(3):195–203.

Knaak, J.B., M.A. Al-Bayati, O.G. Raabe, J.L. Wiedmann, J.W. Pensyl, J.H. Ross, A.P. Leber, and P. Jones. 1992. Mixer-loader-applicator exposure and percutaneous absorption studies involving EPTC herbicide. Biological monitoring for pesticide exposure, measurement estimation, and risk reduction. American Chemical Society. Washington, D.C.

Knarr, R.D., G.L. Cooper, E.A. Brian, M.G. Kleinschmidt, and D.G. Graham. 1985. Worker exposure during aerial application of a granular formulation of Ordram selective herbicide to rice. Arch. Environ. Contam. Toxicol. 14(5):523–527.

Lavy, T.L., J.S. Shepard, and D.C. Bouchard. 1980. Field worker exposure and helicopter spray pattern of 2,4,5-T. Bull. Env. Contam. Toxicol. 24:90–96.

Leung, H. and D.J. Paustenbach. 1994. Techniques for estimating the percutaneous absorption of chemicals due to occupational and environmental exposure. Appl. Occup. and Env. Hyg. 9(3):187–197.

Lioy, P.J., T. Wainman, and C. Weisel. 1993. A wipe sampler for the quantitative measurement of dust on smooth surfaces: Laboratory performance studies. J. Exposure Anal. Environ. Epidemiol. 3(3):315–330.

MacKay, D. 1982. Correlation of bioconcentration factors. Environ. Sci. Technol. 16:274.

McArthur, B. 1992. Dermal measurement and wipe sampling methods: A review. Appl. Occup. Environ. Hygiene 7(9):599–606.

Ness, S.A. 1994. Surface and dermal monitoring for toxic exposures. Van Nostrand Reinhold. New York, New York. 561 pages.

Potts, R.O. and R.H. Guy. 1992. Predicting skin permeability. Pharm. Res. 9:663–669.

Qiao, G.L., J.D. Brooks, R.E. Baynes, N.A. Monteiro-Reviere, P.L. Williams, and J.E. Reviere. 1996. Use of mechanistically defined chemical mixtures (MDCM) to assess component effects on the percutaneous absorption and cutaneous disposition of topically exposed chemicals. I. Studies with parathion mixtures in isolated perfused porcine skin. Toxicol. Appl. Pharmacol. 141:473–486.

Que Hee, S.S., B. Peace, C.S. Clark, J.R. Boyle, R.L. Bornschein, and P.B. Hammond. 1985. Evolution of efficient methods to sample lead sources, such as house dust and hand dust, in the homes of children. Environ. Res. 38:77–95.

Rougier, A., D. Dupuis, C. Lotte, R. Rouget, and H. Schaefer. 1983. In vivo correlation between stratum corneum reservoir function and percutaneous absorption. J. Invest. Dermatol. 81:275–278.

Rougier, A., D. Dupuis, C. Lotte, and R. Rouget. 1985. The measurement of the stratum corneum reservoir. A predictive method for in vivo percutaneous absorption studies: Influence of application time. J. Invest. Dermatol. 84:66–88.

Rougier, A., C. Lotte, and H. Maibach. 1987. In vivo percutaneous penetration of some organic compounds related to anatomic site in humans: predictive assessment by the stripping method. J. Pharm. Sci. 76:451–454.

USEPA (U.S. Environmental Protection Agency). 1985. Development of statistical distributions of ranges of standard factors used in exposure assessments. EPA/600/8-85/010. Office of Health and Environmental Assessment, Exposure Assessment Group. Washington D.C.

USEPA (U.S. Environmental Protection Agency). 1987. Pesticide assessment guidelines. Subdivision U. Applicator exposure monitoring. Publication PB87-133286. National Technical Information Service. Springfield, Virginia.

USEPA (U.S. Environmental Protection Agency). 1989. Methods for assessing exposure to chemical substances. EPA/560/5-85-017. Office of Toxic Substances. Washington D.C.

USEPA (U.S. Environmental Protection Agency). 1992a. Dermal exposure assessment: Principles and applications. Interim report. EPA/600/8-91/011B. Office of Research and Development. Washington, D.C. Page 5.49.

USEPA (U.S. Environmental Protection Agency). 1992b. Laboratory method to determine the retention of liquids on the surface of hands. Final report. Versar, Inc. Springfield, Virginia. EPA-68-02-4254, EPA/747/R-92/003. NTIS PB93-105534. U.S. Environmental Protection Agency, Office of Pollution, Prevention, and Toxics, Washington D.C. 72 pages.

Williams, P.L., D. Thompson, G.L. Qiao, N.A. Monteiro-Reviere, and J.E. Reviere. 1996. Use of mechanistically defined chemical mixtures (MDCM) to assess mixture component effects on the percutaneous absorption and cutaneous disposition of topically exposed chemicals. II. Development of a general dermatopharmacokinetic model for use in risk assessment. Toxicol. Appl. Pharmacol. 141:487–496.

Wilschut, A., W.F. ten Berge, P.J. Robinson, and T.E. McKone. 1995. Estimating skin permeation: the validation of five mathematical skin permeation models. Chemosphere 30:1275–1296.

Zellers, E.T. and R. Sulewski. 1993. Modeling the temperature dependence of N-methylpyrrolidone permeation through butyl- and natural-rubber gloves. AIHA Journal 54:465–478.

6

INGESTION EXPOSURES IN RESIDENCES

The diet or incidental ingestion of nonfood items (*e.g.*, soil) may be a source of exposure to agents or constituents that are present in the residential environment. These agents can include inorganic and organic chemicals as well as microorganisms or toxins produced by microorganisms. They can be naturally or synthetically created.

The mere ingestion of an agent does not imply any adverse health consequence. In fact, some agents are essential nutrients. However, virtually all agents are toxic at some dose. For example, many essential nutrients are toxic at levels higher than the levels that are essential. Therefore, the amount of the agent that is ingested and its toxicity profile should guide the understanding of the significance of the exposure. Furthermore, the amount of the agent that is actually absorbed should be considered rather than the amount that is present in the media. Although the interpretation of the results may be different, the methodologies for estimating exposure are similar for toxins, nutrients and microorganisms.

Residues of agents such as pesticides, antibiotics, and food additives can be found in food and water as a consequence of processing and distribution, or from such activities in the residence or other places where food is consumed as absorption from the household air, surface transfer, or hygienic practices. Food and water can also come in contact with agents outside the home. In addition, agents can be ingested by transfer from the environment to the hands and thence to the mouth. Thus it is possible, and in fact likely, that there will be multiple sources of each agent.

Ingestion will contribute varying amounts to exposure since foods will contain different amounts of each agent on different days. And finally, the diets of individuals vary—both between individuals and in the same individual from day to day.

Additionally, ingestion may include nonfood items (*e.g.*, dust, soil, coins). Agents such as dust and food are commonly ingested through hand-to-mouth behavior, typically seen in infants and small children. Pica behavior (ingestion of nonfood items), which is often associated with ingestion of large amounts of dirt or clay, will also contribute to the ingestion of various agents.

Thus, the goal of the exposure assessment should be to characterize the exposure of the population of concern and to identify the variability in that exposure. This chapter will guide the assessor through the components of ingestion exposure assessments.

OVERVIEW OF INGESTION EXPOSURE METHODS

Typically, the primary objectives are to estimate the absorbed daily dose due to ingestion of the agent and to identify the sources of both variability and uncertainty in

Residential Exposure Assessment, edited by Baker *et al.*
Kluwer Academic/Plenum Publishers, New York, 2000.

the estimate. In addition, the exposure assessment can also be useful in residential assessments to identify the potential importance of diet relative to other pathways of exposure and to indicate where consumption of a particular food commodity or substance or other unique dietary characteristic (*i.e.*, age, regional and ethnic preferences), would indicate the potential for unique exposure patterns.

In order to assess the ingestion pathway to total exposure, four types of data are required:

(1) Potential levels in ingested media (food, water, soil)
(2) Frequency of occurrence of the agent in the ingested media
(3) Absorption factors, and
(4) Amounts of the media that are consumed by the population being evaluated.

It may also be necessary to determine the amount of ingestion due to transfer from the hands to the mouth. This requires estimates of amounts on the hands, and the amount that is subsequently transferred to the mouth. There are many sources of data and a variety of methods that can be used to determine the contribution of ingestion to the total exposure (Petersen and Douglass 1994, Petersen and Barraj 1996). Regardless of the methods and data, all procedures estimate exposure using the following equation:

$$\begin{aligned}\text{Exposure} = \; & \text{concentration in media} \\ & \times \text{consumption of media (quantity and frequency)} \\ & \times \text{proportion absorbed}\end{aligned}$$

The selection of the most appropriate methodology will depend on: (1) the intended application for the exposure assessment, (2) the biological properties, (3) the physical and chemical properties of the agent, (4) the route of entry into media, and (5) the relative contribution of ingestion to overall exposure. Some of the important considerations for each of the five areas are discussed below:

(1) The intended application of the exposure assessment

The purpose of the assessment will play a critical role in determining the most desirable methodology. Different methods will be desirable if the assessment be designed to be conservative (as is often the case for regulatory decision making applications) than when it is designed to be as realistic as possible (as in scientific hypothesis testing).

Some methods, such as the Danish Budget Method that is often used to evaluate intakes of intentional food additives (Hansen 1979), have been designed as "screening" methods. The Budget Method yields a worst-case intake estimate, often called the theoretical maximum daily intake (TMDI) which dramatically overestimates exposure. It can be very useful for preliminary assessments and in designing sampling programs. The Budget Method is an inexpensive means of predicting additive intake because it relies on estimates of physiological requirements for total food and liquid rather than on food consumption survey data.

Screening methods and model diet methods for assessing exposure sacrifice accuracy of estimate for speed, simplicity, and known over-estimation of exposure. In the case of the evaluation of toxic effects, results that predict that intakes will be less than an acceptable daily intake level (ADI) are assumed to mean that exposures will be acceptable, and it can be assumed that there is no need to expend resources to collect better data or to apply more sophisticated techniques in search of greater accuracy. In contrast, a research project that is attempting to evaluate the cause-and-effect relationship of an agent and a disease would require more accurate exposure assessments.

(2) The biological properties of the agent

The length of dosing that is required to elicit a specified biological effect should define the key exposure assessment parameters. That is, the biological effects that are the result of a single or at most few doses will be compared to dietary exposure on a single day. Correspondingly, toxic effects that arise as a result of long-term exposure, will be compared to average dietary exposures (usually over a year).

Other considerations include: any breakdown products of toxicological significance, and the metabolic pathways in plant and animal systems. Potential biological effects must be carefully considered in planning an exposure assessment. Factors of interest include: dose-response relationships, the length of exposure required to produce an adverse effect, potentially sensitive populations, and variability and uncertainty factors.

(3) The physical and chemical properties of the agent

Often when estimating intake of an agent in food, it is necessary to define or characterize the agent in terms of attributes such as structure, volatility, and solubility since there are seldom sufficient data available. For example, a question of concern related to the agent's properties once it is in the food or water is: Does the substance break down during storage, processing, and cooking?

Physical properties of the agent that may be evaluated include: (1) particle size, (2) presence in household air, (3) surface to hand transfer rates, and (4) rates of deposition in food or water.

Our diets are highly processed. Therefore, for most assessments it will be critical to include estimates of the residues in the products as they are consumed (Chin 1991).

(4) The route of entry of the agent into food, water, and other media

The origin of an agent in the media is important to consider when selecting the foods to be analyzed. Is it naturally occurring? Is the agent entering the food prior to the food entering the home or does it enter the food as a result of activities within the home? Does the household water supply contain the agent? Is there a potential for ingestion due to the agent's presence in the air or contact with old paint?

Concentrations of a substance in food may vary due to differences in the cultivar, season, or region of growth, but the substances generally would be expected to occur at some level in all raw plants or uncooked meat of the same type (an exception would be a nutrient or other natural toxin bred into plants by genetic engineering). The extent to which the compound will be present in processed forms of the relevant foods depends on volatility and breakdown characteristics of the chemical and its metabolites.

It is often possible to determine the foods that may contain the agent by evaluating source and transport information for the agent. If it is possible to group foods into categories, it may be possible for data for one food to be extrapolated to foods for which data are not available. For example, if the agent concentration in oranges is expected to be similar to that in grapefruit, it would then be possible to conduct the exposure analysis for "citrus." These food categories can then be used to select the most appropriate food consumption data for the assessment.

Concentrations of the agent in breast milk may need to be considered in evaluating an infant's exposure. It is clear from the literature that a wide variety of agents (*e.g.*, DDT and metabolites, mirex, PCBs, heptachlor, aldrin, dieldrin, chlordane, hexachlorobenzene, and volatile organic chemicals such as tetrachloroethylene and other chlorinated solvents) are found in significant quantities in breast milk. Schreiber (1997) presents a comprehensive analysis of the determinants of the presence of exotic substances in breast milk along with a review of the literature for many agents and a specific case study of tetrachloroethene. There are numerous factors which determine the likelihood that the agent will be found in breast milk. Schreiber identifies three general factors of importance: (1) chemical characteristics, (2) maternal characteristics and (3) magnitude of maternal exposure. For fat soluble compounds, the magnitude of maternal exposure is particularly important in determining the concentrations in breast milk. According to Schreiber, the mother's age is an important determinant since many persistent chemicals are stored in maternal tissues and build up over time. Adipose tissue provides about 60 percent of the fats in breast milk. Therefore, the concentrations of the agent in maternal fat are likely to be good predictors of the concentration in breast milk.

(5) The relative contribution of the ingestion pathway to total exposure

Typically when dietary intake is to be integrated into exposure from other routes, screening methods will be too crude and will need to be replaced with more accurate estimates of intake. In those instances even preliminary (range finding) assessments may require more refined estimates of the concentrations in those foods that appear to be contributing most to the theoretical exposure estimates. It may also be desirable to include information about the effects of processing or cooking. In some instances, it may be necessary to use food consumption data sets that capture the intake patterns of special subgroups such as infants or regional dietary practices.

Range finding exposure assessments can be used for the initial evaluation of "worst case" exposure. The most widely used methods to determine the order of magnitude of exposure are based on theoretical maximum residue concentrations and either average or upper percentiles of consumption. These initial calculations usually do not incorporate distributions of either residue levels or food consumption patterns; similarly the effects of processing or cooking are not included.

An initial decision must be made as to whether samples of the diet will be analyzed or whether calculations will be made based on a dietary intake questionnaire.

Dietary assessments most often rely on survey data on the dietary consumption characteristics of the population combined with data on residues in the food supply. National surveys are normally federally-sponsored and are conducted on a continuing basis. These surveys are described in detail below in the section entitled *Available Sources of Information About Food and Water Consumption in the United States—U.S. National Surveys of Food Intake*. Many other food-related surveys are available that have information which may provide additional information for particular populations, regions, commodities or any number of other factors.

> The methods used to conduct dietary surveys are in a continuing state of development and refinement. The survey instruments and procedures to collect and analyze samples are continually being improved and new compounds are being added to reflect current priorities. The optimal approach for assessing dietary exposures has been debated for years in the U.S. as well as in other countries (FASEB 1988).

In summary, the relative appropriateness, accuracy and reliability of the analysis will depend on the source of the agents in the food, the available data for concentrations of the agent in the media, as well as the toxicological characteristics of the agent. For agents in which adverse effects are observed after a single exposure, it is important to estimate intake by people when they actually ingest media containing the agent. For agents with chronic adverse effects, on the other hand, intake is calculated over days when food, water, or a nondietary substance is and is not consumed to produce an estimate of average daily intake. The USEPA has recently identified two intermediate time periods of interest—short-term and intermediate—in estimating the significance of potential exposures to pesticides (USEPA 1997).

Regardless of the degree of precision of the exposure assessment—from range finding to precise distributional analyses—it is critical that the most appropriate exposure assessment model is applied.

There are four general exposure assessment models:

Average exposure. The most basic models combine data on average intake and average concentration levels of the agent to estimate average exposure. Average chronic exposure is usually estimated on a per capita consumption basis and if compared to the measure of biological/toxicological results from lifetime animal feeding studies or other appropriate test results. Single day or "acute" exposures may be computed using similar methods but using food intake data for a single meal or for the day. Acute exposures are usually compared to results of test results in which the subjects were dosed for a single day.

Simple distribution. A simple distribution of exposure can be calculated by applying a single, average (or worst-case) estimate of the agent's concentration level to a distribution of food consumption. Alternatively, a distribution of concentration levels can be applied to an average or worst-case food consumption level.

Joint distribution/Monte Carlo-type probabilistic assessment. In creating a joint distribution to estimate exposure, the distribution of food consumption is combined with the distribution of agent concentration levels. Joint distribution probabilistic analysis allows the most realistic estimates of exposure.

Commodity contribution analysis. Commodity contribution analyses provide supplemental exposure information. These analyses are conducted to determine the percent of chronic exposure contributed by a specific food or food group.

DEFINING THE EXPOSURE SCENARIOS

The sources of exposure must be defined along with the determination of magnitude and frequency of exposure. It is usually desirable to know how the agent enters the food supply and the time frames that are involved. It is also important to know whether the dietary exposure is correlated with other routes of residential exposure or if it is arising from a separate source.

The following steps help to define the most appropriate exposure scenario:

- Identify the regulatory standards that apply to the agent, if appropriate. For example, the permitted uses of a pesticide on food crops.
- Determine the characteristics of intended or accidental additions of the agent to the food supply, to food contact surfaces, or onto the hands. Determine the impact of processing/cooking and storage of foodstuff on the final levels of the agent.
- Identify and evaluate existing data for both the levels of the agent and for the appropriate consumption estimates.
- Create or apply draft exposure assessment algorithms which define the parameters to be computed and the models to be used to estimate exposure. If appropriate, develop a tiered methodology that allows a sequence of analyses from "range finding" to detailed probabilistic exposure assessments
- Conduct appropriate preliminary exposure assessments including sensitivity analyses to assist in defining the parameters which need the best available data and/or additional data generation.

Sources of the Agent

Agents may be present in or on food, water, and other substances accidentally due to environmental factors such as air pollution, water pollution, accidental contamination, or microbial contamination. The presence of agents from such sources would be expected to vary significantly depending on geographic location or other factors. Biological agents, once entering the food supply may increase or decrease depending on growth characteristics.

Agents present in food may be there incidentally or deliberately. Incidental agents are added indirectly to foods as a result of migration from packaging or a processing step. In general, agents incorporated incidentally are more likely to be present in processed foods than in raw agricultural commodities. Agents in food deliberately are intentionally added during some phase of production or processing.

Agents present in other media, such as soil, may also be there incidentally or deliberately. For example, incidental introduction of substances into soil may occur through migration to soil after foliar pesticide application or through deliberation introduction if the soil itself is being treated. These agents may then be ingested through typical soil ingestion. All people ingest a small amount of soil or dust in their day-to-day behavior. This occurs through a variety of sources including dust deposition on food, ingestion of soil and dust on hands either through transfer to food or other medium, and subsequent ingestion or through direct transfer of soil to the mouth. Infants and small children engage in regular hand-to-mouth behavior particularly by sucking of thumbs, fingers or pacifiers. Additionally, children put their hands in their mouths starting as infants when the sense of taste is used to define their environment.

Quantifying the Concentration of the Agent

The analyst's goal is to characterize the concentration of the agent in each medium that is ingested. Priority should be placed on media that will contribute most to exposure. To assess exposure to an agent, it is important to consider not only the average concentrations, but also the concentration variability.

Existing data can be used to estimate exposure and to guide the collection of future data. A substantial amount of information that is relevant to exposure assessment, and hence risk assessment, is available for some agents (Graham *et al.* 1992, Sexton *et al.* 1992). Such information can be obtained from existing databases on food and drinking water consumption and contaminant residues normally maintained for other purposes (nutritional intake or regulatory surveillance). The Total Diet Study (Pennington and Gunderson 1987) of the U.S. Food and Drug Administration (USFDA), the Nationwide Food Consumption Survey (NFCS) of the U.S. Department of Agriculture (USDA), the National Health and Nutrition Examination Surveys (NHANES) of the U.S. National Center for Health Statistics (NCHS), as well as federal and state regulatory monitoring programs, to name a few, provide information useful for dietary exposure modeling and assessments.

Useful data sources for defining the foods with the highest potential concentration of the agent includes:

- Chemical properties (solubility, heat stability, pH stability, and other properties)
- Metabolites/degradates of potential concern—evaluation of structure of parent
- Differences in levels of raw versus cooked food
- Levels in home prepared versus commercially processed foods
- Time frames for presence of agent in the food supply or in categories of the food supply—continuously or in periodic intervals, and
- Impact of processing/home preparation on potential residues.

Refining the Estimates of the Concentration of the Agent

If additional data are to be collected, the results described above are used to select the food categories, define the representative foods, and focus data collection on the foods contributing most to potential exposure.

There are three common approaches that are appropriate for collecting data on concentrations of agents in food: (1) duplicate diets, (2) market basket or representative sampling, and (3) controlled experimentation. These are discussed below.

Duplicate Diets. Only recently (Davis and Waller 1989, Lioy *et al.* 1988, Manton *et al.* 1991, Berry 1997, Melnyk *et al.* 1997), have direct measurements of the intake of agents in the diets of individuals been made in the U.S. Whereas, other countries (Canada, United Kingdom, Japan, and Sweden) have successfully used daily intake measurements with a high degree of success in both regulatory and assessment programs for a number of years (*e.g.*, Dabeka *et al.* 1987, Sherlock *et al.* 1983, Vahter *et al.* 1990).

To accurately assess dietary exposure and identify sources at the residential level, the actual foods eaten by the residents, both in and away from the residence and in commercial and occupational eating places, must be collected and analyzed. The collection of "duplicate-diets" has proven to be a viable sampling approach. Most extensively, duplicate-diet procedures were the method of choice for the USEPA's National Human Exposure Assessment Survey (NHEXAS) (FASEB 1993) and the World Health Organization's Global Exposure Monitoring System (WHO 1997), where dietary exposure data consistent with measurements of residential exposures from other pathways were required for making assessments of total human exposure. Duplicate diets can include both food and water and can account for residues in either food or water which arise from a wide variety of in-home activities.

In duplicate-diet sampling, a duplicate portion of all foods and beverages consumed during the monitoring period is collected. Foods collected are normally composited for analysis because of economic considerations yielding a single measure of dietary exposure of the participant for the daily or extended monitoring period. Variations of the procedures are used depending on study objectives. Individual food items or food groups and daily food collections may be segregated for independent analysis. Alternatively, it may be necessary to sample foods at the various steps during food preparation in order to identify the reason for excess contamination. Food diaries and questionnaires are used to obtain supporting information useful for identifying sources of contamination and for measuring the completeness and representative nature of the foods collected. Duplicate-diet collections should include all foods consumed, both in and away from the residence (FASEB 1993), and the monitoring period should be extended to the maximum number of days possible without undue burden on the participant. The researcher must determine the level of information required for various exposure scenarios. Often detailed exposure information can only be obtained at considerable expense and participant burden. For example, detailed dietary histories, segregation of individual food items for analysis, and weighing individual foods may not be necessary when a single measure of total intake is the goal.

The USEPA has developed protocols and used duplicate-diet procedures in several residential-based, multimedia human exposure studies (Thomas *et al.* 1997, Berry 1997). Duplicate-diet procedures were used in NHEXAS field studies (Sexton *et al.* 1995) for assessing dietary exposures of over 500 participants in residential environments. Duplicate-diet collections were found to be no more burdensome, costly or time-consuming than breathing-zone air measurements, for example. In pilot studies on farmer exposures to pesticides (Thomas *et al.* 1993, Melnyk *et al.* 1997), duplicate-diet samples verified the persistent nature of several pesticides not applied at one farm in several years, including dieldrin at levels of 10 to 28 ppb in 24-hr composite diets and homegrown food items collected during repeated visits. Bacteria found in drinking water stored in the residences in various containers that were not properly sanitized suggested the need for educational programs on microbial safety.

Duplicate-diet measurement protocols are appropriate when an established or potential route of exposure to an agent exists through ingestion of foods and beverages. Duplicate-diet methods may be used in independent studies of dietary exposure, but are most effectively used to compliment personal measurements for other exposure pathways for evaluating total human exposure. It is not the purpose of the personal monitoring procedures to replace existing regulatory monitoring programs or source-specific assessments that use food consumption and information about the agent's concentration in foodstuffs. A personal monitoring approach for dietary exposure assessment has both strengths and weaknesses (FASEB 1993). A primary strength is its ability to provide data on an individual's total dietary intake of an agent, regardless of source. It is the most accurate approach for measuring the true exposure of an individual since it also measures potential contamination during preparation and consumption in the residence or other eating environment. The personal measurements approach is necessary when direct comparisons to other exposure pathways measured for an individual are required for assessment of that individual's total exposure. However, the personal monitoring approach is not easily interpreted beyond the monitoring period and may not be the most cost-effective approach for long-term assessment of large populations. The collection procedures may alter the eating habits of the individual and, the ability to identify the source of contamination is limited when samples are composited for analysis.

A sample of the mother's breast milk may be essential in estimating infants exposure to an agent. Along with a determination of the concentrations of the agent in the breast milk, it will be necessary to determine whether other foods are also being fed. If other foods are being fed, then estimates of the levels of the agent in those foods need to be determined.

Market Basket and Surveillance Sampling. The U.S. and other countries conduct various types of monitoring or surveillance for agents in commodities or in foods. Monitoring and surveillance studies are conducted to assess compliance with state, national, or international regulations.

At the U.S. federal level, the USDA monitors residue levels in domestic and imported meat and poultry products, and the USFDA monitors residue levels in all other foods. The States of California, Florida, and a number of other states have monitoring programs. A national database, FOODCONTAM, incorporates data from monitoring programs in ten states (Minyard and Roberts 1991). These programs monitor for pesticides, veterinary products, heavy metals, and nutrients. Not all programs monitor for all of these classes of compounds. There is little surveillance for microbiological agents except when there is reported illness. These programs are described in greater detail in the following sections.

Depending on the specific U.S. monitoring program, foods or commodities may be sampled at the point of entry to the country, at the farm gate, at the food processing plant, or at the retail level. Since it is logistically and economically impossible to analyze each and every shipment of every food and commodity, studies are often conducted on target samples suspected to be out of compliance. While the resulting data are not statistically representative, they are useful for "range finding" studies.

Market basket surveys are conducted in the U.S. and other countries to obtain food chemical concentration data that may be used in exposure assessment. A core group of foods representative of national dietary patterns is obtained and analyzed to determine the concentrations of the substances of interest. Generally, samples of food are purchased at retail outlets in different regions of the country and prepared as for consumption.

The USFDA Total Diet Study (TDS). Since 1961, the U.S. FDA has conducted a yearly market basket survey, the Total Diet Study (TDS). Samples of 265 foods chosen to represent the U.S. food supply are collected four times each year, from three cities in each of four U.S. regions. Samples of individual foods from these three cities are composited and prepared as "ready-to-consume" foods in the laboratory prior to analysis. Because the TDS uses only a few hundred foods to represent thousands of foods and is not statistically based, results may be useful for exposure assessment, but must be interpreted cautiously. The multi-residue methods of analysis are capable of detecting over 200 pesticides and industrial chemicals, radionuclides, and essential and toxic elements, including arsenic, lead, cadmium, and mercury. Estimates are made of the intakes of various age and gender groups based on consumption data from USDA's Nationwide Food Consumption Survey (NFCS, USDA 1981) and NHANES (NCHS 1982). The TDS diets are aggregated so that the food items collected are representative of all food consumed in the U.S. For example, apple pie represents not only all the various types of apple pies (homemade, bakery, fast-food, frozen, deep dish, apple with raisin, apple with cheese, and so forth) but all fruit pies, strudels, crisps, turnovers, and pastry with fruit (Pennington 1983). Concentration data on the foods sampled can be used as reference points in designing rough exposure assessments—especially for guiding subsequent studies.

Controlled Experimentation. Controlled experiments are sometimes used to determine the likely levels of agents in foods produced under specified conditions. Controlled trials are conducted for all pesticides that are registered for use on food crops in the U.S. These trials are conducted under the most extreme product label conditions (maximum permitted application rate, maximum number of application, and minimum interval between treatment and harvest). Samples of the edible portions are analyzed to determine the concentration of the pesticide and relevant metabolites. The results of these studies are usually available, at least in a summary form. For some applications, processing studies are also required. These processing studies are conducted in order to determine whether the pesticide is concentrated or diluted/degraded during processing.

Controlled experiments have also been conducted to identify the source of other agents in foods. For example, studies have been conducted to quantify the migration of lead from pottery and glassware into foods and beverages (Bolger *et al.* 1996).

Selection of Appropriate Analytical Methods

All measurements of dietary exposure require appropriate analytical procedures to identify and quantify priority contaminants present in the food samples collected. The complex nature of the food matrix normally requires additional analytical procedures, making them more rigorous than air or water methods. Analytical methods must be capable of assuring proper detection limits and accuracy of the resulting analyses. In general, it is the food analytical procedures that are the ultimate limiting factor in measuring or otherwise assessing dietary exposure at levels comparable to other exposure pathways. Our inability to measure pesticides at low levels in foods limits our ability to accurately estimate exposure of children to pesticides in their diets (NRC 1993).[1]

Historically, food methods for the determination of environmental contaminants have been developed for the purpose of supporting regulatory programs. Whereas agents like pesticides (USFDA 1994) have received considerable methods development efforts, many classes of agents, for example, polycyclic aromatic hydrocarbons (PAHs) and volatile organic compounds (VOCs) have received only limited attention. Most often, food methods are developed, tested, and validated for specific food types or items.

The nature of a duplicate-diet measurement poses several additional challenges, beyond those associated with the complex food matrix itself, that place greater demands on food analytical procedures. Duplicate-diet collections involve many different food types that may be composited into a single sample for analysis. These composites are very diverse and contain varying amounts of fats and other substances that interfere with analysis. Method performance for composite food samples is largely unknown and compositing foods into groups compatible with analytical methods may not be practical or cost efficient. Concentrations of the agent in the composite diet samples may be very low because food items containing the agent will be diluted with food items that do not contain the agent. Thus, very sensitive analytical methods are desirable. The cost of analyzing samples is usually the single most expensive part of an exposure measurement study, often approaching two-thirds of the total study cost. Screening methods (such as

[1] Virtually all exposure assessments assume some concentration in samples in which no residues were detected. Accordingly, the resulting assessment overestimates exposure.

immunoassay procedures, Lopez-Aliva *et al.* 1996) have the potential of greatly reducing analytical costs by detecting the presence of a compound quickly and efficiently, thus eliminating time consuming detailed analyses for all samples, and allow for more efficient use of specific analytical methods.

Currently, no suitable food materials exist that have been sufficiently characterized for use as field and laboratory quality control samples. A food material representative of a composite food sample is needed to evaluate the possible contamination aspects of duplicate-diet and other food sample collection, handling and preparation procedures. Also, standard reference materials (SRMs) are needed for foods. Currently SRMs are available for nutrients and some elements in food samples, but not for most organic agents (NIST 1995).

Detecting microbiological agents requires accurate assays. Currently most assays that are routinely used on foods are qualitative and will need to be modified to yield quantitative estimates. Since microbiological levels will change during processing, transit, cooking, and holding, and are altered by temperature, acidity, and other physical factors, it is also crucial to have approaches that will accommodate such changes.

DIETARY INTAKE

In the U.S. and many other countries (SCF Intake and Exposure Working Group 1994), good food consumption survey data are available. However, these surveys are conducted for purposes other than exposure estimations. Commodity monitoring data are used to track agricultural production and imports. Food consumption data on individuals or groups of individuals are collected to characterize expenditures, dietary intake, or nutritional status of those individuals.

The most appropriate survey for a particular assessment must be determined. There are a number of specific questions to consider when selecting food consumption data to estimate exposure:

- When (how long ago) were the data collected?
- Are current patterns of consumption of the foods of interest similar enough to patterns existing when these data were obtained?
- For which population were the food consumption data collected?
- For which country or countries?
- For which geographical regions?
- Were all population groups included (*e.g.*, children, vegetarians, subsistence fishermen)?
- Were data collected during all seasons?
- How much detail about specific foods was included?
- Was the quantity of each food estimated?

Information about the level of detail in food consumption data available from a particular survey is an important determinant of the potential for accuracy in estimating exposure to a specific agent. For example, if an agent is found in beef liver, consumption data on beef liver should be used in assessing exposure to the agent. If the only available consumption data are on "beef," estimates of an agent's consumption in beef liver will be less accurate. The level of detail available in data from food consumption surveys varies depending, in part, on the survey methodology.

Types of Food Consumption Surveys

There are generally four different broad categories of food consumption data: (1) food supply surveys (market disappearance), (2) household or community inventories, (3) house-hold food use, and (4) individual food intake surveys.

Food Supply Surveys. Food supply surveys, also called food balance sheets or disappearance data, describe a country's food supply during a specified time period. Food balance sheets published by the Food and Agriculture Organization (FAO) of the United Nations describe the food supply in countries on all continents. European food balance sheets are also prepared by the Organization for Economic Cooperation and Development (OECD) and the Statistical Office of the European Communities (EUROSTAT). Food supply data in the U.S. are developed by USDA's Economic Research Service.

Food supply data are derived from economic surveys using the following formula:

$$\frac{\begin{array}{c}[\text{Food production} + \text{beginning inventory}]\\ -[\text{Exports} + \text{ending inventory} + \text{nonfood uses}]\end{array}}{\Delta = \text{Food availability}}$$

Nonfood uses include animal feed, pet food, seed, and industrial use.

Availability is determined at different points for different foods or commodities. Mean *per capita* availability of a food or commodity is calculated by dividing total availability of the food by the total U.S. population.

These surveys provide data on food availability or disappearance rather than actual food consumption, but may be used to indirectly estimate amounts of foods consumed by the country's population. Food supply data may be useful for setting priorities, analyzing trends, developing policy, and formulating food programs. For some countries, food supply data are the only accessible data representing the country's food consumption. Because similar methods are used around the world, these data may be used to make comparisons internationally and also may be useful in some epidemiological studies (Sasaki and Kestelhoot 1992).

However, the value of these data in assessment of exposure to chemicals in foods is limited for several reasons. First, waste at the household and individual levels usually is not considered. Therefore, intake estimates based on food supply data are higher than estimates based on actual food consumption survey data, with the magnitude of the error depending on the quantity of waste produced. Perhaps more importantly for exposure assessment, users of foods cannot be distinguished from nonusers. Therefore, individual variations in intake cannot be assessed nor can exposure of potentially sensitive subpopulations be estimated. Finally, food availability is usually reported in terms of raw agricultural commodities. Processed forms of foods are not considered, nor is there any way to distinguish use of foods as ingredients.

Household or Community Inventories. The design and methodology of food consumption surveys conducted at the household level overlap, to a certain extent, with surveys conducted at the community level and, in the other direction, of the individual. Household surveys generally can be categorized as (a) household or community inventories, or (b) household or individual food use.

Inventories are accounts of what foods are available in the household or, in some cases, in the community. What foods enter the household (or community)? Were they purchased, grown, obtained some other way? What foods are used up by the household? Were they used by household members, guests, tenants? Were they fed to animals?

Inventories vary in precision with which data are collected. Questionnaires may or may not ask about forms of the food (canned, frozen, fresh), source (grown, purchased, provided through food program), cost, preparation, *etc.* Quantities of foods may be inventoried as purchased, as grown, with inedible parts included or removed, as cooked, or raw.

Household or Individual Food Use. Food use studies, usually conducted at the household or family level, are often used to provide economic data for policy development and planning for feeding programs. Survey methods used include food accounts, inventories, records, and list-recalls (Pao *et al.* 1989, Lee and Nieman 1993). These methods account for all foods used in the home during the survey period. This includes foods used from what was on hand in the household at the beginning of the survey period and foods brought into the home during the survey period.

Household food use data may be used in gross estimation of food or food constituent intakes by a population. The USDA has conducted household food consumption surveys approximately every 10 years since 1935, using the results in developing family food plans such as the Thrifty Food Plan, on which Food Stamp Program allotments and other food policies are based. Interviews are conducted with the "household food manager" to assess household characteristics, shopping practices, participation in food assistance programs, income, and expenditures. An inventory is taken of foods used from the household food supply for the previous week. For each food, the amount used, the form, source, unit of purchase, and price are noted.

Although household food use data have been used for a variety of purposes, including exposure assessment, serious limitations associated with data from these surveys should be noted. Food waste often is not accounted for. Food purchased and consumed outside the household may or may not be considered. Users of a food within a household cannot be distinguished, and individual variation cannot be determined. Intakes by subpopulations based on age, gender, health status, and other variables for individuals can only be estimated based on standard proportions or equivalents for age/gender categories.

Individual Intake Studies. Individual intake studies provide data on food consumption by specific individuals. Methods for assessing food intakes of individuals may be retrospective (*e.g.*, 24-hour or other short-term recalls, food frequencies, and diet histories), prospective (food diaries, food records, or duplicate portions), or a combination thereof. The most commonly used studies are those using the recall or record method and the food frequency.

- Recall Method (24-hour recall)

  Recalls are used to collect information on foods consumed in the past. The unit of observation is the individual or the household. The subject is asked to recall what foods and beverages he/she or the household consumed during a specific period, usually the preceding 24 hours. Since this method depends on memory, foods are quantified retrospectively, often with the aid of pictures, household measures, or two- or three-dimensional food models.

  Advantages of this method are the ease of administration and the time necessary for completion (about 20 minutes or less for a 24-hour recall) (Block

1989, Dwyer 1988). Recalls have been used successfully with individuals as young as six years of age, and interviewer-administered recalls are usually the method employed for populations with limited literacy or for individuals whose native language is not English. When individuals are not available for an interview or are unable to be interviewed due to age, infirmity, or temporary absence from the household, surrogate respondents are often used (Samet 1989).

The main disadvantage of the recall method is the potential for error due to faulty memory of respondents. Items that were consumed may be forgotten, or the respondent may recall items consumed that actually were not consumed during the time investigated. To aid recall memories, the interviewer may "probe" for certain foods or beverages that are frequently forgotten, but this probing has also been shown to introduce potential bias by encouraging reporting of items not actually consumed.

- Food Record/Diary Method

  Records are used to collect information about current food intake. The subject is asked to keep a record of foods and beverages as they are consumed during a specific period. Quantities of foods and beverages consumed are entered in the record usually after weighing, measuring or recording package sizes. Occasionally, photographs or other recording devices are employed.

- Food Frequency Questionnaire Method

  The Food Frequency Questionnaire (FFQ), or checklist, determines the frequency of consumption of a limited number of foods, usually fewer than 100. This retrospective method estimates whether and how often foods are usually eaten by an individual during a particular period of time. Subjects indicate how many times each day, week, or month they usually consume each food. Semiquantitative FFQs also estimate amounts consumed by allowing subjects to indicate whether their usual portion size is small, medium, or large compared to a stated "medium" portion. The size of the medium portion is usually standardized for various age/gender groups according to mean intakes of large populations.

  Since FFQs collect data about intake of relatively few foods, a carefully tailored FFQ might provide useful data for assessment of chronic exposure if the chemical in question is concentrated in only a few foods and if the food frequency instrument has been designed to target those foods. Depending on the number of foods and whether or not an attempt is made to quantify intake, the FFQ can be administered in 10 to 15 minutes. Information from FFQs cannot be used to estimate intake of acutely toxic chemicals, since data are collected on single food items or types, not on food combinations eaten at the same time.

In general, data from 24-hour and other short-term recalls and from food diaries, which collect detailed information on the kinds and quantities of foods consumed, are the most accurate and flexible data to use in assessment of exposure to food toxins. Data from these surveys can be used to estimate either acute or chronic exposure. Averages and distributions can be calculated, and exposure estimates can be calculated for subpopulations based on age, gender, ethnic background, socioeconomic status, and other demographic variables, provided that such information is collected for each individual.

Each of the methods discussed above under Individual Intake Studies may also be applied at the family or household levels. This application is most appropriate when household members eat from a common pot.

In general, data from large surveys using recall/record methods provide the most accuracy in assessment of exposure to agents in food when the agent under investigation is found in many foods, and when most of these foods are consumed on a regular basis. It is more difficult to capture intake of infrequently consumed foods using short-term recalls or records. If an infrequently consumed food, such as liver, is a major source of the agent under investigation, exposure estimates for that agent may be low. In this case, a well-designed FFQ that targets the specific foods of concern should be considered instead.

Available Sources of Information About Food and Water Consumption in the United States—U.S. National Surveys of Food Intake

Various survey methods are available for determining consumption of foods and beverages and where these items are consumed: food frequency interviews, food diaries, 24-hour dietary recalls, multiday weighed food record, and household food purchase records. However, the data collected in one of two large, national, food-consumption surveys, conducted by the U.S. government, will be appropriate for most exposure assessments in that country: (1) the Nationwide Food Consumption Survey (NFCS) conducted by the U.S. Department of Agriculture, beginning in 1935, and (2) the National Health and Nutrition Examination Survey (NHANES) undertaken by the U.S. Department of Health and Human Services (USDHHS), beginning in 1971. Both surveys employ multistage area probability sampling procedures to obtain a sample representative of the population. Specific years in which these surveys have been conducted are listed in **Table 6-1**.

Although both surveys contain a dietary intake component, they are quite dissimilar in purpose and in elements other than dietary intake. The purpose of the NFCS is to "measure the food and nutrient content of the diet and the money value of food used by U.S. households and the food and nutrient intakes at home and away from home of individuals" (NRC 1989). The purposes of the NHANES are to: (1) "develop information on the total prevalence of a disease condition or a physical state; to provide descriptive or normative information, and (2) provide information on the interrelationships of health and nutrition variables within the population groups" (NRC 1989). NFCS and NHANES are summarized below.

The Nationwide Food Consumption Survey (NFCS). The NFCS has two major components—the Household component which obtains information on food used by the household over a seven-day period, and the Individual component, in which data on foods and beverages consumed by individuals in the household are collected.

The series of Continuing Surveys of Food Intakes by Individuals (CSFIIs) do not include the lengthy Household Food Use component but do collect detailed data on household demographics. For all of the surveys in the NFCS/CSFII series conducted by the USDA, data are collected for both components by a trained interviewer in the household.

Each household in the NFCS is contacted about one week before the interview to schedule a date for the interview and to ask the member of the household most familiar

Table 6-1. National Food Consumption Surveys Conducted in the U.S.

Year(s)	Survey Description
U.S. Department of Agriculture—Nationwide Food Consumption Surveys	
1965	Individual food intakes collected in Spring quarter. 19,015 individuals.
1977–1978	Individual food intakes collected in all four quarters. 36,142 individuals.
1977–1978	Low Income Survey. 12,042 individuals (Food Stamp Program (FSP) participants and nonparticipants are identified).
1977	Puerto Rico. 7,828 individuals. July–December.
1978	Alaska. Winter quarter (January–March). 2,305 individuals.
1978	Hawaii. Winter quarter (January–March). 3,024 individuals.
1979–1980	Low Income Survey. Differs from 1977–1978 Low Income Survey because the 1979 Food Stamp Act changed the purchase requirements and eligibility standards for Food Stamp Program participation. 9,093 individuals (FSP participants and nonparticipants are included).
1985	Continuing Survey of Food Intakes by Individuals (CSFII). Women 19 to 50 years of age and their children, 1 to 5 years of age. There were 1,503 women and 548 children in the sample for the first dietary recall. This was a longitudinal survey and data is available for many women and children over a period of one year. Four nonconsecutive days of data are available for 1,088 women and 371 of their children.
1985	CSFII. Men 19 to 50 years of age. One day. 1,134 individuals. Summer (July–September).
1985	CSFII. Low-Income. One day. Women 19 to 50 years of age and their children, 1 to 5 years of age. One day. 2,120 women and 1,314 children. Spring (April–June).
1986	CSFII. Women 19 to 50 years of age and their children, 1 to 5 years of age. There were 1,510 women and 547 children in the sample for the first dietary recall. This was a longitudinal survey and data are available for many women and children over a period of one year. Four nonconsecutive days of data are available for 1,164 women and 375 of their children (April–March).
1986	CSFII. Low-Income. One day. Women 19 to 50 years of age and their children, 1 to 5 years of age. 1,329 women and 816 children. Spring (April–June).
1987–1988	All year. 10,172 individuals.
1988	Bridging Study. 697 Women 20 to 49 years of age. Experimental design with methodology bridging 1977 and 1987 surveys.
1989–1991	CSFII. All year. 14,490 individuals.
1994–1996	CSFII. All year.
U.S. Department of Health and Human Services—National Health and Nutrition Examination Surveys (NHANES)	
1971–1975	NHANES I—10,126 individuals, 1 to 74 years of age.
1976–1980	NHANES II—9,983 individuals, 6 months to 74 years of age.
1982–1984	HHANES (Hispanic HANES): Mexican-American—8,554 interviewed Cuban-American—1,766 interviewed Puerto Rican—3,369 interviewed
1988–1994	NHANES III. Proposed sample size of 40,000 over the six years conducted in two national cycles: Cycle I—1988–91 Cycle II—1991–94 Follow-up Supplemental Nutrition Survey to be conducted with examinees 50 years and older to obtain a second 24-hour recall by telephone in Cycle I (1988–91).

with food shopping and meal planning (the "household food manager") to keep receipts and records that would help with the household food use component of the survey. The household food use survey interview is conducted with the household food manager and included questions about the composition and characteristics of the household and its members, shopping practices, participation in food assistance programs, income, and expenditures. An inventory is then made of foods used from the home food supplies during the previous week—the amount used, the form (including fresh, frozen, canned, dried), the source, unit of purchase, if bought, and price.

Individual intake data for both the NFCSs and the CSFIIs are collected by interview with each household member eligible for survey participation. Each individual is asked to recall all foods and beverages and quantities consumed during the previous day (24-hour recall), and then to keep a two-day record of intake for the day of the interview and the following day. Serving sizes are estimated using rulers, household measures such as teaspoons, tablespoons, and cups, and a food instruction booklet, which explained to the respondent how to quantify each type of food. In addition, questions are asked about eating patterns, dietary supplements, activity level, and other areas specific to that particular survey. The household food manager or other responsible adult answers for respondents less than 12 years of age, or individuals unable to answer for themselves.

In summary, the NFCS and CSFII surveys provide data that are quite appropriate and useful for assessment of exposure to food toxins for a variety of reasons:

- Foods coded are numerous and very specific. More than 6,000 different foods have been reported as consumed in at least one year of the surveys between 1987 and 1995. Water consumption is also included.
- Foods are coded in a hierarchical structure, allowing easy aggregation and disaggregation of data in estimating exposure.
- Data on multiple days of intake allow estimation of acute or chronic exposure. (Calculation of usual exposure from data collected on consecutive days is complicated by the nonindependence of intakes over these days. This limitation has been corrected in the 1994 and later years of the CSFII. For these survey years, data were collected on two nonconsecutive days.)
- Distributions of intake may be calculated.
- Data on age, gender, race, region, season, pregnancy and nursing status allow exposure estimation for a wide variety of subpopulations.
- Body-weight data allow exposure estimations per unit body weight.

The National Health and Nutrition Examination Survey (NHANES). NHANES, conducted by the NCHS, includes five major components: (1) a household questionnaire, (2) a medical history questionnaire, (3) a dietary questionnaire, a (4) physical examination, and (5) clinical tests. Parts of the survey are conducted in the home, the rest in specially designed mobile examination centers (MECs), where the dietary and clinical procedures are conducted.

The dietary component consists of a recall of dietary intake for the preceding 24 hours administered by a dietitian in the MEC, a food frequency questionnaire for the previous one to three months (depending on the specific NHANES survey), and questions about special diets, medications, and nutritional supplements. Serving sizes are estimated using food models and geometric shapes. The major limitation of NHANES data is that only one 24-hour recall is obtained for each person, so chronic exposure estimates cannot be made with any accuracy.

Selecting Food Consumption Data for Exposure Analyses

Agents Entering the Food Supply Outside of the Home. In general, data from either of the two major U.S. national surveys provide the most feasible source of food intake data for assessment of exposure to food toxins. Data on the concentrations of agents in food are obtained separately and combined with the food consumption data to estimate intake. This approach can be used to estimate either acute or chronic exposure at the national level. Averages and distributions can be calculated for subpopulations based on age, gender, ethnic background, socioeconomic status, and other demographic variables.

Agents Entering the Food as a Result of Activities within the Home. Dietary intake of agents that are found in food due to household activities such as treatments for pests, and cosmetic usage, can be determined using either one of the national food consumption surveys. The most appropriate survey will depend on what foods are likely to have the residues, how often foods will have residues and on the toxicological characteristics of the agents. The estimates will need to be adjusted by the proportion of the household's food supply that is expected to contain residues.

Infrequently Consumed Foods. Some agents concentrate in liver and other foods that may be consumed infrequently. However, they may be eaten in significant quantities on those occasions when they are eaten. Exposure assessments for those agents will be most accurate when based on surveys in which the frequency of consumption is estimated.

Agents Found in Only a Few Foods. If the agent is found in only a few foods, exposure assessments for that agent will be most accurate if the data used are from surveys which captured very specific information on foods consumed—even if the survey does not capture information about the total diet.

Separating Food Eaten at Home, Versus Away from Home, Versus Eaten at Home but Brought from Outside the Home. If foods from one source—such as institutional foods—have different concentrations of agents than those from other sources, it will be appropriate to use food consumption data that permit exclusion of sources that do not contribute. The USDA CSFII (1989 to 1992) surveys provide extensive detail about the source of each food each respondent reports consuming.

Foods that are Unique to a Specific Population. Some subpopulations may consume foods that come from a unique source and special steps will need to be taken to accurately assess these groups' exposure. Three examples include:

- Subsistence and recreational fisherman who consume fish that were caught from waters in which the agent being studied is present at higher than typical levels.
- Vegetarians, whose diets are limited, to varying degrees, to nonanimal sources.
- Infants and young children eating special foods. Commercial infant foods are prepared with special attention to the needs of infants and children and manufacturers have programs to monitor environmental contaminants. Evaluation of data from the USFDA monitoring programs indicates that levels will be lower for many agents.

The presence of agents in breast milk must be evaluated in assessing potential exposure of nursing infants to an agent in the food supply.

Validity, Reliability, and Sources of Error

The national consumption surveys and food monitoring programs are powerful tools for assessing dietary exposure and evaluating status and trends. One must realize these data are not intended to represent the true diets of all people at all places and that applying these data too broadly may result in false conclusions or poor estimates in dietary exposure for specific populations. The factors affecting the exposure of an individual or small subset of the population to a specific agent are not evaluated by using a national average diet and nationwide residue monitoring data (Lioy 1990). Market-basket surveys reveal little about unique, individual dietary practices which vary from the U.S. norm, such as those of small children living in lead-laden homes (Freeman *et al.* 1997). Care must be taken in the application of these data for specific situations, such as when evaluating residential exposures.

The process of preparation, storage, and consumption of foods in the residence or other eating places can affect an individual's exposure. Such factors are normally not included in risk assessment because they are not well understood or documented. Contamination of food occurs beyond the production and retail distribution system. Additional exposure may come from nonretail food sources such as homegrown food, or fish and other game. Pesticide use in the residence can be an important source of food contamination. Bug bombs or foggers can contaminate both foods and the surfaces they contact. Soiled cooking utensils and dinnerware, food hygiene practices, and food contact with dusty surfaces contribute to excess dietary exposures. A better understanding of the interactions of food with other contaminated media, such as the addition of contaminated water and contacts with contaminants in air and on surfaces during preparation, is needed so that the importance of other contaminated residential media on dietary exposure can be better assessed. It is important to have an overall understanding of the exposure scenarios that lend most to contamination of foods and beverages and, hence, excess dietary exposures. In general, market-basket surveys do not measure levels of contamination in foods "as consumed." In many scenarios, personal dietary measurements are needed when market-level monitoring cannot discern exposures for specific microenvironments, such as a residence, or when comparisons to directly measured exposures from other pathways are required.

Validity. Food consumption data used in assessing exposure to agents in food should be both valid (accurate) and reliable (precise).

Numerous studies attempting to validate one survey method relative to another have been reported (Bingham *et al.* 1982, Blake and Durnin 1963, Block 1982, Bransby *et al.* 1948, Fanelli and Stevenhagen 1986, Grewal *et al.* 1974, Hussain *et al.* 1980, Karvetti and Knuts 1985, Lubbe 1968, Mahalko *et al.* 1985, Meredith *et al.* 1951, Morgan *et al.* 1987, Morrison *et al.* 1949, Nettleton *et al.* 1980, Ramsanen 1979, Russell-Briefel *et al.* 1985, Willett *et al.* 1985, Young *et al.* 1952). For example, the FFQ, a more recent survey method, has been validated by comparing results of dietary intake with repeated multiple day food records, which served as an estimate of usual intake (Block 1989, Pietinen *et al.* 1988a, Willett *et al.* 1988). Results of the validity studies indicated that FFQs could provide useful information about individual nutrient intakes. However, these studies generally have shown better correlations between methods for groups than for individual survey participants. The validity of estimates of intakes of agent in food is less certain.

Validation of the 24-hour recall, diaries, records, and frequencies has also been reported after comparing estimates of dietary intake obtained using one of these survey instruments to subjects' actual intakes. Actual intakes may be based on surreptitious observation in cafeterias, institutional dining centers, or other facilities (Baranowski *et al.* 1986, Gersovitz *et al.* 1978, Greger and Etnyre 1978, Madden *et al.* 1976, Samuelson 1970, Stunkard and Waxman 1981).

Survey methodology has also been validated by use of biological markers associated with dietary intake. Possible sources of biological markers include urine, feces, blood, and tissue samples, but the most easily accessible and therefore most commonly used is urine. Nitrogen content of urine has been used to verify protein intake. If protein intake calculated from the reported food intake is in agreement with protein intake calculated from nitrogen excretion, it is assumed that intake of other nutrients is valid (Bingham and Cummings 1985).

Reliability. Reliability, or reproducibility, is the ability of a method to produce the same or similar estimate on two or more different occasions (Block and Hartman 1989, Pietinen *et al.* 1988b), whether or not the estimate is accurate. However, a method cannot achieve an accurate answer every time unless it gives the same answer every time.

The reliability of food consumption survey data for estimating "usual" intake of a population depends somewhat on the number of days of dietary intake data collected for each individual in the population. The number of days of food consumption data required for reliable estimation of population intakes is related to each subject's day-to-day variation in diet (intraindividual variation) and the degree to which subjects differ from each other in their diets (interindividual variation) (Basiotis *et al.* 1987, Nelson *et al.* 1989). When intraindividual variation is small relative to interindividual variation, population intakes can be reliably estimated with consumption data from a smaller number of days than should be obtained if both types of variation are large. Intake of contaminants can be reliably estimated with fewer days of data when they are present in many foods that are commonly consumed.

In assessing food intake, it is generally accepted that mean intake of a population may be reasonably estimated using a one-day recall or diary if the number of subjects is sufficiently large. However, the percentage of the population estimated to be at risk of toxic effects from an agent will be higher when food intake is assessed using a one-day recall than with a multiday record or dietary history. This is because extreme levels of intake (*e.g.*, 90th or 95th percentiles) are invariably higher for a single day than they are for multiple days. In addition, large intraindividual variation associated with one-day surveys may limit the power to detect differences between different population groups (Liu *et al.* 1978, Beaton *et al.* 1979, van Staveren *et al.* 1985).

Sources of Error. Error in individual food consumption surveys may be due to chance or to measurement factors. Data variability due to chance may be related to the survey sample. Any sample randomly drawn from a population will differ from any other sample, with the degree of difference depending on the size of the sample and the homogeneity of the population from which it was drawn. Error due to chance also arises from data collection at different times of the day, on different days of the week, or at different seasons of the year.

Measurement error may be introduced by the survey instrument, the interviewer, or the respondent. The instrument may bias results if questions are not clear, if probes "lead the subject" to give a desired answer, if questions are culture-specific, or if they do not follow a logical sequence. For self-administered questionnaires, responses will be influenced by the readability level, the use of abbreviations or unfamiliar jargon, clarity of instructions, and amount of space provided for answers. Interviewer bias may be introduced if interviewers make the respondent uncomfortable, are judgmental, or do not use a standard method and/or standard probes.

Respondents may introduce bias if they omit reporting foods they actually ate because they are reluctant to report certain foods or beverages (alcoholic beverages are a good example) or if they are forgetful. Alternatively, they may report the food but understate the quantity consumed. Foods consumed away from home, particularly on occasions when the focus of attention is on the event rather than on the food, are especially difficult for people to remember. Quantities may be underestimated for similar reasons. Foods and beverages that were not consumed may be reported as consumed because of faulty memories, desire to impress the interviewer, or confusion with similar foods.

Measurement errors also include errors in coding due to unclear handwritten records or to erroneous data entry.

ALGORITHMS FOR ESTIMATING EXPOSURE

Dietary exposure modeling consists of combining consumption and residue databases in an appropriate manner and supplementing this information with additional factors that affect exposures of a defined population. One dietary exposure model currently under development, termed the Dietary Exposure Potential Model (EPM), was designed specifically to provide guidance for exposure monitoring programs (Tomerlin *et al.* 1997). Consumption information is derived primarily from USDA survey mean intake values. Highly consumed foods are aggregated into approximately 800 core foods with similar raw agricultural ingredients (*i.e.*, wheat flour with tomato sauce and cheese dishes). The core foods are combined with mean contaminant residue data through the use of standard recipes. Information on residues of specific agents comes primarily from national residue monitoring programs, such as the USFDA's Total Diet Study, the California Pesticide Monitoring Database and the USEPA's Pesticide Residue Information System. The EPM correlates food consumption and agent contamination data in a format that allows exposures to be characterized by highly consumed and highly contaminated foods. The inclusion of consumption factors permits other characterizations.

Dietary models are used to support the risk assessment process and in the design of exposure measurements by identifying potential "hot-spots" of exposure. Such information can be used as guidance in identifying the need for personal dietary measurements and as an aid in selecting populations most likely at risk. Validated dietary models are needed for inclusion in total ingestion (diet, hand-to-mouth ingestion, *etc.*) and total human (multipathway) exposure models.

Modeling the Data: Chronic and Distributional Approaches and Their Associated Statistics

Exposure can be expressed as a single point estimate or a distribution.

Point Exposure. A point estimate of exposure to a specific agent by a particular population is a broad estimate generated using one number to represent concentration of the agent in each food and one number to represent intake of these foods by that population. Table 6-2 presents mean intake quantities for selected foods consumed by the U.S. population and its subpopulations. In estimating chronic exposure, the arithmetic mean of an agent's concentration is most commonly used. However, in the case of pesticides, if the concentration distribution is known to be skewed, use of the median (or 50th percentile) concentration is more appropriate (Mosteller and Tukey 1977).

Simple Distribution. A simple distribution of exposure can be calculated by either of two procedures. A single number chosen to represent concentration of the agent in each of the foods of interest may be applied to a distribution of intake levels for each food. Alternatively, a single number chosen to represent food intakes can be applied to a distribution of agent concentration values. The USEPA currently uses a simple distribution model as a screening tool to assess acute exposure to pesticide residues (Saunders and Petersen 1986).

A distribution of the consumption of tomatoes is presented in Table 6-3. This analysis includes all tomato products that were reported as consumed. Combining this with a single residue value would produce a simple distribution of exposure.

Figure 6-1 is an example of the results of a Monte Carlo analysis. Additional discussions of this topic can be found in Chapter 12.

Joint Distribution. Joint distribution analysis (usually referred to as Monte Carlo analysis) relies on sophisticated calculations to allow combination of representative data using a distribution of agent concentrations and a distribution of food intakes.

Exposure to Multiple Chemicals. Estimates of combined exposure—often referred to as cumulative exposure assessments—involve considerations not required for exposure to one agent. Residues of chemicals cannot be simply summed for use in exposure assessment. The agents may have different toxic effects and/or different potencies. Toxicity must first be standardized to account for differences in biological potencies.

Biological Agents: Special Considerations. Most of the considerations described above apply to biological agents. Ultimately the objective is the same—to estimate the intake of the agent and to compare it to some measure of the biological activity/potency of the agent. Considerations that are unique to microbiological agents include microbial replication patterns, virulence, persistence in individuals and continued risk of spread of infection, and food attributes that may alter pathogenicity. If a quantitative dose-response relationship does not exist or cannot be obtained, ranking schemes may be necessary to characterize severity and/or duration of disease. The potential for outgrowth and die-off

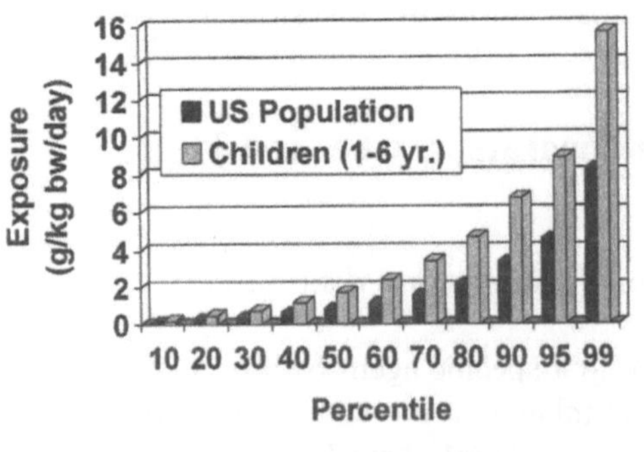

Figure 6-1. Sample exposure analysis using tomato consumption reported in the 1994–1996 CSFII (Appendix 1) and hypothetical residue data.

Table 6-2. Mean food and water intake for selected U.S. populations reported in the 1989–1991 USDA continuing survey for individual intakes

	U.S. Population		Infants (nonnursing)		Children (1–6 years of age)	
Foods	g/kg-bw/day	Percent of diet	g/kg-bw/day	Percent of diet	g/kg-bw/day	Percent of diet
Apples (raw)	0.29	0.79	—*	—	0.72	0.95
Apples (canned)	0.08	0.23	1.60	1.05	0.39	0.59
Apples (juice, cider)	0.47	1.31	3.33	2.19	2.76	3.68
Bananas	0.27	0.75	0.85	0.56	0.77	1.02
Beans (green)	0.15	0.40	0.53	0.35	0.37	0.49
Beef (boneless, fat)	0.20	0.54	—	—	0.43	0.57
Beef (boneless, lean, fat-free)	0.78	2.15	0.35	0.23	1.49	1.98
Broccoli	0.11	0.31	—	—	0.28	0.37
Carrots	0.18	0.50	1.36	0.89	0.32	0.43
Corn sugar	0.61	1.68	1.67	1.10	1.41	1.88
Corn/grain (endosperm)	0.20	0.55	—	—	0.46	0.61
Grapes	0.08	0.23	—	—	0.26	0.35
Grapes (juice)	0.12	0.33	—	—	0.77	1.03
Grapes (wine, sherry)	0.08	0.24	—	—	—	—
Lettuce (head varieties)	0.21	0.57	—	—	0.16	0.22
Milk sugar (lactose)	0.42	1.16	6.37	4.19	1.33	1.77
Milk-based water	5.95	16.5	10.1	6.69	23.0	30.7
Milk-fat solids	0.34	0.93	0.38	0.25	1.06	1.41
Milk-nonfat solids	0.45	1.24	2.17	1.43	1.43	1.90
Oats	0.10	0.228	0.36	0.24	0.40	0.53
Onions	0.17	0.47	—	—	—	—
Orange (juice)	0.86	2.41	0.67	0.44	2.53	3.36
Orange (peeled fruit)	0.09	0.27	—	—	0.24	0.32
Peaches	0.13	0.35	1.12	0.74	0.33	0.45
Peanut butter	—	—	—	—	0.22	0.30
Pears	0.08	0.21	0.73	0.48	0.19	0.25
Pears (nectar)	—	—	1.05	0.69	—	—
Peas (green)	0.12	0.33	—	—	0.27	0.36
Pineapple (juice)	—	—	—	—	0.19	0.25
Pork (boneless, fat)	0.12	0.33	—	—	0.22	0.29
Pork (boneless, lean, fat-free)	0.34	0.94	—	—	0.66	0.81
Potatoes (white, peeled)	0.83	2.30	—	—	1.35	2.00
Rice (milled, white)	0.22	0.61	1.15	0.76	0.51	0.68
Soybeans	0.34	0.95	2.47	1.64	0.62	0.83
Squash (winter)	—	—	—	—	0.86	0.57
Tomato (puree)	0.27	0.74	—	—	0.50	0.66
Tomato (whole)	0.49	1.36	—	—	0.81	1.08
Turkey (boneless, lean, fat-free)	0.10	0.27	—	—	—	—
Water (commercialprocessing)	5.68	15.7	11.0	7.26	8.00	10.7
Water (tap)	9.31	25.8	95.2	62.6	10.5	14.0
Wheat flour	1.35	3.73	0.63	0.42	3.10	4.10

* and—are less than 0.2% of daily food and water intake (on a gram/kg-bw/day basis).

Table 6-3. Distribution of reported tomato consumption by the U.S. population in the 1989–91 USDA Continuing Survey of Food Intake for Individuals (including whole tomatoes, catsup, puree and sauces)

	Consumption (g/kg-bw/day)	
Distribution point or range	U.S. Population	Children (1–6 Years of Age)
Mean user	1.41	2.84
Percentile		
10	0.13	0.25
20	0.28	0.49
30	0.44	0.79
40	0.62	1.19
50	0.85	1.71
60	1.15	2.40
70	1.56	3.42
80	2.18	4.71
90	3.33	6.75
95	4.57	8.89
99	8.27	15.63

of the microorganism and their contributing factors (microbial ecology, time-temperature relationships of processing and storage, *etc.*). In most instances it will be necessary to generate quantitative data about the concentration of microbial agents. There are a host of food-related factors that will determine the final concentration of the biological agent, including the chemical make-up of the food (*e.g.*, pH, water activity, nutrient content), and processing and storage conditions that relate to outgrowth or die-off of the microorganism.

Water Consumption. Data on intake of water is available in the USDA CSFII and NFCS surveys. The data are reported as cups of water consumed. In addition, recipes have been developed (Ershow and Cantor 1989, DEEM 1994) and are used to estimate the source (tap, commercially bottled) of that water. The first assumptions of amount of water ingested daily (2 liters per day for adults and 1 liter per day for children) were published by the Safe Drinking Water Committee of the National Research Council (NRC 1977). These values were further refined by Ershow and Cantor (1989), who statistically analyzed water intake rates by age and divided water intake into: (1) tap water intake (the sum of water drunk as a beverage and water added to food and beverages) and (2) intrinsic water intake (water in foods as purchased).

SENSITIVE SUBGROUPS

Consideration of sensitive subgroups involves their identification and measurement of their exposure patterns:

- *Identification of relevant subgroups with special sensitivity or likely to have unusual exposure levels.* Because of their food intake patterns, some subgroups may have special sensitivities and therefore there is a greater need to accurately measure their exposure. Generally these groups will be identified based on the toxicological information. Also, it may be that there is a desire to provide an extra measure of safety. For example, the U.S. Congress, in passing the 1996 Food Quality Protection Act, specifically identified children as a subgroup for which an extra measure of safety was appropriate. Even groups with similar sensitivities may have substantially different risks because of differences in their exposure patterns—either because of differences in their diets or because of difference in the source of their foodstuffs.

- *Measure potential exposure patterns of subgroups.* Potentially sensitive subgroups include such populations as infants, children, and lactating mothers. For most purposes, it will be desirable to assess both the overall population's intake and the intake for selected populations. As can be seen from Table 6-1, the foods contributing most to the diets of infants, children, and the general population are different. For example, infant diets, on average, contain a higher proportion of ingredients that are often used in commercially prepared infant formulas. These ingredients include soy flour and oil, coconut oil, lactose (milk sugar), and water. Infants have a much higher intake of water per unit body weight than do others in the population. They eat no broccoli. Their raw apple consumption is less than 0.2 percent of their total intake. However, they do eat substantial amounts of processed apples and apple juice. All populations have increased their consumption of corn sugar in the past 10 years due to changes in the sugar content of many commercially prepared foods. Children from 1 to 6 years of age also have a higher intakes of many fruits and vegetables—on a gram per kg body-weight basis—than do older members of the population. They also eat significant quantities of peanut butter. Not surprisingly, they do not consume either wine or beer at levels greater than 0.2 percent of their diet. Both infants and young children consume large quantities of milk. [Note that milk is broken into four categories: nonfat solids, fat solids, milk sugar(lactose), and milk-based water. This was done to avoid double counting when a component of milk is consumed alone. Examples include milk-fat solids in butter, milk sugar in infant formula, and nonfat solids in yogurt.]

  Other subgroups of the population also report differences in the foods they eat. For example, Hispanic subgroups consume more corn and less wheat than other ethnic groups. Fortunately, U.S. national food consumption surveys identify key demographic information so that subgroup analysis is feasible.

  The method used may also depend on whether the population in question is considered to be particularly vulnerable to the agent. Concern over children's exposure to toxins, especially to pesticides, prompted the U.S. Congress to request the National Research Council (NRC) to conduct a thorough review of scientific and policy issues concerning pesticides in the diets of infants and children. The NRC committee recommended that methods be developed for evaluating distribution of exposure in vulnerable populations, for assessing exposure from all foods combined rather than from individual commodities, and for assessing combined exposure from pesticides with similar toxic effects (NRC 1993). A number of sophisticated new methods for assessing exposure have been developed in response to these recommendations.

  For young infants, an important source of exposure to fat-soluble and other environmental agents is human breast milk.

FUTURE RESEARCH NEEDS

- *Ongoing surveys of food consumption of the U.S. population and targeted surveys for subgroups with special sensitivities.* Market-level food programs provide important information for dietary exposure assessments. Although expensive, a high priority should be given to maintaining current information about food consumption. Such programs are essential for establishing trends in food consumption. The U.S. Congress has recognized the importance of quality dietary information and has mandated enhancements to federally-sponsored programs by forming the Interagency Board for Nutrition Monitoring and Related Research (Federal Register 1991).
- *Nationally representative databases of residue information.* Adequate residue information is not readily accessible for most agents. In some cases, data may exist but cannot be accessed except by the scientists who collected the data. Even many of the federally generated data are not available with sufficient documentation regarding methods and results.
- *Additional modeling techniques to better characterize the entire distribution of exposure*, so that both variability and uncertainty can be estimated.

REFERENCES

Baranowski, T., R. Dworking, J.C. Henske, D.R. Clearman, J.K. Dunn, P.R. Nader, and P.C. Hooks. 1986. The accuracy of children's self-reports of diet: family health project. J. Am. Dietet. Assoc. 86:1380.

Basiotis, P.P., S.O. Welsh, J. Cronin, J.L. Kelsay, and W. Mertz. 1987. Number of days of food intake records required to estimate individual and group nutrient intakes with defined confidence. J. Nutrit. 117:1638.

Beaton, G.H., J. Milner, P. Corey, V. McGuire, M. Cousins, E. Stewart, E. de Ramos, D. Hewitt, P., V. Grambsch, N. Kassim, and J.A. Little. 1979. Sources of variance in 24-hour dietary recall data: implications for nutrition study design and interpretation. Am. J. Clin. Nutrit. 32:2456.

Berry, M.R. 1997. Advances in dietary exposure research at U.S. Environmental Protection Agency—National Exposure Research Laboratory. J. Exposure Anal. Environ. Epidemiol. 7(1):3–16.

Bingham, S., and J.H. Cummings. 1985. Urine nitrogen as an independent validatory measure of dietary intake: a study of nitrogen balance in individuals consuming their normal diet. Am. J. Clin. Nutrit. 42:1276.

Bingham, S., H.S. Wiggins, H. Englyst, R. Seppanen, P. Helms, R. Strand, R. Burton, I.M. Jorgensen, L. Poulsen, A. Paerregaard, L. Bjerrum, and W.P. James. 1982. Methods and validity of dietary assessments in four Scandinavian populations. Nutrit. Cancer 4:23.

Blake, E.C., and J.V.G.A. Durnin. 1963. Dietary values from a 24 hour recall compared to a 7-day survey on elderly people. Proc. Nutrit. Soc. 22:I.

Block, G. 1982. A review of validations of dietary assessment methods. Am. J. Epidemiol. 115:492.

Block, G. 1989. Human dietary assessment: Methods and issues. Prevent. Med. 18:653.

Block, G., and A.M. Hartman. 1989. Issues in reproducibility and validity of dietary studies. Am. J. Clin. Nutrit. 50:1133.

Bolger, P.M., N.J. Yess, E.L. Gunderson, T.C. Troxell, and C.D. Carrington. 1996. Identification and reduction of sources of dietary lead in the United States. Food Additiv. Contam. 13(1):53–60.

Bransby, E.R., C.G. Daubney, and J. King. 1948. Comparison of results obtained by different methods of individual dietary survey. Br. J. Nutrit. 2:89.

Chin, H.B. 1991. The effect of processing on residues in foods: The food processing industry's residue database. In: Tweedy, B.G., H.J. Dishburger, L.G. Ballantine, and J. McCarthy (eds.). Pesticide Residues and Food Safety: A Harvest of Viewpoints. American Chemical Society. Washington, D.C. 175 pages.

Dabeka, R.W., A.D. McKenzie, and G.M.A. Lacroix. 1987. Dietary intakes of lead, cadmium arsenic and fluoride by Canadian adults: a 24-hour duplicate-diet study. Food Addit. Contam. 4(1):89–102.

Davis, S., and P. Waller. 1989. Quantitative estimates of soil ingestion in normal children between the ages of two and seven years: analysis of Al, Si, and Ti; field activities and procedures. Final Report. Contract B-Q-1336-A-E. U.S. Environmental Protection Agency. Washington, D.C.

DEEM (Dietary Exposure Evaluation Model). 1994. Novigen Sciences, Inc. Washington, D.C.

Dwyer, J.T. 1988. Assessment of dietary intake. In: Shils, M.E. and V.R. Young (eds.). Modern nutrition in health and disease. 7th edition. Lea and Febiger. Philadelphia, Pennsylvania.

Ershow, A.G., and K.P. Cantor. 1989. Total water and tapwater intake in the United States: Population-based estimates of quantities and sources. Federation of American Societies of Experimental Biology (FASEB). Life Sciences Research Office. Bethesda, Maryland. 98 pages.

Fanelli, M.T., and K.J. Stevenhagen. 1986. Consistency of energy and nutrient intakes of older adults: 24-hour recall vs. 1-day food record. J. Am. Dietet. Assoc. 86:664.

FASEB (Federation of American Societies of Experimental Biology). 1988. Estimation of exposure to substances in the food supply. Life Sciences Research Office. Bethesda, Maryland. 57 pages.

FASEB (Federation of American Societies of Experimental Biology). 1993. National human exposure assessment survey (NHEXAS) dietary monitoring options. Life Sciences Research Office. Bethesda, Maryland. 251 pages.

Federal Register. 1991. Ten-year plan for the national nutrition and related research program. Fed. Reg. 56(209):55716.

Freeman, N.C.G., A. Ettinger, M.R. Berry, and G. Rhoads. 1997. Hygiene and food-related behaviors associated with blood lead levels of young children from lead contaminated homes. J. Exposure Anal. Environ. Epidemiol. 7(1):103–118.

Gersovitz, M., J.P. Madden, and H. Smiciklas-Wright. 1978. Validity of the 24-hour dietary recall and seven-day record for group comparisons. J. Am. Dietet. Assoc. 73:48.

Graham, J., K. Walker, M. Berry, E. Bryan, M. Callahan, A. Fan, B. Finley, J. Lynch, T. McKone, H. Ozkaynak, and K. Sexton. 1992. The role of exposure data bases in risk assessment. Arch. Environ. Health 47(6):408–420.

Greger, J.L., and G.M. Etnyre. 1978. Validity of 24-hour recalls by adolescent females. Am. J. Pub. Health 68:70.

Grewal, T., T. Gopaldas, V.J. Gadre, S.N. Shrivastava, B.M. Pranjpe, B.N. Chatterjee, and N. Srinivasan. 1974. A comparison of weighment and questionnaire dietary survey methods for rural preschool children. Ind. J. Nutrit. Dietet. 11:224.

Hansen, S.C. 1979. Conditions for use of food additives based on a budget for an Acceptable Daily Intake. J. Food Protect. 42:429.

Hussain, M.A., M. Abdullah, N. Huda, and K. Ahmad. 1980. Studies on dietary survey methodology—a comparison between recall and weighing method in Bangladesh. Bangladesh Med. Res. Council Bull. 6:53.

Karvetti, R.L., and L.R. Knuts. 1985. Validity of the 24-hour recall. J. Am. Dietet. Assoc. 85:1437.

Lee, R.D., and D.C. Nieman. 1993. Nutritional Assessment. William C. Brown Publishers. Dubuque, Iowa. 220 pages.

Lioy, P.J., J.M. Waldeman, R. Harkov, C. Pietarinen, and A. Greenburg. 1988. The Total Human Environmental Exposure Study (THEES) to benzo(a)pyrene: Comparisons of the inhalation and food pathways. Arch. Environ. Health 43(4):304–312.

Lioy, P.J. 1990. Assessing total human exposure to contaminants. Environ. Sci. Technol. 24:938–45.

Liu, K., J. Stamler, A. Dyer, J. McKeever, and P. McKeever. 1978. Statistical methods to assess and minimize the role of intra-individual variability in obscuring the relationship between dietary lipids and serum cholesterol. J. Chron. Disease 31:399.

Lopez-Aliva, V., C. Charar, and J. VanEmon. 1996. Quick determination of pesticides in foods by SFE-ELISA. Food Test. Anal. Oct.-Nov. Pages 28–37.

Lubbe, A.M. 1968. A survey of the nutritional status of white school children in Pretoria: description and comparative study of two dietary survey techniques. S.A. Med. Journal Suppl:616.

Madden, J.P., S.J. Goodman, and H.A. Guthrie. 1976. Validity of the 24-hr. recall. J. Am. Dietet. Assoc. 68:143.

Mahalko, J.P., L.K. Johnson, S.K. Gallagher, and D.B. Milne. 1985. Comparison of dietary histories and seven-day food records in a nutritional assessment of older adults. Am. J. Clin. Nutrit. 42:542.

Manton, W.J., K.L. Stanek, A.J. Kuehneman, and C.R. Angle. 1991. The Omaha Duplicate Diet Study. Intrinsic and external sources of dietary lead. Pediatric Res. 29(4):93A.

Melnyk, L.J., M.R. Berry, and L.S. Sheldon. 1997. Dietary exposure from pesticide application on farms in the Agricultural Health Pilot Study. J. Exposure Anal. Environ. Epidemiol. 7(1):61–80.

Meredith, A., A. Matthews, M. Zickefoose, E. Weagley, M. Wayave, and E.G. Brown. 1951. How well do school children recall what they have eaten? J. Am. Dietet. Assoc. 27:749.

Minyard, J.P., Jr., and W.E. Roberts. 1991. FOODCONTAM: A state data resource on toxic chemicals in foods. In: Tweedy, B.G., H.J. Dishburger, L.G. Ballantine, and J. McCarthy (eds.). Pesticide Residues and Food Safety: A Harvest of Viewpoints. American Chemical Society. Washington, D.C. Page 151.

Morgan, K.J., S.R. Johnson, R.L. Rizek, R. Reese, and G.L. Stampley. 1987. Collection of food intake data: An evaluation of methods. J. Am. Dietet. Assoc. 87:888.

Morrison, S.D., F.C. Russell, and J. Stevenson. 1949. Estimating food intake by questioning and weighing: A one-day survey of eight subjects. Proc. Nutrit. Soc. 7:v.

Mosteller, F., and J.W. Tukey. 1977. Data analysis and regression. Addison-Wesley Publishing Co. Reading, Massachusetts. 350 pages.

NCHS (National Center for Health Statistics). 1982. Second national health and nutrition examination survey. 1976–80. Data tape 5704 (24-hour recall, specific food item). NTIS accession number PB82-142639. National Technical Information Service. Springfield, Virginia.

Nelson, M., A.E. Black, J.A. Morris, and T.J. Cole. 1989. Between- and within-subject variation in nutrient intake from infancy to old age: estimating the number of days required to rank dietary intakes with desired precision. Am. J. Clin. Nutrit. 50:155.

Nettleton, P., K.C. Day, and M. Nelson. 1980. Dietary survey methods. 2. A comparison of nutrient intakes within families assessed by household measures and the semi-weighed method. J. Human Nutrit. 34:349.

NIST (National Institute of Standards and Technology). 1995. Standard Reference Materials Catalog. 1995–96. Special Publication 260. National Institute of Standards and Technology. Gaithersburg, Maryland. 40 pages.

NRC (National Research Council). 1977. Drinking water and health. Safe Drinking Water Committee, Commission on Life Sciences. National Academy Press. Washington, D.C. 135 pages.

NRC (National Research Council). 1989. Nutrient adequacy assessment using food consumption surveys. Food and Nutrition Board, Commission on Life Science. National Academy Press. Washington, D.C. 121 pages.

NRC (National Research Council). 1993. Pesticides in the diets of infants and children. Committee on Pesticides in the Diets of Infants and Children, Board on Agriculture and Board on Environmental Studies and Toxicology, Commission on Life Science. National Academy Press. Washington, D.C. 133 pages.

Pao, E.M., K.E. Sykes, and Y.S. Cypel. 1989. USDA methodological research for large-scale dietary intake surveys. 1975–88. Home Economics Research Report Number 49. U.S. Department of Agriculture, Human Nutrition Information Service. Washington, D.C. 100 pages.

Pennington, J.A.T. 1983. Revision of the total diet study food list and diets. J. Am. Dietet. Assoc. 82:166–73.

Pennington, J.A.T., and E.T. Gunderson. 1987. History of the Food and Drug Administration's Total Diet Study—1961 to 1987. JAOAC 70:772–782.

Petersen, B., and L. Barraj. 1996. Assessing the intake of contaminants and nutrients: A review of methods. J. Food Composition Anal. 9:243–254.

Petersen, B., and J. Douglass. 1994. Use of food-intake surveys to estimate exposure to nonnutrients. Am. J. Clin. Nutrit. 59 (Suppl.):2403–2433.

Pietinen, P., A.M. Hartman, E. Haapa, L. Rasanen, J. Haapakoski, J. Palmgren, D. Albanes, J. Virtamo, and J.K. Huttenen. 1988a. Reproducibility and validity of dietary assessment instruments. I. A self-administered food use questionnaire with a portion size booklet. Am. J. Epidemiol. 128:655.

Pietinen, P., A.M. Hartman, E. Haapa, L. Rasanen, J. Haapakoski, J. Palmgren, D. Albanes, J. Virtamo, and J.K. Huttenen. 1988b. Reproducibility and validity of dietary assessment instruments. II. A qualitative food frequency questionnaire. Am. J. Epidemiol. 128:667.

Rasanen, L. 1979. Nutrition survey of Finnish rural children. VI. Methodological study comparing the 24-hour recall and the dietary history interview. Am. J. Clin. Nutrit. 32:2560.

Russell-Briefel, R., A.W. Caggiula, and L.H. Kuller. 1985. A comparison of three dietary methods for estimating vitamin A intake. Am. J. Epidemiol. 122:628.

Samet, J.M. 1989. Surrogate measures of dietary intake. Am. J. Clin. Nutrit. 50:1139.

Samuelson, G. 1970. An epidemiological study of child health and nutrition in a northern Swedish county. 2. Methodological study of the recall technique. Nutrition and Metabolism 12:321.

Sasaki, S., and H. Kestelloot. 1992. Value of Food and Agriculture Organization data on food-balance sheets as a data source for dietary fat intake in epidemiologic studies. Am. J. Clin. Nutrit. 56:716.

Saunders, D.S., and B.J. Petersen. 1986. Introduction to the Tolerance Assessment System. Office of Pesticide Programs. U.S. Environmental Protection Agency. Washington, D.C. 41 pages.

SCF Intake and Exposure Working Group. 1994. Summaries of food consumption databases in the European Union. CS/Int/Gen2 (European Commission). Brussels, Belgium. 150 pages.

Schreiber, J.S. 1997. Transport of organic chemicals to breast milk: tetrachloroethene case study. Pages 95–143 in: Kacew, S. and G. Lambert (eds.). Environmental Toxicology and Pharmacology of Human Development. Taylor and Francis. Washington, DC.

Sexton, K., S.G. Selevan, D.C. Wagner, and J.A. Lybarger. 1992. Estimating human exposures to environmental pollutants: availability and utility of existing databases. Arch. Environ. Health 47(6):398–407.

Sexton, K., D.E. Kleffman, and M.A. Callahan. 1995. An introduction to the National Human Exposure Assessment Survey (NHEXAS) and related Phase I field studies. J. Exposure Anal. Environ. Epidemiol. 5(3):29–232.

Sherlock, J.C., G.A. Smart, B. Walters, W.H. Evans, D.J. McWeeny, and W. Cassidy. 1983. Dietary surveys on a population at Shipham, Somerset, United Kingdom. Sci. Total Environ. 29:121–42.

Stunkard, A.J., and M. Waxman. 1981. Accuracy of self-reports of food intake. J. Am. Dietet. Assoc. 79:547.

Thomas, K.W., L.S. Sheldon, J. Keever, J.M. Roberds, P.W. Geno, and M.R. Berry. 1993. Collection and analysis of duplicate diet samples: a pilot study on farmer exposures to pesticides. In: Proceedings of the EPA/AWMA Symposium: Measurement of toxic and related air pollutants. EPA/600/A93/024. Air and Waste Management Association. Report VIP-34. Pittsburgh, Pennsylvania.

Thomas, K.W., L.S. Sheldon, E.D. Pellizzari, R.W. Handy, J.M. Roberds, and M.R. Berry. 1997. Testing duplicate diet sample collection methods for measuring personal dietary exposures to chemical contaminants. J. Exposure Anal. Environ. Epidemiol. 7(1):17–36.

Tomerlin, J.R., M.R. Berry, N.L. Tran, S.-B. Chew, B.J. Petersen, K.D. Tucker, and K.H. Fleming. 1997. Development of a dietary exposure potential model for evaluating dietary exposure to chemical residues in food. J. Exposure Anal. Environ. Epidemiol. 7(1):81–102.

USDA (U.S. Department of Agriculture). 1981. Nationwide Food Consumption Survey Data Tapes, 1977–78. NTIS Accession numbers PB80-190218, PB80-197429, PB80-200223, and PB81-118853. National Technical Information Service. Springfield, Virginia.

USEPA (U.S Environmental Protection Agency). 1997. Presentation by the Office of Pesticide Programs to the USEPA Science Advisory Board. March 13, 1997. Washington, D.C.

USFDA (U.S. Food and Drug Administration). 1994. Pesticide analytical manual. Third edition. Volume I. Washington, D.C.

Vahter, M., M. Berglund, L. Friberg, L. Jorhem, B. Lind, S. Slorach, and A. Akesson. 1990. Dietary intake of lead and cadmium in Sweden. Var Foda 44 (Suppl. 2). National Food Administration. Uppsala, Sweden. 10 pages.

van Staveren, W. A., J.O. de Boer, and J. Burema. 1985. Validity and reproducibility of a dietary history method estimating the usual food intake during one month. Am. J. Clin. Nutrit. 42:554.

Willett, W.C., L. Sampson, M.J. Stampfer, B. Rosner, C. Bain, J. Witschi, C.H. Hennekens, and F.E. Speizer. 1985. Reproducibility and validity of a semi-quantitative food frequency questionnaire. Am. J. Epidemiol. 122:51.

Willett, W.C., L. Sampson, M.L. Browne, M.J. Stampfer, B. Rosner, C.H. Hennekens, and F.E. Speizer. 1988. The use of a self-administered questionnaire to assess diet four years in the past. Am. J. Epidemiol. 127:188.

WHO (World Health Organization). 1997. Guidelines for predicting dietary intake of pesticides (revised). Geneva, Switzerland. 33 pages.

Young, C.M., G.C. Hagan, R.E. Tucker, and W.D. Foster. 1952. A comparison of dietary study methods. II. Dietary history vs. seven-day record vs. 24-hour recall. J. Am. Dietet. Assoc. 28:218.

TOTAL EXPOSURE

Each of the three previous chapters treated a different exposure pathway (ingestion, inhalation, dermal absorption). This is a reasonable simplification of reality for many, perhaps a majority, of pollutants, since often only one pathway, based on the physical/chemical/biological properties of the agent, dominates exposure. For example, for most volatile organic compounds (VOCs), the compound tends to volatilize from a given medium (*e.g.*, soil or water) into the air and the major exposure is through inhalation of vapors. There are exceptions, however, where multimedia (*e.g.*, lead may be found in ambient air, soil, household dust, drinking water, canned foods) and multipathway (*e.g.*, ingestion of drinking water, inhalation of vapors and dermal contact with water may all be important for a VOC like chloroform) exposures are relevant and a consideration of "total" exposure is necessary in providing a realistic evaluation of residential exposure.

This chapter integrates the three previous chapters under the rubric of "Total Exposure," which we define as exposure to a particular pollutant through all relevant pathways. The main concepts of total exposure have been outlined by several writers (Ott, 1982, 1985, 1990, Duan 1982, 1985, 1988, 1991) and include the use of such tools as personal monitors, human activity patterns, microenvironmental investigations, and body burden measurements. The American Industrial Hygiene Association has published a white paper on total human exposure, supporting the concept and its development (AIHA 1994). An Exposure Factors Manual has also been published (Gephart *et al.* 1994).

This chapter describes the methods that are available to evaluate "total exposure" and, using examples (chloroform, lead, PCBs, DDT), the chapter illustrates approaches taken to integrate and evaluate multi-media and multi-pathway/route exposures. The methods discussed include: (1) large scale exposure studies, where concentrations of target agents in single or multiple media are measured, (2) the use of models to estimate total exposure based on measured concentrations in single-media or microenvironments, and (3) measurement of body burden, which by its nature represents an integration of exposures (more exactly doses) through all pathways.

Although "total" also implies multi-source (*e.g.*, emissions from a number of facilities in a given area); multi-chemical (*e.g.*, mixtures) and cumulative exposures over time, this chapter does not explicitly address these issues. The USEPA Science Policy Council is working towards the development of broad, general guidance on assessing cumulative risks which is defined as "risks from more than one stressor considered in aggregate". One element of this comprehensive, holistic approach to risk assessment is the development of the USEPA's Community-Based Environmental Protection (CBEP) program, an approach that the USEPA is taking to "... assess and manage the quality of air, water,

Residential Exposure Assessment, edited by Baker *et al.*
Kluwer Academic/Plenum Publishers, New York, 2000.

land and living resources in a place as a whole, to better reflect regional and local conditions . . ." The reader is referred to these initiatives for guidance on this approach. Current guidance on mixtures (USEPA 1986) exists but is in the process of being updated.

Exposures to chemical, biological and physical agents in general occur principally because humans engage in normal activities in various microenvironments that bring them into relatively close proximity with a number of agents. These activities and concurrent sources of chemicals occur in outdoor air (*i.e.*, via ambient levels of air pollutants such as nitrogen oxides, carbon monoxide, and particulates such as plant allergens), in the work setting (*e.g.*, exposure to industrial chemicals in factory jobs and exposure to carpet adhesive VOCs in office buildings), pollutant exposures in vehicles while in transit or refueling (*e.g.*, passenger-compartment benzene levels), and exposures in the residence (*e.g.*, disinfectants, personal care products, emissions from heating fuels).

Potential residential exposures result from emission sources that are located both inside and outside the home. Indoor air represents only one of many different "compartments" within a residence. Exposures can also be associated with dislodgeable residues from surfaces (*e.g.*, dermal contact with pesticides applied to carpets, incidental ingestion of dislodged residues following hand-to-mouth transfer), direct dermal contact during application of a variety of consumer products, and ingestion of contaminants in food and water. Further, the agents present in one or more of these compartments are in a "dynamic" state (*i.e.*, there is evidence that particulate contaminants), whether generated inside the residence, tracked in, or infiltrated from the outdoor environment, are resuspended and recycled within the house by walking on floors and rugs, sweeping and dusting, and vacuuming. VOCs that enter the residential environment can be absorbed to surfaces, or "sinks," and then later be released as airborne levels are depleted by various mechanisms, including air exchange with other rooms of the house and with outdoor air, and with physicochemical transformations in residential air. Thus, the residential environment, and potential sources and agents, should be thought of in very dynamic terms, wherein multiple exposure pathways and routes exist. These concepts are discussed in greater detail in Chapter 3.

To better understand the dynamic nature of potential agents (chemical, biological, physical), their sources, and associated exposures in the residential environment, it is helpful to use one or more possible classification systems. Agents can be classified or characterized by the type of source, such as pesticidal consumer products, and by agent group, such as VOCs, particulate matter, bioaerosols, fibers. Sources can be characterized by the agents emitted, by the source and its location, and by the rate and pattern of emissions. These factors may be directly measured or estimated mathematically using mass balance models. The approaches for evaluating sources differ depending on the type of agent, but all consider fate processes such as emission patterns, sinks, and degradation.

A brief summary of the key components in performing exposure studies is presented below (Monitoring Studies) with a discussion of key monitoring studies. This discussion not only highlights the available database but provides insight into the development of the methodologies. This discussion is then followed with a description of body burden studies, including measurement techniques (*e.g.*, for blood, breath, and mother's milk) for determining target agents in body fluids such as blood, urine, and mother's milk, breath and other biological media. A discussion of pharmacokinetic modeling follows with key studies, relating measurements of exposure to burden.

The final section of this chapter illustrates how "total exposure" may be evaluated using specific chemicals. For example, lead is illustrative of exposures from a variety of media (soil, food, dust, drinking water) but primarily via ingestion (although inhalation of gasoline exhaust was also a major exposure pathway) and multipathway/route exposure is illustrated by chloroform in water used in residences where ingestion, inhalation and dermal contact all contribute significantly to exposure. The chloroform example is also indicative of the use of body burden measurement techniques.

PCBs, DDT, and polyaromatic hydrocarbons (PAHs) are also considered in illustrating multimedia, multipathway exposure as well as exposure monitoring methodologies.

MONITORING STUDIES

Principles of Large-Scale Exposure Studies

The basic principles of large-scale exposure studies are:

- Probability-based sample selection, and
- Direct measurement of exposure for each participant.

The first principle ensures that the findings from a sample population can be extrapolated to the entire population. The second principle, which requires development of personal air quality monitors for each of the target pollutants, ensures that the measurements are as close to individual exposure as possible, reducing dependence on models to a minimum.

Probability-Based Sample Selection. The only generally-acccepted way to extrapolate results from a sample to a larger population is by probability-based selection. This means, first, that the entire population of interest must be enumerated in some way; a list must be prepared or accessible that includes every person in the population. Then a method of selection must be devised that assigns a probability of being selected to every person in the entire population.

As a practical matter, the initial "list" is often the most recent U.S. Census for the geographic area being considered. The Census may be the basis for a first stage of selection, in which smaller geographic subunits are selected with a probability proportional to their population. Once these subunits are selected, a more intense phase of "counting and listing" ensues, to detect and document changes that have occurred since the last Census.

The next stage is to gather information from which a final list of target participants can be prepared. If any stratification is desired, a small set of focused questions about the strata can be administered to a larger sample of persons to enable choosing the final sample. For example, in the USEPA TEAM Study of carbon monoxide (CO) (Akland *et al.* 1985), it was desirable to oversample long-distance commuters. Therefore an initial screening questionnaire was administered that included a question about length of commute. With this information from a larger sample, it is possible to assign differing probabilities of selection to the different strata to attain the approximate number of desired persons in each stratum.

The general methods of selecting sample populations are described in textbooks on survey design. A useful paper devoted principally to the methods used in environmental surveys such as the TEAM Studies is Whitmore (1988).

Direct Measurement of Exposure. Direct measurement of individual exposure to airborne pollutants requires use of a personal monitor. Since personal monitors capable of measuring pollutants at environmental concentrations hardly existed in 1980, each of the major USEPA TEAM studies has required development and evaluation of personal monitors. For each study, this required arduous development and testing regimens lasting for several years before final deployment of the monitors. However, once the monitors were developed, they have remained available for others, such as private industry, to employ in later studies.

Key Monitoring Studies

TEAM Studies. The first large-scale studies of human exposure were carried out by the USEPA in the early 1980s (Pellizzari 1987a,b, Wallace 1987, Wallace *et al.* 1982, 1984, 1986ab, 1987ab, 1988, 1989). A personal monitor consisting of a pump and an 8-inch glass tube packed with powdered Tenax was employed to measure personal exposures of 350 persons in New Jersey and another 200 in California to 25 target VOCs. This initial large-scale study resulted in several findings of considerable significance, the most important being that exposure to most of these toxic and carcinogenic compounds came from consumer products, building materials, and personal activities rather than from chemical plants and oil refineries. Subsequent USEPA TEAM studies of carbon monoxide (1982–1983), pesticides (1985–1986) and particles (1989–1990) have been described in Chapter 8.

Other Studies. Private industry has also carried out several large-scale studies, beginning with the NO_2 studies of the SoCal Gas Corporation in Los Angeles. A major study of benzene exposures in Valdez was carried out by the Alyeska Pipeline Service Company in 1992 (Goldstein *et al.* 1992). A study of human exposure to airborne manganese (sponsored by the Ethyl Corporation, makers of the gasoline additive MMT), took place in Toronto, Canada in 1995–96. The tobacco industry sponsored a large study of exposure environmental tobacco smoke (ETS) in 16 cities in the U.S. (Jenkins *et al.* 1996).

No comparable large-scale studies of human exposure have taken place in other countries. However, some large-scale studies of indoor air concentrations have occurred. In Europe, major indoor air studies of VOCs took place in the Netherlands (Lebret *et al.* 1986) and in West Germany (Krause 1987). The German study also included measurements of airborne metals, collected over a period of a year in settling plates (Seifert 1987).

These studies all concentrated mainly on air exposures, although the VOC studies included measurements of drinking water in every home, and the pesticide studies included a few measurements of house dust as well as questions on diet. It is interesting to note that some pesticides meant to be used outside the home, particularly to protect against termites, were found at higher levels indoors, both in air and in house dust. These pesticides included DDT, heptachlor, chlordane, dieldrin, and endrin, all of which had been banned in the U.S. before the study (DDT had been banned about 15 years before). It was hypothesized that they were entering the house either as soil gas (in much the same way as radon) or being tracked in on dirt from shoes. Later studies (Roberts *et al.* 1987) have indicated that track-in is an efficient way to transport both lead and pesticides from outdoor soil to indoor house dust.

Several small-scale personal monitoring studies were carried out during the 1980s. A series of studies in undeveloped countries (Smith 1987, 1993) has been fundamental in increasing our knowledge of the exposures of third-world residents, particularly women and children, to combustion pollutants. Because of the traditional ways of cooking, with very long periods of cooking in homes without chimneys or other proper vents, women and their infants are exposed to extraordinarily high levels of particles, including the carcinogenic PAHs day after day for many years. Although medical records are scarce, and competing causes of death may obscure the likely results of this high exposure, it seems probable that these high particle levels are a serious source of sickness (emphysema, bronchitis, and possibly asthma in children) in these countries. Studies in China have also implicated brown coal used in cooking as an important source of lung cancer in Chinese women nonsmokers.

NHEXAS. A major multimedia study of human exposures was initiated by the USEPA in the mid-1990s. Called the National Human Exposure Assessment study (NHEXAS), it includes measurements of three pollutant groups (VOCs, pesticides, and heavy metals) in air, water, food, soil, and house dust. Field work was completed in 1997. Three geographic areas were included in NHEXAS: (1) the USEPA's Region V—a multi-state Midwest area including Illinois, Wisconsin and nearby states, (2) Arizona, and (3) Baltimore. In the Midwest and Arizona studies, monitoring took place over a single six-day visit to each home. In the Baltimore study, a longitudinal design was implemented, with eight visits to each participant over the course of a year.

16-City ETS Study. A recent major study of environmental tobacco smoke exposure was carried out by the Oak Ridge National Laboratory (Jenkins *et al.* 1996). In addition to measuring respirable suspended particles (RSP, defined as particles smaller than 3.5 μm), the investigators measured a number of other particulate and vapor phase markers, including ultraviolet absorbing particulate matter (UVPM), fluorescent particulate matter (FPM), scopoletin, solanesol, 3-ethenyl pyridine (3-EP), nicotine, and myosmine. Saliva samples were collected at the beginning and end of the monitoring period and analyzed for cotinine (the main nicotine metabolite).

To measure UVPM, the RSP filter is extracted with methanol and an aliquot is injected into a columnless liquid chromatographic (LC) system equipped with an ultraviolet detector (Conner *et al.* 1990). FPM is analyzed using LC with a fluorescence detector. Both methods are calibrated using environmental tobacco smoke (ETS). Results of testing several different types of cigarettes (flue cured, burley, and Turkish tobaccos) show no effect of the particular blend. The resulting values of UVPM and/or FPM in the field are felt to be upper limits on the particle phase of ETS, since other products of combustion can also produce a positive ultraviolet or fluorescence response.

A third possible marker for ETS-related particles is solanesol, a primary terpenoid alcohol present in the dry tobacco leaf at levels up to 4 percent, making it the most abundant constituent of the tobacco lipid fraction. Solanesol is thus more specific to tobacco than either UVPM or FPM. Chamber studies comparing ETS alone to simultaneous cigarette smoking with burning of candles and oil lamps showed that UVPM increased by about a factor of two due to the additional combustion sources, whereas the solanesol levels remained unchanged (Ogden and Maiolo 1989). Provided the solanesol/RSP ratio were known for all cigarettes smoked in a field situation, the observed ratio could be used to correct for other sources of RSP. However, the authors report variations in the ratio from 2 to 4 percent between an ultralow-tar and a Kentucky reference cigarette (K1R4F).

Nelson *et al.* (1992) performed chamber studies to determine the ratio to ETS-related RSP ($PM_{3.5}$) of the three particle-phase markers UVPM, FPM, and solanesol as a function of air exchange rates (0 to $4 h^{-1}$) and measurement times (30 min to 6 h). The ratios were reasonably stable across all rates and times, with the maximum variation in the neighborhood of 50 to 60 percent for each marker. Nicotine was shown to decay faster than particles for approximately an hour and slower thereafter, suggesting an adsorption-desorption phenomenon occurring on the stainless steel walls of the chamber.

Approximately 100 nonsmokers over the age of 18 in each of 16 cities took part. Each participant carried a sampling system during work and another while away from work. A 2×2 design was adopted to study persons exposed to ETS both at work and away from work, only at one or the other location, or not at either location. Although all 1564 participants claimed to be nonsmokers, 66 (4.2 percent) exceeded 15 ng/ml salivary cotinine, a level indicative of active smoking, and were excluded from the data analysis.

A weak point of the study was its selection of subjects. Selection was not by probability-based methods, but rather by use of data bases maintained by market research organizations, random telephone dialing, mall intercepts, *etc.* The result was a sample that was biased in at least two ways, toward more women (68 percent) and higher socioeconomic class (70 percent of respondents exceeded the median U.S. salary). The latter bias, given the strong socioeconomic relation to cigarette smoking, would tend to underestimate the impact of ETS. Exposures to chemical, biological and physical agents in general occur principally because humans engage in normal activities in various microenvironments that bring them into relatively close proximity with a number of agents. These activities and concurrent sources of chemicals occur in outdoor air (*i.e.*, via ambient levels of air pollutants such as nitrogen oxides, carbon monoxide, and particulates such as plant allergens), in the work setting (*e.g.*, exposure to industrial chemicals in factory jobs and exposure to carpet adhesive VOCs in office buildings), pollutant exposures in vehicles while in transit or refueling (*e.g.*, passenger-compartment benzene levels), and exposures in the residence (*e.g.*, disinfectants, personal care products, emissions from heating fuels).

With these caveats, the results confirmed and extended the results of earlier studies. Personal exposures to $PM_{3.5}$ for 808 persons exposed to ETS both at work and away from work averaged 45.4 $\mu g/m^3$ compared to 18.5 $\mu g/m^3$ for 157 persons not exposed at either location. A total of 515 persons exposed to ETS at only one of the two locations had intermediate mean values of about 30 $\mu g/m^3$. Corresponding values for nicotine were 2.98 $\mu g/m^3$ for persons exposed at both locations, about 0.8 $\mu g/m^3$ for persons exposed at only one location, and only 0.12 $\mu g/m^3$ for unexposed persons. The corresponding results for UVPM, thought to represent the actual ETS-related portion of $PM_{3.5}$, were 22.0, 8.3, and 2.0 $\mu g/m^3$.

The study provided a glimpse of gender differences in exposure to ETS. For those persons exposed at home but not at work, 113 women had considerably higher exposures to nicotine (1.6 $\mu g/m^3$) than the 36 men in that category (0.8 $\mu g/m^3$).

The major large-scale studies of human exposure are summarized in Table 7-1.

BODY BURDEN

Another way to do an integrated assessment is to measure the target chemicals in body fluids (breath, blood, or mothers' milk) since the body naturally integrates exposure through all pathways. The body burden at any instant reflects the intake, distribution, metabolism, and excretion rates of the chemical.

Measurement Methods

Breath. In the mid 1970s, a method of sampling exhaled breath for VOCs was developed (Krotoszynski *et al.* 1977, 1979) using what was then a new sorbent—Tenax (a polymer based on 2,6-diphenyl-p-phenylene oxide). This method was capable of measuring sub-ppb levels and represented an improvement in sensitivity over the activated charcoal methods in use until that time. The method was modified and improved for use in USEPA-sponsored studies (Barkley *et al.* 1980, Michael *et al.* 1980, Pellizzari *et al.* 1985, 1990, Gordon *et al.* 1988). The subject inhales pure humidified and charcoal-scrubbed air from a 20-L Tedlar bag that has previously been filled from a pure air cylinder. He then exhales into a second 20-L Tedlar bag. This bag is then emptied by pumping through a glass cartridge containing 1.5 g Tenax. The entire system was mounted in a van to allow "house calls" to the participants in the TEAM Studies of 1980 to 1987. The time to collect a breath sample was about 5 minutes. After the first visit to New Jersey in 1981, it was suspected that the Tedlar bags may have been contaminated by exhaust fumes from the van. All later trips incorporated a helium positive-pressure system to bathe the bags while in storage on the van.

An improvement to this method was developed after completing the TEAM Studies, and was used in a series of field and chamber studies. The method (Pellizzari *et al.* 1990, Raymer *et al.* 1990a, 1992a, Thomas *et al.* 1991) employs a charcoal face mask to allow breathing clean air without the need for a cylinder of clean air and a separate "inhale" bag. Several (2 to 4) breaths are taken through the charcoal filter to flush the alveoli and bronchial tubes of ambient air before beginning to collect the breath sample. The "exhale" bag is replaced by an evacuated electropolished 1.8 L stainless steel cylinder with a critical orifice (although other suitable collecting devices such as Tenax or other sorbents could also be used). The subject exhales through a one-meter long perfluoroethylene tube, which retains the latter part of his breath (the alveolar portion) for a few seconds during the resting and inhalation parts of the respiration cycle, during which time the alveolar air in the tube is pulled into the cylinder through the critical orifice. The timing is such that the canister collects approximately 98 percent alveolar air. About 12 breaths are collected over an 80-second period. The method is readily deployable in the field—electrical power is not needed and the entire set of equipment fits into a metal carrying case about the size of a suitcase. The cycle time (from beginning to collect one breath sample until readiness to collect another) can be as low as three minutes. Besides the chamber studies mentioned above, the system was employed in a field study measuring exposure to benzene while showering with gasoline-contaminated ground water (Lindstrom *et al.* 1994). A miniature version of the system was made and validated for use in space flight (Raymer 1992b).

Recently, a single-breath method has been developed (Pleil and Lindstrom 1995a). The subject breathes directly into a 1 L evacuated cylinder through a straw-like attachment; after wasting the first (dead space) portion of his breath, the subject opens the valve on the cylinder to allow collection of the second (alveolar) portion. The cycle time is reduced to approximately one minute. The new method allows for immediate collection following exposure, thus documenting the maximum breath (and therefore blood) concentration attained during the exposure period. Also, a much more finely detailed

Table 7-1. Major probability-based exposure studies

Sponsor (name)	Pollutant/Media	Location	Season	Year	Participants	Reference
Personal monitoring						
EPA TEAM	VOCs in air, breath, and drinking water	Elizabeth-Bayonne, New Jersey	Fall	1981	355	Pellizzari *et al.* 1987a
			Summer	1982	160	Wallace 1987
			Winter	1983	50	
		Greensboro, North Carolina	Spring	1982	25	
		Devils Lake, North Dakota	Fall	1982	25	
		Los Angeles, California	Winter	1984	120	Pellizzari *et al.* 1989
			Summer	1984	50	Wallace *et al.* 1991
		Antioch-Pittsburg, California	Spring	1984	75	
		Baltimore, Maryland	Spring	1987	150	
EPA TEAM (CO)	Carbon monoxide in air and breath	Washington, DC	Fall-Winter	1983	1161	Johnson 1983
		Denver Colorado	Fall-Winter	1983	485	Hartwell *et al.* 1986
EPA NOPES	Pesticides in air, water, and house dust	Jacksonville, Florida	Summer	1986	65	Immerman and Schaum 1989
			Spring	1987	53	
			Winter	1988	55	
		Springfield, Massachusetts	Spring	1987	49	
			Winter	1988	37	
EPA PTEAM	Particles in air	Riverside, California	Fall	1990	178	Pellizzari *et al.* 1992
EPA NHEXAS	Metals, VOCs, and pesticides in air, water, soil, dust, blood, and urine	Michigan, Minnesota, and other areas in USEPA Region V	All year	1996–97	N/A	Unpublished
		Arizona	All year	1996–97	N/A	
		Baltimore, Maryland	All year	1996–97	65	

CARB	VOCs and SVOCs in air and water	Woodland, California	Spring	1990	128	Sheldon *et al.* 1992
Alyeska	BTEX and 6 other VOCs in air	Valdez, Alaska	Winter Summer	1992	58	Goldstein *et al.* 1992
Ethyl Corporation	Manganese in air	Toronto, Ontario	All year	1995–96	700	Unpublished
CIAR	ETS in air and saliva	16 U.S. cities	All year	1993–94	1548	Jenkins *et al.* 1996
Indoor air only						
AWSH	VOCs and metals	West Germany	All year	1986	488	Krause *et al.* 1987
AUW	VOCs	Ede, Rotterdam	3 Months (?)	1985	350	Lebret *et al.* 1986
BRE	VOCs	Avon, England	All year	1994	170	Brown and Crump 1996
Environment Canada	VOCs	Canada	(?)	1992	750	Fellin and Otson 1993
GRI	CO, VOCs, radon, and NO_2	California	Winter	1991–92	293	Wilson *et al.* 1993

Legend:
TEAM: Total Exposure Assessment Methodology
PTEAM: Particle TEAM
TEAM (CO): TEAM study of carbon monoxide
NOPES: Non-Occupational Pesticide Exposure Study
NHEXAS: National Human Exposure Assessment Survey
CARB: California Air Resources Board
CAIR: Center for Indoor Air Research
AWSH: Air, Water, and Soil Hygiene Office, Germany
AUW: Agricultural University of Washington
BRE: Building Research Establishment
GRI: Gas Research Institute

picture of the decay curve during the first few minutes of very rapid decay can be achieved.

Together with the new sampling method, a three-step sequential analytical approach was developed (Pleil and Lindstrom 1995b). In the first step, only the carbon dioxide level in the sample is quantified. This allows an estimate of the amount of "dead space" air included in the sample. Since CO_2 in alveolar air is about 4–5 percent, compared to only 0.035 percent in outdoor air, a sample including a mixture of alveolar and dead space air will have an intermediate level of CO_2. Determining the CO_2 level allows a quantitative adjustment of the subsequent trace-level determinations to estimate more nearly the alveolar air concentration. Although CO_2 levels vary from person to person according to several factors (such as metabolic rates and the amount of time the breath is held before providing a sample), the CO_2 concentrations should be steady over any given series of breath measurements for a single subject. Deviations from the average established for each person permit correction of the measured VOC levels. In the second step, the sample is analyzed for the very volatile endogenous compounds such as acetone and isoprene, which are found at levels approaching 1 ppm by volume. This step is also capable of quantitating polar compounds such as methanol, ethanol, dimethylsulfide, and 2-propanol at ppb levels. Finally, in the third step the sample is analyzed in detail for the remainder of the VOCs at sub-ppb levels.

A few years ago, Kelly and co-workers (Kelly *et al.* 1989) developed a new exhaled breath interface that allows continuous real-time analysis of undiluted breath. The system takes advantage of the high sensitivity and specificity of tandem mass spectrometry (MS/MS) by coupling the exhaled breath inlet to a direct air sampling source (*e.g.*, atmospheric pressure ionization (API) or glow discharge ionization source) and an MS/MS instrument (*e.g.*, triple-stage quadrupole or ion trap mass spectrometer) (Gordon *et al.* 1992). The subject inhales clean air from a suitable source and exhales directly into the breath interface. The inlet requires no attention from the subject and provides a constant flow of exhaled air into the mass spectrometer. Trace chemicals in the breath are immediately ionized and compounds of interest are isolated according to mass. The selected masses are dissociated and the fragments are identified and quantified. This direct air sampling MS/MS technique thus offers a means of extracting the VOCs directly from the exhaled breath matrix, and eliminates the pre-concentration step that normally precedes exhaled breath analysis by GC/MS.

The method was tested in pilot studies to measure dimethyl sulfide (which has been shown to be elevated in breath levels of person with liver disease) in breath of healthy people, lactic acid in breath during physical exercise (Kelly *et al.* 1989), and the elimination of 1,1,1-trichloroethane from breath of a machine shop worker, sampling at 5.5-second intervals over a 20-minute period immediately following exposure (Gordon *et al.* 1992). The data from the latter experiment were evaluated in terms of a two-compartment model of the body. The residence time for the first compartment was estimated to be 1.2 minutes and for the second compartment to be 17 minutes. The method is capable of analyzing both polar and nonpolar organic compounds. Detection limits for certain compounds measured with the API source, such as dimethyl sulfide, are as low as 5 parts per trillion (ppt). Most other VOCs exhibit detection limits in the low ppb range with either the API/triple quadrupole or glow discharge/ion trap system (Gordon *et al.* 1995).

In a recent study designed to validate a PBPK model, Thrall and Kenny (1996) developed a real-time technique to quantitatively measure the concentration of exhaled breath VOCs using laboratory rats. Breath samples are collected using a specially-designed manifold in which up to four rats can be attached to the manifold via nose-only restraint tubes. The exhaled breath from the rats exit the nose ports and are driven into a common mixing chamber by a continuous flow of supplied breathing air. Samples for analysis by either API/triple quadrupole or glow discharge/ion trap tandem mass spectrometry are continually drawn from the contents of the mixing chamber. Detection limits for the target compound, carbon tetrachloride, are in the low (2 to 10 ppb) range.

The highly compact Teledyne 3DQ Discovery ion trap mass spectrometer is expected to form the nucleus of a field-deployable MS/MS system for real-time monitoring of VOCs in breath (and air) (Gordon *et al.* 1995). The USEPA's National Exposure Research Laboratory (NERL) has instituted a cooperative agreement with Battelle Memorial Institute to refine the technique for applications in assessing human exposure.

Blood. By all measures, blood is a more difficult medium to measure than exhaled breath. It is a complex liquid, containing fats and other constituents with differing affinities for the chemicals of interest. Collection is difficult, partly because it is an invasive procedure, and partly because it is difficult to prevent losses of the volatile components. Cleanup procedures are arduous. Only recently has a broad-spectrum procedure been developed that makes it possible to measure a number of VOCs at normal environmental levels in blood. This procedure (isotope dilution purge and trap) was eventually perfected by a group at the Centers for Disease Control (CDC) (Ashley *et al.* 1994).

Mothers' Milk. As with blood, mothers' milk is a complex mixture of fats, liquids and other constituents that make it difficult to collect and analyze. One method that was developed for PCBs and DDE was widely employed in epidemiological studies in the 1980s (McKinney *et al.* 1984). The procedure employs triple extraction with organic solvents followed by gas chromatography with electron capture detection (GC-ECD).

Pharmacokinetic Models

A pharmacokinetic model is needed to relate measurements of exposure to measurements of body burden. For example, what blood level would result from daily inhalation of 200 μg of benzene? Conversely, what does a measurement of 1 $\mu g/m^3$ in exhaled breath tell us about current blood levels or about previous exposure?

Pharmacokinetic models are all based ultimately on mass-balance considerations. Two types of pharmacokinetic models have been employed to predict the behavior of pollutants (especially VOCs): (1) classical compartment models and (2) physiologically-based models.

Compartment Models. Historically, the first models relating exposure to body burden treated various types of tissue as imaginary "compartments," each with a certain "volume" through which the chemical was supposed to distribute itself instantaneously and homogeneously (Teorell 1937a, 1937b). In the simplest form of the theory, the entire body would be considered a single compartment. In more complex forms of the theory, the first and main "compartment" was the blood (Coburn *et al.* 1965). For pollutants such as carbon monoxide, which are bound in the blood, a blood or breath measurement gives an excellent picture of exposure (Wallace 1983) and no further compartments are needed in the model. For pollutants that distribute throughout the body, additional compartments might be added, such as organs, muscle tissue, fat, or bone (Raymer *et al.* 1990a, Wallace *et al.* 1993, 1997).

The compartment models were constructed by using one or more differential equations to account for the gain of a chemical (through ingestion or inhalation) and its loss (through excretion or metabolism). Differential equations for different "compartments" (liver, other organs, muscle, fat, *etc.*) are developed and solved, usually numerically. Historically, a number of simplifying assumptions were made that allowed analytic solution of the equations for certain simple inputs (a single bolus, or a constant exposure at high concentrations). These solutions were generally in the form of exponential functions with different parameters for each compartment. Such models have been effective in accounting for the main features of exposure to many different VOCs at occupational exposure levels of 50 to 100 ppm.

One important parameter in these "compartment models" is the residence time, τ_i for the ith compartment. This is the time it takes for the chemical to decline to 1/e of its initial concentration in the compartment, assuming all other compartments are at zero concentration. A series of chamber studies sponsored by the USEPA have provided estimates of τ_i for $i = 1 \ldots 4$ for a number of VOCs (Wallace *et al.* 1997).

A second important parameter in the "compartment models" is the fraction (f) of the chemical exhaled under steady-state conditions. This is particularly important for environmental considerations, because it may often be the case that persons under normal conditions are indeed at or near equilibrium with their chemical environments—in such a case, multiplying the breath measurement by 1/(f) is quite a good estimate of their long-term average normal exposure. Estimates of (f) for a number of chemicals have been obtained from USEPA-sponsored field and chamber studies.

Recently, classical compartmental models have been extended to environmental exposure levels of 1 to 100 ppb (Wallace *et al.* 1993). A study of nine VOCs in five subjects suggested that half-lives for the four compartments (blood, vessel-rich tissues, vessel-poor tissues, and fat) were on the order of 2 minutes, 20 minutes, 2 hours, and 2 days, respectively (Wallace *et al.* 1997).

Physiologically Based Models. A weakness of the classical compartment models is their abstract nature. For example, the "volume" of a compartment doesn't have much physical meaning. Also, the factors governing the movement of a chemical into and out of a compartment are "lumped" constants that may contain several physiological processes. Physiologically-based pharmacokinetic (PBPK) models were developed to provide more realistic models (Ramsey and Andersen 1984).

Like classical compartment models, PBPK models also begin with differential equations governing the build-up and decay of chemical concentrations in body tissues and fluids, but they include a number of parameters that correspond to real physiological processes and which have the possibility of being measured. The multiplication of parameters requires extensive computer-aided numerical integration, with some consequent loss of intuitive understanding of relationships. However, these models offer the possibility of ultimately improving our predictive ability, after the values of a sufficient number of parameters have been validated. In particular, they may improve estimates of the actual delivery of the toxicant to the target organ, for which actual measurements may be impossible or severely limited from practical considerations of cost or access.

Key Body Burden Studies

NHANES Studies. Several nationwide studies of body burden have been carried out, particularly by the series of National Human and Nutrition Examination Surveys (NHANES) carried out by the National Center for Health Statistics (NCHS). These surveys provide medical examinations and biomedical measurements for a national sample of some 10,000 to 16,000 participants in a sample of cities over a four-year period. NHANES II (1976 to 80) included measurements of blood carboxyhemoglobin (COHb) in about 8,000 people in 22 cities. Since COHb is due mainly to exposure to carbon monoxide (CO, of which only a small amount, about 0.5 percent, is endogenous), the COHb measurements were a direct reflection of CO exposures for these persons. A nationwide probability distribution of CO exposures could be calculated from these body burden measurements.

NHANES surveys have also provided a continuing record of trends in blood lead levels. A sharp reduction was noted following the removal of lead in gasoline as mandated by the USEPA.

The USEPA's TEAM Study used breath measurements to determine that smokers have about 5 to 10 times the body burden of benzene and styrene as nonsmokers. Blood measurements of children are routinely used to test for lead exposures, which may come from several residential sources including chipped lead paint and household dust containing tracked-in soil containing lead from external house paint or from automobile emissions in earlier decades.

Breath. Measurements of exhaled breath provide an indication of concentrations in the blood, without, however, requiring an invasive procedure. Thus, for exposure studies requiring public participation, breath may often be preferred to blood as the sampling medium.

- *Carbon monoxide.* In the case of CO, which combines with hemoglobin at about 250 times the strength of oxygen, the blood is the first and last stop on its trip inside the body. Thus, the body can be considered a single compartment, and a relatively simple linear relationship exists between CO exposures and COHb concentrations, at least at low (<30 ppm) exposures. This is slightly complicated by the fact that the body produces some COHb, perhaps about at the level of 0.2 to 0.5 percent. Several studies indicate that 1 percent COHb corresponds to about 5 ppm breath CO. Thus, a person with a breath level of A ppm, assuming equilibrium with his or her surroundings, would be exposed to roughly A-1 to A-2 ppm CO. The method of measurement is to collect 1 to 2 L of alveolar air in a container (usually an inflatable plastic bag) and measure it using an electrochemical real-time device. The alveolar air is usually collected by using a breath holding period of 20 seconds followed by wasting the first portion of the breath (to avoid any "dead space" air) and collecting only the last portion: "end-expired air."
- *VOCs.* In the case of the lipophilic VOCs, the blood is a way station on the way to and back from the fat cells. Thus, a VOC breath measurement can no longer be as easily interpreted as a CO breath measurement. However, provided something is known of metabolism of the particular VOC, the level in exhaled breath can be interpreted as a reflection of personal exposure. For example, since less than 10 percent of tetrachloroethylene is metabolized, one can expect

that under equilibrium conditions, the breath level should be approximately 90 percent of the ambient concentration. Measurements of VOCs in breath differ from those in air because of the high levels of carbon dioxide (CO_2) and water vapor in breath. Special analytic techniques must be employed to handle these problems. A quantitative determination of CO_2 levels in breath can be helpful, since some methods require collection of alveolar air, and if some dead space air is mixed in, the CO_2 level will be lower than it should. If the person's normal alveolar CO_2 level is known (typically around 4 to 5 percent), then a sample with lower CO_2 concentration can be corrected to approximate the true (alveolar) concentration.

Blood

- *VOCs*. Only recently has a method been developed capable of measuring multiple VOCs at environmental concentrations in blood. The method involves adding isotopes of each target chemical to the sample, and then adjusting the results by the recovery efficiencies observed for the isotopes (CDC 1990).

  Analysis of blood samples for specific antibodies is standard practice for measuring sensitization to allergens. The presence of IgE antibodies specific for an allergen presumes exposure sufficient to cause sensitization, and is a strong indicator that future exposures will result in symptomatic disease. The presence of specific IgG antibodies is considered to associated primarily with exposure rather than any disease process unless levels are very high.
- *Lead*. Blood measurements have been central in our understanding of lead exposures. Over the years, the blood concentration associated with adverse health effects has been driven steadily downward, from levels on the order of 50 μg/dl to the present guideline of 10 to 15 μg/dl. Methods have included fingertip pricks following alcohol washes, but this method may often be compromised by lead contamination from dirt on the finger. Laboratories often have considerable lead contamination also, leading to uncertainty regarding historical measurements and trends. However, there is no doubt that lead concentrations in blood have been trending generally downward, due to the removal of lead from gasoline.

Skin. Specific IgE sensitization is most often assessed using skin tests. Small amounts of allergen are introduced into the skin either through a scratch or an intradermal injection, and the site is monitored for redness and swelling.

Mothers' Milk. A major study of pesticides and PCBs in mother's milk of about 900 North Carolina women (Rogan *et al.* 1986ab) showed that concentrations of DDE and the many PCB congeners were typically elevated in the milk of mothers, particularly among those (43 percent of the sample) who were giving birth to their first child. Thus consumption of mother's milk could be an important source of early exposure to pesticides and PCBs for the child. Median PCB levels went down from 1.77 ppm (fat basis) at birth to 1.02 ppm at 18 months. The corresponding decline for DDE was from 2.43 ppm to 1.29 ppm. This indicates that the mother's stock of PCBs and DDE was being transferred rather efficiently to the child.

EXAMPLES OF MULTIMEDIA/MULTIPATHWAY EXPOSURES ILLUSTRATED WITH SPECIFIC POLLUTANTS

In the remainder of this section, we will focus in more detail on several examples of pollutants for which multimedia investigations are necessary. One clear example is lead, exposures to which can occur in food (lead solder in cans), drinking water (lead solder in plumbing fixtures), ambient air (from lead in gasoline), or housedust (from tracked-in soil or crumbling leaded paint). One of the few VOCs for which multimedia investigations are necessary is chloroform, to which exposures can occur from drinking chlorinated water, and from inhalation and skin absorption during showers or baths in chlorinated water. An example of a semivolatile compound (or group of compounds) having two or more important routes of exposure is PCBs. An example of a pesticide that requires multimedia investigations is chlorpyrifos, a wide-spectrum pesticide used both externally to the residence (as a termiticide) and inside the residence.

Multimedia Exposure to Lead

Lead is a relatively soft, malleable metal with an atomic weight of 207. In addition to its elemental state, it can also exist as a cation in salts, with a +2 or +4 valence. Naturally occurring lead is found in the earth's crust at concentrations less than 10 ppm and is found at low levels in water, soil, plants, animals and humans. Today lead is used in the manufacture of storage batteries, cable covering, pigments, ammunition, solder, plumbing fixtures, weights and steel (USEPA 1986f).

Childhood lead poisoning has emerged as perhaps the most serious environmental threat to the well-being of the nation's children. The U.S. Public Health Service Centers for Disease Control of the Department of Health and Human Services (USDHHS 1991) Stategic plan for the elimination of childhood lead poisoning concludes that: (1) the effects of exposure to even moderate amounts of lead are more pervasive and long-lasting than previously thought, (2) significant impairment of intelligence and neurobehavioral functions are being reported at increasingly lower levels of lead in blood, and (3) millions of children in the U.S. have blood lead levels in this new range of concern (USDHHS 1991). The ATSDR estimated that between 3 and 4 million children in the U.S. (17 percent of all children) had blood lead levels above 15 ug/dl in 1984 (ATSDR 1990).

Dietary absorption of lead is dependent on age, nutritional status, and food intake (Mahaffey 1985). Adults absorb approximately 7 to 15 percent of dietary lead (ATSDR 1990, Rabinowitz *et al.* 1980), whereas, infants and young children absorb approximately 25 to 50 percent (USEPA 1986f, Ziegler *et al.* 1978). Pulmonary absorption of lead has been estimated at 30 to 50 percent (Fergusson 1986, USEPA 1986f).

Lead is initially distributed to blood (where it is bound to erythrocytes) and soft tissues, then slowly redistributes and accumulates in bone. Skeletal lead accounts for approximately 70 percent of the body burden in children and 95 percent in adults. The half-life of lead in blood and soft tissues is 30 to 40 days and in bone is approximately 25 years. Lead is predominantly excreted in the urine (about 95 percent), and to a lesser degree in bile, sweat, milk, hair and nails (ATSDR 1990, Landrigan *et al.* 1985, USEPA 1986f, NYSLIC 1992). The level of lead in blood reflects relatively recent exposure compared with levels found in teeth and bone, which accumulate over time. However, at steady state, lead cycles between blood and bone, such that blood levels may reflect the body burden of lead, but probably underestimates total body burden (ATSDR 1990a).

Exposure to lead has been extraordinarily difficult to characterize, due to the variety and inhomogeneity of sources. In earlier decades, major exposures were due to inhalation of air containing leaded gasoline exhaust, ingestion of lead in canned food, ingestion or inhalation of leaded paint flakes, ingestion of drinking water from plumbing fixtures containing lead, and ingestion of baby formula. All five of these routes of exposure have been regulated, and exposures have been greatly reduced. Lead was removed from gasoline in the 1970s, and air levels have decreased from 5 to 10 $\mu g/m^3$, to levels well below 0.1 $\mu g/m^3$ today. Use of lead solder in canned foods has been largely prohibited. Leaded paint has been banned for interior use. Leaded plumbing fixtures have been largely replaced. And infant formula must meet strict guidelines on the amount of lead included.

However, despite these advances, too many children still have high blood lead levels. Roadside soils are still contaminated due to half a century of lead fallout from gasoline. Soils near the home are contaminated with washed-out lead from exterior house paint. Both these soil sources contribute to exposure through track-in to the homes, followed by resuspension of the resulting house dust or ingestion by toddlers.

Lead is an important contaminant in house dust, which has been studied by Roberts and colleagues for a decade (Roberts *et al.* 1987, 1993). Their work has succeeded in demonstrating that careful attention to cleaning, taking off shoes on entering households, and reducing the use of carpets and rugs can reduce lead and pesticide concentrations in house dust by factors of 90 percent or more.

Case Study: 3-City Lead Abatement Project. In 1986, Section 111(b)(6) of the Superfund Amendments and Reauthorization Act (SARA) mandated that the USEPA conduct soil lead abatement projects in up to three U.S. cities. The fundamental purpose was to determine whether abatement of lead in soil could reduce the lead in the blood of inner city children. The official name of the resulting study is the Urban Soil Lead Abatement Demonstration Project, but it is also known less formally as the Three City Lead Study (USEPA 1986).

The three cities selected for the project were Boston, Cincinnati and Baltimore. In each city, families in a neighborhood were recruited and sampled during pre-abatement and post-abatement phases. Lead was measured in blood, soil, dust, drinking water, and paint. Questionnaires were administered to determine activity patterns, eating habits, and time budgets.

Research teams from each city included state and local health offices, participating universities, and USEPA regional offices, under the coordination of the USEPA's Office of Research and Development.

Soil lead levels were quite different among the three cities, with Boston being highest and the other cities much lower.

Three forms of intervention were employed: (1) soil abatement, (2) dust removal, and (3) paint stabilization. Soil abatement consisted of excavation and removal, followed by replacement with clean soil (<50 μg Pb/g). Dust was vacuumed and wet mopped, with occasional replacement of rugs and upholstered furniture. Interior dust abatement was not performed in Baltimore. Paint abatement consisted of scraping and smoothing the walls and then repainting.

In Cincinnati, soil was sampled at two depths: the top 2 cm and from 13 to 15 cm deep. If the average of both samples exceeded 500 µg/g, or if the average of the top layer alone exceeded 500 µg/g, the soil was replaced.

- *Boston results*. The Boston study retained 149 of the 152 children enrolled, although 22 children moved to a new location during the study. Only children with blood levels between 7 and 24 µg/dL were included in the study. The key findings from the Boston study were:
  - The median pre-abatement concentration of lead in soil was high, averaging 2,400 µg/g with few samples below 1000 µg/g
  - Abatement reduced the median concentration to about 150 µg/g
  - Lead loadings in house dust were reduced by about a factor of three (from 50 to 75 µg/m^2, to 16 to 21 µg/m^2) after intervention, although one year later they had risen again to about half of the preabatement levels
  - The reductions of lead in both soil and house dust persisted for at least two years
  - Blood lead levels were reduced by 1.86 µg/dL at 10 months after soil abatement, and
  - Additional reductions in blood lead of about 2.0 µg/dL (relative to non-abated groups) were observed at 22 months post-abatement.

The intervention was significant for two of the three study groups. Following the first phase of their study, the Boston group concluded:

> "... this intervention study suggests that an average 1,856 ppm reduction in soil lead levels results in a 0.8–1.6 mg/dL reduction in the blood lead levels of urban children."

Following the second phase of the Boston study, the group concluded (Aschengrau *et al.* 1994):

> "The blood lead reduction in Phase II was somewhat greater than that in Phase I. The combined results from both phases suggest that a soil lead reduction of 2,060 ppm is associated with a 2.2 to 2.7 mg/dL decline in blood lead levels."

- *Baltimore results*. The Baltimore study recruited 472 children, of whom 185 completed the study. The attrition was due to logistical problems resulting in extensive delays between enrollment and abatement. Key findings of the Baltimore study included the following:
  - The soil levels were low, averaging between 500 and 700 µg/g
  - The absolute reduction of lead in soil due to abatement was therefore small, on the order of 400 µg/g
  - Measurements for blood lead showed no decreases relative to the control groups, and
  - Other than exterior paint, there was no control of other sources such as interior paint and house dust.

The Baltimore group stated its conclusions as follows:

> "Statistical analysis of the data from the Baltimore Lead in Soil Project provides no evidence that the soil abatement has a direct impact on the blood lead level of children in the study."

- *Cincinnati results*. The Cincinnati study recruited 307 children, of whom 206 were included in the analysis. Key findings included:

- Soil lead levels ranged from about 300 µg/g in one neighborhood (Pendleton) to about 800 µg/g in Dandridge, and
- Postabatement soil concentrations were below 100 µg/g.

Blood lead concentrations decreased in both the treatment and control areas, and the researchers concluded there was no evidence for an effect of soil abatement on blood lead concentrations.

- *Synthesis*. Although at first glance, the studies appear to have come to opposite conclusions, a closer look suggests that the conclusions are not necessarily inconsistent. One major difference between the Boston study and the other two was that a much greater reduction in soil abatement was made possible by the very high starting values in Boston. The reduction in Boston was more than 2,200 µg/dL, compared with only about 200 to 700 µg/dL in the other two cities. Thus if the reduction in blood lead were linear with the reduction in soil lead, we could expect to see an effect in these two cities only 10 to 30 percent that attained in Boston. Also, a considerable effort was made in Boston to remove or reduce lead in household dust, and this was not done in Baltimore. Although house dust was reduced in Cincinnati, there was evidence of recontamination through contact with exterior sources; in Boston, recontamination was slight.

  Another consideration is the mobilization of blood lead from bone. Autopsy data show that as much as 60 to 70 percent of the lead in a child's body is stored in bone (Barry 1981). (This is even higher—90 to 95 percent—for adults). In adults, lead isotope studies suggest that 30 to 65 percent of the circulating lead is due to skeletal lead (Flegal and Smith 1995, Smith *et al.*1996, Gulson *et al.* 1995). Thus a sudden drop in environmental lead concentrations would not lead to an equally large drop in blood lead levels due to the remobilization of bone lead.

  These findings emphasize the complexity of attempts to abate lead exposure. One conclusion is that soil abatement alone may not be adequate to produce significant declines in blood lead. However, the findings also suggest that soil abatement, *coupled with effective reduction of lead in household dust*, can achieve significant reductions in blood lead, at least in areas with relatively high lead levels in soil.

Multimedia Exposure to Chloroform

Like lead, chloroform exposures can occur in all media and through all major routes, although this fact has only recently been generally accepted. A detailed review of multimedia exposures to chloroform is contained in Wallace (1997). Chloroform is created in water following chlorination. Hypochlorous acid is formed and reacts with organic precursors (*e.g.*, humic or fulvic acids), forming chloroform (Rook 1974, 1977, Schnoor *et al.* 1979, Amy *et al.* 1987). Chloroform has been found to cause cancer in rodents (NCI 1976), and use of chlorinated water has been linked to increased rates of bladder cancer (Cantor *et al.* 1987). Chloroform was discovered in the blood of New Orleans residents in the early 1970s (Dowty *et al.* 1975). It was assumed that exposure was through ingestion of chlorinated drinking water. Therefore, the Safe Drinking Water Act of 1974 resulted in a limit for all trihalomethanes of 100 µg/L in treated water (USEPA 1979). This limit was set assuming ingestion of 2 L of water per day, and did not consider any other pathways of exposure.

Inhalation. In the early 1980s it became evident that inhalation of chloroform might be as or more important than ingestion. These findings were provided by the TEAM Study of VOCs in New Jersey (Pellizzari *et al.* 1987, Wallace 1987, Wallace *et al.* 1982, 1984ab, 1986ab), which documented an average personal exposure of 4 to 8 μg/m^3 over the three seasons of the study. Assuming 10 m^3 of air respired per day (Adams 1993, Layton 1993), this amounted to 40 to 80 μg/day intake. Since the mean water levels varied from 17 μg/L in winter to 70 μg/L in summer, assuming 1 L of water drunk per day (Ershow and Cantor 1989), this would amount to an average daily intake of 17 to 70 μg, comparable to the average inhalation intake.

The finding of elevated chloroform levels in indoor air led to several studies to determine the sources. A series of experiments using model showers have been performed (Andelman 1985a, 1985b, Andelman *et al.* 1986, 1987) to determine the contribution of showers to airborne levels of VOCs contained in the water. These experiments have established that the fraction of a chemical volatilized depends on the physical characteristics of the chemical (particularly the diffusion rate across the water-air interface), the water temperature, the size of the droplet, and the extent to which water remains in the tub without draining. Experiments with chloroform indicated that about 50 percent of the chloroform in the water would volatilize under typical shower conditions (Andelman *et al.* 1986). The end result of these calculations has been to confirm that showers with chlorinated water are an important source of airborne chloroform. Andelman (1985a, 1985b) estimated that inhalation exposure during showers might be comparable to ingestion of one to six liters of drinking water a day. McKone (1987) developed a three-compartment model (shower, rest of bathroom, rest of house) for indoor air concentrations of chloroform due to showering and checked his model against the TEAM Study results. He found that his model was consistent with the TEAM measurements, which did not include measurements in the bathroom during the shower or bath. Thus the exposure during the shower should be *added* to the TEAM Study findings of 4 to 8 μg/m^3 exposure at all locations other than the bathroom.

In 1990, Jo *et al.* (1990a, 1990b) reported that nineteen 10-minute showers using water at chloroform concentrations ranging from 12.9 to 40.0 μg/L resulted in air concentrations in the shower stall ranging from 69 to 327 μg/m^3. The average ratio of chloroform concentrations in air to those in water was 6.8 L/m^3.

Wilkes *et al.* (1990) reported on a model (MAVRIQ) developed to simulate the air concentration of organic chemicals volatilizing from the water supply. The model predicts a peak concentration of about 80 μg/m^3 following a 7.7-min shower using water with a contaminant concentration of 20 μg/L and a volatilization efficiency of 63 percent. A peak concentration of about 28 μg/m^3 was predicted from washing clothes, with a volatilization efficiency of 90 percent estimated for this use. The authors indicated that the unnamed contaminant had characteristics much like TCE, although their transfer efficiencies were actually those estimated for radon (Prichard and Gesell 1981). For example, their estimated shower volatilization efficiency of 63 percent differs from the estimate of 90 percent reported by Giardino.

Wilkes *et al.* (1992) applied the model to a three-person household in which one person took the first shower and left the house, while the second person took the second shower, stayed home, and washed clothes. Contaminant levels in the water are 20 μg/L. Peak concentrations reached 90 μg/m^3 during the first shower and 120 μg/m^3 during the second shower. The second person breathed in about twice as much of the contaminant (58 μg/day compared with about 27 μg/day for the first person).

In a review chapter, Andelman (1990) reported on the results of an experiment in which chloroform had been injected into a laboratory shower at 1840 μg/L, reaching a peak concentration of about 95,000 μg/m^3 after 60 minutes. This experiment indicated that the chloroform volatilization rate falls off with time: 82 percent, 70 percent, and 62 percent at times of 10, 30, and 50 min, respectively.

These studies have led to the conclusion that the ultimate source of most of the chloroform in indoor air in most homes is evaporation from chlorinated water. Major uses of water in the home include showers, baths, toilets, washing clothes, and washing dishes. Sources other than chlorinated water include chlorinated scouring powders and bleaches. A small study using a chlorine-containing scouring powder (Wallace *et al.* 1987a) suggested that such consumer products could be powerful sources of short-term exposure to chloroform. Certain pesticides and rodenticides have been found to emit chloroform (Sheldon *et al.* 1988a,b), and multiple regressions on chloroform concentrations in air and breath of TEAM Study participants often showed increased exposures among persons reporting the use of pesticides (Wallace 1987).

Ingestion. In the New Jersey TEAM Studies, measurements of several types of food using FDA methods showed that chloroform was present in sodas (particularly colas) and dairy products, including milk, cheese, and ice cream (Entz *et al.* 1982).

Dermal Absorption. Jo *et al.* (1990a, 1990b) and Weisel *et al.* (1990) measured exhaled breath of volunteers following showers with and without wet suits to estimate the fraction of total dose absorbed through the skin. Six individuals took 13 normal showers and stood inside the shower stall (with rubber raincoats and boots) during 13 additional showers. During the normal shower, they were exposed both by inhalation and by skin absorption, but the other condition allowed only inhalation exposure. Breath samples were taken both before and after each shower. Breath levels following both types of showers increased linearly with the tap water concentration, but the breath levels were roughly double following the normal shower. The authors interpreted this as evidence that skin absorption during showers (0.22 μg/kg/day for water with a chloroform level of 24.5 μg/L) would be roughly equivalent to the inhalation exposure (0.24 μg/kg/day).

Later, McKone (1993) and Weisel and Shepard (1994) carried out experimental and theoretical studies indicating that the dose of chloroform absorbed through the skin during a shower might be comparable to the dose inhaled. Several studies of indoor swimming pools (Levesque *et al.* 1994, Lindstrom *et al.* 1995) indicate that both inhalation and dermal absorption can provide substantial amounts of chloroform.

Integrated Assessment of Exposure. From the above discussion, it is clear that an adequate assessment of exposure to chloroform would need to consider all three major pathways (inhalation, ingestion, dermal absorption). Although specific assessments would need to consider local geographic and seasonal factors, a general assessment would include consideration of the chloroform levels in drinking water, concentrations in indoor air, length of time taking showers and baths, ingestion of sodas and dairy products, and model calculations to determine inhalation and dermal absorption during showers and baths.

A pilot study carried out in Japan (Toyoda *et al.* 1987) measured the intake of chloroform and other chlorinated hydrocarbons from air, food, and tap water for seven Japanese housewives on three consecutive days in each of two seasons. For all seven subjects in winter, and six out of seven in summer, food contributed the most to their daily intake, accounting for about half of the daily intake of 37 μg in the summer and 70 percent of the smaller winter intake of 14 μg. Air and water intakes were roughly equivalent to each other, but were 4 to 5 times as high in the summer as in the winter.

This study was later enlarged to 30 subjects. Again three consecutive days of air, water, and diet samples were collected, this time from 20 Nagoya residents (10 in summer and 10 in winter) and 10 Yokohama residents (winter). Duplicate portions of all meals consumed were collected and analyzed using a purge-and-trap procedure. Food intakes alone are reported in Toyoda *et al.* (1990). No difference in THM intake was seen between the two cities in the winter. The mean (SD) chloroform concentrations for the Nagoya summer, Nagoya winter, and Yokohama winter food samples were 24.0 (21.6), 18.2 (14.3), and 17.2 (16.4) ng/g (ppb), respectively, leading to estimates of mean daily dietary intake of 47.5 (31.7), 37.3 (21.4), and 35.3 (21.8) μg, respectively. The mean concentrations of the other three THMs never exceeded 0.6 ng/g, with estimated mean daily intakes of 0 to 1.2 μg. The mean daily chloroform intakes from food ranged over about one order of magnitude, from 11 to 108 μg, for the 30 housewives, with an overall grand mean (SD) of 40.0 (25.4) μg. The grand mean daily intakes (SD) of BDCM, CDBM, and bromoform were 0.6 (0.5), 0.3 (0.3), and 0.9 (1.3) μg, respectively. The single case of a Nagoya housewife having a food concentration higher than 100 ppb (the permitted level of THM in drinking water in Japan) was found to be associated with an unusually high indoor air chloroform concentraton of 16.5 $\mu g/m^3$; the authors suggested that the high indoor air concentration may have transferred to food or drink (they did not report on the chloroform concentration in drinking water).

Body Burden. Another way to do an integrated assessment is to measure chloroform in body fluids (breath, blood, or mothers' milk), since body burden naturally integrates exposure through all pathways.

- *Breath.* Early studies of environmentally relevant breath levels of VOCs were carried out by Krotoszynski *et al.* (1977, 1979), who studied 54 healthy nonsmoking volunteers using collection in Teflon bags followed by concentration on Tenax followed by GC-MS analysis. Although the 1977 publication on 28 of the subjects included no mention of chloroform, the 1979 report on all 54 subjects reported that chloroform was prevalent (found in 42 percent of subjects) in exhaled breath. The geometric mean was 23.3 $\mu g/m^3$, with a one-sigma range of 7.4 to 73.7 $\mu g/m^3$. These values seem extremely high compared to subsequent measurements of exhaled breath by others. The authors report low backgrounds of 0.1 to 0.2 $\mu g/m^3$ chloroform in the spirometric system employed, so the high breath values do not seem to be due to contamination at some point with chlorinated water. All subjects lived in the Chicago area in the mid-1970s. Chloroform concentrations in drinking water at that time are unavailable.

The largest existing data set on chloroform concentrations in the body has been provided by the TEAM Study measurements of exhaled breath. About 800 persons, representing a total population of about 800,000, have provided more than 1,250 breath samples. Table 7-2 displays the population-weighted frequency distributions of chloroform in breath for the target populations in the TEAM studies.

Weisel and Shepard (1994) measured air, water, and breath chloroform at an indoor swimming pool on five occasions. The single subject swam for 30 minutes on three occasions, rested in the water for the same length of time on one occasion, and stayed near

Table 7-2. Daily doses (average values) estimated using indirect methods

	First day		Second day	
	μg/kg	Percent	μg/kg	Percent
Inhalation	4.8	23.4	2.4	43.6
Dermal	4.1	20.0	0.9	16.4
Hand/oral	11.6	56.6	2.2	40.0
TOTAL	20.5	100	5.5	100

the pool but out of the water for 30 minutes on the final occasion. Breath samples were collected on Tenax cartridges using a portable sampling apparatus described by Raymer (1990b) for three hours after the subject left the pool area. The chloroform concentrations in the pool water varied from 30 to 150 μg/L, with the air concentrations varying from 23 to 120 μg/m^3. Pre-exposure breath concentrations were less than 2 μg/m^3 on all occasions, rising to 15 to 25 μg/m^3 2.5 minutes after completing the swimming periods, but only to 11 μg/m^3 after the poolside exposure period. The authors found evidence that dermal exposure had occurred, but had insufficient data to estimate the relative contributions of the dermal and inhalation routes.

Levesque *et al.* (1994) measured air, water, and breath chloroform values for 11 swimmers over seven consecutive days at an indoor pool. On each day, the swimmers exercised for 55 minutes at 45 percent of maximum capacity. (On one day, the intensity was increased to 65 percent of maximum). Also, on one day, scuba diving suits were worn to investigate the influence of dermal absorption alone on body burden. Breath values were measured immediately before, during, and after the exercise period. Pool chloroform concentrations were increased each day, from 158 μg/L to over 550 μg/L. Indoor air concentrations also increased, from 507 ppb to 1,630 ppb. Mean breath concentrations went from 100 ppb to about 1,000 ppb. On the day the scuba equipment was used, breath values for nine swimmers averaged only 23 percent of the breath values observed previously at similar air and water concentrations. Therefore, the authors concluded that dermal absorption contributed about 23 percent of the total exposure. Assuming 30 L/min average respiration at 45 percent maximum capacity, the authors calculated that a one-hour exposure in a pool with a typical level of chloroform (160 μg/L) would result in a total dose of about 65 μg/kg/day (50 μg/kg/day from inhalation and 15 μg/kg/day from dermal absorption). This value is about 140 times the dose from a 10-minute shower (0.46 μg/kg/day) and 93 times the dose from ingesting 2 L of tap water at a concentration of 24 μg/L. This result shows that swimming at high intensity in indoor pools can be an important source of exposure to chloroform.

A very recent study of two Olympic-level swimmers in an indoor pool resulted in considerably higher estimates of the dermal contribution (Lindstrom *et al.* 1995). After a two-hour exercise period at very high intensity in a pool containing water at 82°F and chloroform levels of about 70 μg/L, the swimmers had breath concentrations of 340 and 370 μg/m^3, more than double the indoor air concentrations of about 150 μg/m^3. If one assumes that at equilibrium we breathe out about a third of the chloroform we breathe in, then dermal absorption accounted for 6 to 7 times the inhalation uptake. The inhalation uptake could be estimated by assuming 30 L/min respiration rate of 150 μg/m^3 for 120 minutes, or about 540 μg. Adding the dermal contribution of 6 to 7 times that, we arrive at an estimate of close to 4 mg chloroform uptake, or about 100 times the uptake from a shower. For repeated training sessions over many days, one would expect such a high concentration to elevate levels in deep compartments of the body, such as adipose tissue. Supporting this idea, both swimmers had initial breath concentrations of about 3 μg/m^3, which is about 10 times higher than average concentrations noted in the TEAM Study, suggesting elevated long-term body burdens.

In this study, a series of very precise breath concentrations were measured, using a single-breath method developed by Pleil and Lindstrom (1995a). The subject breathes directly into a 1 L evacuated cylinder through a straw-like attachment; after wasting the first (dead space) portion of his breath, the subject closes his lips around the tube and opens the valve on the cylinder to allow collection of the second (alveolar) portion. The cycle time is reduced to approximately one minute. The new method allows for immediate collection following exposure, thus documenting the maximum breath (and therefore blood) concentration attained during the exposure period. Also, a much more finely detailed picture of the decay curve during the first few minutes of very rapid decay can be achieved. A drawback of the method is that the subject does not breathe in clean air, which allows for the possibility of contamination of the sample with "dead space" air. Partly to alleviate this, a sequential analytical approach was developed (Pleil and Lindstrom 1995b). In the first step, only the carbon dioxide level in the sample is quantified. This allows an estimate of the amount of dead space air included in the sample and therefore a quantitative adjustment of the subsequent trace-level determinations to estimate more nearly the alveolar air concentration. In the second step, major endogenous polar VOCs such as acetone and isoprene are quantitated, as well as trace quantities of exogenous polar organics. Finally, the third step allows quantitation of trace quantities of nonpolar organics including the THMs.

- *Blood.* Ashley *et al.* (1994) measured the four individual THMs, together with 28 other VOCs, in the blood of approximately 1,000 people as part of the Third National Health and Nutrition Examination Survey (NHANES), conducted over 4 years between 1988 and 1992. An isotope-dilution purge-and-trap procedure was used (CDC 1990). The levels found in blood appear to be in good agreement with the breath concentrations measured in the TEAM Studies. Both sets of data suggest that a large percentage of the U.S. population is exposed to chloroform, but that very large exposures (*e.g.*, the equivalent of parts per million levels in air or mg/L quantities in drinking water) are rare.
- *Mothers' milk.* As a lipophilic chemical, chloroform should collect in adipose tissue and fatty fluids such as mothers' milk. Chloroform was identified in all eight samples of mothers' milk collected from nursing mothers in New Jersey hospitals (Pellizzari *et al.* 1982). The method employed was purging at 50°C with helium (Michael *et al.* 1980). These samples were part of a larger group of 42 samples collected from nursing mothers in two New Jersey hospitals and from three other hospitals in Pennsylvania, Louisiana, and West Virginia (Erickson *et al.* 1980). Chloroform was detected in 40 of 42 breast milk samples at levels ranging from 0.1 to 65 ng/mL.
- *Risk.* Putting the various exposure and body burden measurements of chloroform together with estimates of potency, it is possible to obtain a rough estimate of cancer risk. Using these methods, chloroform has been identified as one of three VOCs of highest risk (Wallace 1991).

Dietary Exposure to DDT and PCBs Using Body Burden Analysis

DDT. DDT (1,1,1-trichloro-2,2-bis-(p-chlorophenyl) ethane) was one of the most widely used chemicals for controlling insect pests on agricultural crops and for controlling insects that carry such diseases as malaria and typhus (ATSDR 1989). DDT was first used by the military in 1941 to protect American troops and was released for use by the general public in 1945. Dusts and sprays containing DDT drastically reduced the incidence of malaria and despite heavy exposure, there were no verifiable symptoms of poisoning or illness among the troops (ATSDR 1989). DDT was banned from use in the United States in 1972 due to concerns regarding its environmental persistence rather than adverse health effects in humans. Traces of DDT and DDE (1,1-dichloro-2,2-bis (p-chlorophenyl) ethylene) are still found in soil and some agricultural commodities. Because DDT has been sprayed on people, domestic animals, buildings, agricultural crops and forests, it is found widely distributed in the environment. The persistence of DDT and DDE in the environment is mainly due to the fact that these chemicals are virtually insoluble in water and are resistant to chemical, microbial and metabolic degradation (WHO 1979).

General population exposure to DDT and its metabolites DDD and DDE (referred to as total DDT) occurs primarily via the ingestion of food containing trace amounts. Contributions via air and water are much lower (Table 7-2). Although contaminant levels in food have declined, there are measurable amounts in many commodities. DDE was the most frequently occurring pesticide (25 percent incidence in 468 samples) residue found in the total diet study in 1982 to 1984 (the most recent year for which data are available) (Gunderson 1988). Because of their extreme persistence in the environment, residues will continue to be present for decades (ATSDR 1989). The estimated dietary intake of total DDT in the United States was 420 µg/day in 1970, and 8 µg/day in 1974 (Coulston 1985).

Recent market basket surveys indicate that dietary intake of total DDT was 2.4 µg/day in 1980 (Gartrell *et al.* 1985) and 2.2 ug/day in 1981 (Gartrell *et al.* 1986). In a total diet study conducted by the U.S. Food and Drug Administration in 1982, foods representative of the "total diet" of the U.S. population were analyzed. Daily dietary intakes of total DDT were found for different age groups (Gunderson 1988) as follows:

- 6-to-11-month (9 kg) infant, 0.9 ug/day
- 2-year-old (13 kg) infant, 1.3 ug/day
- 25-to-30-year-old (60 kg) female, 1.5 ug/day, and
- 25-to-30-year-old (76 kg) male, 2.5 ug/day.

These studies verify the continuing low level dietary exposure of the general public to total DDT. The acceptable daily intake (ADI) of DDT from food established by WHO/FAO is 5 µg/kg/day or 350 µg/day for an adult (WHO 1979).

Presence of DDT in Biological Tissues and Breastmilk. DDT, DDE, and DDD have very high octanol-water partition coefficients (5.8 × 10:f. sup 5:ef., 5 × 10:f. sup 5:ef, and 1 × 10:f. sup 6:ef., respectively) indicating a strong tendency to partition to fatty tissue. In the Second Health and Nutrition Examination Survey (NHANES II), DDT and its metabolites were detected in the general U.S. population in 99 percent of the samples of blood serum of people aged 12 to 74 years from 64 locations in the U.S. The National Human Adipose Tissue Survey (NHATS) detected DDT and metabolites in 100 percent of the adipose tissue specimens.

In addition to blood and adipose tissue, breastmilk contains total DDT because of the relatively high fat content of milk and due to the continued low level exposure of the mother via her diet. For these reasons, human milk contains higher concentrations of DDT and metabolites compared to concentrations found in cow's milk, infant formula or infant foods. Infant exposure via cow's milk, infant formula and infant foods is negligible (Rogan and Gladen 1991, Rogan *et al.* 1991). As a result, breast-fed infants receive higher dietary exposures than children who are not breast-fed. Breastmilk concentrations of total DDT are shown in Table 7-1. Although the levels of total DDT have declined since the early studies in the 1950s to 1960s (means of 120 to 170 µg/L), mean and median concentrations in the most recent studies (22 to 88 µg/L) are still elevated (Rogan *et al.* 1986ab, Sheldon *et al.* 1985, Shecter 1989, Quinby *et al.* 1965). Exposure of an infant to typical total DDT concentrations in breastmilk results in dietary intakes of 2.1 to 8.6 µg/day (assumes 7.2 kg infant ingests 700 ml breastmilk per day).

DDT and PCBs. The North Carolina Breast Milk Project, begun in 1978 (Rogan and Gladen 1985), has followed a birth cohort of more than 800 children of nonoccupationally-exposed mothers. In the study, PCBs and DDE in milk and other fluids were measured throughout the course of lactation and assessments were made of infant growth, morbidity and development in the children (Rogan and Gladen 1985). The details of the study are discussed later in this chapter. No attempts have been made to assess cancer incidence in children or adults exposed to DDE in breastmilk.

PCBs. PCBs (polychlorinated biphenyls) are a class of chlorinated hydrocarbons whose commercial production began in 1930 (WHO 1979). PCBs were used primarily in capacitors, transformers and other electrical equipment as coolants and lubricants. PCBs were well suited for this purpose since they are chemically inert and resistant to degradation. The manufacture of PCBs stopped in the United States in October 1977 because of evidence that PCBs accumulate in the environment and may cause health hazards for humans (ATSDR 1989). Because of their characteristic stability and lipophilicity and due to their improper disposal, PCBs have accumulated in the environment and in organisms even though they are no longer manufactured (Kimbrough 1987). PCBs are not readily degraded in the environment nor are they easily metabolized and excreted by mammals (Kimbrough 1985). They are predominantly stored in adipose tissue in humans, although measurable amounts have also been detected in serum and breastmilk.

Environmental exposures in the nonoccupationally exposed population are associated with dietary exposures, especially from the consumption of contaminated fresh water fish (Kreiss *et al.* 1981, Humphrey 1983, Fiore *et al.* 1989). However, as with DDT, it is possible that another route of exposure for toddlers is ingestion of soil or house dust.

The concentrations of PCBs in adipose tissue and in blood can be used as biological indicators of exposure and can be used to estimate the concentration of PCBs in breastmilk. Similar to PCE, PCBs are not readily metabolized, are very lipophilic and, therefore, have a high potential for excretion via breastmilk. Unlike PCE, exhalation is not a route of removal of PCBs from the body. The bioaccumulation of individual PCB congeners in animals was correlated with the chemicals' properties (such as partition coefficients) and the degree of chlorination as well as on susceptibility to enzymatic degradation. Unmetabolized PCBs are stored in the fat and are mobilized upon lactation. Elimination in milk is proportional to the level and time of exposure, the lipid content of the diet and milk, the rate of lactation and number of offspring. There is some indication that prenatal exposure to PCBs is more important to developmental delay than breastmilk exposure (NYSDOH 1992). The evaluation of the PCBs in human milk is very complicated since the majority of studies in experimental animals have been performed with commercial mixtures, whose composition varies from that of the mixtures of congeners present in human milk (Jensen and Slorach 1991).

Serum PCB levels in the nonoccupationally-exposed U.S. population range from about 3.6 to 19 ug/l (ATSDR 1989). A recent extensive survey in New Bedford, Mass (Miller *et al.* 1991) found an arithmetic mean serum PCB concentration of 5.9 μg/L among males and 5.7 μg/L among females in 840 random samples collected in 1984 through 1987. Among 110 people in the New Bedford area with known exposure to PCBs via ingestion of contaminated seafood, a mean serum PCB concentration of 13.3 μg/L was reported (Miller *et al.* 1991, ATSDR 1989). Typical concentrations of PCBs identified in breastmilk are shown in Table 7-2. There are no guidelines or standards in the U.S. which indicate an acceptable level of PCBs in breastmilk. A guideline developed by Health and Welfare, Canada advised consultation between general practitioner and mother when the whole milk level of PCB exceeds 50 μg/L (Frank *et al.* 1988). However, it is unclear how women would know the PCB concentrations of their breast milk. The New York State Health Planning Commission (1977) recommended that mothers whose dietary habits regularly include more than two meals per week of fish from water known to be contaminated with PCBs should be instructed to consult with their physicians and consider the advisability of terminating breastfeeding at the end of three months. High levels of PCBs were reported in the breastmilk of Inuit women from the Arctic, Quebec, Canada (Dewailly *et al.* 1989) due to the markedly higher consumption of fish and marine mammals contaminated with PCBs. The 24 Inuit women had milk containing a mean PCB concentration of 111.3 μg/L (range of 16 to 514 μg/L) compared to 48 Caucasian controls whose milk had a mean PCB concentration of 28.4 μg/L (range of 5 to 115 μg/L). Serum samples were not collected and no attempts were made to assess the impact of the PCB contaminated milk on infant health.

Multimedia Exposure to Chlorpyrifos

Vaccaro *et al.* (1991, 1996) estimated inhalation exposure and dosimetry of pesticides in the home, as well as dermal exposure (*e.g.*, skin contact with contaminated surfaces resulting from children's behavior patterns) and incidental oral ingestion (*e.g.*, hand-to-mouth behavior of infants). Although the study is based on the broadcast application of an emulsifiable concentrate formulation by a commercial applicator, the resulting dermal and incidental ingestion exposures may be similar to those that occur following consumer use of other products that result in wide-spread aerosol dispersement and residues on carpets or rugs.

In the Vaccaro *et al.* (1991) study, Dursban LO was used to treat carpet in multiple rooms in two model homes. Post-application, the treated surface was allowed to dry for two hours with normal ventilation and then adult volunteers performed activity patterns with all doors and windows closed.

Eight rooms, four from each of two homes (bedrooms and living rooms, all carpeted, unoccupied and without any furniture) were included in the study. For each house, three rooms were designated for measuring exposure from various activity patterns (*e.g.*, adult volunteers simulating infant activities such as crawling, sitting and lying down) and one room for physicochemical measurements of the rate of dissipation of chlorpyrifos from the treated carpet (*e.g.*, rate of evaporation, amount of dislodgeable residues). Room air concentrations post-application were measured in each house, in two of the three "activity" rooms as well as in the single "physicochemical" room. The total combined surface area from both homes treated with chlorpyrifos was 1,338 ft^2 (Vaccaro 1991). The treated surface area of individual rooms ranged from 113 to 366 ft^2. The average surface area treated per room (*i.e.*, 1,338 ft^2/8 rooms) was approximately 167.3 ft^2.

Total daily dose for the two days of exposure following broadcast application to carpet was estimated for infants (*i.e.*, 10.2 kg, one-year-old) on the basis of environmental measurements and biological monitoring data from six male adult volunteers. Daily (24-hour period) inhalation exposure/dose for children was estimated from a four-hour TWA measurement of chlorpyrifos in air. Oral exposure/dose to children resulting from hand-to-mouth behavior was estimated from hand rinses conducted with adults who has been simulating children's activity patterns (crawling, lying, *etc.*) during a four-hour activity period. Dermal exposure/dose was estimated from measurements of dislodgeable residues that were obtained using "simulated dermal contact" (*i.e.*, a weighted sled) with contaminated surfaces.

Thus, the exposure/dose was estimated by means of both indirect (*i.e.*, passive dosimetry and environmental fate measurements) and direct methods (*i.e.*, biomonitoring). The indirect methods for estimating potential inhalation, dermal and incidental oral) involved the use of several measurement techniques:

- Rate of evaporation was determined from the decline of residues on gauze pads and aluminum foil
- Dislodgeable residues were measured from residues accumulated on a denim-covered block dragged across the floor
- Air samples were taken periodically over a 48-hour period after application (measurements were made at 0, 1, 2, 4, 8, 12, 24, and 48 hours post application), and
- Incidental oral exposure was estimated from hand rinses of volunteer adults simulating children's behavior patterns.

As noted above, a biomonitoring method was also used to measure the daily absorbed dose of chlorpyrifos directly. This method involved measurement of the chlorpyrifos metabolite 3,5,6-trichloro-2-pyridinol in the urine of six adult volunteers, wearing only bathing suits, who performed certain "infant activities" (*e.g.*, sitting and playing with blocks, lying on their back and abdomen, crawling on hands and knees and walking) during a four-hour "activity period." Since the adult volunteers did not actually simulate hand-to-mouth behavior as part of their infant activity pattern simulation, the levels of urinary excretion of the chlorpyrifos metabolite did not include material likely to be absorbed by this route. Consequently, a measure of the total direct exposure/dose estimate was obtained by adding the hand rinse data to the biomonitoring data. Thus, the direct method involved the following techniques:

- Adult volunteers who simulated infant activity patterns
- Hand rinses collected following the activity patterns described above
- Measurement of urinary levels of 3,5,6-trichloro-2-pyridinol, and
- Measurement of blood cholinesterase inhibition.

The results of these direct and indirect exposure assessment methods were then used to estimate the total potential daily dose in exposed children (10.2 kg). The results are summarized as follows:

- Loss of residues from surfaces:
  - Loss of chlorpyrifos from gauze in 48 hours was 32 percent in House #1 and 16 percent in House #2
  - Loss from aluminum foil in 48 hours was 49 percent in House #1 and 93 percent in House #2, and

- Chlorpyrifos residues on both gauze and aluminum foil increased from 24 to 48 hours which was attributed to settling of aerosol particles.
- Air Concentrations:
  - The highest concentration was found to be 28 μg/m^3 in one of the activity rooms 8-hours post application
  - The 4-hour TWA air concentration (T4) measured during the "activity period" ranged from 9.4 to 19.9 μg/m^3 with an average of 15.35 μg/m^3
  - Airborne chlorpyrifos at 48 hours post application ranged from 3.2 to 5.1 μg/m^3
  - The 4-hour TWA air concentration of chlorpyrifos, *i.e.*, 14.5 μg/m^3, was assumed to represent exposure during an active period (8 of 24 hours) and the average air concentration of the T8, T12 and T24 measurements, *i.e.*, 9.46 μg/m^3, was assumed to represent exposure during a rest period (16 of 24 hours), and
  - The average inhalation dose for children on the first day post-application (24-hour period) was estimated to be 4.8 μg/kg body weight.
- Dermal exposures from dislodgeable residues:
  - Based on 4-hour drag values, dislodgeable residues from the carpet were 29.3 μg/ft^2 in House #1 and 15.1 μg/ft^2 in House #2 with an average of 22.2 μg/ft^2
  - Using a percutaneous absorption rate of 3 percent, mean total dose via the dermal route was 4.1 μg/kg for a 10.2 kg child, and
  - The time-averaged transfer coefficient (percent of residue transferred to the body as determined by dragging a piece of cloth with weight) was 0.35 percent for House #1 and 0.27 percent for House #2.
- Oral exposures:
  - Oral/hand contributions averaged 11.64 μg/kg, based on 4-hour activity patterns, and
  - The oral/hand route of exposure resulted in the greatest contribution to the estimated total dose. See Table 7-1 for daily doses (average values).
- Day-1 dose estimates using direct methods:
  - The *maximum* estimated dose of chlorpyrifos, based on urinary metabolite and hand rinse data of an adult male volunteer was 43 μg/kg, and
  - The *average* estimated total (multi-route) dose for a child, scaling from the adult estimate (on the basis of body weight and surface area), was 27 μg/kg.

In summary, the direct (biomonitoring + hand/oral[1]) and indirect (inhalation + dermal + hand/oral) estimates of infant exposure during the first day after application were 27 and 21 μg chlorpyrifos per kg body weight, respectively.

Multimedia Exposure to PAHs

PAHs are a group of compounds with multiple benzene ring structures. Three- and four-ring compounds are generally in the vapor phase at normal ambient temperatures, while five- and six-ring compounds are normally attached to particles. A number of PAHs

[1] As stated previously, because the adult volunteer's infant activity patterns did not include hand-to-mouth behavior, it was necessary to account for the potential contribution of incidental ingestion of chlorpyrifos by conducting hand rinses at the end of the activity period.

are considered probable human carcinogens, with benzo(*a*)pyrene as the most potent of these. PAHs are created by combustion. Major routes of exposure include inhalation of air containing diesel or woodsmoke particles and ingestion of fried, grilled, smoked, or charcoal-broiled foods.

A field study—the Total Human Exposure Environmental Study (THEES)—was conducted in Phillipsburg, New Jersey, a small town with a grey-iron pipe foundry as a major source of benzo(*a*)pyrene emissions (Waldman *et al.* 1991). Fourteen persons in eight homes collected 24-h personal and indoor particle (PM_{10}) samples; concurrent outdoor samples were collected at three locations throughout the city and at the foundry. In addition, each person prepared a "second plate" of food identical to that eaten at each meal. The sampling period was 14 consecutive days during January-February 1987 (no personal samples during this period), January 1988, and September 1988. Participants kept a daily diary of their activities and locations.

A great seasonal difference in airborne benzo(*a*)pyrene concentrations was noted, with the winter personal and indoor means of 1.0 to 1.2 ng/m^3 well above those in summer (0.14 to 0.18 ng/m^3). The mean personal concentrations averaged over winter and summer ranged from about 0.4 to 2.0 ng/m^3 for the 14 participants. The food intake of benzo(*a*)pyrene also varied considerably, from a weeklong average intake of 1 to 1,149 ng/day. The mean intake was 123 ng/day. Assuming alveolar respiration rates of about 10 m^3 per day, even the wintertime mean inhalation intake of benzo(*a*)pyrene (10–12 ng/day) would be only about 10 percent of the intake from food.

A more recent study of personal exposures to airborne benzo(*a*)pyrene for 15 residents of Zagreb, Croatia found similar winter-summer differences, with a mean of 4.34 ($\pm$ 3.07) ng/m^3 in winter and 0.13 ($\pm$ 0.08) ng/m^3 in summer (Sisovic *et al.* 1996). The higher wintertime personal exposures correspond to higher ambient levels in winter due to home heating fuel combustion. However, both THEES and the Croatian study noted the effect of cooking and cigarette smoke on elevating PAH concentrations.

Benzo(*a*)pyrene contents of bacon and hamburger patties cooked by different methods were measured (Creighton *et al.* 1993). Microwave ovens and broiler ovens produced no benzo(*a*)pyrene, but pan frying and especially charcoal broiling produced up to 4 ng benzo(*a*)pyrene for a serving (3 slices) of bacon and up to 40 ng per serving (60 g) of hamburger. Additionally, hamburgers from McDonalds (cooked on metal grills) and Burger King (cooked over a flame) were analyzed with the result that the Whopper had 34 times the benzo(*a*)pyrene level of the Big Mac (195 ng *versus* 5.7 ng).

A study of nine families with incomes below the poverty level in urban and rural North Carolina was undertaken in the winter of 1994–95 (Wilson *et al.* 1996). Each family included a preschool child. Smokers were excluded from the study. Twenty-four-hour samples of indoor and outdoor air were collected using a quartz fiber filter and XAD-2 resin cartridge to collect both particle and vapor phase PAHs. Air exchange rates were monitored using a perfluorotracer (PFT). House dust was measured using an HVS3 vacuum sampler. Yard and entryway soil were also collected, the latter by vigorous banging of a welcome mat that had previously been provided to the family. 24-h urine samples were also collected. Food samples were collected using the duplicate plate method.

The 21 target PAH compounds averaged 4,670 ng/m^3 in the rural homes and 1,890 ng/m^3 in the urban homes, compared to outdoor concentrations of 218 and 277 ng/m^3, respectively. Much of the indoor concentration was contributed by naphthalene, the major constituent of mothballs. The nine carcinogenic PAHs had equivalent indoor and outdoor levels of about 2 ng/m^3 in the rural homes, but higher concentrations of 9 ng/m^3 indoors and 6 ng/m^3 outdoors in the urban homes. Food concentrations of the 21 target PAHs averaged 18.7 ppb for the adults and 14.5 ppb for the children. Concentrations of the nine carcinogenic PAHs were 0.359 ppb for the adults and 0.129 ppb for the children. House dust concentrations of the 21 target PAHs averaged 3.96 ppm and, of the carcinogenic PAHs, 1.57 ppm.

Assuming standard inhalation and ingestion rates, the authors calculated that for adults and children alike, inhalation was the most important route of exposure to total PAH. For adults, inhalation was also the most important route (73 percent of the total) for the carcinogenic fraction. However, for children, ingestion of food (40 percent) and of dust (38 percent) were the most important routes of exposure to carcinogenic PAH.

CONCLUSIONS AND IMPLICATIONS FOR THE FUTURE

Studies of total exposure began in the 1980s and have been extremely important in identifying new important source of exposure. One conclusion that can be drawn from these studies is the importance of *small nearby* sources. These include paints, cigarettes, deodorizers, carpets, glues, adhesives, house dust, solvents, cleansers, and many other everyday objects. Although such sources are miniscule compared to major sources, such as industry and automobiles, the efficiency of delivery can be 10,000 times greater than the more distant industrial sources.

A second lesson from these studies is the importance of body burden measurements in complementing and extending environmental measurements. Without measuring the breath of smokers, it would have been impossible to discover the true extent of their exposure to benzene, styrene, and other agents. Similarly, the slow bioaccumulation of DDT, PCBs, and other semivolatile agents in body fat and breast milk makes a single body burden measurement the equivalent of integrating years of monitoring environmental exposure.

With increasing knowledge of these sources comes greater awareness of actions that can reduce our exposure. In the case of VOCs, we can try to locate products with reduced emissions, and make sure that we use such products under conditions of increased ventilation. Exhaust fans in kitchens and bathrooms can be effective in reducing particle and chloroform concentrations, respectively.

To reduce exposure to lead and pesticides, particularly among infants and toddlers, a number of actions are possible (Roberts *et al.* 1993). Bare floors contain as much as 10 times less house dust than rugs or carpets. If one does not wish to replace rugs or carpets, removing shoes at the entrance will greatly reduce track-in contributions to house dust. Even welcome mats and walk-off mats have been shown to reduce the amount of dirt tracked in. Vacuuming is generally ineffective (about 5 to 15 percent of the total mass of dust is picked up), although if the rug can be turned over and vacuumed from the other side, a series of three such flips with repeated vacuuming on each side has been shown to pick up as much as 85 percent of the total mass of house dust.

In lesser-developed countries, information campaigns on the importance of ventilation and development of a simple chimney system may help reduce the exposure of women and children to particles and PAHs produced by burning biomass fuels.

REFERENCES

Adams, W.C. 1993. Measurement of breathing rate and volume in routinely performed daily activities. Final report. Contract A033-205. California Air Resources Board. Sacramento, California.

AIHA (American Industrial Health Association). 1994. AIHA White Paper on Total Human Exposure. Statement adopted by Board of Directors, December 21, 1994. http://www.aiha.org/papers/totalexp.html.

Akland, G., T.D. Hartwell, T.R. Johnson, and R. Whitmore. 1985. Measuring human exposure to carbon monoxide in Washington, DC and Denver, CO during the winter of 1982–83. Environ. Sci. Technol. 19:911–918.

Amy, G.L., P.A. Chadik, and Z.K. Chowdhury. 1987. Developing models for predicting trihalomethane formation potential and Kinetics. JAWMA 79:89.

Andelman, J.B. 1985a. Human exposures to volatile halogenated organic chemicals in indoor and outdoor air. Environ. Health Perspect. 62:313–318.

Andelman, J.B. 1985b. Inhalation exposure in the home to volatile organic contaminants of drinking water. Sci. Total Environ. 47:443–460.

Andelman, J.B. 1990. Total exposure to volatile organic compounds in potable water. Chapter 20 in: Ram N.M., R.F. Christman, and K.P. Cantor (eds.). Significance and treatment of volatile organic compounds in water supplies. Lewis Publishers. Chelsea, Michigan.

Andelman, J.B., S.M. Meyers, and L.C. Wilder. 1986. Volatilization of organic chemicals from indoor uses of water. Pages 434–440 in: Lester, J.N., R. Perry, and R.M. Sterritt (eds.). Chemicals in the environment. Selper Ltd. London, United Kingdom.

Andelman, J.B., L.C. Wilder, and S.M. Myers. 1987. Indoor air pollution from volatile chemicals in water. Pages 37–42 in: Seifert, B., H. Esdorn, M. Fischer, H. Rüden, and J. Wegner (eds.). Indoor Air '87. Proceedings of the 4th International Conference on Indoor Air Quality and Climate. August 17–21. Volume 1. Institute for Soil, Water, and Air Hygiene. Berlin (West).

Aschengrau, A., A. Beiser, D. Bellinger, D. Copenhafer, and M. Weitzman. 1994. The impact of soil lead abatement on urban children's blood lead levels: Phase II results from the Boston lead-in-soil demonstration project. Environ. Res. 67:125–148.

Ashley, D.L., M.A. Bonin, L. Carkinali, J.M. McCraw, and J.V. Wooten. 1994. Blood concentrations of volatile organic compounds in a nonoccupationally exposed U.S. population and in groups with suspected exposure. Clin. Chem. 40:1401–1404.

ATSDR (Agency for Toxic Substances and Disease Registry). 1989. Toxicological Profile for DDT, DDE and DDD. December. U.S. Public Health Service. Atlanta, Georgia.

ATSDR (Agency for Toxic Substances and Disease Registry). 1990. Toxicological Profile for Lead. (ATSDR/TP-88/17). U.S. Public Health Service. Atlanta, Georgia.

Barkley J., J. Bunch, J.T. Bursey, N. Castillo, S.D. Cooper, J.M. Davis, M.D. Erickson, B.S.H. Harris III, M. Kirkpatrick, L. Michael, S.P. Parks, E. Pellizzari, M. Ray, D. Smith, K. Tomer, R. Wagner, and R.A. Zweidinger. 1980. Gas chromatography mass spectrometry computer analysis of volatile halogenated hydrocarbons in man and his environment—a multimedia environmental study. Biomed. Mass Spectrometry 7:(4)139–147.

Barry, P.S. 1981. Concentrations of lead in the tissues of children. Brit. J. Indus. Med. 38:61–71.

Brown, V.M. and D.R. Crump. 1996. Volatile organic compounds. In: Berry, R.W., V.M. Brown, S.K.D. Coward, D.R. Crump, M. Gavin, C.P. Grimes, D.F. Higham, A.V. Hull, C.A. Hunter, I.G. Jeffery, R.G. Lea, J.W. Llewellyn, and G.J. Raw (eds.). Indoor air quality in homes: Part I. The building research establishment indoor environment study. Construction Research Communications. London, United Kingdom.

Cantor, K.P., R. Hoover, P. Hartge, T.J. Mason, D.T. Silverman, R. Altman, D.F. Austin, M.A. Child, C.R. Key, L.D. Marrett, M.H. Myers, A.S. Narayana, L.I. Levin, J.W. Sullivan, G.M. Swanson, D.B. Thomas, and D.W. West. 1987. Bladder cancer, drinking water sources, and tap water consumption: a case-control study. JNCI 79:1269–1279.

CDC (Centers for Disease Control). 1990. Protocol for measurement of volatile organic compounds in human blood using purge/trap gas chromatography mass spectrometry. U.S. Public Health Service. Atlanta, Georgia.

Coburn, R.F., R.E. Forster, and P.B. Kane. 1965. Considerations of the physiological variables that determine the blood carboxyhemoglobin concentration in man. J. Clin. Invest. 43:1098–1103.

Conner, J.M., G.B. Oldaker III, and J.J. Murphy. 1990. Method for assessing the contribution of environmental tobacco smoke to respirable particles in indoor microenvironments. Environ. Technol. 11:189–196.

Coulston, F. 1985. Reconsideration of the dilemma of DDT for the establishment of an acceptable daily intake. Reg. Toxicol. Pharmacol. 5:332–383.

Creighton, P.J., P.J. Lioy, and A. Greenburg. 1993. Effects of cooking methodology on benzo(a)pyrene exposure from "home-cooked" bacon and hamburgers and "fast-food chain" hamburgers. J. Exposure Anal. Environ. Epidemiol. 2(Supplement 2):27–44.

Dewailly, E., A. Nantel, J-P. Weber, and F. Meyer. 1989. High levels of PCBs in breast milk in Inuit women from arctic Quebec. Bull. Environ. Contam. Toxicol. 43:641–646.

Dowty B., D. Carlisle, J.L. Laseter, and J. Storer. 1975. Halogenated hydrocarbons in New Orleans drinking water and blood plasma. Science 187:75–77.

Duan, N. 1982. Models for human exposure to air pollution. Environ. Internat. 8:305–309.

Duan N. 1985. Application of the microenvironment monitoring approach to assess human exposure to carbon monoxide. Rand Corporation. Report R-322-EPA. Santa Monica, California.

Duan N. 1988. Estimation of microenvironment concentration distributions using integrated exposure measurements. Presented at the Research Planning Conference on Human Activity Patterns. May 10–12, 1988. Las Vegas, Nevada.

Duan N. 1991. Stochastic microenvironment models for air pollution exposure. J. Exposure Anal. Environ. Epidemiol. 2:235–257.

Entz, R., K. Thomas, and G. Diachenko. 1982. Residues of volatile halocarbons in food using headspace gas chromatography. J. Agric. Food Chem. 30:846–849.

Erickson, M.D., B.S.H. Harris III, E.D. Pellizzari, K.B. Tomer, R.D. Waddell and D.A. Whitaker. 1980. Acquisition and chemical analysis of mothers' milk for selected toxic substances. U.S. Environmental Protection Agency. EPA 560/13-80-029. Washington, D.C.

Ershow, A.G. and K.P. Cantor. 1989. Total water and tapwater intake in the United States: population-based estimates of quantities and sources. FASEB. Bethesda, Maryland.

Fellin, P. and R. Otson. 1993. Seasonal trends of volatile organic compounds (VOCs) in Canadian homes. Pages 339–343 in: Jaakola, J.J.K., R. Ilmarinen, and O. Seppänen (eds.). Indoor Air '93: Proceedings of the 6th International Conference on Indoor Air Quality and Climate Volume 1. Helsinki University of Technology. Espoo, Finland.

Fiore, B., H. Anderson, L. Hanrahan, L. Olsen, and W.C. Sonzogni. 1989. Sport fish consumption and body burden levels of chlorinated hydrocarbons: a study of Wisconsin anglers. Arch. Environ. Health 44:82–88.

Frank, R., J. Rasper, M.S. Smout, and H.E. Braun. 1988. Organochlorine residues in adipose tissue, blood and milk from Ontario residents, 1976–1985. Can. J. Pub. Health. 79:150–158.

Fergusson, J.E. 1986. Lead: petrol lead in the environment and its contribution to human blood lead levels. Sci. Total Environ. 50:1–54.

Flegal, A.R. and D.R. Smith. 1995. Measurement of environmental lead contamination and human exposure. Rev. Environ. Contam. Toxicol. 143:1–45.

Gartrell, M., J. Craun, D. Podrebarac, and E. Gunderson. 1985. Pesticides, selected elements, and other chemicals in infant and toddler total diet samples, October 1978–September 1979. JAOAC 68:842–861.

Gartrell, M.J., J.C. Craun, D.S. Podrebarac, and E.L. Gunderson. 1986. Pesticides, selected elements, and other chemicals in infant and toddler total diet samples. October 1980-March 1982. JAOAC 69:138–145.

Gephart, L.A., J.G. Tell, and L.R. Triemer. 1994. Exposure factors manual. J. Soil Contam. 3:47–117.

Goldstein, B.D., R.G. Tardiff, S.R. Baker, G.F. Hoffnagle, D.R. Murray, P.A. Catizone, R.A. Kester, and D.G. Caniparoli. 1992. Valdez air health study. Alyeska Pipeline Service Company. Anchorage, Alaska.

Gordon, S., L.A. Wallace, E.D. Pellizzari, and H. O'Neill. 1988. Breath measurements in a clean-air chamber to determine 'wash-out' times for volatile organic compounds at environmental concentrations. Atmos. Environ. 22:2165–2170.

Gordon, S.M., D.V. Kenny, and T.J. Kelly. 1992. Continuous real-time breath analysis for the measurement of half-lives of expired volatile organic compounds. J. Exposure Anal. Environ. Epidemiol. Supplement 1:41–54.

Gordon, S.M., P.J. Callahan, D.V. Kenny, and J.D. Pleil. 1995. Direct trace analysis of volatile organic compounds in air using ion trap mass spectrometers with filtered noise fields. Presented at the International Symposium on Field Screening Methods for Hazardous Wastes and Toxic Chemicals. Las Vegas, Nevada.

Gulson, B.L., K.R. Mahaffey, K.J. Mizon, M.J. Korsch, M.A. Cameron, and G. Vimpani. 1995. Contribution of tissue lead to blood lead in adult female subjects, based on stable lead isotope methods. J. Lab. Clin. Methods 125:703–712.

Gunderson, E.L. 1988. FDA total diet study. April 1982–April 1984. Dietary intakes of pesticides, selected elements, and other chemicals. JAOAC 71:1200–1209.

Hartwell, T.D., C.A. Clayton, R.M. Ritchie, R.W. Whitmore, and H.S. Zelon. 1984. Study of carbon monoxide exposure of residents of Washington, DC and Denver, Colorado. Final report. Research Triangle Institute. Research Triangle Park, North Carolina. NTIS/PB84-184516.

Hartwell, T.D., C.A. Clayton, R.W. Michie, R.W. Whitmore, H.S. Zelon, S.M. Jones, and D.A. Whitehurst. 1986. Study of carbon monoxide exposure of residents of Washington, DC and Denver, CO. EPA-

600/S4-84-031. PB84-183516. Environmental Monitoring Systems Laboratory. Office of Research and Development. U.S. Environmental Protection Agency. Research Triangle Park, North Carolina. 400 pages.

Humphrey, H. 1983. Population studies of PCBs in Michigan residents. Chapter 21 in: D'Itri, F. and M. Kamrin (eds.). PCBs: Human and Environmental Hazards. Ann Arbor Science. Lansing, Michigan.

Immerman, F.W. and J.L. Schaum. 1990. Nonoccupational pesticide exposure study: Final report. EPA/600-3-90-003. Office of Research and Development. U.S. Environmental Protection Agency. Washington, D.C.

Jenkins, R.A., A. Palausky, R.W. Counts, C.K. Bayne, A.B. Dindal, and M.R. Guerin. 1996. Exposure to environmental tobacco smoke in sixteen cities in the United States as determined by personal breathing zone air sampling. J. Exposure Anal. Environ. Epidemiol. 6(4):473–502.

Jensen, A.A. and S.A. Slosach. 1991. Chemical contaminants in human milk. CRC Press. Boca Raton, Florida.

Jo, W.K., C.P. Weisel, and P.J. Lioy. 1990a. Routes of chloroform exposure and body burden from showering with contaminated tap water. Risk Anal. 10:575–580.

Jo, W.K., C.P. Weisel, and P.J. Lioy. 1990b. Chloroform exposure and the health risk associated with multiple uses of chlorinated tap water. Risk Anal. 10:581–585.

Johnson, T. 1983. A study of personal exposures to carbon monoxide in Denver, Colorado. Final Report. USEPA Contract Number 68-02-3755. Office of Research and Development. U.S. Environmental Protection Agency. Research Triangle Park, North Carolina.

Kelly, T.J., D.V. Kenny, C.W. Spicer, and G.M. Sverdrup. 1989. Continuous analysis of human breath using atmospheric pressure ionization MS/MS with a novel inlet design. In: Proceedings of the 1989 EPA/AWMA International Symposium on Measurement of Toxic and Related Air Pollutants. VIP-13. May 1989. Air and Waste Management Assoc. Pittsburgh, Pennsylvania.

Kimbrough, R. 1985. Laboratory and human studies on polychlorinated biphenyls (PCBs) and related compounds. Environ. Health Perspect. 59:99–106.

Kimbrough, R.D. 1987. Human health effects of polychlorinated biphenyls (PCBs) and polybrominated biphenyls (PBBs). Ann. Rev. Pharmacol. Toxicol. 27:87–111.

Krause, C., W. Mailahn, R. Nagel, C. Schulz, B. Seifert, and D. Ullrich. 1987. Occurrence of volatile organic compounds in the air of 500 Homes in the Federal Republic of Germany. Pages 102–106 in: Seifert, B., H. Esdorn, M. Fischer, H. Rüden, and J. Wegner (eds.). Indoor Air '87. Proceedings of the 4th International Conference on Indoor Air Quality and Climate. August 17–21. Volume 1. Institute for Soil, Water, and Air Hygiene. Berlin (West).

Kreiss, K., M. Zack, R. Kimbrough, L.L. Needham, A.L. Smrek, and B.T. Jones. 1981. Association of blood pressure and polychlorinated biphenyl levels. JAMA 245:2505–2509.

Krotoszynski, B.K., G. Gabriel, and H. O'Neill. 1977. Characterization of human expired air: A promising investigation and diagnostic technique. J. Chromatogr. Sci. 15:239–244.

Krotoszynski, B.K., G.M. Bruneau, and H.J. O'Neill. 1979. Measurement of chemical inhalation exposure in urban populations in the presence of endogenous effluents. J. Analyt. Toxicol. 3:225–234.

Landrigan, P., J. Froines, and K. Mahaffey. 1985. Body lead burden: Summary of epidemiological data on its relation to environmental sources and toxic effects. Pages 421–451 in: Mahaffey, K. (ed.). Dietary and environmental lead: Human health effects. Elsevier. New York, New York.

Layton, D.W. 1993. Metabolically consistent breathing rates for use in dose assessments. Health Phys. 64:23–36.

Lebret, E., H.J. Van de Weil, D. Noij, and J.S.M. Boleij. 1986. Volatile hydrocarbons in Dutch homes. Environ. Internat. 12(1–4):323–332.

Levesque, B., P. Ayotte, A. Leblanc, E. Dewailly, D. Prud'Homme, R. Lavoie, A. Sylvain, and P. Levallois. 1995. Evaluation of dermal and respiratory chloroform exposure in humans. Environ. Health Perspect. 102:1082–1087.

Lindstrom, A.B., V.R. Highsmith, T.J. Buckley, W.J. Pate, and L.C. Michael. 1994. Gasoline-contaminated ground water as a source of residential benzene exposure. J. Exposure Anal. Environ. Epidemiol. 4:183–195.

Lindstrom, A.B., J.D. Pleil, and D.C. Berkoff. 1995. Alveolar breath sampling and analysis to assess trihalomethane exposures during competitive swimming training. Environ. Health Perspect. 105:636–642.

Mahaffey, K.R. 1985. Factors modifying susceptibilty to lead toxicity. Pages 373–419 in: Mahaffey, K. (ed.). Dietary and environmental lead: Human health effects. Elsevier. New York, New York.

McKinney, J.D., L. Moore, A. Prokopetz, and D.B. Walters. 1984. Validated extraction and cleanup procedure for polychlorinated biphenyls and DDE in human body fluids and infant formula. JAOAC 67(1):122–129.

McKone, T.E. 1993. Linking a PBPK model for chloroform with measured breath concentrations in showers: implications for dermal exposure models. J. Exposure Anal. Environ. Epidemiol. 3(3):339–365.

McKone, T.E. and J. Knezovich. 1991. The transfer of trichloroethylene (TCE) from a shower to indoor air: Experimental measurements and their implications. JAWMA 41:832–837.

Michael, L.C., M.D. Erickson, S.P. Parks, and E.D. Pellizzari. 1980. Volatile environmental pollutants in biological matrices with a headspace purge technique. Anal. Chem. 52:1836–1841.

Miller, D., S. Condon, S. Kutzner, D. Phillips, E. Krueger, R. Timperi, V. Burse, J. Cutler, D. Gute. 1991. Human exposure to polychlorinated biphenyls in greater New Bedford, Massachusetts: A prevalence study. Arch. Environ. Contam. Toxicol. 20:410–416.

Nelson, P.R., D.L. Heavner, B.B. Collie, K.C. Maiolo, and M.W. Ogden. 1992. Effect of ventilation and sampling time on environmental tobacco smoke component ratios. Environ. Sci. Technol. 26:1909–1915.

NCI (National Cancer Institute). 1976. Report on carcinogenesis bioassay of chloroform. NTIS PB-264018. Springfield, Virginia.

NYSDOH. New York State Department of Health. 1992. Ambient Air Criteria Document: Polychlorinated Biphenyls (PCBs). Bureau of Toxic Substance Assessment. Albany, New York. 110 pages.

NYSLIC (New York State Joint Legislative Commission on Toxic Substances and Hazardous Waste. 1992. Hearing report: Lead contamination in New York State. March 1992. 51 pages.

New York State Health Planning Commission. 1977. Report of the ad hoc committee on health implications of PCBs in mothers' milk. Health Advisory Council. January 7. Albany, New York.

Ogden, M.W. and K.C. Maiolo. 1989. Collection and determination of solanesol as a tracer of environmental tobacco smoke in indoor air. Environ Sci Technol. 23:1148–1154.

Ott, W.R. 1982. Concepts of human exposure to air pollution. Environ. Internat. 7:179.

Ott, W.R. 1985. Total human exposure: an emerging science focuses on humans as receptors of environmental pollution. Environ. Sci. Technol. 19:880.

Ott, W.R. 1990. Total human exposure: Basic concepts, EPA field studies, and future research needs. JAWMA 40(7):966–975.

Pellizzari, E.D., T.D. Hartwell, B.S.H. Harris III, R.D. Waddell, D.A. Whitaker, and M.D. Erickson. 1982. Purgeable organic compounds in mothers' milk. Bull Environ. Contam. Toxicol. 28:322–328.

Pellizzari, E.D., R.A. Zweidinger, and L.S. Sheldon. 1985. Breath sampling. Pages 399–411 in: Fishbein, L. and I. O'Neill (eds.). Environmental carcinogens: Methods of analysis and exposure measurement. Volume 7. Some volatile halogenated hydrocarbons. IARC Publication Number 68. World Health Organization. Lyon, France.

Pellizzari, E.D., K. Perritt, T.D. Hartwell, L.C. Michael, R. Whitmore, R.W. Handy, D. Smith, and H. Zelon. 1987a. Total exposure assessment methodology (TEAM) study: Elizabeth and Bayonne, New Jersey; Devils Lake, North Dakota; and Greensboro, North Carolina. Volume II. Office of Research and Development. U.S. Environmental Protection Agency. Washington, D.C.

Pellizzari, E.D., K. Perritt, T.D. Hartwell, L.C. Michael, R. Whitmore, R.W. Handy, D. Smith, and H. Zelon. 1987b. Total exposure assessment methodology (TEAM) study: Selected communities in northern and southern California. Volume III. Office of Research and Development. U.S. Environmental Protection Agency. Washington, D.C.

Pellizzari, E.D., K.W. Thomas, J.H. Raymer, D.J. Smith, and S.D. Cooper. 1990. Measurements of exhaled breath using a new portable sampling method. Final Report. Contract Number 68-02-4544. U.S. Environmental Protection Agency. Research Triangle Park, North Carolina.

Pleil, J.D. and A.B. Lindstrom. 1995a. Collection of a single alveolar exhaled breath for volatile organic compounds analysis. Amer. J. Indust. Med. 27:109–121.

Pleil, J.D. and A.B. Lindstrom. 1995b. Measurement of volatile organic compounds in exhaled breath as collected in evacuated electropolished canisters. J. Chrom B: Biomed Applic. 665:271–279.

Prichard, H.M. and T.F. Gesell. 1981. An estimate of population exposures due to radon in public water supplies in the area of Houston, Texas. Health Phys. 41:599–606.

Quinby, G., J. Armstrong, and W. Durham. 1965. DDT in human milk. Nature 207:726–728.

Rabinowitz, M.B., J.D. Kopple, and G.W. Wetherill. 1980. Effect of food intake and fasting on gastrointestinal lead absorption in humans. Am. J. Clin. Nutrit. 33:1784–1788.

Ramsey, J.C. and M.E. Andersen. 1984. A physiologically based description of the inhalation pharmacokinetics of styrene in rats and humans. Toxicol. Appl. Pharmacol. 73:159–175.

Raymer, J.H., E.D. Pellizzari, S.D. Cooper, N.P. Castillo, and K.W. Thomas. 1990a. Evaluation of a pharmacokinetic model for volatile organic compounds in breath and of the application of the analytical method to polar VOCs. Final Report. EPA Contract Number 68-02-4544. U.S. Environmental Protection Agency. Research Triangle Park, North Carolina.

Raymer, J.H., K.W. Thomas, S.D. Cooper, D.A. Whitaker, and E.D. Pellizzari. 1990b. A device for sampling human alveolar breath for the measurement of expired volatile organic compounds. J. Analyt. Tox. 14:337–344.

Raymer, J.H., K.W. Thomas, E.D. Pellizzari, S.D. Cooper, D.J. Smith, P. Kizakevic, and D. Pugh. 1992a. Breath and blood measurements of individuals exposed to volatile organic chemicals: Elimination from the third compartment. Final Report. Contract Number 68-02-4544. Office of Research and Development. U.S. Environmental Protection Agency. Research Triangle Park, North Carolina.

Raymer, J.H., K.W. Thomas, E.D. Pellizzari, D.A. Whitaker, S.D. Cooper, T. Limero, and J.T. James. 1992b. A breath sampling device for measuring human exposure to volatile organic compounds in microgravity. Aviation Space Environ. Med. 65:353–360.

Roberts, J.W., M.G. Ruby, and G.R. Warren. 1987. Mutagenic activity of House Dust. Pages 355–367 in: Sandhu, S.S., D.M. De Marini, M.J. Mass, M.M. Moore, and J.L. Mumford (eds.). Short-term bioassays in the analysis of complex environmental mixtures. Volume V. Plenum Press. New York, New York.

Roberts, J.W., W.T. Budd, J. Chuang, and R.G. Lewis. 1993. Chemical contaminants in house dust: Occurrences and sources. Pages 27–32 in: Jaakola, J.J.K., R. Ilmarinen, and O. Seppänen (eds.). Indoor Air '93: Proceedings of the 6th International Conference on Indoor Air Quality and Climate. Volume 2. Helsinki University of Technology. Espoo, Finland.

Rogan, W.J. and B.C. Gladen. 1985. Study of human lactation for effects of environmental contaminants: The North Carolina breast milk and formula project and some other ideas. Environ. Health Perspect. 60:215–221.

Rogan, W.J. and B. Gladen. 1991. PCBs, DDE, and child development at 18 and 24 months. AEP 1(5):407–413.

Rogan, W.J., B.C. Gladen, J.D. McKinney, N. Carreras, P. Hardy, J. Thullen, J. Tingelstad, and M. Tully. 1986a. Polychlorinated biphenyls (PCBs) and dichlorodiphenyl dichloroethene (DDE) in human milk: Effects of maternal factors and previous lactation. AJPH 76:172–177.

Rogan, W.J., B.C. Gladen, J.D. McKinney, N. Carreras, P. Hardy, J. Thullen, J. Tinglestad, and M. Tully. 1986b. Neonatal effects of transplacental exposure to PCBs and DDE. J. Pediatr. 109:335–341.

Rogan, W.J., P. Blanto, C. Portier, and E. Stallard. 1991. Should the presence of carcinogens in breast milk discourage breast feeding? Reg. Toxicol. Pharmacol. 13:228–240.

Rook, J.J. 1974. Formation of haloforms during chlorination of natural water. J. Water Treat. Exam. 23:234.

Rook, J.J. 1977. Chlorination reactions of fulvic acids in natural waters. Environ. Sci. Technol. 11:478–482.

Schecter, A., P. Purst, C. Furst, H.-A. Meemken, W. Groebel, and D. Quoc Vu. 1989. Levels of polychlorinated dibenzodioxins and dibenzofurans in cow's milk and in soybean derived infant formulas sold in the United States and other countries. Chemosphere. 19:913–918.

Schnoor, J.L., Nitzschke, R.D. Lucas, and J.N. Veenstra. 1979. Trihalomethane yields as a function of precursor molecular weight. Environ. Sci. Technol. 13(9):1134–1141.

Sheldon, L.S., R.W. Handy, T.D. Hartwell, C. Leininger, and H. Zelon. 1985. Human exposure assessment to environmental chemicals—Nursing mothers study. U.S. Environmental Protection Agency and Research Triangle Institute. Research Triangle Park, North Carolina.

Sheldon, L.S., R.W. Handy, T.D. Hartwell, R.W. Whitmore, H.S. Zelon, and E.D. Pellizzari. 1988a. Indoor air quality in public buildings: Volume I. EPA 600/6-88/009a. NTIS PB 89-102503/AS. Office of Research and Development. U.S. Environmental Protection Agency. Washington, D.C.

Sheldon, L.S., H.S. Zelon, J. Sickles, C. Eaton, and T. Hartwell. 1988b. Indoor air quality in public buildings: Volume II. EPA 600/6-88/009b. NTIS PB 89-102511/AS. Office of Research and Development. U.S. Environmental Protection Agency. Research Triangle Park, North Carolina.

Sheldon, L., A. Clayton, B. Jones, J. Keever, and R. Perritt. 1992. Indoor pollutant concentrations and exposures. Final report June 21, 1989—October 1991. NTIS PB-100060. Research Division, California State Air Resources Board. Sacramento, California.

Sheldon, L., D. Whitaker, J. Keever, A. Clayton, and R. Perritt. 1993. Phthalates and PAHs in indoor and outdoor air in a southern California community. Pages 109–114 in: Jaakola, J.J.K., R. Ilmarinen, and O. Seppänen (eds.). Indoor Air '93: Proceedings of the 6th International Conference on Indoor Air Quality and Climate. Volume 3. Helsinki University of Technology. Espoo, Finland.

Sisovic, A., M. Fugas, and K. Sega. 1996. Assessment of human inhalation exposure to polycyclic aromatic hydrocarbons. J. Exposure Anal. Environ. Epidemiol. 6(4):439–447.

Smith, D.R., J.D. Osterloh, and A.R. Flegal 1996. Use of endogenous, stable lead isotopes to determine release of lead from the skeleton. Environ. Health Perspect. 104:60–66.

Smith, K.R. 1987. Biofuels, air pollution and health. Plenum Press. New York, New York.

Smith, K.R. 1993. Fuel combustion, air pollution exposure, and health: The situation in developing countries. Ann. Rev. Energy Environ. 18:529–566.

Teorell, T. 1937a. Kinetics of distribution of substances administered to the body. I. The extravascular modes of administration. Arch. Internat. Pharmacodynam. 57:205–225.

Teorell, T. 1937b. Kinetics of distribution of substances administered to the body. II. The intravascular mode of administration. Arch. Internat. Pharmacodynam. 57:226–240.

Thomas, K.W., E.D. Pellizzari, and S.D. Cooper. 1991. A canister-based method for collection and GC/MS analysis of volatile organic compounds in human breath. J. Analyt. Toxicol. 15:54–59

Thrall, K.D. and D.V. Kenny. 1996. Evaluation of a carbon tetrachloride physiologically based pharmacokinetic model using real-time breath-analysis monitoring of the rat. Inhalation Toxicol. 8:251–261.

Toyoda M., M. Kobayashi, K. Mitsura, A. Saito, Y. Uno, and K. Sakabe. 1987. Estimation of daily intake of chloroform by Japanese housewives. J. Food Safety 8:219–224.

Toyoda, M., K. Takgi, A. Tsurumizu, Y. Saito, H. Kikawa, T. Kawamura, and Y. Sakabe. 1990. Estimation of daily intake of four kinds of trihalomethane by Japanese housewives. J. Food Safety 10:191–200.

USDHHS (United States Department of Health and Human Services). 1991. Strategic plan for the elimination of childhood lead posisoning. Developed for the Risk Management Subcommittee. Committee to Coordinate Environmental Health and Related Programs. Public Health Service. Centers for Disease Control. February.

USEPA (U.S. Environmental Protection Agency). 1979. National interim primary drinking water regulations; control of THMs in drinking water; final rule. Office of Drinking Water. Fed Reg. November 29. 44(231):68624–68707.

USEPA (U.S. Environmental Protection Agency). 1986. Air quality criteria for lead. Draft final (EPA-600/8-83-028aF). Research Triangle Park, North Carolina.

Vaccaro, J.R. 1991. Personal communication. Industrial Hygiene Laboratory. Dow Chemical Company.

Vaccaro, J.R., R.J. Nolan, J.M. Hugo, J.L. Pillepich, P.G. Murphy, and M.J. Bartels. 1991. Evaluation of dislodgeable residues and absorbed doses of chlorpyrifos to crawling infants following indoor broadcast applications of chlorpyrifos-based emulsifiable concentrate. Laboratory Project Study ID: DECO-HEH2.1-1-182(95). The Dow Chemical Company. Midland, Michigan.

Vaccaro, J.R., R.J. Nolan, P.G. Murphy, and D.B. Berbrich. 1996. The use of a unique study design to estimate exposure to adults and children to surface and airborne chemicals. American Soceity for Testing Materials. ASTM STP 1287:166–183.

Waldman, J.M., P.J. Lioy, A. Greenberg, and J.P. Butler. 1991. Analysis of human exposure to benzo(a)pyrene via inhalation and food ingestion in the total human environmental exposure study (THEES). J. Exposure Anal. Environ. Epidemiol. 1(2):193–225.

Wallace, L.A. 1983. Carbon monoxide in air and breath of employees in an underground office. JAPCA 33:678–682.

Wallace, L.A. 1987. The TEAM Study: Summary and Analysis: Volume I. 1987. U.S. Environmental Protection Agency. EPA 600/6-87/002a. NTIS PB 88-100060. Washington, D.C.

Wallace, L.A. 1991. Comparison of risks from outdoor and indoor exposure to toxic chemicals. Environ. Health Perspect. 95:7–13.

Wallace, L.A. 1997. Human exposure and body burden for chloroform and other trihalomethanes. Critical Reviews in Environ. Sci. Technol. 27:113–194.

Wallace, L.A., R. Zweidinger, M. Erickson, S. Cooper, D. Whitaker, and E.D. Pellizzari. 1982. Monitoring individual exposure: Measurement of volatile organic compounds in breathing-zone air, drinking water, and exhaled breath. Environ. Internat. 8:269–282.

Wallace, L.A., E.D. Pellizzari, T.D. Hartwell, M. Rosenzweig, M. Erickson, C. Sparacino, and H. Zelon. 1984. Personal exposure to volatile organic compounds: I. Direct measurement in breathing-zone air, drinking water, food, and exhaled breath. Environ Res. 35:293–319.

Wallace, L.A., E.D. Pellizzari, T.D. Hartwell, R.W. Whitmore, C. Sparacino, and H. Zelon. 1986a. Total Exposure Assessment Methodology (TEAM) study: Personal exposures, indoor-outdoor relationships, and breath levels of volatile organic compounds in New Jersey. Environ. Internat. 12:369–387.

Wallace, L.A., E.D. Pellizzari, T.D. Hartwell, H. Zelon, C. Sparacino, R. Perritt, and R.W. Whitmore. 1986b. Concentrations of 20 volatile organic compounds in the air and drinking water of 350 residents of New Jersey compared with concentrations in their exhaled breath. J. Occup. Med. 28:603–608.

Wallace, L.A., E.D. Pellizzari, B. Leaderer, T.D. Hartwell, C. Sparacino, R.W Whitmore, L. Sheldon, H. Zelon, and R. Perritt. 1987a. The TEAM study: personal exposure to toxic substances in air, drinking water, and breath of 400 residents of New Jersey, North Carolina, and North Dakota. Environ, Res. 43:290–307.

Wallace, L.A., E.D. Pellizzari, B. Leaderer, T.D. Hartwell, R. Perritt, H. Zelon, and L. Sheldon. 1987b. Emissions of volatile organic compounds from building materials and consumer products. Atmos. Environ. 21:385–393.

Wallace, L.A., E.D. Pellizzari, T.D. Hartwell, R. Whitmore, R. Perritt, and L. Sheldon. 1988. The California TEAM study: breath concentrations and personal exposures to 26 volatile compounds in air and drinking water of 188 residents of Los Angeles, Antioch, and Pittsburg, California. Atmos. Environ. 22:2141–2163.

Wallace, L.A., E.D. Pellizzari, T.D. Hartwell, V. Davis, L.C. Michael, and R.W. Whitmore. 1989. The influence of personal activities on exposure to volatile organic compounds. Environ. Res. 50:37–55.

Wallace, L.A., E.D. Pellizzari, and S. Gordon. 1993. A linear model relating breath concentrations to environmental exposures: application to a chamber study of four volunteers exposed to volatile organic chemicals. J. Exposure Anal. Environ. Epidemiol. 3(1):75–102.

Wallace, L.A., T. Buckley, E.D. Pellizzari, and S. Gordon. 1995. Breath measurements as VOC Biomarkers: EPA's experience in field and chamber studies. Environ. Health Perspect. 104(Supplement 5):861–869.

Wallace, L.A., W.C. Nelson, E.D. Pellizzari, and J.H. Raymer. 1997. A four compartment model relating breath concentrations to low-level chemical exposures: Application to a chamber study of five subjects exposed to nine VOCs. J. Exposure Anal. Environ. Epidemiol. In press.

Weisel, C.P., W.K. Jo, and P.J. Lioy. 1990. Routes of chloroform exposure from showering with chlorinated water. In: Measurement of Toxic and Related Air Pollutants: Proceedings of the 1990 EPA/Air and Waste Management Association International Symposium. VIP-17. EPA 600/9-90/026. Air and Waste Management Association. Pittsburgh, Pennsylvania.

Weisel, C.P. and T.A. Shepard. 1994. Chloroform exposure and the body burden associated with swimming in chlorinated pools. Page 47 in: Wang, R.G.M. (ed.). Water contamination and health: Integration of exposure assessment, toxicology, and risk assessment. Marcel Dekker. New York, New York.

Whitmore, R.W. 1988. Design of surveys for residential and personal monitoring of hazardous substances. Atmos. Environ. 22(10):2077–2084.

WHO (World Health Organization). 1979. DDT and its derivatives. Environmental health criteria 9. Geneva, Switzerland.

Wilkes, C.R., M.J. Small, J.B. Andelman, N.J. Giardino, and J. Marshall. 1990. Air quality model for volatile constituents from indoor uses of water. Pages 783–788 in: Walkinshaw, D. (ed.). Indoor Air '90: proceedings of the 5th international conference on indoor air quality and climate. Volume 2. Toronto, Canada. Canada Mortgage and Housing Association. Ottawa, Canada.

Wilkes, C.R., M.J. Small, J.B. Andelman, N.J. Giardino, and J. Marshall. 1992. Inhalation exposure model for volatile chemicals from indoor uses of water. Atmos. Environ. 26A:2227–2236.

Wilson, A.L., S.D. Colome, and Y. Tian. 1993. California residential indoor air quality study. Volume I. Methodology and descriptive statistics. Gas Research Institute, Pacific Gas and Electric Company, San Diego Gas and Electric Company, and Southern California Gas Company. Integrated Environmental Services. Irvine, California.

Wilson, N.K., J.C. Chuang, and C. Lyu. 1996. Evaluation of field methods for estimation of exposure of children in low-income families to polycyclic aromatic hydrocarbons. In: Measurement of toxic and related air pollutants: Proceedings of the 1995 EPA/Air and Waste Management Association International Symposium. AWMA. Pittsburgh, Pennsylvania.

Zeigler, E.E., B.B. Edwards, R.L. Jensen, K.R. Mahaffey, and S.J. Fomon. 1978. Absorption and retention of lead by infants. Pediat. Res. 12:29–34.

III

Evaluating Sources in the Residential Environment

8

CONSUMER PRODUCTS AND RELATED SOURCES

Recent awareness of the generally higher levels of indoor residential exposures relative to outdoor for many chemicals has led to an increasing focus on indoor sources of exposures to chemicals in the residential environment. The TEAM studies have shifted the emphasis to indoor sources of exposures, rather than outdoor ambient sources as the primary contributor to indoor inhalation exposures, as discussed in Chapter 4. This is partly due to (1) the locally important sources of exposure (source terms) provided by some consumer products relative to outdoor levels, (2) time-activity data, which indicate that people spend about 90 percent of their time in indoor environments, and (3) the move over the last couple of decades to energy-efficient residences, which mean less outside air infiltration and more recirculation of indoor releases. Indoor agents or chemicals potentially include a great variety of compounds in consumer products available to the general public (Driver and Whitmyre 1996, Whitmyre *et al.* 1997). Indeed, for certain specific chemicals, consumer products are the main source of emissions and exposure in the indoor residential environment. Assessment of potential consumer exposures has been recognized as a key component of the overall risk evaluation process for consumer products.

Due to a focus of industry on product stewardship, and due to an increased awareness of liability issues, an increasing effort is being made by industry to estimate exposures to new products as part of the premanufacture, premarketing effort. For example, to support and confirm the safety of particular products, several studies of potential residential exposures have been conducted by industry (Hendricks 1970, Wooley *et al.* 1990, Gibson *et al.* 1991). Consumer products are subject to a variety of regulatory pressures from a human health and safety perspective. The Food Quality Protection Act (FQPA) of 1996 provides a legislative mandate for examining exposures to pesticides through multiple use patterns (*e.g.*, dietary and residential). The importance of exposure assessment for consumer products has been established by Proposition 65, and other State and Federal legislation, as mentioned in Section 1. Although the scope of this chapter is limited to assessment of exposures to chemicals in consumer products, it should be recognized that there are a variety of other sources of chemicals in the residence, such as construction materials, combustion sources, vehicles park in attached garages, and dry cleaning brought into the home. These are discussed in Chapter 4.

This chapter provides a discussion of the influence of physicochemical properties on exposure to chemicals in consumer products, and will emphasize the importance of use pattern information on characterizing exposures. This is followed with information

Residential Exposure Assessment, edited by Baker *et al.*
Kluwer Academic/Plenum Publishers, New York, 2000.

on monitoring approaches that can be used to characterize exposure to indoor-use and outdoor-use consumer products, and key issues in designing and conducting such monitoring studies. Next is a discussion on validating emissions models, chemical-specific adjustment of emissions data for surrogate chemicals, and exposure models relevant to inhalation, dermal, and ingestion exposure for consumer products.

ROLE AND SCOPE OF CONSUMER PRODUCT EXPOSURE ASSESSMENT

As discussed in Section 1, exposure assessment plays a pivotal role in risk assessment of chemicals in the residential environment. There are many reasons that an individual may wish to perform an exposure assessment for an indoor use consumer product. These reasons include, for example:

- As part of a determination of whether a proposed product is safe for use by the general public, or which of several proposed formulations of a future product is the safest
- As part of a registration package for a regulatory Agency (*e.g.*, as part of a pesticide registration effort for a new pesticide, or reregistration effort for an existing pesticide)
- In response to potential litigation (*e.g.*, under California's Proposition 65, toluene in nail polish and spray paint), and
- As part of an aggregate risk assessment, such as those required by the Food Quality Protection Act (FQPA) of 1996 for a pesticide.

There are two overriding reasons why accurate characterization of exposure to chemicals in indoor use consumer products may be necessary. First, it is important for consumer product manufacturers and regulators to appropriately characterize and quantify exposure. Failure to meet appropriate risk cutoffs at a screening level should be followed by successive iterations of the exposure assessment, as necessary, until the assessor has assured themselves that all reasonable and scientifically defensible methods have been utilized, and that exposure estimates are not excessively conservative.

Second, characterization of exposure can assist mitigation of risk by providing guidance for appropriate labeling, limiting the use patterns to those that produce acceptable exposures, using protective packaging (as in the case of use of childproof packing on ant bait traps), changing the delivery system (*e.g.*, using microencapsulation to reduce potential exposure), and exploring reformulation of the product such that the problematic chemical is not present in the formulation, is present in reduced amounts, or is less physically available for exposure. Examples of this include reduction in nitrosamines in baby pacifiers through production process changes, (2) reduction in exposures to dioxin in coffee filters by elimination of or changes in the bleaching process for paper products, (3) elimination of exposures to saccharin by substitution with a different artificial sweetening agent, and (4) control of potential sensitization to enzymes in laundry detergents by formulation of laundry detergent as large granules in order to minimize inhalable material that might be suspended in the air during dispensing into automatic washing machines by the consumer.

APPROACHES TO CONSUMER PRODUCT EXPOSURE ASSESSMENT

There is no single standard approach to estimating exposures to chemicals in consumer products. The USEPA is in the process of developing guidance on residential exposure assessment. A number of approaches of varying sophistication and varying data requirements are available for assessment of exposures to consumer product chemicals. There are a number of uncertainties in exposure assessment that can be reduced by methodological or data improvements (Whitmyre *et al.* 1992ab). However, since it would be overly time and resource-intensive to perform the most sophisticated modeling and/or monitoring procedures for every product assessment, including developing or characterizing the required input data for assessment at a detailed level, a "tiered" approach as discussed in Section 1 is recommended. In a tiered approach for an indoor use consumer product, a screening-level exposure assessment might be performed in which a series of simple assumptions could be made in conjunction with the simplest available model for exposure to produce a bounding estimate for exposure. If the resulting exposure estimate, when combined with the available toxicity data (*e.g.*, No-Observed-Adverse-Effect-Levels, or NOAELs) indicate that acceptable Margins of Safety (MOSs) are obtained, then the assessor can end the assessment of the product for the chosen scenario, as the upper bound of exposure has been demonstrated to be without human health consequence. If unacceptable MOSs or estimated excess lifetime cancer risks (if relevant) are obtained, then the assessor will need to more realistically represent exposures.[1] Refinements could include, as in the case of a spray product containing a volatile solvent, (1) use of "typical" use rate values in conjunction with an accepted and validated indoor air model, or (2) use of stochastic modeling in order to obtain the distribution of exposures that would reasonably be anticipated for the specific product and use scenario.

If necessary in cases where available data are limited, and if resources and time limitations permit, actual data may be collected on a key parameter to reduce uncertainties in the risk assessment and to prevent the need to make overly conservative assumptions. This might involve determining analytically the level of dislodgeable residues that can be obtained from a surface treated with pesticides[2], or the breathing zone concentration of a volatile chemical constituent released from the consumer product via collection of samples in a chamber study where the product is released and used under typical anticipated use and reasonable ventilation conditions.

The above description of the "tiered" approach is meant to be illustrative and not to indicate that a rigid procedure must be followed. In fact, for example, if there is

---

[1] An assessor should never end the exposure assessment at the screening level if an unacceptable risk is the result. The consequences of not refining the assessment using more realistic assumptions and better methods could include: (1) a scientifically indefensible decision not to proceed with development of a new product that might have had significant societal benefits and great profit potential for the manufacturer, (2) unnecessary rejection of registration for a product (such as in the case of an indoor consumer-use pesticide), and (3) increased settlement costs resulting from private party or government litigation (such as under Proposition 65 in California).

[2] The level of dislodgeable residues typically drives exposures for consumer-use pesticides applied indoors and outdoors because of the extensive potential for dermal contact and hand-to-mouth ingestion opportunities, particularly for young children.

a strong regulatory need to develop the most realistic measure of exposure for a given consumer product, the assessor may want to immediately proceed to at least a partial stochastic analysis (if the data support such an approach) using a validated exposure model that is not overly conservative. Ultimately, the exact nature of the different levels of the assessment and the basis for the approach and assumptions would vary from case to case. As mentioned in Section I of the book, such an approach saves valuable resources and allows prioritization of the potential exposures from a large number of exposure sources in a time-efficient manner.

Indoor-Use Products as Sources of Exposure

Almost any consumer product is a potential source of exposure to chemicals in the residence. Examples of the types of chemicals present in various products and their associated human exposure pathways are illustrated for some example consumer products in Table 8-1. Toluene is an example of a chemical commonly found in high concentrations in a variety of consumer products as a solvent, such as in nail lacquers and wall paints. By contrast, there are products which contain many chemicals in trace amounts that may find their way into the residential environment to the extent that these compounds can be detected in chamber studies or in indoor air if sufficiently sensitive sampling and analytical methods are employed.

As noted in Chapter 4, consumer products can act as sources of chemicals in the indoor residential environment. These include cleaning agents (Wallace *et al.* 1987), consumer use pesticide products (Wallace *et al.* 1987), floor and wall coverings (Wallace *et al.* 1987, De Bortoli *et al.* 1993, Clausen *et al.* 1993), adhesives and paint removers (Girman *et al.* 1986, Otson *et al.* 1981, Hodgson and Girman 1987), waxes, polishes, and detergents (alkanes, alkenes, alcohols, esters, terpenes) (Knoppel and Schauenberg 1987). As noted in Chapter 4, users of certain products such as mothballs, air fresheners, and other scented products may typically result in indoor/outdoor ratios of key constituents (*e.g.*, naphthalene, *p*-dichlorobenzene, limonene and fragrance components) up to 10:1 to 100:1 (Wallace *et al.* 1989, 1991a).

One example of a consumer product that contributes a multitude of organic compounds to the residential environment is carpet padding, which has undergone extensions emissions testing by the U.S. Consumer Product Safety Commission (CPSC). Using 96-hour small chamber studies, the CPSC detected over 100 VOCs being emitted from 17 samples of carpet padding, including bonded urethane, prime urethane, synthetic fiber, rubberized jute, and sponge rubber padding (Schaeffer *et al.* 1996).

A second example is carpets, which can be a major source of a large number of volatile organic compounds in indoor residential environments (Seifert *et al.* 1989, Bayer and Papanicolopoulos 1990, Pleil and Whiton 1990, Schroder 1990, Hodgson *et al.* 1992). These chemicals include 4-ethenylcyclohexene, styrene, 4-phenylcycohexene, alkyl benzenes, vinyl acetate, 2-ethyl-1-hexanol, propanediol, 2,2,4-trimethylpentane, formaldehyde, 1,1,1-trichloroethane, 1-butanol, toluene, glycol ethers, hexamethylcyclotrisiloxane, and 1,2-dichlorobenzene. An individual compound, 4-phenylcyclohexene (4-PCH) has been associated with the typical "new carpet" odor (Van Ert *et al.* 1987). This compound is a manufacturing byproduct present in the styrene-butadiene rubber (SBR) latex commonly used to bind the secondary backing of the carpet (Hodgson *et al.* 1992). The initial emission rate of 4-PCH from new carpeting was noted to be several times higher than the emission rate measured at 24 hours, and declined approximately 67 percent over a period of one week (Hodgson *et al.* 1992). After four weeks, the emission rate of

Table 8-1. Example consumer products, chemical components, and likely pathways of exposure

Product	Chemical component(s)	Role in product	Predominant exposure pathway(s)	Minor exposure pathway(s)	Comments	Exposure reference(s)
Nail lacquers	Toluene	Solvent	Inhalation	Dermal	Volatilization from nails in breathing zone	Curry *et al.* (1994)
Polyurethane cushions	1,1,1-Trichloroethane	Blowing agent	Inhalation	—	Volatilization from PU foam. Carpet or fabric cover acts as diffusion barrier	Schaeffer *et al.* (1996)
	Butylated hydroxytoluene	Anti-oxidant	Inhalation	—	Slow volatilization from PU foam	
	Siloxanes	Silicone surfactant	Inhalation	—	Slow volatilization from PU foam	
Carpet[1]	4-Vinylcyclohexene	Production residue in latex backing	Inhalation	—		Schaeffer *et al.* (1996)
	4-Phenylcyclohexene	Latex adhesive component in backing	Inhalation	—	Chemical associated with "new carpet" odor	Hodgson *et al.* (1992)
Liquid dishwashing detergent	Ethanol	Solvent	Dermal	Inhalation	Dishwashing detergents can contain up to 10 percent ethanol	Wooley *et al.* (1990)
Liquid laundry detergent	Ethanolamine 1,2-Propanediol	Solvents	Dermal	Inhalation		Wooley *et al.* (1990)

[1] Carpets also provide a reservoir for other important agents, such as dust mites and fungi (see Chapter 4).

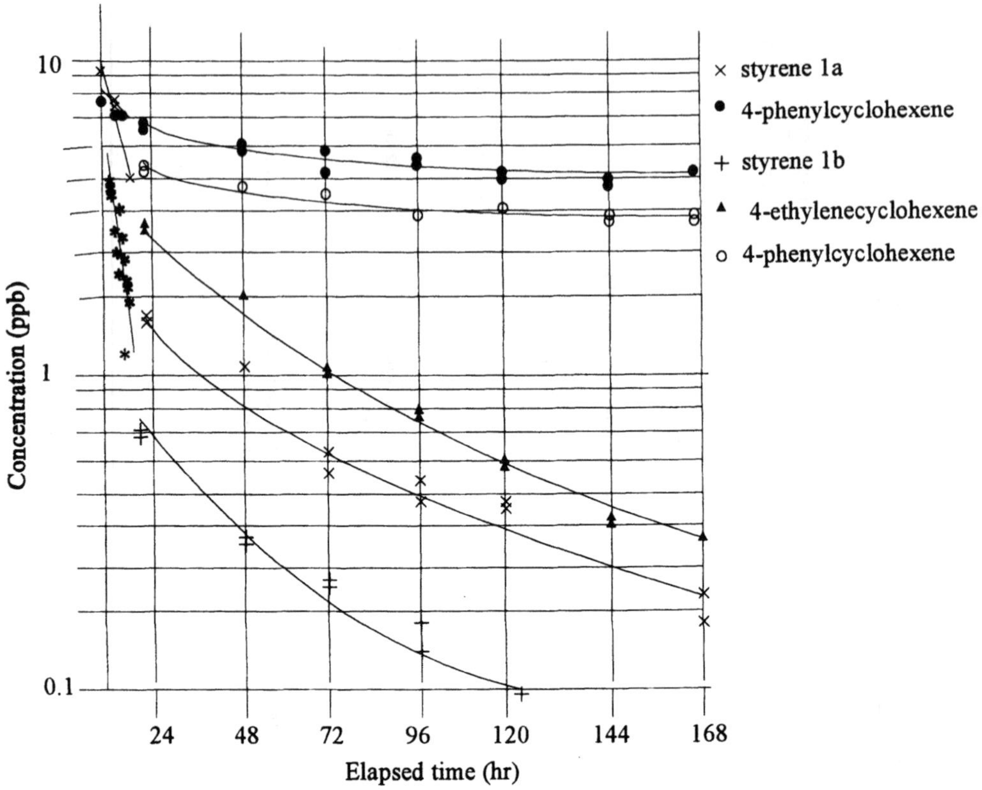

Figure 8-1. Dissipation curves for VOCs emitted from carpeting in large chamber studies (Hodgson *et al.* 1992).

4-PCH was roughly 30 to 60 μg/m^2/hr. The long-term emissions of 4-PCH indicate that there is a relatively large reservoir of this compound present in the backing material. Figure 8-1 shows the decline curve for airborne levels for five organic compounds from duplicate tests in a 20 m^3 environmental chamber with 100 percent nylon cut pile carpeting (SBR latex backing adhesive), conducted at approximately 23 °C, at an air exchange rate of one air change per hour, under approximately 44 to 49 percent relative humidity (Hodgson *et al.* 1992).

With regard to sources of exposure to biological agents, some consumer products can provide a medium for microbial growth and can thus become a source of exposure. When microbially contaminated consumer products are used (*e.g.*, humidifiers), aerosolization during normal use can result in contamination of residential air (Burge *et al.* 1980), and settling of contaminated aerosols can theoretically result in hand-to-mouth transfer of the microbes and incidental oral ingestion of microbes. Carpets and furnishing materials can provide a medium for microbial growth under high humidity conditions, or if water damage occurs in the residence. Some consumer products applied to residential surfaces (*e.g.*, paint) can provide an active surface for microbial growth if an ineffective type or insufficient amount of antimicrobial agent is present in the product. In contrast, many cleaning products, including certain laundry detergents and hard-surface cleaners, contain antimicrobial agents that are intentionally added to control or kill microbes in laundered clothes and on surfaces, respectively. Furthermore, intentionally-added antimicrobial agents are added to a wide variety of consumer products as diverse as contact lens cleaning solutions, fabrics, and coating materials. Importantly, the expressed purpose of some consumer products is to specifically kill a range of microbial agents. Such products include bleach, disinfectants, mouth wash, and certain pool chemicals. There are over 9 million swimming pools and spas in the United States that require regular treatment with chemical biocides to disinfect the water, prevent transmission of disease-causing organisms, and for aesthetic reasons (USEPA 1995a). In the course of normal use of consumer products containing antimicrobial chemicals, residential exposures to these biocides occur through inhalation, dermal, and incidental oral ingestion (*e.g.*, swallowing water while swimming in a public swimming pool). Exposures to biological agents in the residential environment are discussed further in Chapter 9.

The exposure pathways by which consumers may be exposed to chemicals from consumer products may be equally diverse. Exposure to a single chemical may arise from multiple mechanisms and pathways of exposure. The dominant pathway will dependent on the physical chemical properties of the chemical, the formulation type, product delivery system and release characteristics, and anticipated use patterns for the product. Examples of pathways include:

- Inhalation of aerosols and volatile organic compounds (VOCs) dispensed into residential air during product use
- Inhalation of VOCs volatilizing from treated surface (*e.g.*, paint solvents)
- Dermal contact during use
- Post-application dermal exposure to dislodgeable residues, and
- Incidental oral ingestion—settling of aerosols on food and surfaces, incidental oral contact during and after use.

Considering a single factor, such as vapor pressure, may not necessarily result in a correct conclusion regarding the predominant exposure pathway. An example is the case of ethanol in liquid dishwashing detergents. Depending primarily on the amount of product used and the initial water temperature, the amount of ethanol released into the air ranged from 18 to 110 mg per use, as measured by Wooley *et al.* (1990). Given that the total ethanol dispensed per use is on the order of 1,000 to 10,000 mg, it is clear the despite the volatility of ethanol, its high solubility in water and other factors result in most of the ethanol remaining in the water. Therefore, dermal exposure to this component during dishwashing could potentially exceed inhalation exposures. Similarly, only a small fraction of the ethanol contained in liquid laundry detergent is released to residential air during a washing machine cycle (Wooley *et al.* 1990).

Outdoor Use Products as Sources of Exposure

Outdoor use consumer products can also represent sources of exposure to a wide variety of chemicals. These products include lawn and garden pesticides, structural pest control products (*e.g.*, termiticides that are injected or placed in the soil around the outside perimeter of the residence), outdoor paints, wood stains, and other wood treatments (*e.g.*, water-proofing products), antimicrobial products (*e.g.*, pool chemicals, anti-mildew treatments), and combustion product sources (*e.g.*, grills, chain saws, lawn mowers). Some outdoor product uses can potentially result in indoor exposures, such as through tracking indoors of particles contaminated with outdoor use pesticides (Nishioka *et al.* 1996), and infiltration of vapors from outdoors into the residence. Examples of these mechanisms include (1) spraying a pet outdoors with a flea spray, followed by the pet re-entering the residence, where it becomes a source for indoor exposure (*e.g.*, through dermal contact with residues on the dog), (2) outdoor use of paint, resulting in infiltration of solvent vapors into the residence, and (3) tracking of lawn-applied pesticides into the home, resulting in contamination of carpet dust and carpet surfaces (Nishioka *et al.* 1996), which are available for dermal contact and incidental ingestion exposures.

PRODUCT USE CATEGORIES AND FACTORS THAT INFLUENCE EXPOSURE

Consumer products can be categorized in a variety of ways, for example, according to:

- *Purpose/function*. Examples include pesticides, air fresheners, personal care products, disinfectants, hard surface cleaners, laundry/dish washing products, paints and coatings
- *Use pattern*. This includes *intermittent use* (*e.g.*, insecticide space spray) and *continuous release* (*e.g.* fresheners, moth proofing agents)
- *Use category*. For example, do-it-yourself and hobby products, such as auto maintenance/cleaning products and hobbyist supplies (*e.g.*, plastic cement), and
- *Site of use*. For example, outdoor use versus indoor use products.

Impact of Physicochemical Properties and Physical State on Exposure

Physicochemical properties that may influence exposures include vapor pressure, molecular weight, Henry's law constant, and other properties. These physicochemical properties affect distribution among various media and influence the exposure by various routes. Examples of physical state considerations that may influence exposures include formation of aerosols and particles during consumer product use, and the resulting size distributions of the aerosols and particles. The findings of the particle TEAM study (PTEAM) conducted in Riverside California (Clayton *et al.* 1993) suggests that 24-hour personal exposures to 10 μm or less (PM_{10}) particles inside nonsmoking residences may approach or exceed the national ambient air quality standard of 150 μg/m^3, as a result of cooking, consumer product use, and other factors. Many consumer products, by virtue of the nature of their delivery systems (*e.g.*, aerosol spray cans, trigger-pump products), provide a significant source of aerosols in the residential environment. The generation of aerosols provides a vehicle for inhalation of volatile, semivolatile, and nonvolatile materials present in certain consumer products.

For exposure assessment purposes, one needs to know not only the initial concentration of aerosols in residential air from use of a consumer product but also the size distribution (to determine what fraction of the airborne aerosols is actually respirable). For greater realism and accuracy in the exposure assessment, it may be necessary to model the dissipation of aerosols in indoor air as the larger size range of aerosols deposits on the floor and other indoor surfaces, and as fine particles are removed via the air handling system of the residence. For example, Tichenor *et al.* (1990) found that on the order of 20 percent of the particles emitted by kerosene heaters are adsorbed onto indoor residential surfaces.

In one of the first studies conducted to examine respirable particles generated by pressurized consumer products, ITRI (1975) determined the mass median aerodynamic diameter (MMD) of aerosols produced from 10 different aerosol consumer products (air freshener, antiperspirant, carpet cleaner, dusting aid, fabric protector, furniture wax, lubricant, paint, panel wax, and wood panel cleaner) to range from 2.3 to 12 μm. Most of the product MMDs were clustered around 5 μm, with a standard deviation of 1.5 to 2.5. For these studies, most of the delivery rates for the products were in the range of 1 to 2 g/sec.

Ignoring turbulent air flow, the deposition of aerosols in residential environments should follow Stoke's law: the worst-case settling distance for a typical residence without vaulted ceilings would be 8 feet (*i.e.*, from a typical ceiling height) assuming, for example, the use of an aerosol air freshener or insecticide space spray. Based on Stoke's law (Matoba *et al.* 1998), the following settling times can be calculated for various size aerosols: 3.4 minutes for 20 μm, 13.5 minutes for 10 μm, and 54 minutes for 5 μm aerosols. Aerosols significantly less than 5 μm have such long settling times (*e.g.*, 22.4 hours for 1 μm) that they would be more likely to be removed and recirculated in the residence than to deposit. To complicate matters further, aerosols produced from water-based consumer products can grow in size due to hygroscopicity under high humidity conditions (Li and Hopke 1994, Dua and Hopke 1996), which would effectively increase the rate of deposition. Conversely, under low humidity conditions, the distribution of aerosol size can be shifted to lower diameters due to evaporation of water from the aerosols, thus effectively increasing the respirable portion and decreasing the rate of deposition. As a side issue, aerosols produced by indoor consumer products can adsorb radon progeny already present in indoor air and transport them to the surface of the respiratory tract (Li and Hopke 1994, Dua and Hopke undated).

Impact of Use Patterns on Exposure

Product use patterns can have a tremendous impact on individual exposures. As can be imagined, individual use patterns vary greatly. The associated consumer exposures are not only dependent on the total amount of indoor consumer product used, but also the frequency and duration of use of the indoor consumer product, the duration of exposure in the room of use, post-application exposure duration, and the ventilation configuration of the residence. Extensive data are available on these parameters are available in the Westat survey on consumer use of solvent-containing products (USEPA 1987a) and a followup survey on several types of methylene chloride-containing consumer products (USCPSC 1992).

The 1987 survey (USEPA 1987abc) was conducted on 32 categories of common household and automotive products thought to contain methylene chloride or its substitutes. Potential respondents were identified using a random digit dialing procedure, contacted by telephone to obtain their consent, and then sent a questionnaire via mail. Respondents used an average of 5 of the 32 product categories in the year prior to the survey. Most respondents reported having a window or door open during use of a solvent-containing product, and most reported that they read directions on the product labels before use. One finding of the study is that usage of solvent-containing products generally decreases with increasing age (USEPA 1987a).

In contrast, the 1992 survey (USCPSC 1992), which also used random digit dialing to identify potential respondents, employed a telephone interview procedure rather than a written interview, and included a subsample for testing for instruction recall and instruction compliance by mail followup. For products covered by both surveys, central tendency values for these parameters (*e.g.*, arithmetic means, medians) are essentially the same for a given product type, despite expected differences in extreme values (*e.g.*, 1st and 99th percentiles).

An example product covered in both the 1987 and 1992 surveys is aerosol spray paint. The 1987 survey (USEPA 1987a) had a total of 4,917 respondents, 35.4 percent of whom had used spray paints at one time or another (*i.e.*, 64.6 percent of respondents claimed never to have used aerosol spray paint products). The 1992 survey (USCPSC 1992) had a total of 4,979 respondents, 32.4 percent of whom reported using spray paints at one time or another. The mean amount of spray paint applied per use is reported as 13.8 oz in the 1987 survey, and 19.04 oz. in the 1992 survey. However, because the amount of aerosol paint applied per use appears to be lognormally distributed with a geometric mean of 6 oz. per use (see Figure 8-2), the arithmetic mean values will lead to overestimation of exposures if taken as central tendency values. Most of the other use parameters (such as frequency of use, duration of application period, and duration of post-application period) are also closer to being lognormally distributed than normally distributed. Important parameters that affect exposures directly are the room of use (in terms of typical dilution volume available) and ventilation configuration. Figure 8-3 shows the frequency of location for applying spray paint, based on the WESTAT survey (USEPA 1987a). Most of the time (64.1 percent), consumers use spray paint outdoors. The most frequent indoor room of use is the garage, which is also associated with one of the highest ventilation rates in the home relative to other rooms in the residence (USEPA 1989). The 1992 survey (USCPSC 1992) indicates a higher level of awareness among consumers regarding the need for ventilation in using solvent-containing products compared to the WESTAT survey (USEPA 1987a). For use of aerosol spray paint, over 81 percent of respondents in the 1992 survey indicated having a door or window open to the outdoors during use of aerosol paint, whereas only about 63 percent of respondents in the 1986 WESTAT survey indicated that this precaution was taken. Furthermore, 17 percent of respondents from the 1992 survey indicated use of an exhaust fan, whereas only about 10 percent reported doing so in the 1986 WESTAT survey (USEPA 1987a).

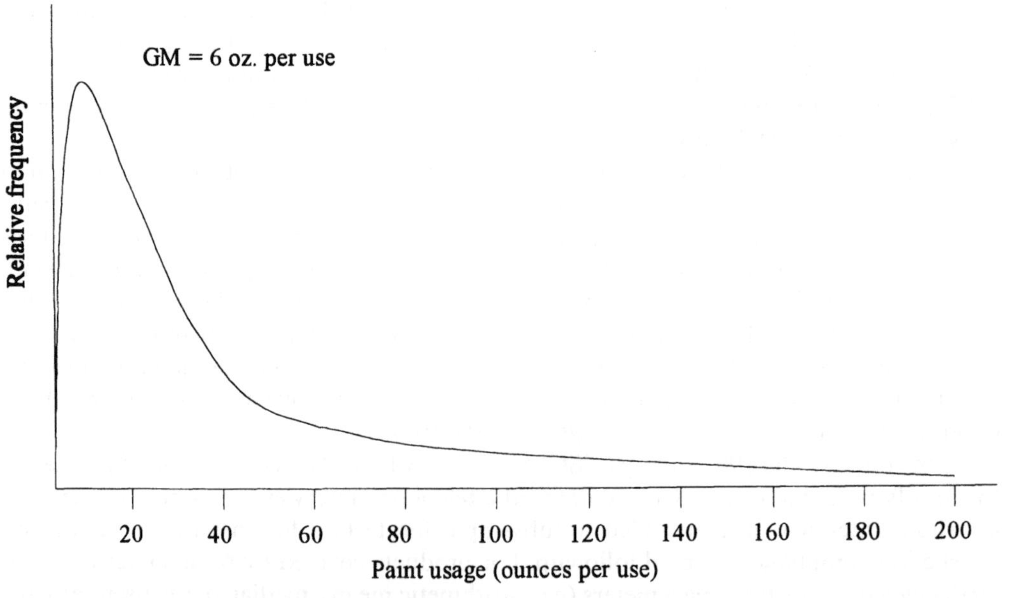

Figure 8-2. Frequency distribution of amount of spray per use (ounces). N = 1118.

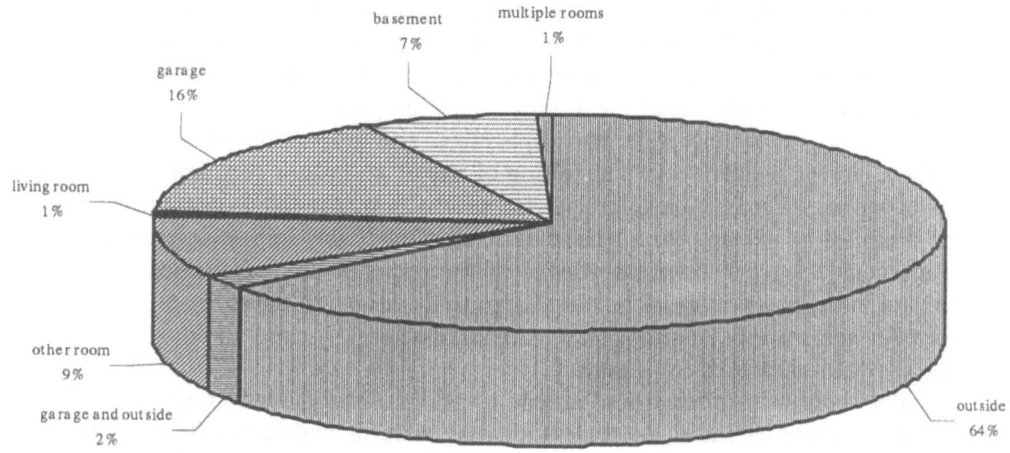

Figure 8-3. Frequency of location of use of aerosol spray paint (N = 1160).

In another survey (USEPA 1987b), national usage data were collected on household cleaning products, using random digit dialing and telephone interviews. This survey tied the frequency and duration of 14 specific cleaning tasks (*e.g.*, cleaning countertop, cleaning bathroom sinks and tubs, doing laundry) to specific products. The survey also collected information on the use of protective measures (including rubber gloves, post-use washing of skin, and appropriate ventilation), likelihood of contact, and the frequency of reading and following label instructions (particularly regarding amount used). The statistical robustness of this survey is limited based on the small number of households interviewed (N = 193).

A major influence on exposures to chemicals in consumer products relates to the amount used by the consumer in terms of its consistency with label instructions (see *Impact of Consumer Psychology on Consumer Product-Related Exposures*). Scenarios should be assessed initially based on typical anticipated use *per label instructions* (including amount per use, frequency of use, and protective measures). This would constitute the base scenario in a consumer product exposure assessment. Next, the "heavy user," perhaps using the 90th percentile use rate and median values for other parameters, could be assessed for acute health benchmark. This would represent a reasonable worst-case scenario.

In performing an assessment, it is important for the assessor to distinguish between the likely user and exposed populations. One way to address this is to break the assessment into application and post-application phases. One of the current concerns of the USEPA is to include an assessment of childhood exposures as part of all exposure assessments for the Agency. This entails not only use of scenarios that may result in increased contact with contaminants (*e.g.*, the "crawling infant" scenario) relative to adults, but also the use of appropriate values for physiological parameters such as body weight, inhalation rate, and skin surface area (see Chapter 10). It is also critical to apply appropriate dose metrics (*e.g.*, acute and subchronic). Long-term health benchmarks such as chronic toxicity and cancer may not be appropriate to assess for children, unless doses are time-integrated through adulthood, including use of incremental changes in values for physiological parameters over time. Furthermore, assessment of children using inappropriate toxicological benchmarks such as reproductive or teratogenic endpoints relevant to chemicals that act only during pregnancy should be avoided.

As with other residential exposures, time-activity patterns as they relate to indoor consumer product use patterns and when people are present in the residence relative to when a product is used in the home, play a key role in estimating and integrating exposures to a chemical. For the use of the product, depending on label instructions, the assessor may wish to finally integrate the application and post-application exposures to chemicals in indoor use products. The timing of use of an indoor consumer product relative to occupancy is particularly crucial for products whose label instructions dictate that the premises be vacated for a period of time immediately following use (*e.g.*, total-release insecticide fogger). Also, knowledge of the presence of odorous/noxious chemicals, or data indicating very brief post-application durations in the room of product use (*e.g.*, as with aerosol spray paints) may lead the assessor to conclude that at least the room of use will be vacated for a period.

Finally, there is an interest and a need to address exposures to multiple products containing the same chemical, to the extent that multiple products will likely be used within the same residence, to determine the total exposure likely to the chemical. As noted above in Section 1, there is a requirement under the Food Quality Protection Act (FQPA) of 1996 to perform "aggregate" risk assessment for pesticides that may occur in food, drinking water, and consumer products. In performing aggregate risk assessment, one must be certain to account for correlations between different product uses. For example, an individual using an insecticide fogger may not need to use a crack and crevice spray frequently, if at all. Distributional analyses, using Monte Carlo techniques (see Chapter 12), may be necessary to successfully develop aggregate exposure assessments and risk assessments that yield a realistic distribution of risks and to prevent gross overstatement of combined risk.

Impact of Consumer Psychology on Consumer Product-Related Exposures

A very real influence on the extent of exposure incurred by a consumer in using a product is consumer behavior. Consumer behavior, including whether cautionary statements on the product label are read and obeyed, have a tremendous impact on exposures. For example, a consumer who does not follow directions to "use in well-ventilated area" or to "use rubber gloves" will likely have a much higher exposure to chemicals in the product than those consumers who read and follow the cautionary language.

In 1996, the USEPA issued the draft Consumer Labeling Initiative (CLI) report. As part of the CLI, a literature search and survey were conducted to document consumers' perception of product labeling and how it does or does not influence perception of risk (USEPA 1996c). The focus of the pilot phase of the CLI project was on indoor insecticides, outdoor pesticides, household antimicrobials/disinfectants, and selected hard-surface cleaners. The objective of this research was not to develop recommendations for development of product labeling standards. Rather the goal was to obtain an accurate view of consumer understanding of environmental health and safety issues as they relate to household product safety. A focus group involved in the CLI initiative concluded that household cleaning product users were not accustomed to reading labels thoroughly because they felt familiar with the products and their associated hazards. Whether or not consumers read cautionary label statements depends on the type of product and their familiarity with it. If the product is (1) labeled "Danger," (2) a pesticide, or (3) being used for the first time, a consumer is more likely to read the label. Consumers with children or pets were more likely to read cautionary labeling for indoor insecticide and outdoor pesticide products, in particular. Studies have found that consumers associate household products with little or no safety threat. Consequently, incidents involving injury usually involve product misuse.

One of the issues highlighted by the survey is the impact of consumer behavior on use. Many followed exact label directions for use. However, most consumers felt that they were familiar enough with dilutable cleaning products that they did not need to follow instructions on product use, often using an arbitrary amount they felt was correct. Also, a large fraction of consumer tended to overuse pesticides (*i.e.*, apply more than specified on the label or take a "Rambo" approach to killing individual weeds and insects). Needless to say, this has a tremendous impact on an important exposure parameter—amount of product per use. This type of behavior would tend to increase the variability for this parameter for the affected product type(s). A separate survey conducted by the Wharton School of Business, University of Pennsylvania (Wall Street Journal 1996) found that consumers often tend to use more of a product than specified on the label if it is purchased in bulk quantities. Regional differences in product use also occur, some due to climate and some due to regional differences in consumer behavior. For example, more consumer insecticide products are used in the U.S. than in Saskatchewan, Canada due in part to the greater occurrence of insect pests in certain parts of the U.S. [*e.g.*, in the southeast (McDuffie and Pahwa 1998)]. Consumers in southern Europe use more grams of laundry and cleaning products per use event than consumers in northern Europe (ECETOC 1994). Some data on the use of consumer products in Europe are located in Appendix 6 of the ECETOC Technical Guidance Document (ECETOC 1994). Weegels (1997) has compiled information on use of consumer products in Europe, comparing questionnaire, diary, and videotape methods; this survey addresses dishwashing products, cleaning products, hair styling products, paints, and pesticides used in and around the residence. Weegels (1997) found that questionnaires tend to overestimate actual product usage, and that key variables such as the amount of product used and duration of use are highly variable within the consumer population.

MONITORING TO CHARACTERIZE EXPOSURE TO INDOOR-USE PRODUCTS

Tichenor *et al.* (1990) has described a three-phased approach for evaluation of indoor air pollution sources. These phases consist of: (1) measurements of emissions in controlled environmental chambers (where the influence of environmental variables such as temperature and sink effects can be evaluated), (2) development of indoor air quality (IAQ) models to predict indoor residential airborne levels over time (incorporating realistic air exchange rates), and (3) simulated residential or full-scale residential testing to collect "real-world" data and to validate the model. Emissions testing and residential monitoring are discussed below.

Characterizing Consumer Product Emissions

Emission Concepts: The Time-Profile Approach and Emission Kinetics. In many cases, the only way to *definitively* quantify the source term, such as for inputting into an indoor exposure model, is to monitor the emission rate under controlled conditions. This is often conducted in small test chambers. A major source of the uncertainties in estimating exposures to chemicals released from consumer products is in defining the emission rate of the chemical of interest. The comparability of airborne concentrations estimated by inputting these emissions data into indoor air models versus actual measured airborne levels associated with product use in an actual residence has been determined in only a few cases (Sparks *et al.* 1990, Nagda *et al.* 1995). One such validation study focusing on toluene emitted by tile adhesives (Nagda *et al.* 1995), involving multichamber modeling of toluene based on emission measurements from small chamber studies indicated relatively good correspondence between modeled and measured airborne levels in the room of use (see Figure 8-4).

Emission rates modeled for many consumer products that are applied to a porous matrix, or that are in contact with a porous matrix often represent biphasic time courses. Thus, there is a "fast" (evaporation-dominated) component that primarily governs the early stage of emissions and a "slow" (diffusion-limited) component that has the greatest influence on the emission curve after the first few hours of coating application (USEPA 1996a). That is, the time-profile of the indoor air level is represented by a linear log-log dissipation curve that suggests a biphasic or multiphasic release mechanism. In simplest terms, this would mean that an initially rapid release occurs from a readily available pool (*e.g.*, chemicals released from the surface of a coating, such as paint), followed by a more slowly-available pool of chemical (*e.g.*, chemicals released through a "skin" formed on the outside of the coating as the coating dries or "cures" and/or from product retained within the interstices of the complex matrix to which it has been applied. This biphasic emission pattern has been observed in small-chamber studies of VOC emissions following application of wood stain to hardwood boards (Chang and Guo 1992). Another case in point is the release of toluene from tile adhesive measured after it is covered with floor tile (Nagda *et al.* 1995). The best fit was obtained by applying a nonlinear least squares fit test (see Figure 8-5). Another case where a nonlinear type of emissions curve might be anticipated is the release of pesticides from a treated surface. This applies to termiticides injected into the soil under or around a residence as well as pesticide applied by the consumer to certain indoor surfaces such as carpeting, for example. This biphasic pattern is more subtle in the application of a coating such as paint to a nonporous matrix (*e.g.*, glass or steel), where the emissions are evaporation-controlled but eventually become film-limited, thus leading to adequate approximation of the time-profile of emissions using a single-exponential curve (Clausen *et al.* 1991, USEPA 1996a).

Based on 96-hour chamber studies conducted on carpet padding, Schaeffer *et al.* (1996) noted a time-weighted average VOC emission rate of 10 to 500 mg per m^2 of product area, or approximately 0.10 to 5.0 mg/m^2/hr. A decline in emission rates of VOCs from carpet padding were observed over time. Declines in product emission rates are often product and chemical specific in nature. In the case of carpet padding, the decline in emission rate was most rapid for the more volatile compounds. For example, the emission rate for N,N-dimethylformamide and trichloropropane at 96 hours was only 2 percent of the emission rates at 6 hours. In contrast, the emissions of less volatile components declined more slowly over time. For example the emission factor at 96 hours for BHT was still 85 percent of that at 6 hours (Schaeffer *et al.* 1996). Thus, for a complex source such as carpet padding, the profile of major individual VOCs emitted within the first 24 hours can be quite different than those emitted several days or weeks later. The European Collaborative Action on indoor air quality and its impact on human health has produced two reports on testing emissions (2: EUR 12196 EN and 13: EUR 15054 EN).

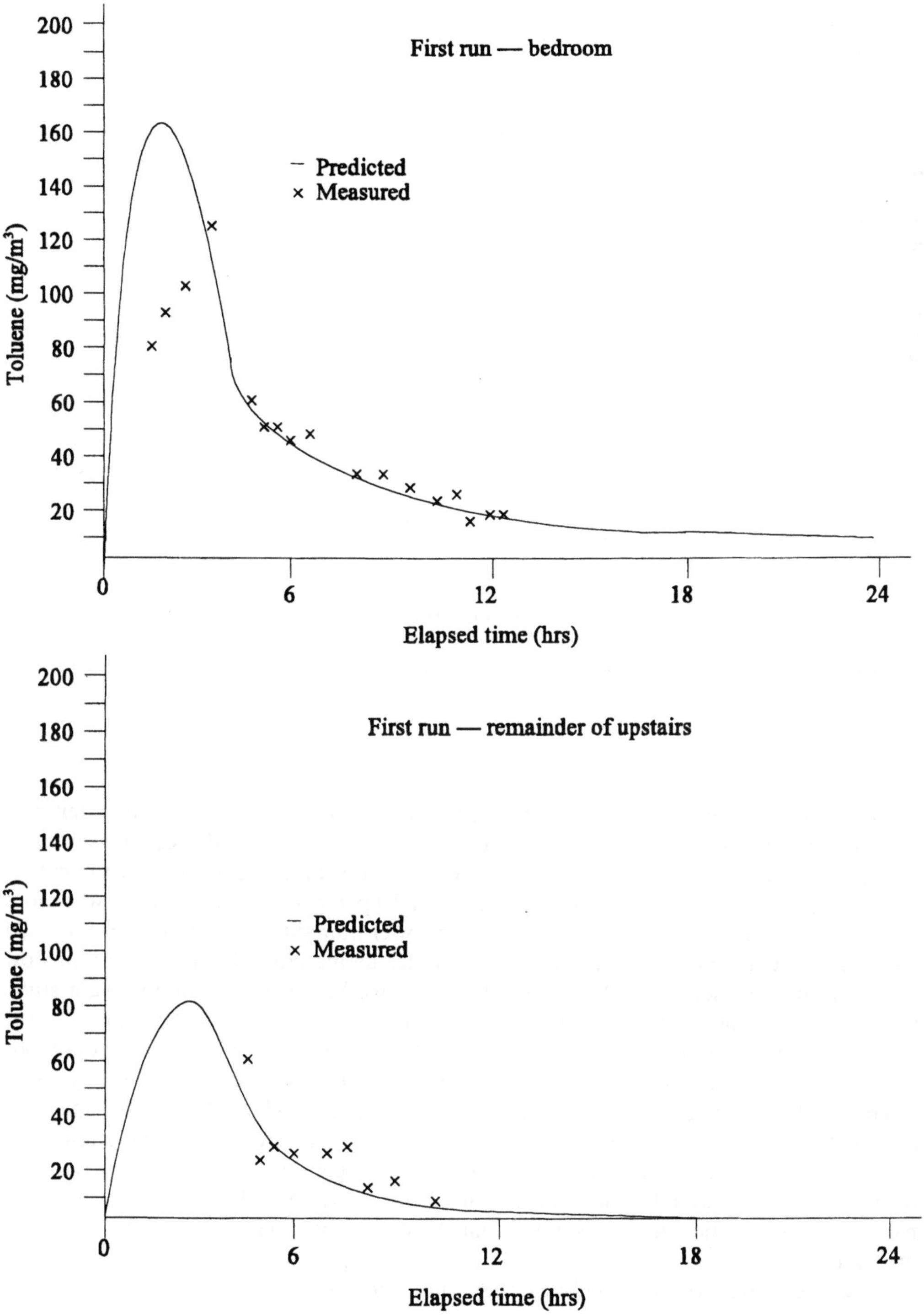

Figure 8-4. Predicted and measured toluene concentration in room of use (bedroom) and adjacent zone (remainder of upstairs) resulting from use of tile adhesive (Nagda *et al.* 1995).

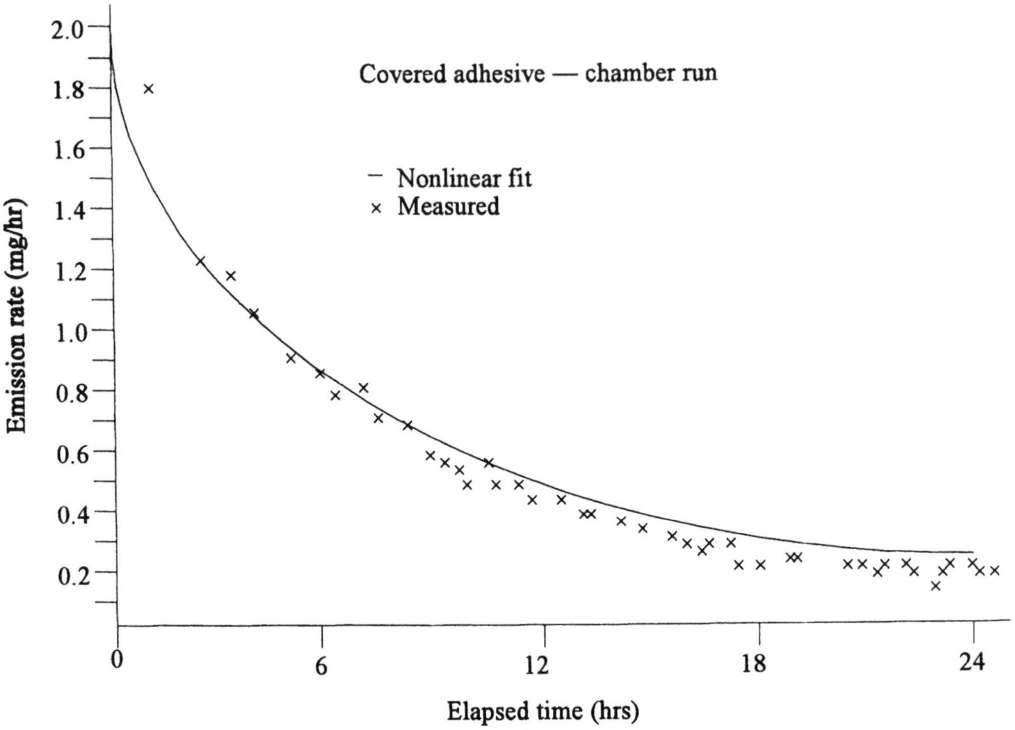

Figure 8-5. Nonlinear fit test of toluene concentration-time profile of tile adhesive in room of use (Nagda *et al.* 1995).

Influences on Emission Rate Measurements and Limitations. Relative humidity, temperature, and air flow rate can have a significant impact on the time-profile of emission. For example, Schaeffer *et al.* (1996) found that the amount of acetic acid emitted from carpet cushions (padding) during chamber testing under 50 percent relative humidity was more than 10 times higher than that occurring under lower humidity conditions. It is critically important to conduct small chamber testing under temperature, humidity, and air flow rates that are representative of residential conditions. For example, conducting a study under too high a flow rate may lead to a gross overestimate of emission rate due to the enhanced volatilization occurring, particularly if the substance of interest is volatile or semivolatile in nature. ASTM standard guide D5116-90 is applicable to these types of small chamber studies, and provides some basic guidelines (ASTM 1990). VOC emissions from materials have been observed in other tests to be different when the studies are carried out under nitrogen compared to air. For example, Jensen *et al.* (1996) reported higher carboxylic acid/aldehyde ratios emitted from linoleum in the presence of air compared to nitrogen and, thus, concluded that the emissions were influenced by oxidation processes.

Serious limitations of small-chamber studies include:

- *Sorption on chamber wall surfaces.* It has been observed in numerous studies that many organic vapors adsorb onto surfaces (*e.g.*, walls) in test chambers and in real indoor environments (Dunn and Tichenor 1988, Tichenor *et al.*

1991, Colombo *et al.* 1993, Levsen and Sollinger 1993, Kjaer *et al.* 1996). This problem is likely to become more severe as the surface area-to-volume ratio of the chamber increases (for smaller chambers) and as the emission rate of material decreases, resulting in greater fractional loss to the walls of the chamber. As an example, compare a 52 liter cubic test chamber approximately 0.373 m in each dimension, which has a surface area-to-volume ratio of 16.1 to a cubic room size chamber 27 m^3 in volume and 3.0 m in each dimension, which has a surface area-to-volume ratio of 2.0. The smaller 52 liter cubic test chamber has 8 times more surface area per unit volume as the room-size chamber.

- *High air velocity and turbulence.* The air velocity in the immediate vicinity of the material being tested in a chamber is an important factor that affects the emission rate of VOCs from the surface of the material. This effect has been noted by numerous researchers (ASTM 1990, Tichenor 1996, Zhang and Haghighat 1996, Zhang *et al.* 1996). The major effect of air velocity in affecting emission rate seems to be that higher velocity increases the mass transfer rate coefficient by reducing the thickness of the diffusional boundary layer adjacent to the material's surface (Zhang and Haghighat 1996). The amount of turbulence in the air over the material surface has also been found to affect emission rates, with higher turbulent intensity resulting in higher emission rates (Zhang and Haghighat 1996, Zhang *et al.* 1996).
- *Ventilation rate and mixing.* The ventilation rate is usually expressed as the air exchange rate in air changes per hour (ach), which is the number of chamber volumes of fresh (*i.e.*, outside) air that passes through the chamber per hour. For a given emission rate, the air exchange rate will be a factor in establishing the airborne VOC concentration. The airborne concentration in turn can affect the emission rate. In general, lower air concentrations in highly-ventilated chambers will result in higher emission rates because the concentration gradient between the test material's surface and air is reduced (ASTM 1990). Another parameter that is related but not the same as air exchange rate is mixing. A test chamber should be well-mixed (ASTM 1990) in order to provide uniform air concentrations so that the sampled VOC concentrations are representative of the air throughout the chamber.

Thus, results reported from small-chamber studies should be considered screening-level or qualitative in nature, rather than accurate quantitative emission factors suitable for exposure assessments of product use in actual residences.

Comparison of airborne concentrations of semivolatile compounds such as BHT emitted from new carpeting tested in 4 liter versus 20 m^3 chambers suggests that greater sorptive losses of semivolatile compounds may occur during small chamber tests due to the higher surface area-to-volume ratio, thus possibly leading to underestimates of emission rates and airborne concentrations for this class of compounds (Hodgson *et al.* 1992). The existence of barriers may reduce the emission rate of VOCs from a product. For example, placement of a carpet over a carpet cushion reduces the peak VOC concentration in air above the carpet, and delays the occurrence of the peak concentration. Thus, carpets appear to act as a diffusion barrier to chemicals present in the carpet cushion (Schaeffer *et al.* 1996). Product emission rates may also be affected by storage time prior to purchase. For example, carpet cushions sealed and shipped directly from the manufacturer were associated with substantially higher emission rates of VOCs than those obtained at a retail outlet (Schaeffer *et al.* 1996). Small chamber testing has also been conducted to determine the nature of emissions from a variety of products, including wallpaper and vinyl wall coverings (De Bortoli *et al.* 1993) and vinyl flooring (Clausen *et al.* 1993).

Monitoring to Characterize Indoor Air Concentrations for Consumer Products

Full-scale monitoring of indoor air levels of contaminants in actual or research residences has been conducted for almost 15 years, as part of large-scale studies (such as TEAM) or small privately-funded efforts. Typically, room area or personal monitoring has been conducted, usually representing an integrated sample (*e.g.*, over 8 or 12 hours). However, currently available instrumentation allows essentially continuous real time or near-real time monitoring of many contaminants. A caution in conducting studies involving monitoring of indoor air levels is to select a measurement method that will have adequate sensitivity. Too often, occupational hygiene screening methods are applied and can result in inconclusive studies in which most of the samples are below the detection limit. With high detection limits, even quantifying the exposures as one-half the detection limit, the typical practice, can lead an assessor into areas of unacceptable risk, particularly if the no-observed-effect-level (NOEL) is low, as in the case of inhalation toxicity benchmarks for certain compounds.

Collection of monitoring data can focus on area concentrations in the room of use and/or adjacent areas, and monitoring of breathing zone concentrations (personal monitoring, see Chapter 4). Monitoring data are available for a limited number of consumer products during actual or simulated use, representing application and/or post-application periods. Examples of these data are shown in Table 8-2. The information in Table 8-2 illustrates that consumer exposures can be highly variable due to differences in individual use patterns and behaviors, and that exposures are frequently reported in a variety of different units.

Exposure Monitoring Case Studies

- *Laundry detergents*. One of the first studies of personal exposures to a consumer product was the measurement of laundry detergent dust levels in the air during dispensing of laundry detergent into an automatic washing machine. In this study by Hendricks (1970), use of an Anderson dust sampler with an air inlet held near a consumer during in-home product use indicated an average exposure of 0.27 μg detergent dust inhaled per cup of laundry detergent dispensed using the "double-pour" technique (Hendricks 1970). This represents a very low exposure, and reflects the intentional formulation of laundry detergent as large granules with minimum fine dust in order to reduce the likelihood of adverse respiratory reactions to enzymes present in the formulation.
- *Pesticides*. Limited monitoring data on inhalation exposures to pesticides delivered from a hand-held aerosol can are contained in the Pesticide Handlers Exposure Database (PHED). PHED is being commonly used by registrants and government agencies to supplement and validate field exposure studies, and as an evaluation tool for analysis of field exposure data. PHED contains over 1,700 records of data on measured dermal and inhalation exposures, as well as accompanying data on parameters that may affect the magnitude of exposures

Table 8-2. Examples of monitoring data for residential inhalation exposures to chemicals in selected consumer products

		Sample duration (hrs)		Mean air	Exposure		
Product	Chemical	Application	Post-application	concentration (mg/m^3)	Range (mg/day)	Central tendency	Data sources and comments
Aerosol can	Pesticide	0.47	—	—	—	0.605 mg/lb a.i. (median)	USEPA (1995b). Assumes inhalation rate of 13.8 l/min. Specifics on room volumes and air exchange rates are not available.
Nail polish	Toluene	0.43–0.77	1.0–4.0	3.2–5.65 (appl.) 0.2–1.7 (post-appl.)	1.03–2.82	1.89 mg/day (mean)	Curry *et al.* (1994). Based on personal monitoring in California residences, representing various housing types and ventilation conditions. Exposures include application and post-application periods. Assumes 10.7 m^3/day inhalation rate.
Tile adhesive	Toluene	10–12	30.8–39.6	—	—		Nagda *et al.* (1995). Toluene content of 4.8 percent by weight in adhesive, application of tiles 75 minutes after start of application. Room volume 20.5 m^3. Air recirculation rate of 9.4 ACH. Exchange with outdoor air of 1.8 ACH.

(USEPA 1995b). Exposure data from PHED are usually reported in normalized form (*e.g.*, μg exposures per lb of active ingredient [a.i.] handled or applied per day or event). In the case of aerosol cans, the median (50th percentile) inhalation exposure is 0.605 mg per lb a.i. handled, assuming an inhalation rate of 13.8 liters/min and an average exposure duration during use of approximately 30 minutes.

A limited number of studies have been conducted to characterize the airborne concentrations of specific pesticides in residential indoor air following use of a consumer pesticide product intended for indoor use. Because many pesticides are semi-volatile, pesticides can exist in both aerosol (respirable and nonrespirable diameters) and vapor phases. For example, chlorpyrifos is a cholinesterase-inhibiting semivolatile organic insecticide with a vapor pressure of 2×10^{-5} mm Hg at STP. The post-application airborne levels declined from approximately 17 μg/m^3 immediately after use to near 14 μg/m^3 1 to 6 hours after use, and around 6.5 μg/m^3 7 to 16 hours after use, for a fogger product containing 0.5 percent chlorpyrifos (Ross *et al.* 1992) used in a room of approximately 52 m^3. Extensive air monitoring collected as part of the U.S. National Occupational Pesticide Exposure Survey (NOPES) study indicates that chronic inhalation exposures to pesticides in residences are usually in the low μg/kg dose range (Ross *et al.* 1992).

- *Toluene*. Toluene occurs widely in consumer products, such as nail lacquers. Therefore, one might anticipate that toluene would be found in the residential indoor air environment associated with use of particular consumer products. A study in 161 California homes indicated median background concentrations of toluene (48-hr samples) of 27 μg/m^3 for indoors versus 17 μg/m^3 for outdoors (Wilson *et al.* 1993). A monitoring study on inhalation exposures to toluene associated with use of nail lacquers in California residences was conducted by Curry *et al.* (1994). The study involved collection of personal, area, and background samples (charcoal tubes) before, during, and after application of nail lacquer products by 5 subjects in their homes. The total amount of nail polish applied by subjects ranged from 511 to 833 mg (mean of 701 mg). The toluene content of the lacquers ranged from 21.0 to 33.1 percent. Personal exposures (encompassing both the application and post-application periods) ranged from 1.03 to 2.82 mg/day, with an arithmetic mean of 1.89 mg/day. Time-weighted average toluene concentrations in indoor air ranged from 3.2 to 9.2 mg/m^3 during application. Airborne levels of toluene declined relatively rapidly after application. For the poorest-ventilated residence, a log-linear dissipation curve was observed, with a post-application half-life of 67 minutes in the subject's breathing zone. Although the study captured a variety of residential housing stock types, room volumes, and ventilation conditions, the highest observed exposure was not associated with the residence having the poorest ventilation conditions. Rather, it was experienced by a woman who held her hand closest to her face (due to poor vision) during application of the nail lacquer. This case study illustrates the importance of personal behavior on exposures to this product category. A study by Ceschin-Roques *et al.* (1991) suggests that certain chemicals may have the potential to penetrate the nail and provide a dermal exposure.
- *Perchloroethylene*. Perchloroethylene (tetrachloroethylene) is a good example of a chemical used outside the home that can find its way into the home through indirect means. Although there are still a large number of consumer products

containing perchloroethylene, one of the major routes of this compound into the residential environment appear to be (1) through dry-cleaned clothes brought into the home, (2) offgassing of perchloroethylene from workers' clothing and from workers' breath into residential air, and (3) infiltration of perchloroethylene from dry cleaning facilities into residential air in concurrent or adjoining building.

Wallace *et al.* (1991b) measured elevated levels of perchloroethylene in the homes of dry cleaning workers. Thompson and Evans (1993) used a physiologically-based pharmacokinetic model to verify that the workers' body burdens were sufficient to explain elevated indoor residential airborne levels (on the order of $100\,\mu g/m^3$), presumably attained by workers unintentionally exhaling the compound into the home environment after working hours. In a monitoring study by Aggazzotti *et al.* (1994), alveolar air samples were taken from dry cleaning workers, family members of dry cleaning workers not employed at the facilities, and control group individual not working in or living near a dry cleaning facility. In addition, residential indoor air samples were taken from the respective households. Indoor air levels of perchloroethylene in dry-cleaners' homes were roughly two orders of magnitude higher than those in the homes of the nonexposed control group (Aggazzotti *et al.* 1994).

The effect of bringing dry-cleaned clothes into the residence on perchloroethylene levels in indoor residential air has been investigated by Thomas *et al.* (1991) in several New Jersey homes. Elevated indoor levels of perchloroethylene were measured in most of the homes into which dry-cleaned clothes were brought. Indoor/outdoor concentration ratios exceeded 100 for all four of the homes with both dry-cleaned clothes and outdoor concentration data present. Indoor air levels in homes with dry-cleaning present were as high as $300\,\mu g/m^3$, and levels remained elevated in the affected homes for at least 48 hours after the dry-cleaned clothes were introduced. Maximum source strengths ranged from 16 to 69 mg/hr. The 2- to 6-fold higher breath levels of perchloroethylene in individuals living in residences where dry-cleaned clothes were present verified the anticipated higher exposures. Indoor air, personal air, and breath levels were found to be related to the number of dry-cleaned garments introduced into the residence divided by the total house volume (Thomas *et al.* 1991).

Several investigations by the New York State Department of Health have found very elevated levels of perchloroethylene in the air of residences near or above dry cleaning facilities (Schreiber *et al.* 1993). In one investigation, elevated levels were identified in second and third floor apartments located directly above a first floor dry cleaning facility and next to an adjacent building that contained a laundromat with dry cleaning equipment. In this case, the highest perchloroethylene level detected in one apartment was $197{,}000\,\mu g/m^3$ measured over a 12-hour period. This level exceeds the OSHA 8-hour TLV of 170,000 $\mu g/m^3$ (Schreiber *et al.* 1993). In at least one home adjacent to a dry cleaning facility, a level of $85\,\mu g/m^3$ was obtained, reflecting the locally high outdoor level of $140\,\mu g/m^3$. A controlled study by Schreiber *et al.* (1993) in which perchloroethylene levels in residences in the same building as dry-cleaning operations were compared to those in control homes verified the suspected relationship between co-location of dry-cleaning operations and elevated levels of perchloroethylene in residential air. The type of dry-cleaning machines present in the building was significantly associated with the concentration of perchloroethylene found in the residential air of apartments located above the facility. Perchloroethylene has been found in the breast milk of nonoccupationally exposed nursing mothers; Schreiber (1997) has used a PBPK model to predict concentrations of perchloroethylene in the breast milk of nursing mothers exposed to airborne levels monitored in (1) the dry cleaning work environment, and (2) apartments located above dry cleaning establishments.

Monitoring to Characterize Dislodgeable Residues on Surfaces

A major parameter in assessment of dermal exposures in the residence relates to the fraction of residues on a surface (such as in the case of a floor treated with pesticides) that can be transferred to skin. Dermal exposure can occur through direct skin contact with residues of chemicals on surfaces and subsequent absorption through the skin. Furthermore, transfer of residues on the skin via hand-to-mouth activity, such as in toddlers, may provide a significant *oral* dose from a *dermal* exposure. On an absorbed dose basis, the hand-to-mouth exposure pathway often "drives" exposure/risk assessments for many products in the home, particularly pesticide carpet spray and other pesticide products that lead to significant deposition of product on floor surfaces, such as total-release foggers and space sprays.

A variety of direct (*i.e.*, skin surface sampling) and indirect (*e.g.*, environmental surface sampling, videography) techniques for measuring dislodgeable residues on a surface or residues transferred to skin have been developed, including: (1) surrogate skin surface methods, such as cotton dosimeter patches, gloves, garments and whole body dosimeters (Roberts and Camann 1989, Ross *et al.* 1990), (2) removal methods, such as hand rinse or wipes (Lewis *et al.* 1994, Geno *et al.* 1996, Fenske and Lu 1994), (3) videotaping of human volunteers and analysis of dermal contact variables, such as duration of frequency of contact of each body part (Zartarian *et al.* 1995, 1997ab, 1998), (4) videoimaging of skin surfaces exposed to fluorescent tracer compounds (Thompson and Roff 1996, Fenske *et al.* 1986, Fenske 1993), and (5) surface sampling methods, such as surface extraction, surface wipes, dust collection, drag sleds, roller methods, and hand press (USEPA 1990, Lioy *et al.* 1993, Freeman *et al.* 1996, Lewis *et al.* 1994, Black and Fenske 1996, Ross *et al.* 1991, Vaccaro *et al.* 1996, USEPA 1996b).

Total theoretical surface residues in the residential environment may be estimated based on (1) label use rates or based on actual use trials in which the amount of product used is measured by weight loss, or (2) measurements of amounts applied by analysis of foil-backed denim coupons placed prior to application. In the latter case, an adequate time period should be allowed (usually on the order of 1 or 3 hours) to permit deposition of the majority of the mass of the aerosols released from the product delivery system (*e.g.*, aerosol can).

The percent dislodgeability of the sprayed consumer product after application can be determined by several methods including the drag sled method, the polyurethane foam (PUF) roller, cloth roller, manual wipe methods, and hand press. USEPA (1996b) did a comparison of dislodgeability methods for a variety of pesticides (chlorpyrifos, pyrethrin I, piperonyl butoxide, and methoprene), focusing primarily on the drag sled, PUF roller, and hand press techniques. For all pesticides tested, the drag sled removed the highest percentage of residues. The drag sled provided a 3- to 4-fold higher transfer of chlorpyrifos, PBO and pyrethrin I from plush carpet than did a single pass of the PUF roller. The drag sled produced about a 9-fold higher dislodgeability than the hand press method. An order of magnitude or higher proportion of residue could be removed from vinyl flooring than carpet. Significant reductions over approximately a month in the amount of chemical that can be dislodged were observed for both the drag sled and PUF roller methods. For example, the dislodgeability from plush carpeting was reduced by approximately 40-fold and 130-fold on 31 days after application relative to dislodgeability 1 day after application for PBO and pyrethrin I, respectively. This reduction in dislodgeability is thought to be due to decreased availability from the matrix of the surface rather than due to volatilization (although this may play a role for some compounds). For a given method, dislodgeability was more strongly related to flooring surface than chemical identity, although there were clearly chemical-specific differences in the rate of decline of dislodgeable residues for each surface type and method tested.

How the actual product is applied to the surface is critical in determining the usefulness of the data. Hand-held application of a product can lead to high variability in the data, such that time course information may be obscured by the variability. Total release foggers result in nonuniform distributions of released material, reflected in spatial heterogeneity in the deposition coupons (Ross *et al.* 1990).

Home pesticide foggers typically result in deposition of 55 to 69 percent of its contents onto the floor of the residence, thus providing considerable potential for transfer of pesticide residue to a person in contact with the floor. A number of case studies are available in the literature that provide useful examples of indoor surface residue measurements. These represent useful surrogate data for performing assessments of exposure to dislodgeable residues of pesticides in the indoor environment. Most of these studies are the work of Ross *et al.* (1990, 1991, 1992), Vaccaro (1992), and Vaccaro *et al.* (1996) on the widely used insecticide chlorpyrifos. In addition to using standard dislodgdeability methods, some of these studies have used structured activities (such as a Jazzercise routine) to standardize the contact regime with the treated floor surface (Ross *et al.* 1990), followed by quantifying exposure using gauze patches or clothing as external dosimeters, or using biomonitoring (*e.g.*, via blood and urine samples) to characterize absorbed dose.

Using the CDFA cloth roller method (see above), Ross *et al.* (1991) investigated the dislodgeability of chlorpyrifos and d-trans allethrin from nylon carpeting, following application of chlorpyrifos from a commercial total release fogger. Only about 1 to 3 percent of the total residue on the nylon carpet was transferrable by this method 6 to 12.5 hours after application. Consistent with the USEPA (1996b) study, dislodgeability declined over time (Ross *et al.* 1990, 1991). A These results suggest that the potential reservoir of dislodgeable residues is limited, and that the available "pool" of chemical decreases following initial contact so that subsequent contacts will not transfer nearly as much as the initial contact. The transfer from carpet to cotton cloth correlated highly with transfer to cotton clothing worn by individuals exercising on the treated carpet.

Several reasons for the observed decline in transferability of residues from indoor surfaces over time have been proposed. These include (1) entry of pesticide molecules into the matrix of the flooring material where they are no longer accessible, (2) chemical degradation, (3) volatilization of the pesticide, or (4) migration and adsorption to particulate matter deposited near the carpet backing (Ross *et al.* 1991). For stable nonvolatile pesticides, incorporation into or association with flooring material can be postulated as a possible loss mechanism responsible for declining dislodgeability of low volatility organic compounds over time.

Several studies indicate that dislodgeable residues can contribute a significant oral exposure component, particularly in infants and toddlers, through hand to mouth activity (Ross *et al.* 1992, Vaccaro 1992, Vaccaro *et al.* 1996). Potential hand to mouth exposures in adults is significantly increased by smoking activity (Krieger *et al.* undated). Ross *et al.* (1992) estimated the following route-specific attribution of total exposure to chlorpyrifos after treatment with a fogger product: adult—70 percent dermal, 16 percent oral, and 14 percent inhalation; infant—29 percent dermal, 67 percent oral, and 4 percent inhalation. The major difference in the Ross *et al.* (1992), Vaccaro (1992), and Vaccaro *et al.* (1996) estimates for oral attribution of total exposure relates to the assumption of less hand area entering the mouth (5 percent for adult, 50 percent for the infant) in the Ross *et al.* (1992) study. The dermal exposure component resulting from contact with dislodgeable residues on indoor residential surfaces is substantially reduced if dermal absorption is limited, as in the case of disodium octaborate tetrahydrate used as a carpet flea treatment (Krieger *et al.* undated). In contrast oral exposure to boron compounds via oral exposure is of great concern in children due to moderate to high oral toxicity.

Methods for Characterizing Particle-Bound Contaminants on Surfaces

Consumer products used indoors or tracked indoors from outdoor use can associate with house-dust or soil particles in the indoor environment. Some residues of pesticides may volatilize or be resuspended into indoor air and associate with airborne particles as well. Carpet fibers and carpet backing receiving this contaminant loading from indoor consumer product use may act as long-term reservoirs that transfers chemicals (*e.g.*, pesticides) to house dust. Indoor activities such as walking or vacuuming may resuspend the particles into indoor air.

Characterization of particle-associated chemicals in the home originating from consumer product use is a relatively new area of research. Several vacuum-based devices for collecting a sample of particle-bound contaminants from a floor surface, such as a carpet have been developed, such as the Small High Volume Surface Sampler (HVS3), now memorialized as ASTM Standard Practice D 5483–93 (Camann 1994). In several California homes, carpet dust collected from heavily trafficked carpeted areas of the homes (*i.e.*, family rooms) had on the order of approximately 100 ppm Dieldrin and DDT, 14 ppm methoxychlor, and 1 to 6 ppm benzo(a)pyrene (Camann 1994), when sampled in 1993 using this sampling device. More remote rooms, such as the bedroom, were reported to have several orders of magnitude lower pesticide levels in carpet dust, but levels of benzo(a)pyrene comparable to that in family room carpet dust (Camann 1994). These results are notable, given that DDT and dieldrin were banned more than a decade prior to the study. Low concentrations of pesticides in bedroom carpet dust indicates that these pesticides were not disseminated evenly throughout the residence. Low concentrations of pesticides in entry mats and foundation surface soil suggest that an indoor source, possibly involving long-term recycling of particles in the indoor environment, are responsible for the current levels. This suggests that once contaminated particles are introduced to the indoor residential environment, considerable recycling of particles in the residence may occur for very long periods of time, potentially measured in years.

Additional Case Studies

Furtaw *et al.* (1996a) conducted a large chamber study of VOC emissions from marker pens, including dry-erase markers and permanent felt-tip markers. The permanent felt-tip markers were determined by GC-MS to emit ethylbenzene and xylenes, and the dry-erase markers were determined by GC-MS to emit methyl isobutyl ketone. The total VOC emissions from the marker pens were in the range of 24 to 130 mg per minute of use. Personal air monitoring was conducted for the subject who uncapped and wrote with the pen on a whiteboard for a defined period of time inside a 9 ft × 12 ft. × 8 ft (24.5 m^3) stainless steel chamber. Samples were directed into a GC/MS with a cryogenic sample preconcentration trap at 10 minute intervals. By comparing the levels of methyl isobutyl ketone (MIBK) measured in the personal air samples to MIBK levels collected at more distal points in the chamber, Furtaw *et al.* (1996a) were able to determine that the breathing zone samples were elevated relative to the corner of the chamber, thus verifying a "source-proximate" effect. Furtaw *et al.* (1996a) were also able to verify that sinks inside the chamber were reversible and became a source of emissions after airborne levels of MIBK from pen usage start declining by comparing the results of the actual airborne concentration *versus* time curve to that predicted by a well-mixed model using the actual SF_6-based air exchange rate for the chamber.

MONITORING TO CHARACTERIZE EXPOSURE TO OUTDOOR-USE PRODUCTS

Several monitoring studies have measured homeowner mixer/loader/applicator and residential reentry exposures to lawn and garden pesticides (*e.g.*, Harris *et al.* 1992, Harris and Solomon 1992, Solomon *et al.* 1993, Stephenson *et al.* 1996, Vaccaro *et al.* 1996). In a study by Harris *et al.* (1992), homeowner exposures associated with application of 2,4-D to lawns was measured for liquid and granular formulations with and without the use of protective equipment. Normalized absorbed doses of 2,4-D (mg per lb. active ingredient handled) were estimated from this study based on biomonitoring (*i.e.*, post-application urinary levels of 2,4-D). On average, 2,4-D exposures associated with use of granular products were almost 20 times lower than exposures associated with liquid products, when normal clothing was worn (Harris *et al.* 1992). In a study by Stephenson *et al.* (1996), homeowner mixer/loader/applicator exposures to chlorpyrifos associated with use of a granular formulation applied at 1 kg a.i./hectare via a cyclone spreader, versus a liquid formulation (hose-end application) applied at 5 kg a.i./hectare were compared, using biomonitoring for the metabolite 3,5,6-trichloropyridinol (TCP). When adjusted for application rate, the mean normalized absorbed doses of chlorpyrifos were roughly 15 times lower for application of granular formulation using a cyclone spreader compared to mean exposures for hose-end application of a liquid formulation (Stephenson *et al.* 1996).

Harris and Solomon (1992) measured exposures to 2,4-D for 10 individuals engaged in 1 hour of structured activities (walking, sitting, laying on turf) on residential lawns starting at 1 hr and 24 hr after application. The highest absorbed dose measured by Harris and Solomon (1992) was 5.36 µg per kg body weight for an individual with minimal clothing (*i.e.*, shorts only). Vaccaro *et al.* (1996) measured residential reentry exposures to chlorpyrifos following application at 4 lbs a.i./acre by a professional lawn care operator; at four hours after application, human volunteers engaged in structured activities consisting of touch football, sunbathing, frisbee, weeding and/or picnicking for 4 hours. Dislodgeable residues were measured using a weighted drag sled device; biomonitoring for the metabolite TCP indicated a mean total absorbed dose of approximately 7 µg/kg for adults and 13.5 µg/kg when scaled to children. Use of physicochemical measurements and predictive models yielded a strikingly similar absorbed dose estimate of 10.5 µg/kg for adults.

Important considerations in assessing exposures to outdoor products applied to a surface are: (1) the extent of dislodgeable residues, and (2) the rate at which the constituent chemicals will dissipate from the surface (*e.g.*, lawn, painted wall). Clearly, the lower the amount of chemical dislodged by contact and the longer the delay in initial contact after application, the lower the dermal exposure will be. A number of studies have examined the dislodgeability of pesticide residues from turf. Examples of these include Vaccaro *et al.* (1996) and Schneider *et al.* (1994). Several studies have examined the rate of dissipation of pesticide residues on turf, including Hurto and Prinster (1993). Dislodgeable residue measurements taken using wipe samples indicated a rapid decline in transferable residues from approximately 8 percent at 1 hr after application of 2,4-D to turfgrass, to 1 percent at 24 hours following application (Harris and Solomon 1992, Solomon *et al.* 1993). The dislodgeability results are largely consistent with those of Thompson *et al.* (1984), who found that 5 percent of the applied 2,4-D residues were dislodgeable by wipe techniques shortly after application. Hurto and Prinster (1993) measured the dislodgeability of 5 pesticides (chlorpyrifos, DCPA, diazinon, isophenfos, and pendimethalin) from turfgrass. At 2 hrs post-application, the dislodgeability (based on a detergent stripping procedure) ranged from a low of 0.6 percent for chlorpyrifos, to a high of 10.7 percent for isophenfos. Total residue on the turfgrass declined significantly within 2 days post-application. For example, DCPA levels decreased by approximately 66 percent and 25 percent for irrigated and nonirrigated plots, respectively (Hurto and Prinster 1993). Schneider *et al.* (1994) measured dislodgeable residues of diazinon from treated turf using "shoe shuffling" (cheesecloth as the dosimeter) or the CDFA roller (which uses a section of percale sheet as a dosimeter). Less than 1 percent of the diazinon was transferable based on the CDFA roller method. Both the Hurto and Prinster (1993) and the Schneider *et al.* (1994) studies indicated that dislodgeable residues can be substantially reduced when the applied formulation is "watered-in" immediately following application. Outdoor-use turf chemicals can become a source of indoor exposures by tracking of residues from residential turf to carpets, as shown in the case of 2,4-D by Nishioka *et al.* (1996).

Only a limited number of studies have measured exposures to chemicals in products that are applied by brushing outdoors (Roff and Thompson 1995). Dermal exposures during outdoor application of a water-based wood preservative containing 0.2 percent permethrin or a solvent-based formulation containing 0.16 percent permethrin were visualized and quantified using a fluorescent tracer method. This study concluded that: (1) most of the exposure occurred on the hands of the painters, (2) the covering of arms and legs reduced exposures by a factor of 20, (3) subjects using solvent-based formulation had higher dermal exposures (by a factor of almost 4) than subjects using water-based formulation, and (4) the maximally exposed subject received 8 mg of permethrin from 30 minutes of painting, corresponding to dermal exposure to 5.5 ml of formulation.

A few studies have examined indoor residential exposures that occur following application of termiticides outside of the structure of the residence (*e.g.*, under crawl spaces, under concrete slab foundations, soil injection outside around the perimeter of the residence). Wright and Leidy (1982) found detectable levels of chlordane and heptachlor in the indoor air of 6 houses for a period of one year following termiticide application, with chlordane levels ranging from 0.3 to 5.8 $\mu g/m^3$ and levels of chlordane ranging from 0.01 to 1.8 $\mu g/m^3$. Louis and Kisselbach (1987) conducted airborne sampling of indoor air in 13 houses at several points (1 week, 3 months, 1 year) during the year following treatment of the residences with termiticides. Levels of chlordane ranged from nondetect to approximately 1 μg/m^3 in the living areas of the residences over the one-year monitoring period. Some of the pre-treatment levels were equivalent to the post-treatment levels, even at 1 week post-application. Post-application air levels in the basement and crawl spaces of treated homes were substantially higher than levels in living areas of the residences by a factor of approximately 2- to 7-fold. Vaccaro (1992) measured air levels of chlorpyrifos and solvent following termiticide treatment of residences with Dursban®. Air levels for house with crawl spaces or slab construction were generally less than 1 μg/m^3, with the exception of the 24 hr post-application levels in some houses. Houses with basements had somewhat higher levels, with maximum levels ranging up to approximately 5 μg/m^3 at 24 hours post-application, dropping to approximately 3 μg/m^3 at 30 days post-application and up to 0.5 μg/m^3 at 1 year post-application. A peak level of xylene-range aromatic compounds, consisting of trimethylbenzene, 1-methyl-ethylbenzene, N-propylbenzene, and xylene, at 0.6 to 4.5 ppm occurred in the first 24 hours in homes treated with Dursban. This level fell rapidly in 2 or 3 days after application to on the order of 0.1 ppm, where it remained for several months. The persistent compounds, chlordane and heptachlor, are no longer used as termiticides. They have largely been replaced by synthetic pyrethroid termiticides such as permethrin (Vaccaro 1992), that have 100- to 1,000-fold lower vapor pressures than chlorpyrifos, and chemicals that act as growth regulators to prevent maturation of the worker termites. These developments will likely lead to reductions in residential exposures following termiticide applications.

MODELING (PREDICTIVE EXPOSURE ASSESSMENT METHODS) AS A WAY OF ESTIMATING EXPOSURES TO CHEMICALS IN INDOOR-USE PRODUCTS

Modeling of Inhalation Exposures to Chemicals in Indoor Consumer Products

Estimation of human inhalation exposures to chemicals in indoor residential environments as a result of use of consumer products requires integration of time-varying chemical emissions and associated airborne levels with information on individuals locations in the residence, activities, and associated inhalation rates (USEPA 1997). Often the airborne concentration term and its time profile may be estimated using an indoor air quality (IAQ) model. The accuracy of the concentration term is greatly enhanced by a realistic depiction of the emissions profile for the product use scenario, either through emissions monitoring or modeling. The timing of the product release relative to the presence of the user or other individuals in the residence, as well as key factors that are reflective of product labeling (*e.g.*, ventilation conditions) are critical to obtaining a scientifically defensible and meaningful exposure assessment.

Adjustment of Emission/Concentration Data for Surrogate Chemicals. Often, emissions or airborne concentration data for a chemical of interest in a consumer product are not available. It is often possible to find relevant surrogate emissions data or airborne concentration data for chemicals applied to surfaces for the initial (volatilization-driven) phase of the emission from the surface. Hartley (1969) empirically determined a relationship that allows for correction of surrogate emissions or airborne concentration data based on physicochemical properties for the chemical of interest, assuming all other conditions relating to volatilization of a compound from a surface are the same, including initial loading of chemical on the surface, air exchange rate and type of surface. To obtain the corrected value, the emission rate (E) or airborne concentration, C, is multiplied by the ratio of the products of the vapor pressure (VP) and square root of the molecular weight (MW) of the compound in which the assessor is interested (Compound 1, Equation 1) to that for which data are available (Compound 2, Equation 2). In other words:

$$E_1 = E_2\left(\frac{VP_1\sqrt{MW_1}}{VP_2\sqrt{MW_2}}\right) \tag{1}$$

$$C_1 = C_2\left(\frac{VP_1\sqrt{MW_1}}{VP_2\sqrt{MW_2}}\right) \tag{2}$$

Typically, the assessor will also need to adjust the airborne concentration for the surrogate compound proportionally to account for differences in use rate (UR, for instance, g/m^2), and area of surface treated (A) for the chemical of interest versus that for which surrogate data are available. In this case, the above equation is expanded as follows:

$$C_1 = C_2\left(\frac{VP_1\sqrt{MW_1}}{VP_2\sqrt{MW_2}}\right)\left(\frac{UR_1}{UR_2}\right)\left(\frac{A_1}{A_2}\right) \tag{3}$$

Use of the Theoretical Saturation Concentration as a Screening-tool for Inhalation Exposures to Vapor-chemicals. For semivolatile and relatively nonvolatile chemicals, the vapor inhalation exposure pathway can often be eliminated as a significant contributor to indoor residential exposures by calculating a theoretical saturation concentration for the chemical of interest based on its physicochemical properties, and calculating the inhalation exposure that would be associated with this maximum worst-case vapor concentration. The saturation concentration in air is calculated using the ideal gas law:

$$PV = nRT \tag{4}$$

where

P = pressure (atm)
V = volume (cm^3)
n = number of moles = g/molecular weight
R = universal gas constant (82.1 cm^3-atm/mole-°K)
T = temperature (°K)

Solving for mass:

$$\text{Mass(g)} = \frac{\text{Vapor pressure (mm Hg)} \times \text{Room volume (m}^3) \times \text{MW (g/mole}}{\text{Temp}(^\circ\text{K}) \times \dfrac{82.1\,\text{cm}^3\cdot\text{atm}}{\text{K}\cdot\text{mole}} \times \dfrac{1\,\text{m}^3}{10^6\,\text{cm}^3} \times \dfrac{760\,\text{mm Hg}}{\text{atm}}} \tag{5}$$

Then, the saturation concentration (C_{sat}) is calculated by dividing the mass by the room volume. The intake via inhalation (I_a) corresponding to this saturation concentration, which can be scaled by time to represent either the application exposure, the post-application exposure, or both depending on the exact nature of the product use scenario, can be calculated as follows:

$$I_a(\mu g/day) = C_{sat}(\mu g/m^3) \times \text{Duration}(hr/day) \times \text{Inhalation rate}(m^3/hr) \quad (6)$$

Indoor Air Models for Consumer Products Used When the Emission Rate Is Known. As noted in Chapter 4, the general form for a single-chamber indoor air model when the emission rate is known is:

$$C_t = \frac{S}{aV} + \left(C_{t-1} - \frac{S}{aV}\right)e^{-\Delta t} \quad (7)$$

where

C_t = Concentration in the chamber at time t (mg/m^3)
C_{t-1} = Concentration in the chamber for the measurement immediately **prior to** time t (*i.e.*, t − 1)
S = Emission rate (mg/hr)
a = Air exchange rate (hr^{-1})
V = Volume of chamber (m^3)
Δt = Time increment of interest (hr)

For many indoor emission involving one-time use or intermittent use of a product, the emission declines over time in an exponential or double exponential manner. Thus, for a chemical such as toluene in nail polish, which exhibits an exponential decline in emissions over time, the time-dependency of the emission can be built into the indoor air model using the following emission term:

$$S_t = S_0 e^{-rt} \quad (8)$$

where

S_t = Emission rate at time t (mg/hr)
S_0 = Initial emission rate (*i.e.*, at time 0)
r = Rate constant to account for decrease in the emission rate over time (hr^{-1})
t = Time (hr)

The nonlinear solution for concentration at time t then becomes (Nagda *et al.* 1995):

$$C_t = \frac{S_0}{V(a-r)}\left(e^{-rt} - e^{-at}\right) \quad (9)$$

A number of general indoor air models are available (see Chapter 4) that can be applied to determination of airborne levels of vapors or particles. A number of validated

USEPA modeling tools exist to address indoor airborne levels of chemicals from many types of emission sources. Some of the desirable features one would look for in such a model would be:

- The capability to represent air flows between two or more rooms (zones) with each other and with outside air
- The ability to turn sources on and off in a specified way in one or more rooms (zones)
- The ability to build in a decay term to represent chemical transformation/sink adsorption or deposition (in the case of aerosols or particles)
- Data files containing typical measured room volumes and air exchange rates in different regions of the country, and
- The option to artificially construct a hypothetical residence with user-specified room volumes and air exchange rates.

One example of a model that has all of these capabilities is the *Multi-Chamber Concentration and Exposure Model* (MCCEM) developed for the Environmental Monitoring Systems Laboratory, U.S. Environmental Protection Agency, Las Vegas (USEPA 1991). MCCEM is a user-friendly computer program that estimates indoor concentrations for, and inhalation exposures to, chemicals released from products or materials used indoors. Concentrations can be modeled in as many as four zones (*e.g.*, rooms) in a building. The user provides values for emission rates, the zone where the source is located, the zone where exposure occurs, duration of exposure, air exchange rates, the nature of the building, and whether a short-term model (including average and maximum peak values) or long-term model is desired. The model contains room volume data and measured air flow rate data between different rooms for different building configurations and different geographic locations, or the user may build a hypothetical house or building, assigning the desired room (zone) volume and air exchange rates. Other examples of similar modeling tools include several USEPA models as well as the CONTAM model developed and updated regularly by the National Institute of Standards and Technology (NIST 1994).

A new database/model management tool developed by the University of Nevada at Las Vegas for the Environmental Monitoring Systems Laboratory, U.S. Environmental Protection Agency, Las Vegas, is anticipated to revolutionize the modeling of indoor air exposures. This software tool is called the Total Human Exposure Risk Database and Advanced Simulation Environment (THERdbASE). This software integrates a number of indoor air models (as well as other pathway models) with distributional data on variables such as demographics, time-activity, food consumption, and physiological parameter data that can be subset according to the needs of the assessment (Pandian *et al.* 1995). THERdbASE can also be used for estimating dermal and ingestion exposures and total human exposure via multiple agents and pathways (*i.e.*, multiple agents present in more than one media and coming into contact with humans via multiple exposure pathways and routes).

Nagda *et al.* (1995) compared the results obtained with MCCEM, using exponential decay of emission rates, to those obtained by direct monitoring. While the overall correspondence is reasonably good for the room of product use (*e.g.*, bedroom), the modeled peak concentration values during the first several hours of exposure in the room of use appear to overestimated somewhat compared to monitored concentrations for toluene, and the correspondence in the adjacent zone is not as good as in the zone of use (Nagda *et al.* 1995). It has been suggested by Nagda *et al.* (1995) that the correspondence between measured and predicted (*i.e.*, modeled) concentrations be tested routinely using the following statistical benchmarks borrowed from ASTM (1991):

- Correlation coefficient of 0.9 or greater
- Regression slope between 0.75 and 1.25
- Regression intercept of 25 percent or less of the average measured concentration
- Normalized mean square error of 0.25 or lower
- Fractional bias of the mean concentrations of 0.25 or lower, and
- Fractional bias of the variance of the concentration of 0.25 or lower.

Jayjock (1994) has developed a model that estimates airborne concentrations indoors for chemicals that are released from indoor sources primarily through volatilization. For large sources where volatilization drives the airborne levels (*e.g.*, rugs, painted walls, spills), the airborne concentration can become a significant portion of the equilibrium partial pressure, thus reducing the net volatilization rate. This significant retarding pressure is termed "back pressure". Jayjock (1994) presents a back pressure model for estimating airborne levels over time for indoor volatilizing sources greater than a few hundred square meters of surface area in large rooms, or about for sources covering about 10 percent of the floor surface area for small rooms (*i.e.*, rooms smaller than 30 m^3. A reversible sink model has been developed by Jayjock *et al.* (1995) to describe the behavior of isothiazolone biocide volatilized from treated wood into glass chambers and into indoor residential environments. In a reversible sink model, the chemical is assumed to adsorb to sinks such as walls, and then desorb back into the air as the airborne level falls. Adsorption and apparent degradation of isothiazolone on the chamber walls resulted in equilibrium airborne concentrations that were about one-third of those predicted without consideration of these mechanisms. This model has been modified to reflect actual residential settings, based on monitoring studies, and overestimates actual room concentrations within an acceptable factor of 2 to 4 (Jayjock *et al.* 1995). A reversible sink model that describes the airborne levels of alkanes emitted from wood stain applied indoors was described by Chang and Guo (1994). An important consideration is that model development should be followed by model validation with actual monitoring data, whenever possible, such as in the case of Nagda *et al.* (1995), Jayjock *et al.* (1995), and Chang and Guo (1994).

Indoor Air Models for Consumer Products When the Emission Rate Is Unknown. To date, only a few indoor air modeling tools have been developed specifically for indoor use consumer products where the emission rate term is unknown. One example is the Screening-Level Consumer Inhalation Exposure Software (SCIES) developed by the U.S. Environmental Protection Agency (USEPA 1994). The SCIES model calculates screening-level estimates of time-weighted average inhalation exposures to chemicals in consumer products in 11 product categories for individuals in the room of use and for individuals in an adjacent zone. The model uses physicochemical properties of the chemical of interest, specifically molecular weight, vapor pressure, and density along with use rate and weight fraction of chemical in the product to estimate an emission rate for the chemical. SCIES also contains default values for volume and air exchange rate of the anticipated room of product use. Example default values used in SCIES are shown in Table 8-3. Many of the default values used in SCIES are "conservative," (*i.e.*, likely to overestimate actual exposures). An example of this is that some of the values for frequency of use represent "high-end" (*e.g.*, 90th percentile) values, and some of the air exchange

Table 8-3. Example product-specific default values used in the screening-level consumer inhalation exposure software (USEPA 1994)

Product category	Frequency of use (events/year)	Mass product used per event(g)	Product density (g/cm^3)	Volume of room of use (m^3)	Whole house volume (m^3)	Exchange with outdoor air(ACH)	Duration of exposure (hr)
Liquid fabric softeners	169 (semivolatiles) 338 (volatiles)	60	—	20	292	0.2	0.33 (semivolatiles) 0.117 (volatiles)
Automobile vinyl upholstery cleaner	2	163	0.99	2.4	—	12.5	1.18
Aerosol paint	6	133	0.88	40	292	0.5	0.66
Latex wall paint	6	2,267	1.24	40	292	0.2	4.9
Oil-based paint	6	1,118	1.35	40	292	0.2	3.2

rates are less than the national average (summarized in Pandian *et al.* 1998; refined in Murray and Burmaster 1995) and inconsistent with product labeling for those products (*e.g.*, paints) that require use with adequate ventilation. Thus, the results obtained from SCIES should not be assumed to represent "typical" airborne concentrations from use of specific consumer products in residences. Rather, the airborne concentrations predicted by SCIES are likely to represent significant overestimations of actual exposures. The reader should refer to Chapter 4 for more generally applicable IAQ models.

The "Personal Cloud" Effect. There are many types of indoor use consumer products that are used in the immediate vicinity of humans and which could lead to high levels of chemicals in the breathing zone of an individual. Examples include volatile chemicals in cleaning products, cosmetics, grooming/personal hygiene products, art/hobby supplies, and aerosols/particulate matter generated by use of such products. By their nature, these types of products are used by consumers in a way that results in emissions being released in very close proximity to the user's breathing zone. Assessments of exposure to chemicals in indoor residential air typically use models that assume complete mixing and homogeneous concentrations throughout a building or room (Furtaw *et al.* 1996b). However, practical experience and experimental data suggest that airborne concentrations of chemicals in residential air tend to be greater in close proximity to the location of the source than at a distal point in the same room (Furtaw *et al.* 1996b). This increased airborne concentration in the immediate vicinity of the consumer is referred to as the "personal cloud" effect or "source-proximate effect" (SPE). This effect is particularly strong during the initial period of product use when emission rates are maximal. This period is referred to as the "alpha" period. When emissions decline, airborne concentrations become closer to the well-mixed situation. This is referred to by some researchers as the "gamma" period (Furtaw *et al.* 1996b).

Furtaw *et al.* (1996b) validated the existence of the "personal cloud" effect for vapor phase materials using the tracer gas sulfur hexafluoride (SF_6). They found that the extent of the discrepancy between personal and distal airborne concentrations in the same room increased with increasing distance from source and with increasing room air flow rate. For SF_6 at "typical" air flow rates (air recirculation rate of 10 room volumes per hour), the average airborne concentration at arm's length (approximately 0.4 meters) exceeded the theoretical well-mixed concentration by a ratio of approximately 2:1 (Furtaw *et al.* 1996b). This effect has been observed for a number of settings, including personal toluene exposure from nail polish use in residences, where breathing zone concentrations of toluene during application were 1.2 to 11 times higher than in other areas in the room of application located 0.61 to 2.6m away from the test subject (Curry *et al.* 1994). The airborne concentration data from the study of Furtaw *et al.* (1996b) fit a two-compartment model reasonably well, where the source is located in a virtual small compartment, and the source compartment is connected with the room compartment via a stochastic air transfer rate parameter, reflecting the variability of air flow, flow direction, turbulence, and mixing.

The Particle Total Exposure Assessment Methodology (PTEAM) study, which focused on measuring personal exposures to inhalable particles (PM_{10}) of approximately 200 Riverside, California residents support the finding of a "personal cloud" effect for particulate material in the residence as well (Clayton *et al.* 1993). A major finding from this work is that personal exposures to particles in the daytime are approximately 50 percent greater than either general indoor or outdoor concentrations, on average. These data have been hypothesized to suggest that individuals are exposed to a "personal cloud" of particles as they go about their daily activities (Rodes *et al.* 1991). Resuspension of household dust via walking in the residence, such as contaminated soil particles tracked into the home, and certain household activities such as vacuuming and cooking, or sharing a home with a smoker lead to significant particle exposures. Rodes *et al.* (1991) reported typical ratios of 1.2 to 3.3 for personal concentration of a contaminant to that for area concentration in the residence. These ratios appear to be lognormally distributed and depend on the proximity of the source to the consumer. This subject is further discussed in Chapter 4.

Attempts have been made to develop nonsteady-state fugacity-based models to provide improved ways of describing the time-dependent application and post-application indoor air exposures to chemicals released from consumer products (Matoba *et al.* 1993, 1994, 1995, 1998). Fugacity is the tendency of a chemical to escape from one medium (*e.g.*, wall surface) and to transfer to another medium (*e.g.*, air). Thus, taking a fugacity-based approach to modeling is a unified way of quantitatively describing the dynamics and connections of the various compartments in the residential environment. Fugacity-based models have been developed for estimation of application and post-application exposures associated with space-spraying of pesticides (Matoba *et al.* 1993), electric vaporization of pesticides (Matoba *et al.* 1994), and broadcast spraying of pesticides (Matoba *et al.* 1995). These models have been validated based on residential monitoring data and have been condensed into a Microsoft Excel-based simulation package known as *InPest* (Matoba *et al.* 1998). *InPest* is able to simulate the time-dependent concentrations of pesticide in the air as well as time-dependent residue levels on residential surfaces such as walls, floors, and ceilings. Thus, combining these predicted residue levels with certain assumptions, inhalation, dermal, and incidental oral exposures for applicators and occupants (including children) inside the residence can be estimated. *InPest* incorporates considerations of generation, deposition, and horizontal movement of aerosols (for 3 aerosol size ranges), adsorption and desorption, volatilization, degradation, and dissipation for water-based and oil-based formulations. The model can be run for different ventilation conditions (windows open or closed), and with a variety of flooring materials (Matoba *et al.* 1998).

A technical guidance document (TGD) prepared by the European Centre for Ecotoxicology and Toxicology of Chemicals contains a number of screening-level models (Appendices 1, 4, and 7 of the TGD) and product use data (Appendix 6 of the TGD) that are applicable to estimating residential exposures associated with consumer products (ECETOC 1994). Other European screening-level models include those of Vermeire *et al.* (1993) and Van de Meent *et al.* (1995). Van Veen (1995, 1997) has developed mechanistic models applicable to estimating exposures associated with consumer product use. This set of models is known as CONSEXPO (Van Veen 1997), which combines contact, exposure, and uptake models for inhalation, dermal and oral routes of exposure. CONSEXPO 2.0 (CONSumer EXPOsure) is being further developed at the RIVM, the Netherlands, to provide estimation routines for exposure to consumer products, based on the modeling framework described by Van Veen (1996).

Modeling of Dermal Exposures to Chemicals in Indoor Consumer Products

Dermal exposure of adults and children in the indoor residential environment (*e.g.*, following the application of a consumer or professional pesticide product indoors) may result from contact with treated surfaces, such as carpets, vinyl and hardwood floors, and treated pets. Dermal exposure represents a dynamic complex process affected by multiple rate-limiting phenomena and highly variable factors (USEPA 1992, USEPA 1996b, Zartarian and Leckie 1998). In indoor residential environments, this complexity can be attributed, in part, to:

(1) Time-dependent dislodgeability and transferability or residues (*e.g.*, pesticides from fogger treatment) for different surface types (*e.g.*, carpet, hardwood floors, upholstery), and
(2) Related differential dermal "body part" surface area loading (Ross *et al.* 1990, 1991) as a function of human time-activity patterns and biomechanical variables (*e.g.*, frequency and duration of body part surface area contacted, pressure exerted by body region on residential surface contacted).

Dermal exposure assessment for certain subpopulations, such as infants and children, must also address secondary pathways, such as incidental ingestion associated with hand-to-mouth activity (Zartarian *et al.* 1998).

There are numerous opportunities for dermal exposure to chemicals in the residential environment. These include, but are not limited to: direct contact with cleaning/laundry products (*e.g.*, cleanser, laundry detergent) during use, indirect contact with cleaning product residues (*e.g.*, laundry detergent residues in washed clothing), contact with dislodgeable residues of a chemical after use (*e.g.*, crawling infant contact with pesticide residues on a carpet), and direct contact with materials that are intentionally applied to the skin (*e.g.*, soap, cosmetics). There are basically three types of approaches to assessing dermal exposures—(1) the film-thickness approach, (2) dermal-permeability based approaches (USEPA 1992), and (3) transfer-factor based approaches.

The film-thickness approach assumes that a uniform layer of a material (*e.g.*, liquid consumer product) is present on a certain area of the skin, and that all of the material in that layer is available for absorption. Default film thickness data, in the absence of data on the actual product of interest, are available in USEPA (1987c). Other variables that are unique to the film thickness approach are the density of the product (g/cm^3) and the percent dermal absorption anticipated during the per-event exposure period. Absorption can be assumed to be 100 percent for screening-level assessments, but severe overestimation of dermal exposure is likely to occur.

In contrast, dermal permeability-based methods recognize the fact that dermal absorption is a time-dependent process, and that under controlled conditions, the dermal penetration can be expressed as a time-dependent parameter known as the dermal permeability coefficient (K_p). Measured and estimated dermal flux ($\mu g/cm^2/hr$) and/or permeability coefficients (cm/hr) have been published for various substances that may occur in consumer products (USEPA 1992, Driver *et al.* 1993). Additional discussion/information regarding dermal exposure assessment and percutaneous absorption kinetics can be found in USEPA (1992), Kasting and Robinson (1993), and Wilschut *et al.* (1995).

Dermal exposure modeling approaches in the residential environment have traditionally relied on the use of conservative assumptions (such as 100 percent transfer from the contacted surface to the skin), surface residue levels, and worst-case time-activity assumptions for subpopulations of interest (USEPA 1997, ILSI 1998, Ross *et al.* 1990, 1991, Vaccaro *et al.* 1996). The development of more realistic models of dermal exposure in the U.S. is being facilitated by a variety of research and regulatory compliance programs initiated, in part, due to the requirements of the U.S. Food Quality Protection Act of 1996 (FQPA). FQPA establishes a single, consistent, health-based standard for pesticide exposures to residues in food and residential use products, which must be considered jointly. Relevant research programs that will provide a better understanding of residential exposures, including dermal exposures, are the USEPA National Human Exposure Assessment Survey (NHEXAS), and industry-sponsored residential exposure data development activities, such as the Outdoor Residential Exposure Task Force (ORETF), and the Pyrethrin/Piperonyl Butoxide/MGK-264/Deltamethrin Non-Dietary Exposure Task Force (NDETF), formed to respond to the USEPA's call for improved estimates of residential exposures.

An example of a more advanced approach to modeling post-application dermal exposures, in particular, involves the use of "transfer factors" (TFs). The unitless generic TFs represent an activity-specific basis for estimating dermal loading of residues on the skin (μg/cm²) for various anatomical body regions in conjunction with compound-specific transferable residue (TR) data. The general equation for estimating potential dermal exposure can be described as follows:

$$\text{Post-application dermal exposure}_{(\text{body parts (1-n)})} = \sum (\text{TF} \times \text{TR} \times \text{SA}) \qquad (10)$$

where

Transfer Factor (TF)	=	unitless "generic" body-part specific factor relating transferable residue values to dermal exposure associated with a specific activity
Transferable Residue (TR)	=	chemical-specific mass of residue per unit surface area transferring from a treated surface to a "surrogate skin" collection media, *e.g.*, cotton dosimeter (g/cm²)
Surface Area (SA)	=	skin surface area represented by a specific body part (cm²)

The transfer factor-based dermal modeling approach can be based on dermal transferability studies, such as that of Ross *et al.* (1990, 1991). In these studies, post-application transfer factors derived from human volunteer dermal dosimetry and transferable residue measurements for carpet were made following treatment of a carpeted room with a total release indoor fogger pesticide product. During the studies, the human subjects engaged in a high-contact structured activity known as Jazzercise® for approximately 20 minutes, while wearing dosimeter clothing that could later be extracted and analyzed for the active ingredient. Transferrable residue samples were obtained by use of an indoor roller method (Ross *et al.* 1990, 1991). The roller device was a cylinder that was manually rolled over a cotton cloth dosimeter placed on the treated carpet. The method was shown to transfer approximately 1 to 3 percent of the potentially available pesticide residues from the nylon carpeting to the collection media (cotton cloth). Transfer from the carpet to the cloth was found to be highly correlated with transfer to the cotton clothing worn by the subjects who exercised on the carpet.

It is important to note that transfer factor-based modeling methods inherently considers only the activities and time domain included in the dermal monitoring study from which the factors are derived. Therefore, if the design of the selected monitoring study from which the transfer factors are derived represents a high-contact activity over a short duration, the estimated exposures must be placed in perspective with more representative time-activity profiles (USEPA 1996b, ILSI 1998). For example, dermal exposures associated with 20 minutes of Jazzercise may be similar to or greater than an entire day of exposure associated with more typical everyday activities on the same type of floor surface (*e.g.*, sitting, walking, playing on floor surface), with all other product use and environmental conditions being equivalent (ILSI 1998).

The subject of dermal exposure assessment is discussed in further detail in Chapter 5.

Assessment of Incidental Ingestion Exposures to Chemicals in Consumer Products

Ingestion of chemical residues can occur in the home beyond chemical residues (*e.g.*, pesticides) consumed in food derived from nominally contaminated raw agricultural commodities (RACs) from spraying in the field. Primary examples of incidental residues include ingestion of cleaning agent and pesticide residues on plates and silverware following product use, and ingestion of trace levels of organics (*e.g.*, haloforms) in drinking water entering the home. Another important pathway for incidental ingestion exposure is hand-to-mouth behavior in infants and toddlers in particular. Vaccaro (1992) has shown this to be actually the predominant exposure pathway (for this age group) for exposures to pesticides applied to carpets either directly or incidentally (*e.g.*, through insecticide fogger use, such as a flea bomb), more so than inhalation or dermal contact with contaminated surfaces. For food-related incidental contact, it will be necessary to consider the nature of the toxicological endpoint (*e.g.*, short-term versus long-term health effects) to determine which type of dietary consumption data is most appropriate (*e.g.*, an upper bound on the amount eaten on one day in which the commodity is consumed, or long-term averages, which would include days on which the commodity is not consumed). This area is discussed in further detail in Chapter 6.

Modeling of Exposures to Chemicals in Outdoor Consumer Products

Some of the modeling methods for dermal exposure indicated above can be used or modified for outdoor consumer products. For example, the transfer factor based approach can be used for estimation of post-application exposures to pesticide residues on treated residential lawns. The more traditional transfer-coefficient (TC) approach, which has its origins in application to estimating dermal exposures for agricultural reentry workers (Popendorf 1985, Krieger *et al.* 1990), can be used to estimate dermal exposures for individuals contacting pesticide-treated foliage in the outdoor residential environment, whether it be treated turf, ornamentals, or residential flower and vegetable gardens. The transfer coefficient (TC) can be thought of as a the surface area of foliage contacted per hour for a specific activity (*e.g.*, weeding, fruit harvesting, Jazzercise on turf). The TC is believed not to be pesticide-specific, but rather dependent on the type of plants contacted, the nature of the activity, and the foliage height or degree of foliage at the time of contact. One of the key variables associated with estimation of outdoor post-application dermal exposures is the dislodgeable foliar residue (DFR), which is the mass of residue that is potentially available for contact per unit area of treated foliage. The DFR (and its decline curve) is pesticide-specific. The relationship between TC, DFR and dermal exposure, in its simplest terms, is expressed in Equation 11:

$$\text{Transfer Coefficient}\,(\text{cm}^2/\text{hr}) = \frac{\text{Dermal Exposure}\,(\mu\text{g/hr})}{\text{DFR}(\mu\text{g/cm}^2)} \qquad (11)$$

Thus, "generic" transfer coefficients for specific activities can be developed by applying a pesticide and determining post-application exposures and measuring DFR.

Equation 12 represents a general dermal exposure modeling approach that incorporates the concepts presented in Popendorf (1985):

$$\text{Dermal Exposure } (\mu g) = K_d \times t \times R^{(-kr)(T)} \tag{12}$$

where

k_d = activity-specific transfer coefficient (cm^2/hr)
t = duration of activity (hrs)
R_0 = initial dislodgeable foliar residue level ($\mu g/cm^2$)
k_r = dislodgeable foliar residue decay coefficient (fraction/day)
T = reentry interval, or number of days after application that contact activity occurs (days)

Thus, if the dislodgeable residue level is known for the day on which contact with treated foliage occurs (R_T), the dermal exposure is simply:

$$\text{Dermal Exposure } (\mu g) = K_d(cm^2/hr) \times R_T(\mu g/cm^2) \times t(hr) \tag{13}$$

One caution for exposure assessors is that transfer coefficients (TCs) obtained from agricultural worker reentry studies must be scaled downward based on productivity rate (*e.g.*, based on pounds strawberries harvested per hour) to obtain residential-equivalent TCs (*e.g.*, for harvesting strawberries from a backyard garden).

Modeling approaches are also available for a limited number of outdoor use consumer products. Dang (1995) has developed a model called SWIMODEL to estimate exposures to pool chemicals, including disinfectants, and chlorine reaction by-products, such as chloroform.

ACKNOWLEDGMENTS

The authors would like to acknowledge several individuals who provided assistance in completion of this chapter, including: (1) Dr. Yoshihide Matoba of Sumitomo Chemical Company, Environmental Health Science Laboratory, Hyogo, Japan, who provided extensive text on fugacity-based modeling, (2) Dr. Mark Van Veen of RIVM, the Netherlands, who provided text on exposure modeling for consumer products in Europe, and (3) Edwin Furtaw of the Environmental Monitoring Systems Laboratory/USEPA/Las Vegas, who provided text on the limitations of small-chamber studies.

REFERENCES

Aggazotti G., G. Fantuzzi, G. Predieri, E. Righi, and S. Moscadelli. 1994. Indoor exposure to perchloroethylene (PCE) in individuals living with dry-cleaning workers. Sci. Total Environ. 156:133–137.

ASTM (American Society for Testing and Materials). 1990. Standard guide for small-scale environmental chamber determinations of organic emissions from indoor materials/products. ASTM Standard D 5116-90. Philadelphia, Pennsylvania.

ASTM (American Society for Testing and Materials). 1991. Guide for statistical evaluation of indoor air quality models. ASTM Standard D 5157-91. Philadelphia, Pennsylvania.

Bayer C.W. and C.D. Papanicolopoulos. 1990. Exposure assessments to volatile organic compound emissions from textile products. Pages 725–730 in: Indoor Air '90. Proceedings of the 5th International Conference on Indoor Air Quality and Climate. Toronto, Canada. July 29–August 3, 1990. Volume 3.

Black, K.G. and R.A. Fenske. 1996. Dislodgeability of chlorpyrifos and fluorescent tracer residues on turf: comparison of wipe and foliar sampling techniques. Arch. Environ. Contam. Toxicol. 31:563–570.

Burge, H.A., W.R. Solomon, and J.R. Boise. 1980. Microbial prevalence in domestic humidifiers. Appl. Environ. Microbiol. 39(4):840–844.

Camann, D.E. 1994. Investigating the effects of pesticide exposure in the home. Technology Today. June 1994.

Ceschin-Roques, C.G., H. Hanel, S.M. Pruja-Bougaret, J. Luc, J. Vandermander, and G. Michel. 1991. Ciclopirox nail lacquer 8%: *In vivo* penetration into and through nails and *in vitro* effect on pig skin. Skin Pharmacol. 4:89–94.

Chang, J.C.S. and Z. Guo. 1992. Characterization of organic emissions from a wood finishing product—wood stain. Indoor Air 2:146–153.

Chang, J.C.S. and Z. Guo. 1994. Modeling of alkane emissions from a wood stain. Indoor Air 4:35–39.

Clausen, P.A., P. Wolkoff, E. Holst, and P.A. Nielsen. 1991. Long-term emission of volatile organic compounds from waterborne paints—methods of comparison. Indoor Air 4:562–576.

Clausen, P.A., B. Laursen, P. Wolkoff, E. Rasmusen, and P.A. Nielsen. 1993. Emission of volatile organic compounds from a vinyl floor covering. In: N.L. Nagda (ed.). Modeling of Indoor Air Quality and Exposure. ASTM STP 1205. American Society for Testing and Materials. Philadelphia, Pennsylvania. Pages 3–13.

Clayton, C.A., R.L. Perritt, E.D. Pellizzari, K.W. Thomas, R.W. Whitmore, L.A Wallace, H. Ozkaynak, and J.D. Spengler. 1993. Particle total exposure assessment methodology (PTEAM) study: Distributions of aerosol and elemental concentrations in personal, indoor, and outdoor air samples in a southern California community. J. Exposure Anal. Environ. Epidemiol. 3(2):227–250.

Colombo, A., M. De Bortoli, H. Knoppel, E. Pechchio, and H. Vissers. 1993. Vapour deposition of selected VOCs on indoor surface materials in test chambers. In: K. Saarela, P. Kalliokoski, and O. Seppanen (eds.). Volume 2. Chemicals in Indoor Air, Material Emissions. Pages 407–412. Proceedings of the 6th International Conference on Indoor Air Quality and Climate. Helsinki, Finland. July 4–8, 1993.

Curry, K.K., D.J. Brookman, G.K. Whitmyre, J.H. Driver, R.J. Hackman, P.J. Hakkinen, and M.E. Ginevan. 1994. Personal exposures to toluene during use of nail lacquers in residences: description of the results of a preliminary study. J. Exposure Anal. Environ. Epidemiol. 4(4):443–456.

Dang, W.T. 1995. The swimmer exposure assessment model (SWIMODEL) and its use in estimating risks of chemical use in swimming pools. Antimicrobial Division, Office of Pesticide Programs, U.S. Environmental Protection Agency. Washington DC.

De Bortoli, M., E. Pecchio, H. Schauenburg, H. Schlitt, and H. Vissers. 1993. Emission of formaldehyde, vinyl chloride, VOCs, and plasticizers from different wallcoating materials. In: K. Saarela, P. Kalliokoski, and O. Seppanen (eds.). Volume 2. Chemicals in Indoor Air, Material Emissions. Pages 413–418. Proceedings of the 6th International Conference on Indoor Air Quality and Climate. Helsinki, Finland. July 4–8, 1993.

Driver J.H., R.G. Tardiff, L. Sedik, R.C. Wester, and H.I. Maibach. 1993. In vitro percutaneous absorption of [^{14}C] ethylene glycol. J. Exposure Anal. Environ. Epidemiol. 3(3):277–284.

Driver, J.H. and G.K. Whitmyre. 1996. Assessment of residential exposures to pesticides and other chemicals. Pesticide Outlook. August 1996. Pages 6–10.

Dua, S.K. and P.K. Hopke. 1996. Hygroscopic growth of assorted indoor aerosols. Aerosol Sci. Technol. 24:151–160.

Dua, S.K. and P.K. Hopke. Undated. Hygroscopicity of indoor aerosols and its influence on the deposition of inhaled radon decay products. Unpublished paper. Department of Chemistry, Clarkson University. Pottsdam, New York. Pages 1–18.

Dunn, J.E. and B.A. Tichenor. 1988. Compensating for sink effects in emissions test chambers by mathematical modeling. Atmos. Environ. 22:885–894.

ECETOC (European Centre for Ecotoxicology and Toxicology of Chemicals). 1994. Assessment of non-occupational exposures to chemicals. Technical report 58. Brussels, Belgium.

Fenske, R.A. 1993. Dermal exposure assessment techniques. Ann. Occup. Hyg. 37(6):687–706.

Fenske, R.A. and C. Lu. 1994. Determination of handwash removal efficiency: Incomplete removal of the pesticide chlorpyrifos from skin by standard handwash techniques. AIHA Journal 55(5):425–432.

Fenske, R.A., S.M. Wong, J.T. Leffingwell, and R.C. Spear. 1986. A video imaging technique for assessing dermal exposure. I. Instrument design and testing. AIHA Journal 47:764–770.

Freeman, N.C.G., T. Wainman, and P.J. Lioy. 1996. Field testing of the LWW dust sampler and association of observed household factors with dust loadings. Appl. Occup. Environ. Hyg. 11(5):476–483.

Furtaw, E.J., R.M. Gonzalez, and M.D. Pandian. 1996a. VOCs from marker pens in exposure scenarios. Paper presented at the Air and Waste Management Association Symposium "Measurement of Toxics and Related Air Pollutants." May 7, 1996. Research Triangle Park, North Carolina.

Furtaw, E.J., M.D. Pandian, D.R. Nelson, and J.V. Behar. 1996b. Modeling indoor air concentrations near emission sources in imperfectly mixed rooms. JAWMA 46:861–868.

Geno, P.W., D.E. Camann, H.J. Harding, K. Villalobos, and R.G. Lewis. 1996. Handwipe sampling and analysis procedure for the measurement of dermal contact with pesticides. Arch. Environ. Contam. Toxicol. 30:132–138.

Gibson, W.S., F.R. Keller, D.J. Foltz, and G.J. Harvey. 1991. Diethylene glycol monobutyl ether concentrations in room air from application of cleaner formulations to hard surfaces. J. Exposure Anal. Environ. Epidemiol. 1:369–383.

Girman, J.R., A.T. Hodgson, A.S. Newton, and A.W. Winkes. 1986. Emissions of volatile organic compounds from adhesives with indoor applications. Environment International 12:317–321.

Hartley, G.S. 1969. Evaporation of pesticides. Pages 115–134 in: Pesticide formulation research: Physical and colloidal chemical aspects. American Chemical Society. Washington, D.C.

Harris, S.A. and K.R. Solomon. 1992. Human exposure to 2,4-D following controlled activities on recently sprayed turf. J. Environ. Sci. Health B27(1):9–22.

Harris, S.A., K.R. Solomon, and G.R. Stephenson. 1992. Exposure of homeowners and bystanders to 2,4-dichlorophenoxyacetic acid (2,4-D). J. Environ. Sci. Health B27(1):23–38.

Hendricks, M.G. 1970. Measurement of enzyme laundry product dust levels and characteristics in consumer use. J. Am. Oil Chemists' Soc. 47:207–211.

Hodgson, A.T. and J.R. Girman. 1987. Exposure to methylene chloride from controlled use of a paint remover in residences. Paper presented at the 80th Annual Meeting of the Air Pollution Control Association. New York, New York. June 21–26, 1987. Lawrence Berkeley Laboratory Report No. LBL-23078.

Hodgson, A.T., J.D. Wooley, and J.M. Daisy. 1992. Emissions of volatile organic compounds from new carpets measured in a large-scale environmental chamber. Paper presented at the 85th Annual Meeting of the Air & Waste Management Association. Kansas City, Missouri. June 21–26, 1992. Paper No. 92-79.15. Pages 1–19.

Hurto, K.A. and M.G. Prinster. 1993. Dissipation of turfgrass foliar dislodgeable residues of chlorpyrifos, DCPA, diazinon, isophenfos, and pendimethalin. Chapter 9, pages 86–99 in: Pesticides in Urban Environments. American Chemical Society, Washington, DC.

ILSI (International Life Science Institute). 1998. Aggregate Exposure Assessment for Pesticides. Proceedings of the Aggregate Exposure Assessment Workshop. February 9–10, 1998. ILSI Risk Science Institute. Washington DC.

ITRI (Inhalation Toxicology Research Institute). 1975. Annual Report 1974–1975. Albuquerque, New Mexico. Pages 57–62.

Jayjock, M.A. 1994. Back pressure modeling of indoor air concentrations from volatilizing sources. AIHA Journal 55(3):230–235.

Jayjock, M.A., D.R. Doshi, E.H. Nungesser, and W.D. Shade. 1995. Development and evaluation of a source/sink model of indoor air concentrations from isothiazolone-treated wood used indoors. AIHA Journal 56:546–557.

Jensen, B., P. Wolkoff, and C.K. Wilkins. 1996. Characterization of linoleum: identification of oxidative emission process. Pages 145–152 in: B.A. Tichenor (ed.). Characterizing sources of indoor air pollution and related sink effects. ASTM STP 1287. American Society for Testing and Materials.

Kasting, G.B. and P.J. Robinson. 1993. Can we assign an upper limit to skin permeability? Pharmaceut. Res. 10:930–931.

Kjaer, U.D., P.A. Nielsen, K.V. Vejrup, and P. Wolkoff. 1996. A method for determination of the sink effect of VOCs from building materials. Pages 123–133 in: B.A. Tichenor (ed.). Characterizing sources of indoor air pollution and related sink effects. ASTM STP 1287. American Society for Testing and Materials.

Knoppel, H. and H. Schauenberg. 1987. Screening of household products for the emission of volatile organic compounds. Pages 27–31 in: Proceedings of the 4th International Conference on Indoor Air Quality and Climate. Volume 1.

Krieger, R., C. Blewett, S. Edmiston, H. Fong, D. Gibbons, D. Meinders, L. O'Connell, J. Ross, F. Schneider, J. Spencer, and T. Thongsinthusak. 1990. Gauging pesticide exposure of handlers (mixer/loader/applicators) and harvesters in California agriculture. Med. Lav. 81:474–479.

Krieger, R.I., T.M. Dinoff, and J. Peterson. Undated. Human disodium octaborate tetrahydrate (DOT) exposure following carpet flea treatment is not associated with significant dermal absorption. Personal

Chemical Exposure Program, Department of Entomology, University of California. Riverside, California.

Levsen, K. and S. Sollinger. 1993. Textile floor coverings as sinks for indoor air pollutants. Pages 395–400 in: K. Saarela, P. Kalliokoski, and O. Seppanen (eds.). Volume 2. Chemicals in Indoor Air, Material Emissions. Proceedings of the 6th International Conference on Indoor Air Quality and Climate. Helsinki, Finland. July 4–8, 1993.

Lewis, R.G., R.C. Fortmann, D.E. Camann. 1994. Evaluation of methods for monitoring the potential exposure of small children to pesticides in the residential environment. Arch. Environ. Contam. Toxicol. 26:37–46.

Li, W. and P.K. Hopke. 1994. Hygroscopicity of consumer spray product aerosols. J. Aerosol Sci. 25:1341–1351.

Lioy, P.J., T. Wainman, and C. Weisel. 1993. A wipe sampler for the quantitative measurement of dust on smooth surfaces: Laboratory performance studies. J. Exposure Analysis and Environmental Epidemiology. 3:315–330.

Louis, J.B. and K.C. Kisselbach. 1987. Indoor air levels of chlordane and heptachlor following termiticide applications. Bull Environ. Contam. Toxicol. 39:911–918.

Matoba Y., J. Ohnishi, and M. Matsuo. 1993. A simulation of insecticides in indoor aerosol space spraying. Chemosphere 26(6):1167–1186.

Matoba Y., J. Ohnishi, and M. Matsuo. 1994. Indoor simulation of insecticides supplied with an electric vaporizer by the fugacity model. Chemosphere 28(4):767–786.

Matoba Y., J. Ohnishi, and M. Matsuo. 1995. Indoor simulation of insecticides in broadcast spraying. Chemosphere 30(2):345–365.

Matoba Y., J. Yoshimura, J. Ohnishi, and N. Mikami. 1998. Development of the simulation model *InPest* for prediction of the indoor behavior of pesticides. JAAWMA 48:969–978.

McDuffie, H.H. and P. Pawha. 1998. Exposure to home fumigation, indoor pesticide use, and insect repellents as reported by Saskatchewan men and women. Poster presented at the Eighth conference of the International Society of Exposure Analysis. Boston, Massachusetts. August 15–18, 1998.

Murray, D.M. and D.E. Burmaster. 1995. Residential air exchange rates in the United States: empirical and estimated parametric distributions by season and climatic region. Risk Anal. 15(4):459–465.

Nagda, N., M.D. Koontz, and P.W. Kennedy. 1995. Small-chamber and research house testing of tile adhesive emissions. Indoor Air 5:189–195.

Nishioka, M.G., H.M. Burkholder, M.C. Brinkman, S.M. Gordon, and R.G. Lewis. 1996. Measuring transport of lawn-applied herbicide acids from turf to home: Correlation of dislodgeable 2,4-D turf residues with carpet dust and carpet surface residues. Environ. Sci. Technol. 30:3313–3320.

NIST (National Institute of Standards and Technology). 1994. CONTAM93. User manual. Building and Fire Research Laboratory, NIST, U.S. Department of Commerce. Gaithersburg, Maryland.

Otson, R., D.T. Williams, and P.D. Bothwell. 1981. Dichloromethane levels in air after application of paint removers. AIHA Journal 42:56–60.

Pandian, M.D., E.J. Furtaw, H. Reed, and D. Gonzales. 1995. THERdbASE: Total human exposure relational database and advanced simulation environment. Harry Reid Center for Environmental Studies, University of Nevada at Las Vegas. U.S. Environmental Protection Agency, Office of Research and Development, Environmental Monitoring Systems Laboratory. Las Vegas, Nevada.

Pandian, M.D., J.V. Behar, W.R. Ott, L.A. Wallace, A.L. Wilson, S.D. Colome, and M. Koontz. 1998. Correcting errors in the nationwide data base of residential air exchange rates. J. Exposure Anal. Environ. Epidemiol. 8(4):577–586.

Pleil, J.D. and R.S. Whiton. 1990. Determination of organic emissions from new carpeting. Appl. Occup. Environ. Hyg. 5:693.

Popendorf, W.J. 1985. Advances in the unified field model for reentry hazards. In: R.C. Honeycutt, G. Zweig, and N.N. Ragsdale (eds.). Dermal exposure related to pesticide use: Discussion of risk assessment. ACS Symposium Series 273. Washington, DC: American Chemical Society.

Roberts, J.W. and D.E. Camann. Pilot study of a cotton glove press test for assessing exposure to pesticides in house dust. Bull. Environ. Contam. Toxicol. 43:717–724.

Rodes, C.E., R.M. Kamens, and R.W. Wiener RW. 1991. The significance and characteristics of the personal activity cloud on exposure assessment measurements for indoor contaminants. Indoor Air 2:123–145.

Roff, M.W. and J. Thompson. 1995. Dermal exposure of amateur users of pesticide products: Application by brushing outdoors. Health and Safety Executive, Health and Safety Laboratory, Sheffield, England, UK. HSL Internal Report No. R 41.75 IR/L/A/95/15.

Ross, J., T. Thongsinthusak, H.R. Fong, S. Margetich, and R. Krieger. 1990. Measuring potential dermal transfer of surface pesticide residue generated from indoor fogger use: An interim report. Chemosphere 20:349–360.

Ross, J., H.R. Fong, T. Thongsinthusak, S. Margetich, and R. Krieger. 1991. Measuring potential dermal transfer of surface pesticide residue generated from indoor fogger use, using the CDFA roller method. Interim report II. Chemosphere 22:975–984.

Ross, J., H.R. Fong, T. Thongsinthusak, and R. Krieger. 1992. Experimental method to estimate indoor pesticide exposure to children. Pages 227–241 in: Guzelian, P.S., C.J. Henry, and S.S. Olin (eds.). Similarities and differences between children and adults. Implications for risk assessment. ILSI Press. Washington DC.

Schaeffer, V.H., B. Bhooshan, S.B. Chen, J.S. Sonenthal, and A.T. Hodgson. 1996. Characterization of volatile organic chemical emissions from carpet cushions. JAWMA 46:813–820.

Schneider, F., B. Hernandez, C. Benson, M. Dong, and J. Ross. 1994. Dislodgeable residues from treated soil and turf. Report No. HS-1693. California Department of Pesticide Regulation, Worker Health and Safety Branch. Sacramento, California.

Schreiber, J.S. 1997. Transport of organic chemicals to breast milk: Tetrachloroethylene case study. Chapter 5, pages 95–143 in: S. Kacew and G.H. Lambert (eds.). Environmental toxicology and pharmacology of human development. Taylor & Francis Publishing. Washington DC.

Schreiber, J.S., S. House, E. Prohonic, G. Smead, C. Hudson, M. Styk, and J. Lauber. 1993. An investigation of indoor air contamination in residences above dry cleaners. Risk Anal. 13(3):335–344.

Schroder, E. 1990. Textile floor coverings and indoor air quality. Pages 719–723 in: Walkinshaw, D. (ed.). Indoor Air '90: Proceedings of the 5th international conference on indoor air quality and climate. July 29–August 3. Volume 3. Toronto, Canada. Canada Mortgage and Housing Association. Ottawa, Ontario.

Seifert, B., D. Ullrich, and R. Nagel. 1989. Volatile organic compounds from carpeting. Pages 253–258 in: Brasser, L.J. and W. Mulder (eds.). Man and his ecosystem: Proceedings of the 8th World Clean Air Congress. September 11–15. Volume 1. The Hague, Netherlands.

Sparks, L.E., M. Jackson, B. Tichenor, J. White, J. Dorsey, and R. Stieber. 1990. An integrated approach to research on the impact of sources on indoor air quality. Pages 219–224 in: Walkinshaw, D. (ed.). Indoor Air '90: Proceedings of the 5th international conference on indoor air quality and climate. July 29–August 3. Volume 4. Toronto, Canada. Canada Mortgage and Housing Association. Ottawa, Ontario.

Solomon, K.R., S.A. Harris, and G.R. Stephenson. 1993. Applicator and bystander exposure to home garden and landscape pesticides. Chapter 22, pages 262–273 in: Proceedings of the 203rd American Chemical Society National Meeting. San Francisco, California.

Stephenson, G.R., C.S. Bowhey, S.R. Lang, and K.R. Solomon. 1996. Homeowner exposure to granular and liquid formulations of the turfgrass insecticide chlorpyrifos. Preliminary Joint Report of the Ontario Ministry of Environment and Energy, the Turfgrass Research Foundation, and the Ontario Ministry of Agriculture and Food and Rural Affairs.

Thomas, K.W., E.D. Pellizzari, R.L. Perritt, and W.C. Nelson. 1991. Effect of dry-cleaned clothes on tetrachloroethylene levels in indoor air, personal air, and breath for residents of several New Jersey homes. J. Exposure Anal. Environ. Epidemiol. 1(4):475–490.

Thompson, K.M. and J.S. Evans. 1993. Worker's breath as a source of perchloroethylene in the home. J. Exposure Anal. Environ. Epidemiol. 3(4):417–430.

Thompson, J. and M. Roff. 1996. Dermal exposure of occasional users of pesticide products: Flea spray in aerosol cans. Sheffield, UK: Health and Safety Laboratory, Health and Safety Executive. HSL Internal Report No. IR/A/96/10.

Thompson, D.G., G.R. Stephenson, and M.K. Sears. 1984. Persistence, distribution, and dislodgeable residues of 2,4-D following its application to turfgrass. Pesticide Sci 15:353–360.

Tichenor, B.A. 1996. Overview of source/sink characterization methods. Pages 9–19 in: B.A. Tichenor (ed.). Characterizing sources of indoor air pollution and related sink effects. ASTM STP 1287. American Society for Testing and Materials.

Tichenor, B.A., L.A. Sparks, J.B. White, and M.D. Jackson. 1990. Evaluating sources of indoor air pollution. JAWMA 40:487–492.

Tichenor, B.A., Z. Guo, J.E. Dunn, L.E. Sparks, and M.A. Mason. 1991. The interaction of vapor phase organic compounds with indoor sinks. Indoor Air 1:23–35 (Danish Technical Press, Copenhagen, Denmark).

USCPSC (U.S. Consumer Product Safety Commission). 1992. Methylene chloride consumer use study survey findings. Abt Associates, Inc. Cambridge, Massachusetts. Contract C-89-2032. Consumer Products Safety Commission. Bethesda, Maryland. Pages 1–120.

USEPA (U.S. Environmental Protection Agency). 1987a. Household solvent products. A national usage survey. WESTAT, Inc. Rockville, Maryland. Contract Number 68-02-4243. Exposure Evaluation Division, Office of Pesticides and Toxic Substances. Washington DC. NTIS No. PB88-132881.

USEPA (U.S. Environmental Protection Agency). 1987b. National usage survey of household cleaning products. WESTAT, Inc. Rockville, Maryland. Contract Number 68-02-4243. Exposure Evaluation Division, Office of Pesticides and Toxic Substances. Washington DC.

USEPA (U.S. Environmental Protection Agency). 1987c. Methods for assessing exposure to chemical substances. Volume 7. Methods for assessing consumer exposure to chemical substances. USEPA Publication Number 560/5-85-007. Exposure Evaluation Division, Office of Pesticides and Toxic Substances. Pages 1–189.

USEPA (U.S. Environmental Protection Agency). 1989. Consumer inhalation exposure estimates for solvents. Versar Inc. Springfield, Virginia. Contract Number 68-02-4254. Exposure Assessment Branch, Office of Toxic Substances. Washington DC.

USEPA (U.S. Environmental Protection Agency). 1990. Non-occupational pesticide exposure study (NOPES). Research Triangle Park, NC: USEPA. Publication No. 600/3-90/003. Pages 2–3.

USEPA (U.S. Environmental Protection Agency). 1991. MCCEM. Multi-chamber concentration and exposure model. User's guide, version 2.3. Geomet Technologies, Inc. Germantown, Maryland. Geomet Report Number IE-2130. Office of Research and Development, Environmental Monitoring Systems Laboratory. Las Vegas, Nevada.

USEPA (U.S. Environmental Protection Agency). 1992. Dermal exposure assessment: Principles and applications. EPA 600/8–91–011. Exposure Assessment Group, Office of Health and Environmental Assessment, Office of Research and Development. Washington DC.

USEPA (U.S. Environmental Protection Agency). 1994. Screening-level consumer inhalation exposure software (SCIES): Description and user's manual Version 3.0. Geomet Technologies, Inc. Germantown, Maryland. USEPA Contract Number 68-D3-0013. Exposure Assessment Branch, Office of Pollution Prevention, Pesticides, and Toxic Substances. Washington DC.

USEPA (U.S. Environmental Protection Agency). 1995a. Antimicrobial pesticides, uses, human exposures, and risk assessments. Draft report. Office of Pesticide Programs, Occupational and Residential Exposure Branch. Washington DC. Pages 165–214.

USEPA (U.S. Environmental Protection Agency). 1995b. Pesticide handlers exposure database (PHED), Version 1.1. Occupational and Residential Exposure Branch, Office of Pesticide Programs. Office of Pollution Prevention, Pesticides, and Toxic Substances. Washington DC.

USEPA (U.S. Environmental Protection Agency). 1996a. Estimation of emission profiles for latex and alkyd paints. Geomet Technologies, Inc. Germantown, Maryland. USEPA Contract Number 68-D3-0013. Economics, Exposure and Technology Division, Office of Pollution Prevention, Pesticides, and Toxic Substances. Washington DC.

USEPA (U.S. Environmental Protection Agency). 1996b. Comparison of methods to determine dislodgeable residue transfer from floors. USEPA Publication No. 600/R-96/089. National Exposure Research Laboratory, Office of Research and Development, Research Triangle Park, North Carolina. Pages 1–85.

USEPA (U.S. Environmental Protection Agency). 1996c. Consumer labeling initiative interim report. August 21, 1996 Draft Report. Office of Prevention, Pesticides, and Toxic Substances. Washington DC.

USEPA (U.S. Environmental Protection Agency). 1997. Exposure factors handbook. NCEA-W-0005. Exposure Assessment Group, National Center for Environmental Assessment, Office of Research and Development. Washington DC.

Vaccaro, J.R. 1992. Residential exposure to pesticides. Presentation at the Reference House Workshop. Annual meeting of the Society for Risk Analysis. December 6, 1992. San Diego, California.

Vaccaro, J.R., R.J. Nolan, P.G. Murphy, and D.B. Berbrich. 1996. The use of unique study design to estimate exposure of adults and children to surface and airborne chemicals. Pages 166–183 in: B.A. Tichenor (ed.). Characterizing sources of indoor air pollution and related sink effects. ASTM STP 1287. West Conshohockan, Pennsylvania.

Van de Meent, D., J.H.M. De Bruijn, F.A.A.M. De Leeuw, A.C.M. De Nijs, J.T. Jager, and T.G. Vermeire. 1995. H.4. Exposure modeling. Pages 103–145 in: C.J. Van Leeuwen and J.L.M. Hermens (eds.). Risk assessment of chemicals: An introduction. Kluwer Academic Publishers. Dordrecht, The Netherlands.

Van Ert, M.D., J.W. Clayton, C.L. Crabb, and D.W. Walsh. 1987. Identification and characterization of 4-phenylcyclohexene—an emission product from new carpeting. Office of Toxic Substances, U.S. Environmental Protection Agency. Washington DC.

Van Veen, M.P. 1995. CONSEXPO: a program to estimate consumer product exposure and uptake. RIVM Report No. 612810.002. Bilthoven, The Netherlands.

Van Veen, M.P. 1996. A general model for exposure and uptake from consumer products. Risk Analysis 16:331–338.

Van Veen, M.P. 1997. CONSEXPO 2.0, RIVM. Bilthoven, The Netherlands.

Vermeire, T.G., P. Van der Poel, R.T.H. Van der Laar, and H. Roelfzema. 1993. Estimation of consumer exposure to chemicals: application of simple models. Sci. Total Environ. 136:155–176.

Wallace, L.A., E. Pellizzari, B. Leaderer, H. Zelon, and L. Sheldon. 1987. Emissions of volatile organic compounds from building materials and consumer products. Atmos. Environ. 21(2):385–393.

Wallace, L.A., E.D. Pellizzari, T.D. Hartwell, V. Davis, L.C. Michael, and R.W. Whitmore. 1989. The influence of personal activities on exposure to volatile organic compounds. Environ. Res. 50:37–55.

Wallace, L.A., W.C. Nelson, R. Ziegenfus, and E.D. Pellizari. 1991a. The Los Angeles TEAM study: Personal exposures, indoor-outdoor air concentrations, and breath concentrations of 25 volatile organic compounds. J. Exposure Anal. Environ. Epidemiol. 1(2):37–72.

Wallace, L.A., E.D. Pellizzari, L. Sheldon, T. Hartwell, R. Perritt, and H. Zelon. 1991b. Exposures of dry cleaning workers to tetrachloroethylene and other volatile organic compounds: Measurements in air, water, breath, blood, and urine. Presented at the annual meeting of the International Society for Exposure Analysis and Environmental Epidemiology. November 15–18, 1991. Atlanta, Georgia.

Wall Street Journal. 1996. News summary of Wharton School of Business study on consumer behavior. Thursday, October 24, 1996. Page A1.

Weegels, M.F. 1997. Exposure to chemicals in consumer products. Report of the Technical University. Delft, The Netherlands.

Whitmyre, G.K., J.H. Driver, M.E. Ginevan, R.G. Tardiff, and S.R. Baker. 1992a. Human exposure assessment I: Understanding the uncertainties. Toxicol. Indust. Health 8(5):297–320.

Whitmyre, G.K., J.H. Driver, M.E. Ginevan, R.G. Tardiff, and S.R. Baker. 1992b. Human exposure assessment II: Quantifying and reducing the uncertainties. Toxicol. Industrial Health 8(5):321–342.

Whitmyre, G.K., J.H. Driver, and P.J. Hakkinen. 1997. Assessment of residential exposures to chemicals. Pages 125–141 in: Molak, V. (ed.). Fundamentals of Risk Analysis and Risk Management. CRC Lewis Publishers. Boca Raton, Florida.

Wilschut, A., W.F. ten Barge, P.J. Robinson, and T.E. McKone. 1995. Estimating skin permeation: The validation of five mathematical skin permeation models. Chemosphere 30:1275–1296.

Wilson, A.L., S.D. Colome, and Y. Tian. 1993. California residential indoor air quality study. Volume I. Methodology and descriptive studies. Integrated Environmental Services, Irvine, California.

Wooley J, W.W. Nazaroff, and A.T. Hodgson. 1990. Release of ethanol to the atmosphere during use of consumer cleaning products. JAWMA 40(8):1114–1120.

Wright, C.G. and R.B. Leidy. 1982. Chlordane and heptachlor in the ambient air of houses treated for termites. Bull. Environ. Contam. Toxicol. 28:617–623.

Zartarian, V.G. and J.O. Leckie. 1998. Dermal exposure: the missing link. Environ. Sci. Technol. March. Pages 134–137.

Zartarian, V.G., J. Streicker, A. Rivera, C. Cornejo, S. Molina, O. Valdez, and J.O. Leckie. 1995. A pilot study to collect micro-activity data of 2–4 year-old farm labor children in Salinas Valley, California. J. Exposure Anal. Environ. Epidemiol. 5(1):21–24.

Zartarian, V.G., A.C. Ferguson, C.G. Ong, and J.O. Leckie. 1997a. Quantifying videotaped activity patterns: Video translation software and training methodologies. J. Exposure Anal. Environ. Epidemiol. 7(4):535–542.

Zartarian, V.G., A.C. Ferguson, and J.O. Leckie. 1997b. Quantified dermal activity data from a four-child pilot field study. J. Exposure Anal. Environ. Epidemiol. 7(4):543–552.

Zartarian, V.G., A.C. Ferguson, and J.O. Leckie. 1998. Quantified mouthing activity data from a four-child pilot field study. J. Exposure Anal. Environ. Epidemiol. 8(4):543–553.

Zhang, J.S. and F. Haghighatt. 1996. A small air velocity-controlled test chamber for emission studies. Pages 23–33 in: B.A. Tichenor (ed.). Characterizing sources of indoor air pollution and related sink effects. ASTM STP 1287.

Zhang, J.S., C.Y. Shaw, J.M. Kanabus-Kaminska, R.A. MacDonald, R.J. Magee, E. Lusztyk, and H.J. Weichert. 1996. Study of air velocity and turbulent effects on organic compound emissions from building materials/furnishings using a new small test chamber. Pages 184–199 in: B.A. Tichenor (ed.). Characterizing sources of indoor air pollution and related sink effects. ASTM STP 1287.

9

BIOLOGICAL AGENTS

The major classes of biological agents are infectious agents, allergens, and toxins. Biological agents are unique among other environmental agents of disease in that they either are or were derived from living organisms that can grow, change, and reproduce themselves over time. Exposure to biological agents, especially as aerosols, continues to be of increasing interest to the community of those interested in indoor air quality and human health. The major classes of biological agents are infectious agents and toxins.

Infectious diseases (which are *only* caused by biological agents) have plagued mankind since the beginning, and (more than once) have changed the course of history. Infections continue to be the single most important cause of morbidity (ill heath) throughout the world. Millions of people each year lose work or school time with viral infections (influenza, colds, measles, chicken pox, *etc.*) (Burge 1989, 1995). Tuberculosis is increasing in most of the world, with multiply drug resistant strains posing significant challenges for the medical community. Currently however, a major focus of research on the residential environment concerns the role of biological-source allergen exposures in the increasing incidence and severity of asthma. As much as 10 to 30 percent of some populations have asthma, and the disease affects (overall) about 8 percent of the world population (Pope *et al.* 1993). In addition to these diseases that are unique to biological agents, some microorganisms produce potent toxins that can have serious health effects including cancer, birth defects, damage to the immune system, and central nervous system malfunctions (Betina 1989). The organisms that produce these toxins are common in the residential environment.

The primary sources for biological agents of disease in houses are the human and animal occupants, outdoor air, settled dust, and growth of microorganisms on or in reservoirs.

EXPOSURE SCENARIOS FOR BIOLOGICAL AGENTS

Human occupants of indoor environments carry infectious agents that are readily spread both through direct contacts that occur in residences (some common colds), but also through the air (*e.g.*, measles, whopping cough, chicken pox, tuberculosis) in (often) poorly ventilated houses. Human occupants also produce volatile organic compounds that, if they accumulate, lead to poor air quality.

Outdoor air penetrates residences through doors, windows, and inadvertent openings. Along with the air come fungal spores, bacteria, and (rarely) other biological materials that may result in exposure to allergens (*e.g.*, *Coccidioides immitis*, an infectious fungus common in southwestern soils).

Residential Exposure Assessment, edited by Baker *et al.*
Kluwer Academic/Plenum Publishers, New York, 2000.

Particles that enter from outdoor air, as well as those released indoors, settle and become a part of house dust which can then be reaerosolized. Included in dust are human skin scales, fibers and hairs, skin, urinary, and respiratory proteins from animal occupants, food fragments, arthropod fragments and excrement, bacteria, fungi, pollen, and many other biological and nonbiological particles. House dust is considered to be a primary reservoir for indoor allergens that play a primary role in the development of asthma.

Finally, organisms may grow in indoor reservoirs and release disease-causing particles and gases. Thus, fungi growing on wall surfaces may release allergen and toxin-containing spores. *Legoinella pneumophila* may grow in residential hot water systems, and aerosols released during showering may lead to Legionnaires' disease.

This chapter will discuss the nature of the organisms and the disease agents they produce, and specific factors within residences that lead to exposure.

THE ORGANISMS

The primary sources for biological agents of disease are viruses, bacteria, fungi, arthropods, birds, and mammals (that is, most forms of life on earth). This section provides a brief overview of the types of organisms of concern.

Viruses

Viruses are obligatory intracellular parasites that can reproduce only within living cells because they need the metabolic machinery of cells (*e.g.*, human or animal cells) to generate new viruses. They are classified based on their structure, and are usually named with reference to the diseases they cause (Tortora *et al.* 1992). Viruses are found only in association with higher host organisms, and cause only infectious disease.

Bacteria

Although small and relatively simple, the bacteria are considerably more complex than viruses, and contain all the cellular machinery that is necessary for them to reproduce on their own. They are prokaryotic, meaning that they have no organized nucleus or mitochondria. Bacteria are classified on the basis morphology (shape, size and arrangement of cells), staining reactions, physiology, and (where appropriate) diseases they cause. Of particular importance in bacterial classification is the Gram stain (developed by Hans Christian Gram), which divides bacteria into two large groups. Gram positive bacteria have a thick, rigid cell wall composed of many layers of peptidoglycan with embedded teichoic acids which, in addition to many other functions, confer antigenic specificity to the cell and are used for identification. The Gram negative bacteria have only a thin layer of peptidoglycan surrounded by an lipid-rich outer membrane that includes lipoproteins, lipopolysaccharides, and phospholipids. The polysaccharides in the outer membrane serve as antigens and can be used to distinguish species of Gram negative bacteria. The lipid portion of the lipopolysaccharide molecule is known as endotoxin. Bacteria occupy nearly every imaginable niche on earth. They are present in hot springs where temperatures are near boiling, and in the arctic ice. They are present on all living and dead plant and animal surfaces, and everywhere where water is consistently present. Exposure to bacteria or their products can cause infection, hypersensitivity diseases, or toxicoses (Tortora *et al.* 1992).

Fungi

Fungi are formed of chains of cells (hyphae), in mass called mycelium. Fungal cells are bound by a rigid cell wall composed (primarily) of acetyl-glucosamine polymers (chitin) and β-glucans. (Sietsma and Wessels 1981). Fungi excrete extracellular polysaccharides (EPS) which aid in adhesion of fungal cells to substrates, and may confer antigenic specificity and be useful in identification. Fungi excrete enzymes designed to facilitate degradation of complex food sources (usually carbohydrates such those in plants). Mycotoxins are secondary metabolites produced by some fungi. The purpose of their production is not well understood. Fungi have complex life cycles that include both sexual and asexual modes of reproduction. Both modes usually produce spores that are more or less resistant to environmental stresses, and most travel through the air to inoculate new environments. Fungi are classified on the basis of their mode of sexual reproduction (where known), and (secondarily) their mode of asexual reproduction. Within these large categories, classification is primarily based on morphology. Because they are often visible, fungi have many common names that cause some confusion. Mold and mildew are both names for fungi that are growing (usually) where they are not wanted (*i.e.*, these are the weeds of the fungal kingdom). Yeasts are unicellular fungi commonly used in making bread and fermented beverages (beer, wine). Mushrooms are fungi that produce large fruiting bodies that may be delicious and either nutritious or contain deadly poisons (Kendrick 1992). Fungi are ubiquitous in both outdoor and indoor environments. Exposure to fungal spores, fragments, or metabolites may cause infections, hypersensitivity diseases, or toxicoses.

Arthropods

The phylum Arthropoda (jointed leg animals with exoskeletons) includes the classes Arachnida (*e.g.*, dust mites, spiders) with four pairs of legs, Crustacea (*e.g.*, lobsters and shrimp) with five pairs of legs, and Insecta (*e.g.*, cockroaches) with three pairs of legs. The most common species of house dust mites are *Dermatophagoides pteronyssinus* and *D. farinae*, with *Blomia tropicalis, D. microceras, Euroglyphus maynae, Cheyletus spp, Tarsonemus spp., Typhagus spp.* also locally abundant. Dust mites are small (~100 μm) creatures that feed on human skin scales and fungi in dust, and decaying organic matter found on the surfaces of leaves, mosses, lichens, and fungi. They produce membrane-bound fecal balls that contain glycoprotein enzymes that are highly allergenic (Arlian 1989). Cockroaches are members of the class Insecta, and are common throughout all parts of the world. Some of the most common species of cockroach are *Blatella germanica* (German cockroach), *Periplaneta americana* (American cockroach), *Blatella orientalis* (Oriental cockroach), and *Supella supelledium* (brown-banded cockroach). In the United States, while the American cockroach is abundant outdoors in humid regions, the German cockroach produces the majority of indoor infestations. Cockroaches are nocturnal and live in damp, dark places indoors. They venture out into lighted interiors only when population pressure forces excursions for food (Mathews 1989). Cockroach feces and body parts contain potent allergens.

Mammals and Birds

Mammals and birds are familiar to all. Mammals that are important sources for disease-causing agents in the indoor environment include cats, dogs, small furry pets (hamsters, Guinea pigs, rabbits), and pests such as mice, rats and bats. Rats and mice require food, water, and shelter like other animals, and among the safest environments for them are human residences. These animals can multiply several times before a resident realizes the home is infested with rodents. Birds are kept as pets, and may indirectly affect the indoor environment when they colonize attics or eaves near open windows. These animals release skin scales (dander), saliva, and urinary proteins, all of which contain allergens.

Plants

Plants differ from animals and fungi in the nature of their cell walls (cellulose as opposed to chitin for fungi) and in their ability to transform CO_2 (using chlorophyll) into carbohydrates for use as food. The plant kingdom includes algae, mosses and liverworts, club mosses, ferns, gymnosperms (pine, spruce, *etc.*), and seed plants. All of these except gymnosperms and seed plants reproduce by spores, most of which travel through the air. Gymnosperms and seed plants reproduce by seeds that each contain a small plant. Seeds are produced by fusion of male and female cells. The male cells are called pollen, and are an important allergen source. Plants may be herbaceous with cell walls composed primarily of cellulose, or woody, with lignin-impregnated cell walls. Many plants produce complex compounds ("sap") that travel through vessels within stems. Saps of important to human health include the sensitizing oils produced by poison ivy and related plants, the sap of the rubber tree (latex) which contains potent allergens, and many potent toxins (*e.g.*, digitalis).

BIOLOGICAL AGENTS AND HUMAN DISEASE

Infections

Infectious disease is unique to biological agents, and infection results when one living organism invades another and reproduces to cause disease. By definition, in order to cause disease an infectious agent must be alive and able to grow in the host. The capacity of being able to grow in a host, which involves overcoming host defenses, is called virulence. There are three general types of infectious diseases: contagious, vector-borne, and environmental. The contagious diseases are transmitted directly from one person to another, either by direct contact (*i.e.*, touching) or through the air. Most airborne contagious diseases are caused by viruses (*e.g.*, influenza, common colds, chicken pox and measles). A few result from bacterial infection (*e.g.*, tuberculosis and whooping cough). Susceptibility to contagious disease is controlled by the presence or absence of antibodies specific for the disease agent. Such antibodies are acquired either through natural infection or by artificial immunization. (Most agents of contagious disease are highly virulent, and exposure to very few agents will cause disease in susceptible individuals.)

Vector-borne diseases are transmitted from animals to humans, either by a direct contact (*e.g.*, bite) or through an insect vector. Some of the most devastating of human diseases are in this category, including the black plague and malaria. Important vector-borne diseases that may be associated with residential environments include Lyme disease (transmitted via deer ticks) and equine encephalitis (transmitted via mosquitoes). Hanta virus respiratory syndrome is caused by a mouse-borne virus and excreted in mouse urine. When dry urine is disturbed, the virus may become airborne and cause infection (Farr 1994).

A third category of infectious disease are those that are caused by organisms that can grow in the environment as well as the human body. A few of these environmental-source diseases are caused by virulent agents, and all exposed people without specific immunity generally develop disease (or at least support the organism long enough so that antibodies can be produced). Included in this category are the fungal diseases histoplasmosis, blastomycosis, and coccidiomycosis, all resulting from exposure to fungi that occur only in outdoor reservoirs. Most environmental-source diseases are considered "opportunistic," which means that the amount of exposure required to significantly increase the risk of disease is dependent on the status of the host's natural disease resistance. Healthy people with normal immune system function are rarely affected. Included in this category is Legionnaires' disease, which requires some damage to lung immunity before infection is likely to occur (Eichoff 1979). As many as 25,000 cases of Legionnaires' disease may occur annually in the United States primarily from residential exposures.

Allergies

Allergies are the result of an inappropriate response of the immune system to environmental stimuli. The agents that cause allergies are almost all high molecular weight proteins or polysaccharides. All of the allergic diseases require one or more sensitizing exposures before symptoms can occur. The two types of allergic disease that commonly result from indoor exposure are the atopic diseases which are mediated by blood-borne antibodies, and hypersensitivity pneumonitis, which is mediated (primarily) by the cellular immune system.

Hay fever and asthma result when exposure to environmental allergens causes the production of specific antibodies of the class immunoglobulin E (IgE). IgE molecules attach to cells in the body that contain histamine. When exposure to the sensitizing allergen occurs, the allergen binds to the IgE molecules and stimulates the cell to release the histamine, resulting in inflammation. If the inflammation is confined to the upper airway and eyes, the familiar symptoms of hay fever result. If lower airway inflammation occurs, asthma (reversible airway constriction) is the result. The development of atopic disease is controlled by genetic factors as well as exposure to allergens. Somewhere between 30 and 50 percent of the population is genetically susceptible to the atopic diseases (Pope *et al.* 1993).

Common sources of allergens that cause IgE mediated diseases are listed in Table 9-1. Levels of allergen in dust is the most commonly used indicator for exposure, and estimates of sensitizing and symptom-eliciting levels are also listed.

The prevalence of fungal-induced allergy is unknown, primarily because of the poor quality of extracts used for skin testing, and because few diagnosticians are familiar with patterns of exposure. Up to 10 percent of unselected populations and 60 percent of atopic populations have reacted positively to specially prepared fungal extracts (Verhoeff and Burge 1997).

Dust mite allergens are clearly a major cause of sensitization and symptom development for childhood asthma. As many as 60 percent of atopic children may develop sensitization to dust mites, often with asthma as the outcome (Platts-Mills and Chapman 1987). Cockroach allergy is also extremely common (Squillace 1995), and sensitivity to cockroach allergens among asthmatic patients has been reported as 51.3 percent in Kentucky. Recently, an association between exposure to cockroach allergen and asthma in residences of children living in the inner city has been reported (Rosenstreich *et al.* 1997).

Table 9-1. Sources for the most common allergens causing hay fever and asthma

Living organism	Nature of the allergen	Reservoirs	Sensitizing exposure	Symptom exposure
Fungi	Secreted enzymes Cell wall components	Air, dust, surface growth	(?)	(?)
Dust mites	Digestive enzymes in feces	Dust	2 μg/g dust	10 μ/g dust
Cockroaches	Digestive enzymes (?)	Dust	2 U/g	10 U/g
Cats	Skin secretions	Dust, air	1 μg/g	8 μg/g
Dogs	Skin secretions, urine	Dust, air	(?)	(?)
Pollen	Recognition proteins	Air	(?)	(?)

The actual amounts of cockroach allergen exposure required for sensitization or symptom development are unknown.

Exposure to cat allergens has been linked to allergic sensitization and the development of childhood asthma. The prevalence of cat allergy has been estimated to be as low as 2 percent in the United States, and as high as 13.3 percent in New Zealand (Luczynska 1995). Pollen-related allergies are extremely common, with as many as 30 percent of some populations experiencing symptoms (Chiu *et al.* 1993).

Hypersensitivity Pneumonitis

While specific antibodies are also produced in hypersensitivity pneumonitis, they are of the class immunoglobulin G (IgG) and (apparently) play only a peripheral role in the disease. Rather, the disease process is related to sensitization of specific cells in the lung that release inflammatory agents when exposed to the sensitizing allergen. These mediators cause flu-like symptoms as well as inflammation of lung tissue. While initially reversible, continuing exposure to allergens responsible for this disease results in permanent changes (fibrosis) in lung tissue. Risk factors other than exposure are unknown for hypersensitivity pneumonitis (Salvaggio 1987). The kinds of exposures necessary to cause disease are also generally unknown and co-exposure with allergen *and* a nonspecific immune stimulant (known as an adjuvant) may be necessary. Natural adjuvants that are often a part of sensitizing exposures include endotoxin, and cell wall components of some kinds of bacteria (*Mycobacterium*, thermophilic actinomycetes). Sources for antigen exposures include bacteria (especially the thermophilic actinomycetes which causes farmer's lung disease), fungi (cheese workers disease), wood dusts (wood-trimmers disease), and birds (pigeon breeder's disease). The incidence of hypersensitivity pneumonitis is unknown, but the disease-causing exposures associated with residential environments are probably very rare. Many of these rare residential cases result from exposure to humidifier aerosols (Burke *et al.* 1977, Shiue *et al.* 1990).

Toxicoses

Biological toxicoses are similar to those caused by nonbiological toxins. Specific chemicals cause direct damage to cells and tissues, and the kind of damage is related to the kind of chemical. Bacteria and fungi that are abundant in indoor environments produce potent inflammatory agents (*e.g.*, endotoxin), cytotoxins (*e.g.*, the fungal trichothecene toxins), or carcinogens (*e.g.*, the aflatoxins). Endotoxin is the most familiar of the environmental bacterial toxins. It stimulates the immune system and exposure may be necessary to normal human development. In excess, it causes fever, other flu-like symptoms, chest tightness, and, as mentioned above, may enhance the development of sensitization leading to hypersensitivity pneumonitis. Other bacterial cell components have not been studied for their toxic effect. Bacteria do release gases (*e.g.*, hydrogen sulfide) and volatile organic compounds (*e.g.*, aldehydes, alcohols and ketones) during metabolism that have well-known toxic effects. The prevalence of diseases related to exposure to bacterial toxins in the residential environment is unknown (Milton 1995, 1996).

Thus far, satratoxin (produced by the fungus *Stachybotrys chartarum (atra)* and other fungi) is the only mycotoxin to have been reasonably implicated in human respiratory disease in a residential environment (Croft *et al.* 1986), although the toxin is often suspected to play a role in indoor air-related disease. *Stachybotrys* is rarely found in air surveys of the residential environment in levels likely to cause disease. *Stachybotrys* spp were found in only 10 percent of air samples, and 2 percent to 3 percent of dust samples from 100 Danish homes (Gravesen 1987). However, the fungus grows poorly on standard culture media, and further study using spore counting techniques are necessary to document is actual prevalence in residences.

PATHWAYS OF EXPOSURE TO BIOLOGICAL AGENTS

The Airborne Route

Agents may become airborne as whole organisms, spores or other reproductive units, fragments, dried residuals of dissolved materials, or as vapors or gases.

Droplet Nuclei

Especially important in the transmission of infectious disease are droplet nuclei which are what remains when small droplets of water that contain cells, fragments or solutes dry (Knight 1973, Riley 1982). Water droplets may be released from the human respiratory tract or from environmental reservoirs. Those released from the respiratory tract may contain one or more virions and perhaps a bacterium or two imbedded in a matrix of proteins and polysaccharides. The spread of truly airborne contagious infections depends on droplet nuclei. Droplet nuclei may also be released from environmental reservoirs such as cooling towers, water-spray humidifiers, fountains, and other sites where water is actively disturbed. When such reservoirs contain microorganisms, either the organisms themselves or their fragments or metabolites become airborne as droplet nuclei. Legionnaires' disease is transmitted by droplet nuclei containing viable *Legionella* bacteria. Humidifier fever results when endotoxin-containing droplet nuclei are released from reservoirs contaminated with Gram negative bacteria. Hypersensitivity pneumonitis may result when aerosol-releasing water reservoirs are contaminated with *Mycobacterium* or with fungi.

Cells and Spores

Some organisms are small enough so that they can individually become airborne. The bacteria, and the unicellular fungi (*e.g.*, yeasts) are in this category. Although little studied, it appears likely that, in outdoor air, bacteria and yeasts are released from plant surfaces in dry form (*i.e.*, not as droplet nuclei).

The fungi and actinomycetes (filamentous bacteria) that are common indoors reproduce by means of spores which may be produced singly, in chains or in clumps.

Individual fungal spores range in size from about 1.5 μm to 50 μm and may be unicellular and more or less spherical, or multicellular with complex shapes. Actinomycete spores are usually near 1 μm in diameter. Spore walls may be hydrophilic and coated in mucoid polysaccharide material, or hydrophobic, containing waxes in the outer walls. Once formed, spores are released into the air by internal mechanisms usually related to changes in water activity or (more commonly) by mechanical actions such as air movement or other mechanical disturbance. In the latter case, the levels of disturbance required are related to the strength of attachment of the conidia to the conidiophore, and (probably) relative humidity. Reported threshold speeds required for fungal spore release range from <1 to 3 m/s with wind gusts as the primary mechanism for dispersal (Aylor and Waggoner 1980). Required air speed varies with the kind of fungus. For example, *Aspergillus* and *Penicillium* species required 0.5 mm/s while *Cladosporium* species needed 1.0 m/s for spore release (Pasanen *et al.* 1991). Once airborne, the influencing factor in flight through air and deposition onto surfaces is the aerodynamic diameter of the fungal spore or cluster (Madelin and Johnson 1992).

Pollen, spores of primitive vascular plants (ferns) and nonvascular plants such as mosses, and algal cells also are transported via air. The most important of these particles with respect to human disease is pollen. To ensure that at least a few pollen grains reach their target, plants that produce windborne pollen generally produce massive quantities of small, dry, relatively smooth grains in inconspicuous flowers. Pollen grains are large in comparison with many other biological aerosols. Most are between 18 and 50 μm in diameter when viewed wet under the microscope.

FRAGMENTS AND METABOLITES

Fragments of bacteria and fungi, dust mites, cockroaches and plants may also become airborne. In addition, metabolites of some may be transmitted through the air. Much of the allergenicity attributed to dust mites is contained in the 10 μm to 40 μm fecal pellets (Tovey *et al.* 1981). Not "respirable" in the usual sense, the mite fecal particles are thought to impact at the bifurcations in larger airways, slowly releasing the large amounts of contained allergen. This theory is supported by the relatively long onset time for symptoms following exposure.

Fel d 1 (the primary cat allergen) is often associated with small particles ≤2.5 μm in diameter (Luczynska 1995). Rapid air changes apparently differentially remove small particles, and low air exchange rates and presence of carpeting and cloth sofa have been shown to favor smaller Fel d 1 particles in air compared to high air exchange rates and no soft furnishings. The small particle size is considered the reason why cat-allergic patients develop an almost immediate response upon entering a room where a cat is or has been. Because of the small size of the particles, cat allergens may remain airborne and available for inhalation for hours. The major dog allergen, Can f 1, is borne on particles that range from <5 μm to >9 μm. Custovic *et al.* (1996) reported that approximately 20 percent of airborne Can f 1 collected on filters was associated with particles <5 μm. As is the case for cat allergen, these small particles, which can penetrate into the lower airways, can remain airborne for hours and pose a threat to dog-allergic individuals who enter a dog owner's residence even after the dog has been removed from the room or home. Particle sizes for other allergens have not been reported.

All living organism release volatile organic compounds during metabolism. Human body odor is, in part, natural volatile compounds, and, in part, volatile compounds produced by bacterial degradation of skin secretions. Bacterial and fungal growth in indoor environments is often first recognized by the presence of odors, indicating the presence of active growth. The nature and amounts of volatile compounds produced by these organisms depends on the type of organism, the nutrients that are being digested, and the conditions under which growth is occurring (Batterman *et al.* 1991). Carbon dioxide and ethanol are the most abundant carbon-containing compounds released during microbial growth. The odoriferous compounds are released in much lower amounts and include a variety of alcohols, aldehydes, and ketones, often in the 8 to 9 carbon range.

Vector-Borne Diseases

A large proportion of infectious diseases are transmitted via vectors. A good example of this pathway is the black plague. The bacterial agent (*Yersinia pestis*) resides in rats (which congregate near human residences), and is transmitted to humans when fleas bite and ingest rat blood (including the bacterium), then bite people, injecting the bacterium into the human blood stream. Black plague is endemic in some parts of the Unites States (Tortora *et al.* 1992). The devastating epidemics of the past are now prevented by the use of highly effective antibiotics.

Hanta viruses could also be considered to be vector-borne, and infection with these agents may be related (rarely) to indoor exposures. In North America, the Hanta virus resides in rodents and is excreted in rodent urine. The virus can apparently survive in the excreted urine which, when dry, can become airborne with disturbance (Farr 1994). The actual incidence of Hanta virus respiratory syndromes in North America is unknown.

ROUTES OF EXPOSURE

Exposure to biological agents can occur through all of the primary routes of exposure, either directly or indirectly. For example, airborne agents can be inhaled directly or contacted on surfaces or foods on which they settle.

Inhalation

The most common route of exposure to airborne biological agents is inhalation (see Chapter 4).

Dermal

The most familiar biological sources for dermal exposure and subsequent symptoms are poison ivy, poison oak, and poison sumac. These plants produce oils that are potent sensitizers, activating the cellular immune system in the skin and mucosa, and (with repeated exposure) causing severe dermatitis. Other plants have hairs or spines that readily penetrate human skin and release irritants that cause immediate irritant reactions such as burning and itching. In this category are the stinging nettles (species of *Urtica* and *Laportea*). Some plants also produce extremely toxic materials in the sap (the complex liquid that flows through the plant vascular system). These materials usually do not affect the skin directly but cause extremely uncomfortable and sometimes dangerous irritation of the mucous membranes. Some common house-plants contain these toxins (Poinsettias, other members of the Euphorbiaceae).

As discussed above, most fungi produce toxins, and some are best known for their dermal effects. Thus, the toxins produced by *Stachybotrys chartarum* (*atra*) were first recognized by the severe dermatitis that can result when extensively contaminated materials are handled.

Ingestion Exposure

Water, food, and other ingested materials can be exposure sources for biological agents. Some of the most important infectious diseases are water-borne. Recent outbreaks of cholera, cryptosporidiosis, and Giardia infections have been related to ingestion of contaminated water (Eisenberg *et al.* 1996). In addition, some *Legionella* transmission may occur via ingestion and subsequent aspiration of contaminated water (Fraser 1985). Many outbreaks of water-borne infectious disease occur when human fecal material contaminates drinking water systems. Public water supplies in the developed countries are treated to control these enteric pathogens, and such water is monitored for *Escherichia coli*, an organism that is considered to be an indicator for fecal contamination. Prevention of water-borne infectious disease is the basis for laws regulating septic systems, sewage treatment, and all disposal of human waste.

Many of our staple foods contain potent allergens (*e.g.*, peanuts, milk, shell fish, wheat). Some people are so exquisitely sensitive to some food proteins that life-threatening anaphylactic reactions occur when the slightest hint of the material is ingested.

Some processed foods contain microorganisms that are intrinsic to the nature of the processing. For example, all cheeses, beer, wine, vinegar, yogurt, soy sauce, tempeh, and many other food types are produced by purposely inoculating plant-based materials with specific microorganisms, and the microorganisms and their metabolites (which may include allergens or, rarely, toxins) are ingested along with the food item (Pitt 1975). Food may also become inadvertently contaminated with microorganisms. Recent outbreaks of infection with virulent strains of *Escherichia coli* contaminating imported fruits are good examples of invisible contamination. Botulism results when an anaerobic bacterium (*Clostridium botulinum*) contaminates canned foods and produces an extremely potent toxin that is often fatal when ingested. In addition, we have all seen molds growing on food in the cupboard or refrigerator, and have been tempted to cut off the moldy part and eat the rest (not a good idea!).

Finally, some extremely toxic plants, animals, and fungi are eaten on the Russian Roulette system. The best examples of this are the Japanese custom of eating an extremely toxic fish that, if improperly prepared, causes rapid death, and the collection of wild mushrooms, some of which contain toxins that are so potent that only a single mouthful of the mushroom will cause inevitable death. Inadvertently ingested materials in residences can contain biological materials, although the health effects from this source of exposure are essentially unexplored. Ingestion of dust by small children is a good example, and the possibility is being considered that some sensitization to mite and cockroach allergen is the result of ingestion of dust containing high levels of these allergens.

RESIDENTIAL RESERVOIRS

Outdoor Sources

Outdoor air is an important source for bacterial and fungal aerosols and the only source for pollen. Airborne pollen prevalence is determined by the kinds of source plants in the environment. Most plants produce pollen seasonally, at least in temperate climates. Most tree pollen is released during the early months (spring). Pollen of the perennial grasses is released beginning in May and peaks in June, while annual grass pollen may be released throughout the summer. The common annual weeds generally produce pollen in the late summer to early fall (AAAAI 1994).

Most bacteria in outdoor air are derived from living and dead plant surfaces, with the most frequently isolated types being Gram negative species of genera such as Pseudomonas, and members of the Gram positive genus *Bacillus*. All nonliving plant surfaces are colonized by fungi, providing the principal source for outdoor aerosols. Fungi are also the primary agents of plant disease. Fungi are always present in outdoor air, with levels depending on the concentrations in sources, and factors that disturb the source sufficiently to cause spore release. When the ground (and ground-level sources) are covered by snow, spore levels in air can be virtually undetectable. When snow is not present, sudden changes in humidity, wind, and mechanical disturbance influence aerosol concentrations and composition. Seasonal changes in spore aerosols are related to weather, with snow having the largest effect, and the seasonal plant cycle which causes large increases in available dead organic material as winter approaches. Diurnal patterns of variation are generally related to changing water content of the air and changes in wind speed and turbulence. Geographic variation is related both to weather patterns and to the kinds of plants that are available for colonization.

The outdoor biological aerosols penetrate the home in the same way as other relatively large (>2 μm) particles (Alzona *et al.* 1979). To construct a structure where all particles from outdoors are excluded is possible (*e.g.*, space ships) but not practical and certainly would not be convenient for its occupants. Of course, the ventilation system of this structure would have to actively remove pollutants generated indoors. In general, houses are not constructed specifically to exclude outdoor airborne particles, but, due to other goals (*e.g.*, energy efficiency), residences are often quite efficient in isolating occupants from most of these particles. Filtration of intake air is deemed necessary to protect heating/ventilation/air-conditioning systems, reduce cleaning costs, and to protect from penetration of agents of human disease (in that order of importance). Filtration systems vary from extremely coarse filters that remove only very large objects (leaves, insects, *etc.*) to highly efficient filters that move up to 99 percent of all particles down to sizes well below 1 μm. Most residential systems have only coarse filtration.

Other outdoor sources of biological agents include ground and flood water that may carry infectious agents and may penetrate into residences. Flooding is especially of concern with respect to the potential for sewage contamination (see above). Finally, poison ivy and poisonous mushrooms are not restricted to wild lands, but readily occupy residential gardens. In fact, most cases of mushroom poisoning in children result from ingestion of "lawn" mushroom (although not all are poisonous) and most cases of poison ivy dermatitis from gardening activities.

Indoor Living Sources

As has been stated above, nearly all viral agents of disease are shed directly from infected humans. The same is generally true for the airborne bacterial diseases with only a few exceptions. Indoors, most bacteria are released from human skin, and Gram positive cocci such as Staphylococcus and Micrococcus tend to be abundant in air and are dominant bacteria usually collected during air sampling studies. These bacteria rarely present a risk for disease. Their concentration depends primarily on the numbers of people in the space, and the activity of each person. Although ventilation also probably plays some role, the skin scales on which the bacteria are borne are relatively large and settling is probably the largest factor in removal.

The primary sources for many allergens in the residential environment are pets and pests. Removal of cats and dogs from the residential environment is the only way to significantly reduce exposure to these allergens, although washing of the animals, and restricting entry to specific rooms may cause slight decrements in exposure. Control of dust mites through killing (washing bedding in hot water, benzoyl peroxide treatments, humidity control) and control of exposure (plastic encasings for bedding, removal of carpeting) can reduce mite allergen exposure to a significant extent. Cockroach control can also be effective in single-family homes if rigorously addressed. However, significant problems are encountered in multiple family dwellings unless the control measures are applied to the entire building.

Inanimate Indoor Reservoirs

Generally, three types of inanimate indoor reservoirs exist: (1) dust, which is an accumulation of organisms that settle from the air and other sources as well as a complex, living ecosystem, (2) water and associated surfaces, and (3) solid materials that may be supporting active microbial growth. In all of these reservoirs, water activity and temperature play an important role in determining source strength.

Water activity. Water activity (a_w) is the ratio of the water vapor pressure of a substance to the vapor pressure of pure water at a given temperature (note the important role that temperature plays in determining water activity). Water activity of a substance is measured by enclosing the substance, allowing it to come to equilibrium with the surrounding air, and measuring RH in the surrounding air. If the absolute amount of water does not change, water activity increases as temperature decreases. Water activity requirements for mites and for microorganisms are defined at specific temperature levels. As temperatures decrease, metabolism of most organisms slows, and growth rate decreases regardless of increasing water activity. On the other hand, as temperatures increase, the metabolism of the organism tends to increase as long as water activity (which is simultaneously decreasing) remains within a range that is adequate to support growth. The majority of studies that have evaluated the ability of organisms to colonize and grow in dust (and on other substrates in indoor environments) have ignored the role of temperature.

Generally bacteria require relatively warm temperatures and high water activities (>0.95) to grow. The dominance of *Bacillus* species in dust may relate to accumulation over time of the very resistant spores that these organisms produce. For fungi, the minimum a_w required for spore germination is lower than that to sustain linear growth which was usually lower than the minimum required for asexual sporulation (Adan 1994). In a case study of fungal species prevalence in an indoor environment, colonization of papered surfaces by different fungal species changed with degree of saturation. *Penicillium* species and *Aspergillus versicolor* appeared when the paper was comparatively dry, while *Ulocladium* species and *Stachybotrys chartarum (atra)* appeared in later stages when conditions were very wet. When there was direct ingress of water, *Stachybotrys chartarum* was found nearest to the point of ingress with *Penicillium* and *Aspergillus versicolor* in the drier areas (Grant *et al.* 1989). The xerophilic fungi (*e.g.*, members of the genus *Eurotium*) require relatively dry conditions for growth (a_w ~0.8), and many will not grow under wetter conditions. The xerotolerant fungi (*e.g.*, common species of

Cladosporium and members of the *Aspergillus versicolor* group) can grow over a wide range of water activities. Most fungi, however, require water activities in excess of 0.95 for growth. On the other hand, many fungi may be able to maintain metabolic activity (but not grow) at water activities as low as 0.65 (65 percent RH).

Dust mites absorb water directly through their exoskeleton. Ambient relative humidity between 70 and 85 percent is conducive to mite proliferation. How this translates to a_w within mite reservoirs is unknown. Apparently an RH >85 percent inhibits population growth either through direct effects from humidity, or (more likely) because other organisms such as fungi may flourish and compete for nutrients or parasitize the mites.

Dust. Dust is composed of (1) human skin scales, (2) silica and other materials tracked in from outdoors, (3) fibers, (4) microorganisms such as bacteria, yeasts, and other fungi that may have settled from air or other sources, or that may be growing within the dust, (5) whole bodies, fragments and metabolites of arthropods, and (6) many other materials that may be purposely added (*e.g.*, carpet fresheners, pesticides) or may accumulate from other sources (*e.g.*, fragments of paint).

Levels of cultural fungi in dust are often in the range of 10^3/g to 10^4/g dust (Chew *et al.* 1996). Mean dust levels of Fel d 1 have been reported as 2.5 Units/g of bed dust in the US (Lintner and Brame 1993), and are strongly dependent on whether or not there is a cat in the home. A similar relationship holds for dog allergens. It should be noted that houses without cats or dogs can still have elevated allergen levels. Dog allergen has been reported in levels that have been associated with allergic asthma in homes without dogs (Custovic *et al.* 1996).

The mean levels of Bla g 1 in homes from Charlottesville, Virginia were 10,755 U/g of dust (kitchen) and 204 U/g of dust (bedroom floors, sofas, and bedding). In dust samples obtained from homes in Atlanta, GA, the levels of the cockroach allergen ranged from 4 to 1340 U/g of dust (Squillace 1995).

Water Reservoirs. Standing or recirculated water always supports bacterial growth, and may encourage the growth of fungi as well. Common stagnant and recirculating home water reservoirs include humidifiers, fish tanks, and poorly draining refrigerator drip pans (among others). Levels of bacteria may exceed 10^6/ml of liquid in these reservoirs, while fungal levels often reach 10^4 (Burge *et al.* 1980). Gram negative bacteria (with associated endotoxin) are frequent water reservoir contaminants. Yeasts and wet-spored fungi such as species of *Acremonium*, *Fusarium*, and *Phialophora* are often found in water reservoirs as well as following flooding (Patterson *et al.* 1981).

Microbial Growth on Building Materials and Contents. Fungi will grow wherever condensation is consistently present, and on materials that are wetted from leaks and floods. Species of *Penicillium* and *Aspergillus* are the fungi most commonly isolated from air in colonized homes, although many other organisms (including the notorious fungus *Stachybotrys chartarum (atra)* are frequently isolated directly from colonized substrates. It is often assumed that surface fungal contamination is direct evidence of airborne contamination. However, fungal spore release and dispersal are dependent upon many factors as discussed above. In the indoor environment these include human activity, turbulence at the surface of the substrate, and strength of attachment to the conidiophore (Ingold 1971). In the indoor environment, high spore counts have coincided with human activity such as cleaning, construction, and high occupancy load (Hunter *et al.* 1988,

Buttner and Stetzenbach 1993, Lehtonen and Reponen 1993). Reported concentrations of culturable fungal propagules in air (measured as colony forming units) depend on the method of sample collection and the culture medium chosen. Levels ranging from 0 to >5000 cfu/m^3 have been reported in residences depending on season and type of ventilation in the home.

SAMPLING CONSIDERATIONS

Sampling for biological agents depends primarily on the analytical method to be used. The volatile organic compounds produced by biological agents are sampled and analyzed using methods similar to those used for other kinds of volatile compounds (see Chapter 4). Other categories of analysis that determine sampling approach include culture for viable cells, direct counting using microscopy, immunoassays, bioassays, gene probes, and chemical assays.

Infectious Agents

Agents that must be alive to cause disease (*e.g.*, infectious agents) are appropriately detected using culture methods. All are particles, so that samples can be collected from the air by applying the relevant principles of aerosol physics while providing protection for fragile organisms. Samples from environmental reservoirs can also be cultured. It should be noted that culture always underestimates actual concentrations of the organisms. Culture conditions are never ideal and even organisms that might cause disease may be missed. Fungal pathogens, in particular, are difficult to culture. Organisms that are difficult to recover in culture may often be identified and sometimes counted using microscopy (often combined with immunostaining), bioassays, gene probes, or chemical assays. However, it should be noted that such data are biased in the opposite direction to culture (*i.e.*, such methods provide an overestimation of potential risk, since both living and nonliving cells are counted).

In practice, sampling for infectious agents is rarely done, except for water-borne agents and to identify kinds of organisms in human tissues and secretions. Both are routinely tested using culture. Water samples are concentrated and, for example, a count of coliform bacteria (bacteria representative of contamination with fecal material) is made. For *Legionella pneumophila*, water samples are concentrated and cultured, and resulting colonies checked for identity using fluorescent immunostaining and microscopy. Counts of specific viruses are made by inoculating appropriate cell types grown in culture. Human-source samples may be cultured and the isolates identified using mechanized systems that consist of arrays of biochemical tests, or in some cases, direct staining of the material using special dyes. Chemical methods (GC/MS) are also becoming more widely available for identification of human-source microorganisms, particularly bacteria.

Allergens

Allergens are best quantitated using immunoassays, which directly detect the proteins responsible for health effects. Such assays are available for many animal allergens (*e.g.*, dust mite, cockroach, cat, dog, mouse), for some pollen allergens (*e.g.*, ragweed, birch, grass), and several fungal allergens (*e.g.*, *Alternaria*, *Aspergillus fumigatus*). When such assays are available, they are performed on (usually) settled dust and results reported as allergen/gram of dust. For cat allergen, air samples can also be analyzed.

Presence and levels of allergens for which immunoassays have not been developed must be estimated indirectly. Thus, pollen and fungal spore types and levels in outdoor air are estimated using microscopic identification of air samples collected by impaction onto optically suitable surfaces. Indoors, exposure to fungal and bacterial allergens is usually estimated using culture. This obviously leads to gross underestimation of exposure since many allergen-bearing cells are likely to be nonliving, and many cannot be grown in culture. Microscopy can used to accurately estimate fungal levels as for outdoors, although some of the fungal types common indoors can be identified only to group (*e.g.*, *Aspergillus* and *Penicillium*) are not differentiable by microscopy alone).

Some investigators have advocated the use of chemical assays (for ergosterol or glucans) or immunoassays (for glucans) to estimate total fungal burden (and hence allergen exposure) in houses. Although these methods may in the future provide useful tools in evaluating the overall condition of a house, they do not provide the evidence necessary to make the connection between exposure and sensitization to specific fungi.

Toxins

Assessment of risk from exposure to toxin is best accomplished by measuring the specific toxin of concern. Bacterial endotoxin is routinely measured using the *Limulus* bioassay, which measures the actual effect of the toxin. GC/MS is also used, but total endotoxin is measured, rather than just the biologically relevant fraction. Some mycotoxins can also be readily measured, at least in bulk samples. On the other hand, many microorganisms produce many different kinds of toxins, and assessing the potential exposure to all is, at present, not possible using chemical methods. Instead, bioassays such as the brine shrimp assay for cytotoxins or the Ames assay for mutagens can be used to assess overall toxicity in a sample. However, most often investigators resort to cultural (or occasionally, microscopic) identification of specific organisms from air or bulk samples and infer toxin exposure if a known toxin-producing organism is found to be present. This is (essentially) a worst-case approach, since many organisms produce toxins only under a restricted set of environmental conditions.

REFERENCES

Adan, O.C.G. 1994. On the fungal defacement of interior finishes. Thesis at Eindhoven University of Technology.

Alzona, J., B.L. Cohen, H. Rudolph, H.N. Jow, and J.O. Frohkiger. 1979. Indoor-outdoor relationships for airborne particulate matter of outdoor origin. Atmos. Environ. 13:55–60.

AAAI (American Academy of Allergy, Asthma, and Immunology). 1994. Pollen and Spore Report. Milwaukee, Wisconsin. 248 pages.

Arlian, L. 1989. Biology and ecology of house dust mites, *Dermatophagoides* spp. and *Euroglyphus* spp. Immunol. Allergy Clin. North Amer. 9(2):339.

Aylor, D.E. and P.E. Waggoner. 1980. Aerial dissemination of fungal spores. Ann. N.Y. Acad. Sci. 641:116–122.

Batterman, S., N. Bartoletta, and H. Burge. 1991. Fungal volatiles of potential relevance to indoor air quality. Paper 91-62.9, presented at the annual mMeeting of the Air and Waste Management Association. Vancouver, Canada. June. Air and Waste Management Association. Pittsburgh, Pennsylvania.

Betina, V. 1989. Mycotoxins: Chemical, biological and environmental aspects. Bioactive molecules Volume 9. Elsevier. New York, New York.

Burge, H. 1989. Indoor air and infectious disease. Occupational medicine: State of the art reviews 4(4):713–721.

Burge, H. 1995. Airborne contagious disease. Pages 25–48 in: Burge, H.A. (ed.). Bioaerosols. CRC/Lewis Press. Boca Raton, Florida.

Burge, H.A., W.R. Solomon, and J.R. Boise. 1980. Microbial prevalence in domestic humidifiers. Appl. Environ. Microbiol. 39(4):840–844.

Burke, G.W., C.B. Carrington, R. Strauss, J.N. Fink, and E.A. Gaensler. 1977. Allergic alveolitis caused by home humidifiers. JAMA 238:2705.

Buttner, M.P. and L.D. Stetzenbach. 1993. Monitoring airborne fungal spores in an experimental indoor environment to evaluate sampling methods and the effects of human activity on air sampling. Appl. Environ. Microbiol. 59(1):219–226.

Chew, G., M. Muilenberg, D. Gold, and H.A. Burge. 1996. Is dust sampling a good surrogate for exposure to airborne fungi? J. Allergy. Clin. Immunol. 97(1)pt 3:419 (Abstract).

Chiu, J.T., B. Nelson, W. Sokol, M. Ellis, S. Galant, and H.S. Novey. 1993. Extended evaluation of diagnostic skin testing practices in Orange County, California. Allergy Proc. 14(4):283–286.

Croft, W.A., B.B. Jarvis, and C.S. Yatawara. 1986. Airborne outbreak of trichothecene toxicosis. Atmos. Environ. 20:549–552.

Custovic, A., R. Green, S.C. Taggart, A. Smith, C.A. Pickering, M.D. Chapman, and A. Woodcock. 1996. Domestic allergens in public places. II: Dog (Can f 1) and cockroach (Bla g 2) allergens in dust and mite, cat, dog and cockroach allergens in the air in public buildings. Clin. Exper. Allergy 26(11):1246–1252.

Eichoff, T.C. 1979. Epidemiology of Legionnaires' disease. Ann. Intern. Med. 90:499–502.

Eisenberg, J.N., E.Y. Seto, A.W. Olivieri, and R.C. Spear. 1996. Quantifying water pathogen risk in an epidemiological framework. Risk Anal. 16(4):549–563.

Farr, R.W. 1994. Hantavirus pulmonary syndrome. West Virginia Medical Journal 90(10):422–425.

Fraser, D.W. 1985. Potable water as a source for Legionellosis. Environ. Health Perspect. 62:337–341.

Grant, C., C.A. Hunter, B. Flannigan, and A.F. Bravery. 1989. The moisture requirements of moulds isolated from domestic dwellings. Intern. Biodeterior. 25:259–284.

Gravesen, S. 1987. Microbial and dust pollution in non-industrial work places. Pages 279–282 in: Boehm, G. and R. Leuschner. Advances in aerobiology. Proceedings of the 3rd international conference on aerobiology. Birkhauser Verlag. Basle.

Hunter, C.A., C. Grant, B. Flannigan, and A.F. Bravery. 1988. Moulds in buildings: The air spora of domestic dwellings. Intern. Biodeterior. 24:81–101.

Ingold, C.T. 1971. Fungal spores: Their liberation and dispersal. Clarendon Press. Oxford, England. 302 pages.

Kendrick, B. 1992. The Fifth Kingdom. 2nd edition. Mycologue Publications. Newburyport, Massachusetts. 406 pages.

Knight, V. 1973. Airborne transmission and pulmonary deposition of respiratory viruses. Pages 1–9 in: Knight, V. (ed.). Viral and mycoplasmal infections of the espiratory tract. Lea and Febiger. Philadelphia, Pennsylvania.

Lehtonen, J. and T. Reponen. 1993. Everyday activities and variation of fungal spore concentrations in indoor air. Internat. Biodeterior. Biodegrad. 31:25–39.

Lintner, T.J. and K.A. Brame. 1993. The effects of season, climate, and air-conditioning on the prevalence of *Dermatophagoides* mite allergens in household dust. J. Allergy Clin. Immunol. 91(4):862–867.

Luczynska, C.M. 1995. Mammalian aeroallergens. Pages 19–162 in: Burge, H.A. Bioaerosols. CRC/Lewis Press. Boca Raton, Florida.

Madelin, T.M. and H.E. Johnson. 1992. Fungal and actinomycete spore aerosols measured at different humidities with an aerodynamic particle sizer. J. Appl. Bacteriol. 72:400–409.

Mathews, K. 1989. Inhalant insect-derived allergens. Immunol. Allergy Clin. North Amer. 9(2):321–338.

Milton, D. 1995. Endotoxin. Pages 77–86 in: Bioaerosols. CRC/Lewis Press. Boca Raton, Florida.

Milton, D.M. 1996. Bacterial endotoxins: A review of health effects and potential impact in the indoor environment. Pages 179–196 in: Gammage, R.B. and B.A. Berven (eds.). Indoor air and human health. Second edition. CRC/Lewis Press. Boca Raton, Florida.

Pasanen, A.L., P. Pasanen, M.J. Jantunen, and P. Kallioski. 1991. Significance of air humidity and air velocity for fungal spore release into the air. Atmos. Environ. 25A:459–462.

Patterson, R., J.N. Fink, W.B. Miles, J.E. Basich, D.B. Schleuter, D.G. Tinkelman, and M. Roberts. 1981. Hypersensitivity lung disease presumptively due to Cephalosporium in homes contaminated by sewage flooding or by humidified water. J. Allergy Clin. Immunol. 68:128–132.

Pitt, J.I. 1975. Xerophilic fungi and the spoilage of foods of plant origin. Pages 273–307 in: Duckworth, R.G. (ed.). Water relations of food. Academic Press. London, United Kingdom.

Platts-Mills, T.A.E. and M.D. Chapman. 1987. Dust mites: Immunology, allergic disease, and environmental control. J. Allergy Clin. Immunol. 80:755–775.

Pope, A., R. Patterson, and H. Burge. 1993. Indoor allergens. National Academy Press. Washington, DC. 308 pages.

Riley, R.L. 1982. Indoor airborne infection. Environ. Internat. 8:317–320.

Rosenstreich, D.L., P. Eggleston, M. Kattan, D. Baker, R.G. Slavin, P. Gergen, H. Mitchell, K. McNiff-Mortimer, H. Lynn, D. Ownby, and F. Malveaux. 1997. The role of cockroach allergy and exposure to cockroach allergen in causing morbidity among inner-city children with asthma. New England J. Med. 336(19):1356–1363.

Salvaggio, J. 1987. Hypersensitivity pneumonitis. J. Allergy Clin. Immunol. 79:558–571.

Shiue, S.T., H.H. Scherzer, A.C. DeGraff Jr., and S.R. Cole. 1990. Hypersensitivity pneumonitis associated with the use of ultrasonic humidifiers. N.Y. State J. Med. 90:263–265.

Sietsma, J.H. and J.G.H. Wessels. 1981. Solubility of (1 → 3)-β-D/(1 → 6)-β-D-glucan in fungal walls: Importance of presumed linkage between glucan and chitin. J. Gen. Microbiol. 125:209–212.

Squillace, S.P. 1995. Allergens of arthropods and birds. Pages 133–148 in: Burge, H.A. (ed.). Bioaerosols. CRC/Lewis Press. Boca Raton, Florida.

Tortora, G.J., B.R. Funke, and C.L. Case. 1992. Microbiology. An Introduction. Fourth edition. Benjamin/Cummings Publ. Co. Inc. Redwood City, California. 810 pages.

Tovey, E.R., M.D. Chapman, C.W. Wells, and T.A.E. Platts-Mills. 1981. The distribution of dust mite allergen in the houses of patients with asthma. Amer. Rev. Respir. Dis. 124:630–635.

Verhoeff, A.P. and H.A. Burge. 1997. Health risk assessment of fungi in home environments. Ann. Allergy. 78(6):544–559.

IV

Exposure Factors

10

HUMAN EXPOSURE FACTORS

Data Sources and Principles for Their Use

Accurate assessment of human health risks associated with exposure to environmental agents and stressors depends on realistic estimates of human exposure, as well as adequate knowledge and understanding of dose-response relationships. Exposure assessment is one of the four major steps in the standard risk assessment frame work, along with hazard identification, dose-response assessment, and risk characterization (NRC 1983, 1994). From the perspective of regulatory agencies, such as the U.S. Environmental Protection Agency (USEPA), exposure assessment is a formalized process aimed at identifying important sources and pathways of exposure, estimating the number of people exposed to specific pollutant concentrations for the time period of interest, and determining the amount of the pollutant that actually enters the human body (dose) (USEPA 1992, 1996). A crucial and sometimes controversial aspect of most exposure assessments is the need to make well-reasoned decisions about human exposure factors, which are the critical parameters used to calculate exposure and dose.

This chapter provides the details on how to obtain and use information on human exposure factors. General issues on scenario-based exposure assessment, and on uncertainty and variability are discussed in Sections 1 and 5 of this book, respectively.

KEY ISSUES FOR EXPOSURE ASSESSORS

No matter why an exposure assessment is conducted, whether for screening or for site- and situation-specific reasons, the assessor's primary goal must be to estimate exposure and dose as accurately as possible. Success in attaining this goal depends on careful, informed decisions about several key issues related to presenting, interpreting, and communicating the results of residential exposure assessment. These issues, discussed in detail in Section 5, are:

- Distinguishing between uncertainty and variability
- When to use point estimates or distributions of exposure factors, and
- Communicating exposure and dose estimates.

Additionally, exposure assessors need to consider measured versus modeled exposure concentrations and sources of exposure factors.

The American Industrial Health Council (AIHC 1994) defines what it calls "Good Exposure Assessment Practices," as general principles that should guide assessors in preparing exposure assessments. The AIHC recommends that eight practices should apply to all exposure assessments:

- *Protocol*—Every exposure assessment should have a written protocol before other work begins. It should first state the purpose of the assessment and also discuss the variables to be evaluated (*i.e.*, a clearly defined assessment endpoint), the necessary level of detail, methods for addressing uncertainty, and the relationship between uncertainty and conclusions. The protocol should also describe each of the subsequent seven practices in sufficient detail to document that the assessment is adequate for the stated purpose.
- *Organization, Personnel, and Resources*—The organization sponsoring the assessment should accept the protocol, have the commitment to carry it out, and draw only those conclusions warranted by the data and the analysis. The assessors and other scientific and technical staff should have the necessary education, experience, and certification necessary to complete the assessment successfully. The resources committed to the assessment should be adequate to complete it and, where necessary, commitments for the cooperation of others should be obtained in advance.
- *Study Model*—The model or set of models that relate the presence of a substance to dose should be explicitly stated, and the basis for each model (*e.g.*, deterministic, empirical, statistical) should be described. It should be made clear which model parameters are to be measured and which assumed, and how uncertainty in parameters and the model itself is to be evaluated and communicated. Validation of the model and underlying statistical assumptions should be described.
- *Study Design*—For the studies on which the assessment is based, sampling statistics, data collection methods, analytical methods, and data analysis methods should be documented and shown to be adequate for making inferences with stated levels of confidence and power.
- *Quality Assurance*—For the studies on which the assessment is based, the assessor should demonstrate that procedures are adequate to ensure an acceptable level of data quality in sampling, transport, analysis, compilation, recording, and storage of the data. Independent quality audits should be done periodically.
- *Statement of Overall Uncertainty*—The results of the assessment should include a statement of overall uncertainty, including consideration of both variability and uncertainty, and, at a minimum, provide an estimate of the range between specified percentiles.
- *Archiving*—The protocol and all raw data collected during the assessment, including questionnaires, reports, instrument calibrations, quality control records, and other important documents, should be retained in such a way so that they are retrievable for some specified period.
- *Communication and Confidentiality of Results*—Results of personal exposure studies should be reported to the individual on whom the data were collected, while other, less direct measurements (*e.g.*, area samples) should be reported to relevant individuals if data related substantially to their exposure. The identity of persons participating in exposure studies should be kept confidential.

AVAILABLE INFORMATION ON HUMAN EXPOSURE FACTORS

A critical challenge confronting all exposure assessors is the need to choose suitable values for important human exposure factors. Basically, there are two general types of information sources to which the assessor can turn:

- Primary sources, which are studies or collections of studies reported in the scientific literature (*e.g.*, peer-reviewed journals), and
- Secondary sources, which are compilations that summarize existing data from primary sources and recommend values for important human exposure factors.

The following discussion summarizes information about human exposure factors contained in important secondary sources.

The USEPA's Exposure Factors Handbook

Historically the most widely used secondary source has been USEPA's Exposure Factors Handbook, updated in 1996. The purpose of the Exposure Factors Handbook is to: (1) summarize data on human behaviors and characteristics which affect exposure to environmental contaminants, and (2) recommend values to use for these factors.

These recommendations are not legally binding on any EPA program and should be interpreted as suggestions which exposure assessors can consider and modify as needed. Most of these factors are best quantified on a site or situation-specific basis. The Handbook strives to include full discussions of the issues which assessors should consider in deciding how to use these data and recommendations. The Handbook is intended to serve as a support document to EPA's Guidelines for Exposure Assessment (USEPA 1992). The Guidelines were developed to promote consistency among the various exposure assessment activities that are carried out by the various EPA programs. The Handbook assists in this goal by providing a consistent set of exposure factors to calculate dose.

To develop the Exposure Factors Handbook, USEPA reviewed the majority of the literature pertaining to a factor and then distinguished between "relevant" and "key" studies. Scientific studies chosen for inclusion in the Handbook were selected based on the considerations listed in Table 10-1. The recommended values for most human exposure factors are based on results of designated "key" studies, which are those deemed to best exemplify the attributes listed in Table 10-1. Other studies, which are not considered to be as important but are still deemed to be pertinent and potentially useful, are designated as "relevant." The USEPA points out that "The recommended values were derived solely from USEPA's interpretation of the available data. Different values may be appropriate for the user to select in consideration of policy, precedent, strategy, or other factors such as site-specific information."

The USEPA used a five-step procedure to develop its recommendations:

- Key studies were identified and evaluated according to the criteria listed in Table 10-1.
- If only one key study was identified for a particular factor, the mean value was selected as the recommended central value for that population. If it was determined there were multiple key studies, all reasonably equal according to quality, relevance, and study design, then a weighted mean of the studies was designated as the recommended mean value. If the key studies were deemed to

Table 10-1. Considerations in selecting studies for the USEPA's Exposure Factors Handbook

Characteristic	Explanation
Level of peer review	Studies were selected predominantly from the peer-reviewed literature and final government reports. Internal and interim reports were therefore avoided.
Accessibility	The preferred studies were those that the use could access in their entirety if desired.
Reproducibility	Sought-after studies were those that contained sufficient studies so that the methods could be reproduced, or at least so that details of the author's work could be accessed and evaluated.
Focus on exposure factor of interest	Selected studies were those that directly addressed the exposure factor of interest, or which addressed related factors which have significance for the factor under consideration. As an example of the latter case, a selected study contained useful ancillary information concerning fat content in fish, although it did not directly address fish consumption.
Data pertinent to the U.S.	The selected studies were those that addressed the U.S. population. Data on populations outside the U.S. were sometimes included if behavior patterns and other characteristics of exposure were similar.
Primary data	Studies were deemed preferable if based on primary data, but studies based on secondary sources were included where they offered an original analysis. For example, the Handbook cites studies of food consumption based on original data collected in the USDA National Food Consumption Survey.
Current information	Studies were chosen only if they sufficiently recent to represent current exposure conditions. This is an important factor for those factors that change with time.
Adequacy of data collection period	Because most users of the Handbook are addressing primarily chronic exposures, studies were sought that utilized the most appropriate techniques for collecting data to characterize long-term behavior.
Validity of approach	Studies utilizing experimental procedures or approaches that more likely or closely capture the desired measurement were selected. In general, direct exposure data collection techniques, such as direct observation, personal monitoring devices, or other known methods were preferred where available. If studies utilizing direct measurement were not available, the selected studies were those that rely on validated indirect measurement methods such as surrogate measures (e.g., heart rate for inhalation rate) and questionnaires. If questionnaires or surveys were used, proper design and procedures must have included adequate sample size for the population under consideration, a response rate large enough to avoid biases, and avoidance of bias in design of the instrument and interpretation of results.
Representativeness of the population	Studies seeking to characterize the national population, a particular region, or subpopulation were selected, if appropriately representative of that population. In cases where data were limited, studies with limitations in this are were included and limitations were noted in the Handbook.

be unequal in quality, relevance, or study design, then the range of means was presented and the user must select the most appropriate value. For the national population, the midpoint of the range is usually considered to be the most appropriate value.
- The variability of the factor across the population was discussed and, if adequate data were available, the variability was described as either a series of percentiles or a distribution.
- The uncertainty in the recommended value was discussed in terms of data limitations, the range of circumstances over which the estimates are (or are not) applicable, and possible biases in the values themselves. A statement was included about parameter uncertainties (*e.g.*, measurement and sampling error), and model or scenario uncertainties.
- Each recommended value was assigned a confidence rating according to the general guidelines summarized in Table 10-2, and a rationale was provided for the rating.

A summary of USEPA's recommendations and associated confidence ratings for human exposure factors are provided in Table 10-3.

American Industrial Health Council's Exposure Factors Sourcebook

The American Industrial Health Council, an industry association, has published an Exposure Factors Sourcebook (AIHC1994) that summarizes recommended values for commonly used human exposure factors. The report addresses three major types of factors:

- Receptor physiologic parameters (*i.e.*, adult and child body weight, adult height, skin surface area)
- Time-activity patterns (*i.e.*, years in one residence, time spent indoors and outdoors, weekly hours at and away from home for both adults and children, weekly hours at work, working tenure, shower duration, time-use patterns), and
- Receptor contact rates (*i.e.*, adult and child soil ingestion, total dietary intake, consumption of homegrown fruits and vegetables, consumption of fish and shellfish, consumption of meat and beef, consumption of drinking water, and daily inhalation rate for adults and children).

Recommended values are summarized in Table 10-4. For each exposure factor a rationale was provided for the recommended of a single value and, where available, information is included on the range of values as well as a statistical treatment of the data.

The AIHC also recommends a list of fourteen principles that define "good exposure assessment practices" for probabilistic risk assessment by Monte Carlo simulation (Burmaster and Anderson 1994):

- *Principle 1*—Show all formulas used to estimate single-point values for exposure, dose, potency, hazard indices, or risk.
- *Principle 2*—Calculate and present the single-point estimates of exposure and risk according to the existing deterministic risk assessment guidelines for the appropriate regulatory agency. The calculation of point estimates using standard techniques is a desirable first step in Monte Carlo-based risk assessment.

Table 10-2. Considerations used to rate confidence in recommended values

Considerations	High confidence	Low confidence
Study elements		
Level of peer review	Studies received high level of peer review (*e.g.*, appear in peer-reviewed journals)	Studies received limited peer review
Accessibility	Studies are widely available to public	Studies are difficult to obtain (*e.g.*, draft reports, unpublished data
Reproducibility	Results can be reproduced or methodology can be followed and evaluated	Results cannot be reproduced, methodology is hard to follow, and author(s) cannot be located
Focus on factor of interest	Studies focused on the exposure factor of interest	Purpose of the studies were to characterize a related factor
Data pertinent to the U.S.	Studies focused on the U.S. population	Studies focused on population outside the U.S.
Primary data	Studies analyzed primary data	Studies based on secondary sources
Currency	After 1990	Before 1990
Adequacy of data collection period	Study design captures the measurement of interest (*e.g.*, usual consumption patterns of interest)	Study does not capture the measurement of interest very accurately
Validity of approach	Studies used the best methodology available to capture the measurement of interest	There are serious limitations to the approach used
Study sizes	$n > 100$	$n < 20$
	Sample size depends on how target population is defined. As size of the sample relative to total size of the target population increases, estimates are made with greater statistical assurance that the sample results reflect actual characteristics of the target population.	
Representativeness of the population	Study population same as population of interest	Study population very different from the population of interest
Variability in the population	Studies characterized variability in the study population	Characterization of variability is limited
Lack of bias in study design (a high rating is desirable)	Potential biases in the studies are stated or can be determined from the study design	Study design introduces biases in results
Response rates		
In-person interviews	>80[a]	<40[a]
Telephone interviews	>80[a]	<40[a]
Mail surveys	>70[a]	<40[a]
Measurement error	Study design minimizes measurement error	Uncertainties in the data exist due to measurement error
Other elements		
Number of studies	>3	1
Agreement among researchers	Results of studies from different researchers are in agreement	Results of studies from different researchers are in disagreement

[a] Differences include age, sex, race, income, or other demographic parameters.

Table 10-3. Summary of exposure factor recommendations and confidence ratings

Exposure factor	Recommendation	Confidence rating
Drinking water intake rate	21 ml/kg-day (average)	Medium
	34 ml/kg-day (90th percentile)	
	Percentile and distribution also included	
Total fruit intake rate	3.4 g/kg-day (average)	Medium
	12.4 g/kg-day (95th percentile)	Low
	Percentiles also included	
	Means presented for individual fruits	
Total vegetable intake rate	4.3 g/kg-day (average)	Medium
	10 g/kg-day (95th percentile)	Low
	Percentiles also included	
	Means presented for individual vegetables	
Total meat intake rate	2.1 g/kg-day (average)	Medium
	5.1 g/kg-day (95th percentile)	Low
	Percentiles also included	
	Means presented for individual meats	
Total dairy intake rate	8.0 g/kg-day (average)	Medium
	29.7 g/kg-day (95th percentile)	Low
	Percentiles also included	
	Means presented for individual dairy products	
Breast milk intake rate	742 ml/day (average)	Medium
	1,033 ml/day (upper percentile)	Medium
Fish intake rate	*General population*	
	20.1 g/day (total fish) average	Medium
	13.5 g/day (marine) average	Medium
	6.6 g/day (freshwater estuarine) average	Medium
	63 g/day (total fish) 95th percentile long-term	Medium
	Serving size	
	123 (average)	High
	305 g (95th percentile)	High
	Recreational marine anglers	
	2–7 g/day (fintfish only)	Medium
	Recreational freshwater anglers	
	8 g/day (average)	Medium
	25 g/day (95th percentile)	Medium
	Native american subsistence population	
	70 g/day (average)	Medium
	170 g/day (95th percentile)	Low
Home produce food intake	*Total fruits*	Medium (for means and short-term distibutions)
	2.7 g/kg-day (average)	
	11.1 g/kg-day (95th percentile)	
	Total vegetables	
	2.1 g/kg-day (average)	
	7.5 g/kg-day (95th percentile)	
	Total meats	
	2.2 g/kg-day (average)	
	6.8 g/kg-day (95th percentile)	
	Total dairy products	
	14 g/kg-day (average)	
	44 g/kg-day (95th percentile)	
	Percentiles also included	
	Means presented for individual food items	
	Low (for long-term distributions)	
Inhalation rate	*Children (<1 year)*	
	4.5 m^3/day (average)	High
	Children (1–12 years)	
	8.7 m^3/day (average)	High

Table 10-3. *(Continued)*

	Adult females	
	11.3 m^3/day (average)	High
	Adult males	
	15.2 m^3/day (average)	High
Surface area	*Water contact (bathing and swimming)*	
	Use total body surface area for children in USEPA 1996 Tables 6-6 to 6-8, for adults Tables 6-2 to 6-4 (percentiles are included)	High
	Soil contact (outdoor activities)	
	Use whole body part area for children in USEPA 1996 Tables 6-6 to 6-8, for adults in Tables 6-2 to 6-4 (percentiles are included)	
Soil adherence	Use values presented in USEPA 1996 Table 6-16 depending on activity and body part (central estimates only)	Low
Soil ingestion rate	*Children*	
	100 mg/day (average)	Medium
	400 mg/day (upper percentile)	
	Adults	
	50 mg/day (average)	Low
	Pica child	
	10 g/day	Low
Life expectancy	75 years	High
Body weight	71.8 kg	High
Showering/Bathing	*Showering time*	
	8 min/day (average)	Medium
	12 min/day (95th percentile)	
	(percentiles are also included)	
	Bathing time	
	20 min/event (median)	High
	45 min/event (90th percentile)	
	Bathing/showering frequency	
	1 shower event/day	High
Swimming	*Frequency*	
	1event/month	High
	Duration	
	60 min/event (median)	High
	180 min/event (90th percentile)	
Time indoors	*Children (ages 3–11)*	Medium
	19 hr/day (weekdays)	
	17 hr/day (weekends)	
	Adults (ages 12 and older)	Medium
	21 hr/day	
	Residential	High
	16.4 hr/day	
Time outdoors	*Children (ages 3–11)*	Medium
	5 hr/day (weekdays)	
	7 hr/day (weekends)	
	Adults	Medium
	1.5 hr/day	
	Residential	High
	2 hr/day	
Time spent inside vehicle	*Adults*	
	1 hr 20 min/day	Medium
Occupational tenure	6.6 years (16 years of age and older)	High
Population mobility	9 years (average)	Medium
	30 years (95th percentile)	

Table 10-4a. Reverence *versus* default point value: physiologic parameters and their distributions. Adapted from AIHC 1994. All data sources reported in AIHC 1994

Receptor physiologic parameters	Reference point value[a]	Commonly used default value (Average case, worst case)	Data quality	Exposure factor	Representative distribution	Units
Adult body weight	72 kg	70 kg, NA	High	Adult body weight	@CUMUL(44,107,52,3,.05,57.6,.15,68.7,.5,84.4,85,97,.95,5)	kg
Child body weight ages 1–4	13 kg	16 kg (ages 1–6), NA	High	Child body weight	@CUMUL(7,20,10.6,.05,11.4,.15,12.9,.5,14.7,.85,16,.95,5)	kg
Adult height	66.5 in	NA	High	Adult height—men Adult height—women	@NORMAL(69.12,2.85) @NORMAL(63.68,2.60)	in in
Adult male body surface area	1.93 m^2	1.94 m^2, 2.20 m^2	Moderate	Skin surface area—men	@CUMUL(1.5,2.3,1.66,.05,1.53,.15,1.69,.5,1.91,.85,2.09,.95,5)	m^2
Adult female body surface area	1.69 m^2	1.69 m^2, 1.98 m^2	Moderate	Skin surface area—women	@CUMUL(1.4,2.1,1.45,.05,1.53,.15,1.69,.5,1.91,.85,2.09,.95,5)	m^2

[a] Reference point values represent measures of central tendency (median, mode, mean) from the best available source(s) of data.

Table 4b. Reverence *versus* default point value: time-activity parameters. Adapted from AIHC 1994. All data sources reported in AIHC 1994

Time activity	Reference point value[a]	Commonly used default value (average case, worst case)	Data quality	Exposure factor	Representative distribution	Units
Years at one residence	8.1 yr	25 or 30 yr (9 yr, 30 yr)	High	ROP	@CUMUL(1,75,4,.25,8,.5,15, .75,26,.9,33,.95,47,.99,6)	yrs
Weekly hr spent: indoor-outdoor	156 hr–12 hr	168 hr–0 hr (156 hr–12 hr, NA)	High			
Weekly hr: adult home and away	108 hr–60 hr	168 hr–0 hr (108 hr–60 hr, NA)	Moderate			
Weekly hr: child home and away	138 hr–30 hr	168 hr–0 hr	Moderate			
Weekly hr at work	23 hr	40 hr	Moderate	Hours at work	@CUMUL(0,107,.34,.3,8.31, .4,20.22,.5,32.08,.6,37.68,7, 41.33,.8,46.88,.9,7)	hrs/week
Working tenure	4 yr	25 or 30 yr	High			
Shower duration	7.6 min	7 min/day, 12 min/day	High	Shower duration	@CUMUL(1,20,2,.008,4,.10,5,.2 3,7,.53,9,.73,12,.9,15,.96,19, .99,8)	min

Table 10-4c. Representative distributions for receptor: contact rate exposure factors. Adapted from AIHC 1994. All data sources reported in AIHC 1994

Exposure factor	Reference point value[a]	Commonly used default value (average case, worst case)	Data quality	Exposure factor	Representative distribution	Units
Adult soil ingestion	0.1–10 mg/day[b]	100 mg/day[c]	Low	Adult soil ingestion	@CUMUL(0,216,0,.17,0,.33,0,.5,17,.67,148,.83,216,1,6)	mg/day
Child soil ingestion	16 mg/day	200 mg/day, 800 mg/day	Moderate	Child soil ingestion	@CUMUL(0,1391,0,.05,0,.10,16,.5,67,.9,110,.95,5)	mg/day
Total dietary intake	1.6 kg/day	1.6 kg/day	Moderate	Vegetable	@LOGNORMAL(62,1800)	g/day
Adult consumption of homegrown fruits vegetables	28–50 g/day	28–50 g/day	Moderate	ingestion		
Adult consumption of fish/shellfish: total nonmarine	14 g/d–6.5 g/day	14–6.5 g/day	Moderate	Freshwater finfish ingestion	@CUMUL(.4,15.,4,.02,1,.13,1.6,.36,2,.49,3,.7,5,.9,10,.98,7)	g/day
				Saltwater finfish ingestion	@CUMUL(.2,10,.2,.03,.8,.25,1.4,.53,2,.74,4,.93,10,1,6)	g/day
				Shellfish ingestion	@CUMUL(.2,10,.2,.12,.8,.45,1.4,.66,2,.79,4,.96,5)	g/day
Adult beef intake: total-homegrown	88 g/d–40 g/d	100–44 g/day, 100–75 g/day	Moderate	Beef ingestion	@NORMAL(75,56)	g/day
Adult water intake	1.4 L/day	1.4–2 L/day	High	Water ingestion—adult	@CUMUL(.4,2,.4,.192,.96,.396,1.28,.6,1.7,.8,1.96,1, 5)	L/day
				Water ingestion—child	@CUMUL(.6,3,.676,.025,1.046,.25,1.316,.5,1.655,.75,2.562,.975,5)	L/day
Adult inhalation rate	18 m^3/day	20–30 m^3/day	Moderate			
Child inhalation rate	12 m^3/day	15–20 m^3/day	Moderate			

[a] Reference point values represent measures of central tendency (median, mode, mean) from the best available source(s) of data.
[b] Range of values is recommended over single point value because of the shortage of reliable data.
[c] USEPA makes no recommendation for this factor because data are lacking.

- *Principle 3*—Present results from sensitivity analyses of the deterministic calculations to identify inputs suitable for probabilistic treatment, and discuss variables not included in analyses.
- *Principle 4*—Restrict the use of probabilistic techniques to the exposure pathways and compounds of regulatory importance, saving time, money, and other resources.
- *Principle 5*—Provide detailed information on the input distributions selected, including a graph showing the full distribution and location of the point value used in deterministic assessment, a table showing the mean, standard deviation, minimum value, 5th percentile, 50th percentile (median), the 95th percentile, and the maximum value, and a written justification for the distribution selected.
- *Principle 6*—Show, to the extent possible, how the input distributions and their parameters capture and represent both variability and uncertainty.
- *Principle 7*—Use measured data to inform the choice of input distributions whenever possible, after making sure the data are relevant to and representative of the population, place, and time, which are the focus of the assessment.
- *Principle 8*—For parametric distributions, report the goodness-of-fit statistics and related methodological issues for input variables fit quantitatively to measured data.
- *Principle 9*—Discuss the presence or absence of moderate to strong correlations between or among input variables.
- *Principle 10*—Provide detailed information and graphs for each output distribution.
- *Principle 11*—Perform probabilistic sensitivity analyses for all key inputs represented by a distribution in the Monte Carlo analysis so as to distinguish the effects of variability from the effects of uncertainty.
- *Principle 12*—Investigate and report the numerical stability of the central moments (mean, standard deviation, skewness, kurtosis) and the tails of the output distribution from the simulation.
- *Principle 13*—Report the name and statistical quality of the random number generator used for the simulation.
- *Principle 14*—Discuss limitations of the methods and caveats related to interpretation of results.

TIME-ACTIVITY PATTERNS

Because time-activity patterns are such an important part of all scenario-based exposure assessments, it is useful to take a closer look at the data underlying USEPA-recommended activity factors. In general, activity factors refer to:

- The time people spend in various locations (*e.g.*, indoors at home, commuting in a vehicle, working in an office) during their normal daily routine
- The various activities in which they engage (*e.g.*, eating, exercising, smoking, cooking, sleeping) or are affected by (*e.g.*, presence of active smokers), and
- Their occupational and residential mobility patterns over a lifetime (*e.g.*, years at one residence or job).

Unfortunately, activity patterns are generally not well characterized and appropriate data are sparse.

The Role of Activity Patterns in Exposure Assessment

Exposure is defined as the simultaneous occurrence of two events: a person or persons present in location *x, y, z* at time *t*, and agent concentration *c* is present at location *x, y, z* at time *t* (Ott 1982). In the absence of personal measurements of exposure and/or dose, integrated exposure can be calculated by summing the product of the pollutant concentration in each microenvironment and the time spent by each person in each microenvironment, across all relevant microenvironments. For purposes of this discussion, a microenvironment is defined as a three-dimensional space where the pollutant level at some specified time (or period of activity) is uniform or has constant static properties. The mathematical model for calculating the integrated exposure (E_i) for a population of N people over time period T is summarized in Equation 1 (Duan 1988).

$$E_i = c_j t_{ij} \tag{1}$$

where

E_i = time-weighted, integrated exposure for person i
i = 1, 2, 3, 4. . . . N people
c_j = pollutant concentration in microenvironment j
t_{ij} = time spent by person i in microenvironment j
J = number of microenvironments through which person i moves over the time period of interest
T = time period of interest

Equation 1 implies that the complete understanding of exposure requires explicit knowledge of important microenvironments and related activities, time spent in these microenvironments and associated exposure-related activities, and agent concentrations in important microenvironments. This kind of information allows us to estimate exposures indirectly and to begin to understand how, why, when, and where exposure occur. Only by grasping the full context within which exposures take place can we hope to make informed choices about which exposures (risks) are unacceptable, which control strategies are likely to be effective in preventing or reducing exposures (risks), and how alternative approaches compare according to cost effectiveness.

Activity Pattern Data

Exposure-related studies to discern and describe human activity patterns typically involve administering questionnaires to a defined study population, which can be selected based on statistical design (*e.g.*, stratified random sample of the U.S. population) or as a matter of convenience (*e.g.*, volunteers in a pollution monitoring study). The six most fundamental exposure-related questions that might be included in a study of human activity patterns are summarized in Figure 10-1 (Klepeis *et al.* 1996). These six questions can be grouped into three general categories:

- *Questions about microenvironments*—What are people doing (*e.g.*, activities, background activities) and where are they doing it (*i.e.*, microenvironments)?

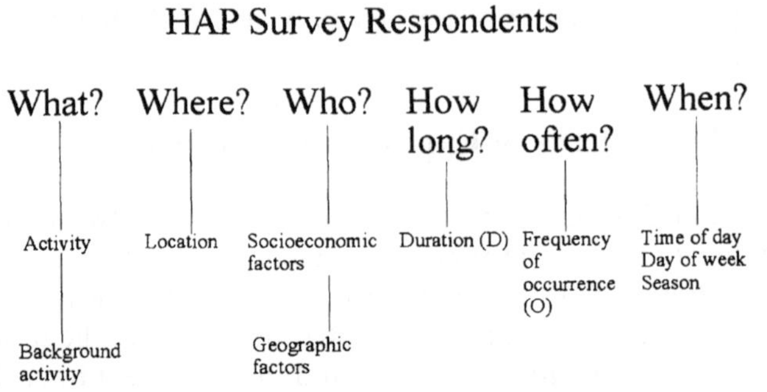

Figure 10-1. The most basic question posed in human activity pattern analysis: The What? and Where? questions specify the microenvironmental factors, Who? and When? determine the background (gender, age, census region) and time (time-of-day, day-of-week, season) factors, respectively, and How long? and How often? determine the duration D and frequency of occurrence, O, of the microenvironments, respectively. Based on Klepeis *et al.* (1996).

- *Questions about people*—Who are the (*e.g.*, sociodemographic characteristics, geographic location)?
- *Questions about time*—How long were they doing it (*i.e.*, duration), how often were they doing it (*i.e.*, frequency of occurrence), and when were they doing it (*e.g.*, time of day, time of week, season)?

As part of the analysis of the National Human Activity Pattern Survey (NHAPS), Klepeis *et al.* (1996) expand the microenvironmental concept by defining the *generalized n-factor microenvironment*, which incorporated three basic ideas. First, whereas the term "environment" encompasses all factors that influence an individual life, a microenvironment is some fraction of all possible discrete events or episodes that occur during a defined portion of time in an individual life. Second, a *microenvironmental factor* is a discrete event related to human behavior such as location, an activity, some other simultaneous activity, or a background activity. Third, microenvironments are fully microenvironmental factors and the corresponding time in which each microenvironmental factor or combination of factors occurs.

The *generalized n-factor microenvironment* concept implies the existence of four types of microenvironmental factors, any one or combination of which may comprise the n-factor portion of the generalized microenvironment for an individual (Klepeis *et al.* 1996). The four kinds of environmental factors are:

- Location of respondent (*e.g.*, kitchen, office, restaurant, airliner)
- Primary activity of the respondent (*e.g.*, cooking, driving, eating, cleaning)
- Simultaneous activities of the respondent (*e.g.*, smoking while driving, playing with pets while eating, applying pesticides while cleaning, using air freshener while watching television), and
- Background activities that the respondent is not directly controlling (*e.g.*, fireplace in operation, guest smoking in the living room, dishwasher running, custodial staff cleaning in the office).

The detailed mechanisms of exposure for each microenvironment, which are derived for the relevant microenvironmental factors, are experimentally and/or theoretically determined. For example, a mass-flow model might be used to calculate typical inhalation exposures to environmental tobacco smoke for a person in a room where several people are smoking. Once the microenvironmental factor(s) of interest has (have) been specified, the characterization of the microenvironment is completed by assigning the appropriate starting and ending times for which the microenvironmental factor(s) occur(s) over the time period of interest (Klepeis *et al.* 1996).

The USEPA Exposure Factors Handbook (Volume III) recommends values for human activity factors including (see Table 10-5):

- Time spent on indoor activities (children and adults)
- Time spent on outdoor activities (children and adults)
- Time spent inside vehicles (adults)
- Time spent taking baths and showers and swimming
- Occupational tenure, and
- Population mobility (years at one residence).

As a guide to locating information on human activity patterns, the USEPA Exposure Factors Handbook (Volume III) also provides a summary of published studies that are appropriate for estimating activity factors for the U.S. population and various subpopulations. These studies used diaries to collect information on time spent at various activities and locations for children, adolescents, and adults, and for certain sociodemographic groups. As shown in Table 10-6, the studies are divided into "key studies," those most useful for deriving activity factors, and "relevant studies," those applicable or pertinent, but not necessarily the most important.

Table 10-5. Summary of recommended values for activity factors. Data sources cited in AIHC 1994

Type	Value
Indoor activities	*Children*
	19 hr/day (weekdays)
	17 hr/day (weekends)
	Adults
	21 hr/day
Outdoor activities	*Children*
	5 hr/day (weekdays)
	7 hr/day (weekends)
	Adults
	1.5 hr/day
Time spent inside vehicle	*Adults*
	1.3 hr/day
Taking baths	20 min/event
Taking showers: duration	7.5 min/day
Occupational tenure	6.6 yr (16 yrs and older)
Population mobility	9 yr (average)
	30 yr (95th percentile)
Swimming	1 event/month
	60 min/event
Residential	
Indoors	16.4 hr/day
Outdoors	2 hr/day

Table 10-6. Summary of consumer use studies. All studies cited in AIHC 1994

Study	Study size	Approach	Relevant population	Comments
Key studies				
USCPSC 1992	4.997 product interviews, 527 mailed questionnaires	Direct: interviews and questionnaires	Adults	Random digit dialing method used to select sample. Information on use of 3 products containing methyl chloride was requested.
USEPA 1987a	4,920 individuals	Direct: questionnaire	18 yrs selected to be representative US population	Waksberg Method (random digit dialing) used to select sample. Respondents asked to recall use in past 2 months of 32 categories of household products containing methyl chloride.
USEPA 1987b	193 households	Direct: telephone survey, two post-survey validation efforts: 30 reinterviewed, then another 50 reinterviewed	Adult household members who do cleaning tasks in household	Waksberg Method (random digit dialing) used to select sample. Household use of cleaning products requested. Phone survey during end-of-year holidays may reflect biased useage data. Two validation resurveys conducted 3 months after survey.
USEPA 1987c	777 households	Direct: one post-survey validation effort conducted with 30 reinterviewed	Household members who do painting tasks in household	Waksberg Method (random digit dialing) used to select sample. Painting product use information in past 12 months was requested. One validation resurvey conducted 3 months after survey.
Klepeis *et al.* 1996	9,386 individuals	Direct: interviews and questionnaires	Representative of US general population	National Human Activity Pattern Survey (NHAPS). Participants selected using random digit dialing and computer assisted telephone interviewing. 24-hr diary data, and followup questions, nationally representative, represents all seasons, age groups, and genders.
Relevant study				
CTFA 1983	*Survey 1*: 47 women employees and relatives or employees *Survey 2*: 1,129 cosmetic purchasers *Survey 3*: 19,035 females	*Survey 1*: Direct 1-week prospective survey *Survey 2*: Direct prospective survey *Survey 3*: Direct 9.5 months prospective survey	*Survey 1*: 16–61 yr-old-females *Survey 2*: Customers of cosmetic manufacturer *Survey 3:* Market research company sampled female consumers nationwide	

Human Activity Factors for Consumer Products

Exposure assessment is often focused on chemicals released during the use of consumer products. To estimate specific pathway exposures for consumer products or their constituents, information is needed about amount of the product used, concentration of the product in each type of activity, percent weight of the chemical in the product, duration and frequency of product use and, for dermal exposure, the amount of solution (material) on the skin after contact (Hakkinen *et al.* 1991, USEPA 1996). The USEPA Exposure Factors Handbook (Volume III) presents a survey of studies that collected human activity data on the frequency, duration, and amount of use for cleaning products, painting products, household solvent products, and cosmetic products. As shown in Table 10-6, studies were classified as either key or relevant based on their utility for estimating activity factors for consumer products.

In addition, two documents developed by the USEPA Office of Toxic Substances were identified as important sources of further information. The first document (USEPA 1986), *Standard Scenarios for Estimating Exposure to Chemical Substances During Use of Consumer Products*, presents data and supporting information to assess consumer exposure to constituents in household cleaners and components of adhesives. Information is also included describing standard scenarios selected to represent upper-bound exposures for each product. The second document (USEPA 1987d), *Methods for Assessing Consumer Exposure to Chemical Substances*, presents detailed information about databases, tools, and a systematic approach for estimating exposures to a given chemical in consumer products. It also includes a comprehensive list of consumer product categories with associated consumer products and potential exposure pathways and mechanisms.

The USEPA has decided "it not feasible to specify recommended exposure values" for consumer products because of the large range and variation among consumer products and related exposure pathways (USEPA 1996). Instead, assessors are referred to the "contents and references" included in the consumer products section of the USEPA Exposure Factors Handbook (Volume III).

REFERENCES

AIHC (American Industrial Hygiene Council). 1994. Exposure factors sourcebook. Washington DC. 118 pages.

Burmaster, D.E. and P.D. Anderson. 1994. Principles of good practice for the use of Monte Carlo techniques in human health and ecological risk assessments. Risk Anal. 14(4):477–481.

Duan, N. 1988. Estimating microenvironment concentration distributions using integrated exposure measurements. Pages 15–114 in: Proceedings of the research planning conference on human activity patterns. T.H. Starks (ed.). EPA report number EPA-600/4-89/004. U.S. Environmental Protection Agency, Office of Research and Development. Washington DC.

Hakkinen, P.J., C.K. Kelling, and J.C. Callender. 1991. Exposure assessment of consumer products: Human body weights and total body surface areas to use, and sources of data for specific products. Vet. Human Toxicol. 33(1):61–65.

Klepeis, N.E, M.A. Tsang, and J.V. Behar. 1996. Analysis of the national human activity pattern survey (NHAPS) respondents from a standpoint of exposure assessment. Final report. Lockheed Martin. Las Vegas, Nevada. Contract Number 68-01-7325. U.S. Environmental Protection Agency. USEPA/600/R-96/074.

NRC (National Research Council). 1983. Risk assessment in the federal government: Managing the process. National Academy Press. Washington DC. 196 pages.

NRC (National Research Council). 1994. Science and judgment in risk assessment. National Academy Press, Washington, D.C. 652 pages.

Ott, W.R. 1982. Concepts of human exposure to air pollution. Environ. Internat. 7:179.

USCPSC (U.S. Consumer Product Safety Commission). 1992. Annual report. Washington DC.

USEPA (U.S. Environmental Protection Agency). 1986. Standard Scenarios for Estimating Exposure to Chemical Substances During Use of Consumer Products. Exposure Evaluation Division, Office of Pesticides and Toxic Substances.

USEPA (U.S. Environmental Protection Agency). 1987a. Household solvent products. A national usage survey. WESTAT, Inc. Rockville, Maryland. Contract Number 68-02-4243. Exposure Evaluation Division, Office of Pesticides and Toxic Substances. Washington DC. NTIS No. PB88-132881.

USEPA (U.S. Environmental Protection Agency). 1987b. National usage survey of household cleaning products. WESTAT, Inc. Rockville, Maryland. Contract Number 68-02-4243. Exposure Evaluation Division, Office of Pesticides and Toxic Substances. Washington DC.

USEPA (U.S. Environmental Protection Agency). 1987c. National household survey of interior painters. WESTAT, Inc. Rockville, Maryland. Contract Number 68-02-4243. Exposure Evaluation Division, Office of Pesticides and Toxic Substances. Washington DC.

USEPA (U.S. Environmental Protection Agency). 1987d. Methods for assessing exposure to chemical substances. Volume 7. Methods for assessing consumer exposure to chemical substances. USEPA Publication Number 560/5-85-007. Exposure Evaluation Division, Office of Pesticides and Toxic Substances. Pages 1–189.

USEPA (U.S. Environmental Protection Agency). 1992. Guidelines for exposure assessment. Fed. Reg. 57 (104):22888-22938. May 29.

USEPA (U.S. Environmental Protection Agency). 1996. Exposure factors handbook (Draft). EPA Publication Number Volume I 600/P-95/002Ba, Volume II 600/P-95/002Bb, and Volume III 600/P-95/002Bc.

11

RESIDENTIAL EXPOSURE FACTORS—DATA SOURCES AND PRINCIPLES FOR THEIR USE

Information on residential exposure factors presented in this chapter will not have much relevance unless their use in the exposure assessment process is demonstrated. The good exposure practices and principles discussed in Chapter 10 are also applicable to residential exposure factors. From a human exposure perspective, the environment of concern is the residence and the exposed receptors of concern are humans occupying the residences.

Exposure is a component in the following continuous process:

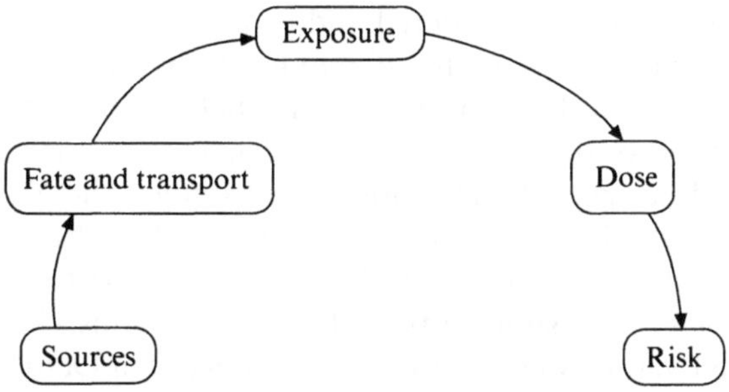

In this process, the human element is introduced into the exposure component through activity patterns. Activity patterns are characterized by the following three factors:

- Where people are (*location*)
- What they do (*activity*), and
- For how long (*duration*).

The first factor, location, comprises the following three broad categories:

- Residential indoor environments
- All other indoor environments, and
- All outdoor environments.

The first category, residential indoor environments, is the focus in this chapter.

In their residences, humans are exposed to pollutants resulting from their actions, actions of their co-residents, and sources uncontrolled by the residents.

Residential Exposure Assessment, edited by Baker *et al.*
Kluwer Academic/Plenum Publishers, New York, 2000.

Examples of exposures from one's actions include exposure to primary and secondary cigarette smoke while smoking and exposure to volatile organic compounds while taking a shower. Examples of exposures from actions of co-residents include exposure to secondary cigarette smoke while someone else is smoking and exposure to volatile organic compounds while in a bathroom that was just used for showering by a co-resident. Examples of exposures from sources uncontrolled by residents include exposure to outdoor pollutants from infiltration into residences and exposure to emissions from building materials. To characterize exposures in the above-mentioned scenarios and numerous others that occur in residences, human activity patterns, the residential environment, contaminant sources, and the factors influencing the residential environment must be well understood.

Human Activity Patterns

Human activity patterns were addressed in Chapter 10. Since the exposure receptor of importance is humans, defining what they do in residences puts in perspective those details relevant to the assessment of human exposures to contaminants in residential environments. Referring to the three factors of human activity patterns listed in the previous section, *location* requires knowledge of where people spend their time in the residence. It is important to know how often such locations as kitchens, bathrooms, basements, and garages are used. The second factor, *activity*, addresses what is actually being done in the different locations. Knowing what activities take place in the different locations deduces what contaminant sources are encountered and/or used. The last factor, *duration*, elucidates for how long people are present in different locations and for how long they indulge in different activities while in those locations. Information on human activity patterns is gathered through surveys or diaries where respondents are queried for their activity patterns, typically for 24-hour periods. The most recent study is the National Human Activity Pattern Survey (NHAPS) that was conducted by the U.S. Environmental Protection Agency. It is the largest and most current human activity pattern survey available (see Chapter 10). Data for 9.386 respondents in the 48 contiguous United States, randomly chosen, were collected via minute-by-minute diaries between October 1992 and September 1994. Detailed data were collected for a maximum of 82 different possible locations, and a maximum of 91 different activities. The results of this study are summarized in a report by USEPA (1996).

Residential Environments

Residences where humans live include single family homes, attached family homes such as duplexes, apartments in single or double level complexes, apartments in multiple level complexes, and mobile homes. Individual residences may span multiple levels, may include basements, and may have attached garages.

Residences are constructed by using combinations of wood, brick, rock, cement, sand, steel, glass, and plastic products. Air handling systems are built to exchange indoor air with outdoor air and circulate indoor air. This system also cools or heats the indoor air depending upon the season. Insulation is inserted between the interior and exterior walls to preserve the thermal conditions of indoor environments from outdoor weather conditions. Their design is a compromise between energy efficiency that demands minimum infiltration and flushing of indoor contaminants that demands maximum exfiltration.

A comfortable residence is possible only if the structure tempers the prevailing outdoor temperature, humidity, and air speed. This can be achieved by one of two ways. One way is by constructing a tight box with an air-conditioning system that is independent of outdoor conditions. Tight boxes have the advantage that they can be designed and engineered for use anywhere without regard to the local climate. Examples of such structures are skyscrapers, sport arenas, hotels, shopping malls, and mobile homes. The other way is by designing an envelope that interacts with the outdoors and where the flux of heat, air, and moisture can be modified as necessary to attain a comfortable level. Most residences include some elements of both ways.

Any indoor environment exchanges air by one of the following three forms: natural ventilation, infiltration, and forced ventilation. Natural ventilation includes air movement through open windows, the fireplace, and open doors. All indoor environments are subject to involuntary infiltration through cracks, doors, windows, and chimneys. Most of the infiltration occurs through floors, walls, windows, and ducts. Other infiltration points include doors, kitchen exhausts, fireplace, light fixtures, dryer exhausts, and bathroom vents. Infiltration of outdoor air depends on wind speed and temperature, the location of the residence and its design. Forced ventilation is provided by central mechanical ventilation systems.

Each building material has different physical and thermal properties, and thus react differently to wind, temperature, humidity, and air impurities.

According to Reeves (1996), amenities have changed in new houses in the last 25 years. These changes are reflected in Table 11-1, and reflect the diversity of residences in the U.S. This diversity is extended to the factors that influence human exposure to contaminants in residential environments, particularly indoors.

Sources—Location and Use

Humans in residential environments are exposed to contaminants from one or more of the following contaminant source categories:

- Sources used by self
- Sources used by co-inhabitants, and
- Sources external to humans.

Table 11-1. What amenities have changed in new houses in the last 25 years

	Percent	
Amenity	1971	1996
Central air	36	80
$2^1/_2$ Baths or more	15	48
4 Bedrooms or more	24	30
1 Fireplace or more	36	63
Full or partial basement	36	39
Slab	38	42
No garage or carport	26	14
2-Car garage or more	39	76
1 Story	73	49
2 Stories or more	17	48
1,200 Square feet or fewer	36	10
2,400 Square feet or more	9	28
Median square footage	1,375 sq ft	1,920 sq ft

Source: National Association of Home Builders 1996.

Sources include consumer products, building materials, combustion related products, bioaerosols (see Chapter 9), and outdoor contamination. Depending upon residential factors, agents present in these sources are transported to humans.

Models for Human Activity Patterns, (Indoor) Environments, and Contaminant Sources

Residential environments can be characterized for their quality either by monitoring or by modeling. It is impossible to monitor all residential environments. One reasonable compromise is to develop probabilistic distributions based on a statistically-designed monitoring study. The number of residences to be monitored increases with the increase in the number of residential variables involved. Another approach is to model residential environments to predict their quality. Model results have to be validated by using monitoring results. One advantage in developing models is that the physics and chemistry of the problem can be better understood during the process.

Residential environment models take three factors into consideration:

- Location of residences
- Structure characteristics, and
- The activity patterns of inhabitants.

The *location of a residence* determines the quality and the quantity of air that enters it. *Structure characteristics* include building materials used, the number of levels, including the presence of a basement, the number of rooms and their volumes, the air flow rates in and out of the rooms, and the nature of surface coverings. Information on air exchange rates, whole house and individual room, can be used to determine the air flow rates. The *activity patterns of inhabitants* determine the source characteristics of contaminants.

Contaminants in indoor air adhere to surfaces. Particulate matter deposits on surfaces. Inorganic and organic gases adsorb onto material surfaces. (Renes *et al.* 1985, Seifert and Schmahl 1987, Matthews *et al.* 1987, Borazzo *et al.* 1990, Tichenor *et al.* 1990a, 1990b, 1991, Jorgensen *et al.* 1993, Chang and Guo 1992, Colombo *et al.* 1993, Levsen and Sollinger 1993).

The sink terms are frequently based on first-order kinetics. In this case, the rate of adsorption per unit area of the adsorbing surface is directly and linearly proportional to the mass of adsorbed contaminant per unit area of the sink. Tichenor *et al.* (1991) have shown that this first-order rate model can be derived from the Langmuir adsorption concept. Slightly different sorption terms, which can be shown to correspond to the Freundlich adsorption model, have sink rates proportional to concentrations raised to some arbitrary power, n. Thus, the Langmuir-based model can be seen as a special case of the Freundlich model, in which the value of n is 1.0. Several of the cited studies have reported that the Langmuir model provides a good fit to experimental data. However, some researchers have also found that the Langmuir model does not provide as good a fit to data for complex or "fleecy" sink surfaces such as carpets, or dust, as it does for relatively flat, smooth surfaces (Tichenor *et al.* 1991, Kjaer and Nielsen 1993).

In each of these sink models, the proportionality factor between air concentration and rate of adsorption per unit area of sink is referred to as the adsorption rate constant, frequently denoted as k_a. Similarly, the proportionality factor between desorption rate per unit area of sink and mass in the sink is referred to as the desorption rate constant, k_d. The rate, k_a/k_d, referred to as k_e, is the equilibrium capacity of a sink surface to adsorb a certain chemical at a given air concentration. These constants (k_a, k_d, and k_e) can be estimated from concentration versus time data, such as is usually measured in environmental chamber tests (Dunn and Tichenor 1988, Tichenor *et al.* 1991, Chang and Guo 1992, Colombo *et al.* 1993, Levsen and Sollinger 1993). The rate constants are dependent on the type of surface and environmental conditions like temperature, humidity, and air movement.

According to Tichenor *et al.* (1991) a sink model can be represented as:

$$Rs = (kaCA \sin k) - (kdMs^n A \sin k) \tag{1}$$

where:

A_{sink} = area of the sink, m^2
C = concentration of gaseous contaminant, mg/m^3
k_a = adsorption constant, m/hr
k_d = desorption constant, m/hr
M_s = mass collected in the sink, mg/m^2
n = exponent, and
R_s = the rate at which mass accumulates in the sink, mg/hr.

In this model, for typical gaseous organic pollutants of interest in indoor air, k_a ranges from about 0.1 to 0.5 m/hr, k_d is about 0.001 m/hr, and n equals 1.

HOUSING STOCK

Numerous studies that characterize residences on a local basis exist. Most of these studies limit their focus to less than 50 residences and the residences are not randomly chosen. Since 1974, the U.S. government has been studying the characteristics of American houses through a survey called the American Housing Survey. This survey also tracks the demographics of residents.

The American Housing Survey (AHS) is a collaborative effort by the U.S. Department of Commerce (Bureau of the Census) and the U.S. Department of Housing and Urban Development. It is conducted by field representatives who obtain information from occupants of homes. They obtain information on vacant homes from informed people such as landlords, rental agents, or knowledgeable neighbors. The most recent survey—the 1993 metropolitan survey—was conducted from April through December 1993 for all areas surveyed. Each metropolitan area had a sample of about 4,600 interviews.

Data are obtained on apartments, single family homes, mobile homes, vacant housing units, age, sex, and race of householders, income, housing and neighborhood quality, housing costs, equipment and fuels, and size of the housing units. Additional information on homeowner's repairs and mortgages, rent control, rent subsidies, previous unit of recent movers, and reasons for moving is also collected.

Geographic Distribution

Table 11-2, based on the 1993 metropolitan survey, indicates the:

Table 11-2. Household characteristics: tenure, urban versus rural, region, and year built

Characteristic	Number	Percent
1 **Total number of households in the U.S.**	94,724,000	100
2 **Tenure**		
Owner Occupied	61,252,000	64.7
Renter Occupied	33,472,000	35.3
3 **Metropolitan/nonmetropolitan areas**		
Inside Metropolitan Statistical Areas	73,898,000	78.0
In Central Cities	29,838,000	31.5
Suburbs	44,060,000	46.5
Outside Metropolitan Statistical Areas	20,826,000	22.0
4 **Geographical regions**		
Northeast	18,906,000	20.0
Midwest	23,031,000	24.3
South	32,936,000	34.8
West	19,850,000	20.9
5 **Year structure built**		
1990 to 1994	4,576,000	4.8
1985 to 1989	7,969,000	8.4
1980 to 1984	7,171,000	7.6
1975 to 1979	10,708,000	11.3
1970 to 1974	10,110,000	10.7
1960 to 1969	14,405,000	15.2
1950 to 1959	12,360,000	13.0
1940 to 1949	7,539,000	8.0
1930 to 1939	5,853,000	6.2
1920 to 1929	5,047,000	5.3
1919 or Earlier	8,986,000	9.5

Source: American Housing Survey 1996.

- Total number of households in the U.S.
- Division of households between owner- and renter-occupied
- Number of households in metropolitan and nonmetropolitan areas
- Number of households by geographical region, and
- Division of households by the year structure was built.

Table 11-3 through Table 11-6 indicate, respectively: (1) household characteristics with respect to number of units, number of levels, and square footage by type of dwelling: apartment, single-family home, mobile home, *etc.*, (2) housing characteristics with respect to number of rooms, number of bedrooms, and number of complete bathrooms, (3) lot size, and (4) human characteristics with respect to persons per room, square feet per person.

ENVIRONMENT CHARACTERISTICS

Building Materials

The choice of building material has a substantial influence on indoor environment quality, either by direct interaction or by defining the temperature and humidity of buildings. Traditionally, buildings have been built with locally available materials. In North America, wood is still far more abundant than any other building construction material. Whole wood or timber has largely disappeared from the building market as floor or wall material except in structural applications. It has been replaced by plywood panels, which are made from veneer obtained by continuous peeling of a tree trunk along its axis with a large, sharp blade. The veneer is flattened and then sprayed with 2.5 wt percent glue, usually a phenolic aldehyde resin or urea-formaldehyde resin.

Three, five, or more layers are cross-stacked and bonded at 350°F (Meyer 1983).

During the last few decades, particleboard has replaced some of the plywood stock used as building material. Particleboard is made by spraying wood chips with 7–8 wt percent urea-formaldehyde resin and pressing panel-shaped cakes on huge cookie sheets, called cauls, at 350°F for 3–10 minutes. Particleboard makes an excellent flooring or underlayment on dry floors, and it has unexcelled properties for cabinet work and furniture. Almost all desks, tables, doors, and furniture are now made from particleboard covered with melamine-impregnated paper with an imprinted wood-grain design.

The heat transmission of structural and wall materials is so large that the building envelope must be equipped with separate layers of insulation material if the flow of heat between indoors and outdoors is to be significantly reduced. The U.S. Department of Energy has approved and established standards for 11 thermal insulating materials: cel-

Table 11-3. Household characteristics: number of units, number of levels, and square footage

Characteristic	Number	Percent
1 **Units in structure**		100
1 (Detached)	58,918,000	62.2
1 (Attached)	5,375,000	5.7
2 to 4	9,279,000	9.8
5 to 9	4,724,000	5.0
10 to 19	4,190,000	4.4
20 to 49	3,154,000	3.3
50 or More	3,429,000	3.6
Mobile home or trailer	5,655,000	6.0
2 **Number of levels** (Limited to multiunit structures)		
1	2,424,000	—
2	9,101,000	—
3	7,137,000	—
4 to 6	3,829,000	—
7 or More	2,294,000	—
3 **Square footage of unit** (Median = 1,725 sq ft)		
Single-family detached and mobile homes	64,574,000	100
Fewer than 500	697,000	1.1
500 to 749	2,381,000	3.7
750 to 999	5,704,000	8.8
1,000 to 1,499	15,084,000	23.4
1,500 to 1,999	13,414,000	20.8
2,000 to 2,499	9,653,000	15.0
2,500 to 2,999	5,374,000	8.3
3,000 to 3,999	4,799,000	7.4
4,000 or More	2,688,000	4.2
Not reported	4,780,000	7.4

Source: American Housing Survey 1996.

Table 11-4. Housing characteristics: number of rooms, number of bedrooms, and number of complete bathrooms

Characteristic	Number	Percent
1 **Number of rooms in house** (median = 5.5)	94,725,000	100
1 Room	611,000	0.6
2 Rooms	989,000	1.0
3 Rooms	7,959,000	8.4
4 Rooms	17,221,000	18.2
5 Rooms	21,030,000	22.2
6 Rooms	19,870,000	21.0
7 Rooms	13,083,000	13.8
8 Rooms	7,683,000	8.1
9 Rooms	3,738,000	3.9
10 or More Rooms	2,541,000	2.7
2 **Number of bedrooms in house**	94,725,000	100
(Median = 2.7)		
None	1,129,000	1.2
1	11,279,000	11.9
2	28,965,000	30.6
3	38,504,000	40.6
4 or More	14,848,000	15.7
3 **Number of complete bathrooms in house**	94,725,000	100
None	526,000	0.6
1	43,944,000	46.4
1½	14,740,000	15.6
2 or More	35,515,000	37.5

Source: American Housing Survey, 1996.

lulose fibers, mineral fibers, mineral blankets and batts, vermiculite, perlite, cellular polystyrene, polyurethane and isocyanate, urea-formaldehyde foam, and aluminum foil. In March 1982, the U.S. Consumer Product Safety Commission passed a ban (47 FR 14366) on urea-formaldehyde foam insulation, because formaldehyde levels in some homes had been reported at, or close to, occupational levels. In the United States, about 1.5 million homes have been insulated with urea-formaldehyde foam.

Table 11-5. Housing characteristics: lot size (relevance to outside)

Characteristic	Number	Percent
Lot size (Median = 0.43)	69,934,000	100
Less than ⅛ acre	6,608,000	9.4
⅛ up to ¼ acre	11,976,000	17.1
¼ up to ½ acre	9,830,000	14.1
½ up to one acre	6,958,000	10.0
1 to 4 acres	10,519,000	15.0
5 to 9 acres	1,633,000	2.3
10 acres or more	3,621,000	5.2
Do not know	13,186,000	18.9
Not reported	5,603,000	8.0

Source: American Housing Survey 1996.

Table 11-6. Human characteristics: persons per room, square feet per person

Characteristic	Number	Percent
1 **Persons per room**	94,724,000	100
0.50 or Less	64,611,000	68.2
0.51 to 1.00	27,727,000	29.3
1.01 to 1.50	1,940,000	2.0
1.51 or More	446,000	0.5
2 **Square feet per person** (Median = 689 sq ft)		
Single-family detached and mobile homes	64,572,000	100
Less than 200	1,752,000	2.7
200 to 299	4,283,000	6.6
300 to 399	6,222,000	9.6
400 to 499	6,623,000	10.2
500 to 599	5,897,000	9.1
600 to 699	5,756,000	8.9
700 to 799	4,731,000	7.3
800 to 899	3,607,000	5.6
900 to 999	3,394,000	5.3
1,000 to 1,499	9,798,000	15.2
1,500 or more	7,729,000	12.0
Not reported	4,780,000	7.4

Source: American Housing Survey 1996.

Building Appliances

Table 11-7 through Table 11-9 illustrate basic housing characteristics related to appliances, type of fuel used and frequency, and purpose of fuel use.

Sources—Location, Use, Release Rates

Three national surveys have been conducted for the USEPA (USEPA 1987abc) that provide usage data for household solvent products, household cleaning products, paints, and other paint-related products. The primary purpose of these surveys was to gather usage data needed to assess consumer exposures to chemicals in common household products. The data that can be obtained from these studies are frequency of use, duration of use, and amount used. Individual questionnaires were completed by respondents who recalled product usage behaviors over the previous 12 months.

Ventilation Parameters—Air Exchange Rate, Air Recirculation Rate, Interzonal Air Flows, Intrazonal Circulation

Residential air exchange is the exchange of air between the inside and the outside of a residence. It is composed of three processes:

- *Infiltration.* Air leakage through random cracks, interstices, and other unintentional openings in the building envelope
- *Natural ventilation.* Airflows through open windows, doors, and other designed openings in the building envelope, and
- *Forced or mechanical ventilation.* Controlled air movement driven by fans.

Table 11-7. Housing characteristics: building appliances

Characteristic	
1 **Equipment**	
Lacking complete kitchen facilities	1,107,000
Complete kitchen (sink, refrigerator and burners)	93,617,000
Kitchen sink	94,119,000
Refrigerator	94,423,000
Burners and oven	94,146,000
Burners only	123,000
Oven only	104,000
Neither burners nor oven	350,000
Dishwasher	49,332,000
Washing machine	73,146,000
Clothes dryer	67,464,000
Disposal in kitchen sink	40,091,000
2 **Air Conditioning**	
Central	42,183,000
1 Room unit	16,790,000
2 Room units	6,818,000
3 Room units or more	2,482,000
3 **Main Heating Equipment**	
Warm-air furnace	51,248,000
Steam or hot water system	13,657,000
Electric heat pump	8,422,000
Built-in electric units	6,722,000
Floor, wall, or other built-in hot air units without ducts	4,746,000
Room heaters with flue	1,766,000
Room heaters without flue	1,597,000
Portable electric heaters	682,000
Stoves	2,831,000
Fireplaces with inserts	580,000
Fireplaces without inserts	304,000
Other	1,257,000
None	911,000

Source: American Housing Survey 1996.

The evaluation criteria for assessing the ventilation quantities include (Bearg 1993):

- Cfm of outdoor air
- Percentage of outdoor air in the supply air
- Cfm of outdoor air per person
- Cfm of outdoor air per square foot of building area, and
- Air changes per hour of ventilation.

The term "cfm of outdoor air" is concerned with the absolute quantity of outdoor air. The use of "percentage of outdoor air in the supply air" reflects the fact that the supply air is made up of a mixture of both the outdoor air for ventilation and air that is recirculated from the building which are then combined to achieve a total volume needed to facilitate thermal regulation. The term "cfm of outdoor air per person" is an approach for quantifying ventilation rates. Expressing ventilation rate in this unit is a very important evaluation criterion because the ASHRAE Standard 62 (Ventilation for

Table 11-8. Housing characteristics: type of fuel used and frequency

Type of fuel	Number of units
Electricity	94,691,000
All-electric units	19,667,000
Piped gas	58,254,000
Bottled gas	7,370,000
Fuel oil	13,475,000
Kerosene or other liquid fuel	2,360,000
Coal or coke	456,000
Wood	12,589,000
Solar energy	383,000
Other	1,195,000

Source: American Housing Survey 1996.

Acceptable Indoor Air Quality) lists minimum recommended ventilation rates expressed in these units. There is no ASHRAE standard for "cfm of outdoor air per person" for residences. However, for offices, this criterion is 20 cfm per person based on an estimated maximum occupancy of 7 people per $100\,m^2$ ($1000\,ft^2$). For smoking lounges, it is 60 cfm per person based on an estimated maximum occupancy of 70 people per $100\,m^2$ ($1000\,ft^2$), and for classrooms it is 15 cfm per person based on an estimated maximum occupancy of 50 people per $100\,m^2$ ($1000\,ft^2$). Recognizing that furnishing and building components also have the potential to be sources of air contaminants, besides people, the ventilation rate is expressed in terms of "cfm of outdoor air per square foot of building area". This criterion is a ventilation requirement of the California Title 24 regulations. Expressing the ventilation rate in terms of the air changes per hour (which is equal to the quantity of outdoor air divided by the building volume) is another approach that is frequently used. Total supply air rates are typically in the range of 5 to 7 ach (total air), while minimum ventilation rates are typically around 1.0 ach of outdoor air. These typical values correspond to a range of 14 to 20 percent outdoor air in the supply air.

ASHRAE Standard 62-1989 (Ventilation for Acceptable Indoor Air Quality) lists minimum outdoor air requirements which are expected to be deemed capable of providing an acceptable level of indoor air quality. Acceptable indoor air quality is defined by

Table 11-9. Housing characteristics: purpose of fuel use

	Main house heating	Cooking fuel	Water heating	Central air conditioning	Clothes dryer fuel
Total	93,813,000	94,363,000	94,517,000	42,183,000	67,464,000
Electricity	25,107,000	55,887,000	35,242,000	39,234,000	51,487,000
Piped gas	47,669,000	33,813,000	49,415,000	2,605,000	15,007,000
Bottled gas	3,922,000	4,183,000	3,135,000		
Fuel oil	11,168,000		5,594,000		
Kerosene or other liquid fuel	1,021,000	303,000	318,000		
Coal or coke	297,000	14,000	44,000		
Wood	4,104,000	17,000	45,000		
Solar energy	30,000		281,000		
Other	496,000	145,000	442,000	343,000	970,000

Source: American Housing Survey 1996.

ASHRAE as "air in which there are no known contaminants at harmful concentrations as determined by cognizant authorities and with which a substantial majority (80 percent or more) of the people exposed do not express dissatisfaction" upon survey.

U.S. Department of Housing and Urban Development (USHUD) ventilation requirements are incorporated into its minimum property standards for residences (USHUD 1990a) which are part of its mortgage insurance and low rent public housing program and in construction requirements for manufactured housing (USHUD 1990b). Ventilation for the construction of manufactured housing can be met in two ways. The rule specifies that an area equivalent to not less than 8 percent of the floor area must be available for natural ventilation (windows or doors), or alternatively, a mechanical system must be capable of a ventilation rate of 2 ach of outdoor air. Bathrooms and toilet compartments require either 1.5 ft^2 of openable glazed area or a mechanical system capable of producing 5 ach. The mechanical system must exhaust directly outside the house. It should be noted that since this requirement for mechanical ventilation is only for a given capability, it does not say anything about actual operational requirements.

The primary building codes that are in use in the United States include those written by the Building Officials and Code Administrators International (BOCA), the Southern Building Code Congress International (SBCCI), the Council of American Building Officials (CABO), and the American Public Health Association (APHA) model code. The CABO code for one- and two-family dwellings, specifies ventilation requirements for habitable rooms in terms of openable windows or mechanical ventilation systems that can provide 2 ach of outdoor air. For bathrooms, toilet compartments, and similar rooms, the ventilation requirements can again be achieved by openable windows or a mechanical ventilation system that can provide 5 ach of outdoor air.

Air conditioning is the simultaneous control of temperature, humidity, cleanliness, odor, and air speed as required by the occupants of a space. Due to age, sex, clothing, and the level of activity of each person not everyone within a given space can be made completely comfortable by one set of conditions. ASHRAE comfort standard 55 (11) defines acceptable thermal comfort as an environment that at least 80 percent of the occupants will find thermally acceptable.

A complex regulating system in the body acts to maintain the deep body temperature at approximately 36.9°C (98.6°F). If the environment is maintained at suitable conditions so that the body can easily maintain an energy balance, a feeling of comfort will result. Comfort in environments can be defined through indices. Indices include dry and wet bulb temperatures, dew point temperature, relative humidity, air movement, mean radiant temperature, operative temperature, humid operative temperature, heat stress index, and skin wettedness.

The dry bulb temperature is an important index of comfort, particularly when the relative humidity is between 40 percent and 60 percent, and particularly in colder regions. The dew point temperature is a good single measure of the humidity of the environment and is directly related to the water vapor pressure in saturated air. The wet bulb temperature is useful in describing comfort conditions in the regions of high temperature and high humidity where dry bulb temperature has less significance. For example, the upper limit for tolerance of the average individual with normal clothing is a wet bulb of about 30°C (86°F) when air movement is in the neighborhood of 0.25 to 0.38 m/s (50 to 75 ft/min). Relative humidity has no real meaning in terms of comfort unless the accompanying dry bulb temperature is also known. Very high or very low relative humidity is generally associated with discomfort, however. Occupants are comfortable in a warm humid environment if the air movement is high. Yet, if the temperature is too low, they become uncomfortable in the air movement is too high. Generally, when air motion in the vicinity of an individual is in the neighborhood of 0.25 m/s (50 ft/min), the average person will be comfortable.

The mean radiant temperature is the uniform surface temperature of an imaginary black enclosure with which a person, also assumed to be a black body, exchanges the same heat by radiation as in the actual environment. This parameter is important when surrounding surfaces are at a temperature different from the body. The operative temperature is the uniform temperature of an imaginary enclosure with which an individual exchanges the same heat by radiation and convection as in the actual environment. For practical applications, the operative temperature is the mean of the dry bulb and the mean radiant temperatures at a given location in the space. The humid operative temperature is the uniform temperature of an environment at 100 percent relative humidity with which a person will exchange the same heat from the skin surface by radiation, convection, conductance through clothing, and evaporation as in the actual environment. The heat stress index is the ratio of the total evaporative heat loss required for thermal equilibrium of the body to the maximum possible evaporative heat lost to the environment. Closely related to this is the index of skin wettedness, which is the ratio of the observed skin sweating to the maximum sweating possible for the environment as defined by the skin temperature, air temperature, humidity, air movement, and clothing.

Clothing insulation can be described in terms of its clo value (1 clo = 0.155 (m^2-C)/W = 0.88 (ft^2-hr-F)/Btu). A heavy 2-piece business suit and accessories has an insulation value of about 1 clo, whereas a pair of shorts is about 0.05 clo. During the summer months, typical clothing in commercial establishments have insulation values ranging from 0.35 to 0.6 clo. During the winter months, typical clothing have insulation values ranging form 0.8 to 1.2 clo. At other times, between seasons, the clothing may have an insulation value in the range of 0.6 to 0.8 clo.

The comfort conditions discussed so far relate to sedentary activity, which accounts for most applications. In spaces where the occupants may be engaged in active work, the comfort zone temperatures should be decreased.

Smell plays an important role in human survival (although, in some cases, for odorless substances such as carbon monoxide, odor is not a good indicator). Through it, a person is able to detect danger, food, and other things relating to safety and health. To eliminate or reduce an odor or gas so that it is undetectable, it is necessary either to remove the offending material or to dilute the air until the concentration is too low for perception. Humidity has an effect on odor perception. If the conditioned space is held at a relative humidity of about 45 percent to 60 percent, odor perception and irritation will be at a minimum.

Ventilation air is the combination of outdoor air and recirculated air from the conditioned space that after passing through the air-conditioning unit becomes supply air. The term makeup air may be used synonymously with outdoor air and the terms return and recirculated air are often used interchangeably. ASHRAE Standard 62 (10) defines acceptable air quality as ambient air in which there are no known contaminants at harmful concentrations and with which a substantial majority of the people exposed do not express dissatisfaction. Standard 62 defines ventilation requirements and specifies allowable contaminant concentrations for ventilation air. Table 11-10, from Standard 62, gives the minimum outdoor air requirements for ventilation of residential buildings.

A minimum supply of outdoor air is necessary to dilute the carbon dioxide produced by metabolism and expired from the lungs. This value, 7.5 L/s (15 cfm), allows an adequate factor of safety to account for health variations and some increased activity levels. The air that is recirculated must pass through a filtering system. The filtering system, which may be located in the recirculated air stream or in the plenum which mixes the outdoor air and return air, should be designed to remove both particulate and gaseous contaminants. Air cleaning filters are characterized by their efficiency, air flow resistance, and dust-holding capacity.

ASHRAE 62-1989 specifies that the combined mechanical and natural ventilation rate should be no less than 0.35 ach, while the CSA (Canadian Standards Association) standard specifies that a fan must supply mechanical ventilation at a rate of 0.3 ach. The CSA standard is directed at new energy-conserving Canadian housing with very tight envelope construction and negligible natural infiltration.

Canadian standard CSA F326.1 is based on a required air flow rate of 5.0 L/s per room, with master bedrooms and basements requiring a higher rate of 10.0 L/s. In contrast, ASHRAE Standard 62-1989 sets a requirement based on the number of building occupants rather than the number of rooms. The required ASHRAE level of 7.5 L/s per person is converted to a building ventilation rate by assuming that there are N + 1 occupants in a house with N bedrooms. Table 11-11 shows minimum ventilation rates calculated using these per-person and per-room requirements (Wilson and Walker 1992).

Based on experimental and simulation efforts on bungalows and two-story detached and townhouse dwellings, Wilson and Walker came to the following conclusions. Most detached single-story and two-story houses are adequately ventilated in winter by natural air infiltration with no passive vent openings. Townhouses, with two walls common to adjacent units, are usually not adequately ventilated by natural infiltration. Even with several large ground-level passive ventilation openings, houses cannot be adequately ventilated in summer, spring, and fall seasons in light winds without using large open-area vents that cause overventilation in winter. The implications of these conclusions is that even for passively ventilated houses, some supplementary form of ventilation is required

Table 11-10. Ventilation guidelines for different environments from Standard 62

Environment	Outdoor air Requirements	Comments
Living areas	0.35 ach but not less than 7.5 L/s (15 cfm) per person	For calculating the ach, the volume of the dwelling shall include all areas within the conditioned space. The ventilation is normally satisfied by infiltration and natural ventilation. Dwellings with tight enclosures may require supplementary ventilation supply for fuel-burning appliances, including fireplaces, and mechanically exhausted appliances. Occupant loading shall be based on the number of bedrooms as follows: first bedroom—two persons, each additional bedroom—one person. Where high higher densities are known, they shall be used.
Kitchens	50 L/s (100 cfm) 25 L/s (50 cfm)	Intermittent mechanical exhaust, natural ventilation (opened windows) for control of moisture, odors, gases, vapors, and particles.
Garages separate for each dwelling unit	50 L/s (100 cfm) per car	Normally satisfied by infiltration or natural ventilation
Common for several units	7.5 (L/s)/m^2 (1.5 cfm/ft^2)	

Table 11-11. Required minimum air exchange rates for typical single-family dwellings with full ventilated basements

Number of bedrooms (N)	Floor ft^2	Area* m^2	ASHRAE Standard 62-1989 7.5 lps per occupant for (N + 1) occupants		CSA Standard F326.1** 5 lps per room (except 10 lps for master bedroom, basement, and)	
			1-story ach	2-story ach	1-story ach	2-story ach
3	1000	93	0.24	0.32	0.44	0.58
3	1300	121	0.18	0.25	0.34	0.45
4	1800	167	0.17	0.22	0.27	0.35
4	2000	186	0.15	0.20	0.24	0.32

* Excluding basement, wall height of 2.44 m for all levels assumed for active air volume.
** Kitchen, living room, family room, two bathrooms, basement, N bedrooms.

in spring, fall, and summer when indoor-outdoor temperature differences are less than 10°C and winds are "light" to "moderate".

Effective Volumes

Room volumes have not been well characterized for the United States housing stock. The research in Murray and Burmaster (1996) reports empirical distributions and estimated univariate parametric probability distributions for house volumes and certain zone volumes within households for residential structures in the United States. The author conducted analyses on two separate databases. The first database contains 7,041 households was derived from the U.S. Department of Energy Residential Energy Consumption Survey (RECS). While this database is very representative of the whole United States, measurements are in square footage of floor area and the author assumed an 8-foot ceiling height. The second database, frequently called the PFT database, compiled by the Brookhaven National Laboratory, contains volume data, as well as air exchange measurements, on 1,751 households.

There is information on several well-characterized houses that have been used for energy conservation and indoor air quality research. Four examples are given in Table 11-12. All houses were built in the late 1970s or early 1980s.

Surface Characteristics—Type and Area

Amounts of selected products or materials used in constructing or finishing residential surfaces have been estimated (Tucker 1991). Many of these estimations, found in the USEPA Exposure Factors Handbook, were made by assumption. Products used for floor surfaces, and for which usage amounts were estimated, include adhesives, varnish and wood stains, and materials used for walls include paneling, gypsum board, and wallpaper. Particleboard and chipboard are most likely to be used for interior furnishings such as cabinets, shelves, decking, and underlayment.

Water Use

The residential water supply may convey certain chemicals to which occupants can be exposed through dermal contact, ingestion, and/or inhalation. Among residential

Table 11-12. Housing characteristics: typical volumes, floor areas, and wall areas

Room / zone	Volume (m^3)	Floor area (m^2)	Wall area (m^2)
NIST two-story style			
Total	420		
Living room	61	25	49
Dining room	42	17	40
Kitchen/family room	75	31	55
Bedroom 1 (master)	69	28	52
Bedroom 2	35	15	45
Bedroom 3	29	12	34
Bedroom 4	29	12	34
Hall	11	4	28
Bathroom 1 (master)	11	4	20
Bathroom 2	11	4	20
Bathroom 3 (half)	8	3	18
Closet 1 (master)	22	9	30
Closet 2	12	5	22
Closet 3 (utility)	5	2	14
Attached garage	131	54	73
NIST Ranch Style			
Total	250		
Living/dining room	75	31	60
Kitchen	28	12	34
Bedroom 1 (master)	44	18	42
Bedroom 2	33	14	36
Bedroom 3	31	13	36
Hall	13	5	35
Bathroom 1 (master)	11	5	24
Bathroom 2	15	6	24
Attached garage	89	37	62
EPA ranch style			
Total	293		
Living/dining room/kitchen	150	63	81
Bedroom 1 (master)	42	18	36
Bedroom 2	34	14	27
Bedroom 3	33	14	27
Bathroom 1 (master)	10	4	8
Bathroom 2	12	5	7
Hall	12	5	24
GEOMET split foyer style			
Total	311		
Living/dining room/kitchen	100	41	36
Bedroom 1 (master)	35	14	36
Bedroom 2	21	8	28
Bedroom 3	23	9	34
Bathroom 1 (master)	8	3	18
Bathroom 2	9	4	20
Hall	11	5	24
Downstairs	104	43	59
Attached garage	108	44	65

Table 11-13. Housing characteristic: water use

Characteristic	Number
1 **Source of Water**	
Public system or private company	81,028,000
Well serving 1 to 5 units	11,967,000
Drilled	10,229,000
Dug	1,163,000
Not Reported	575,000
Other	1,730,000
2 **Plumbing**	
With all plumbing facilities	93,345,000
Lacking some plumbing facilities	227,000
No hot piped water	57,000
No bathtub or shower	181,000
No flush toilet	131,000
No plumbing facilities for exclusive use	1,152,000
3 **Means of sewage disposal**	
Public sewer	72,797,000
Septic tank, cesspool, or chemical toilet	21,807,000
Other	121,000

Source: American Housing Survey 1996.

water uses, showering, bathing, and other hygiene-related washing, preparing foods, handwashing of dishes and clothes, and watering other objects (washing surfaces, watering the lawn, *etc.*) provide opportunities for dermal contact. All water uses, primarily those indoors, will result in some volatilization of chemicals, leading to inhalation exposure. The exposure potential for a given situation will depend on the source of water, the types and extent of water uses, and the extent of volatilization of specific chemicals. Table 11-13 and Table 11-14 illustrate water-use characteristics by water source and by use activity, respectively.

Table 11-14. Water use characteristics in residences (gallons per day)

Study	Total (all uses)	Shower/bath	Toilet	Laundry	Dishwashing	Other
Bennett	45	9	15	11	4	6
Cohen	52	6	17	11	18	
EBMUD	67	20	28	9	4	6
Laak	42	9	20	7	4	2
Ligman (rural)	46	11	18	14	3	
Ligman (urban)	43	10	18	11	4	
Milne	70	21	32	7	7	3
MWD	93	26	30	20	5	12
Partridge	52–86	20–40	4–6	20–30	8–10	
Reid	59	20	24	8	4	3
USDHUD	40	15	10	13	2	
USEPA	40	10	9	11	5	5
Mean (across studies)	59	17	18	13	6	5
Median (across studies)	53	15	18	11	4	5

REFERENCES

American Housing Survey. 1996. 1993 national data chart. http://www.census.gov/hhes/www/ housing/ ahs/93dtchrt/ahs93hmtl.

Bearg, D.W. 1993. Indoor Air Quality and HVAC Systems. Lewis Publishers/CRC Press, Inc. Boca Raton, Florida.

Borazzo, J.E., C.I. Davidson, and J.B. Andelman. 1990. Sorption of organic vapors to indoor surfaces of synthetic and natural fibrous materials. In: Walkinshaw, D. (ed.). Indoor Air '90: Proceedings of the 5th international conference on indoor air quality and climate. July 29–August 3. Volume 2. Toronto, Canada. Canada Mortgage and Housing Association. Ottawa, Ontario.

Chang, J.C.S. and Z. Guo. 1992. Characterization of organic emissions from a wood finishing product—wood stain. Indoor Air 2:146–153.

Colombo, A., M. De Bortoli, H. Knoppel, E. Pechchio, and H. Vissers. 1993. Vapour deposition of selected VOCs on indoor surface materials in test chambers. In: Saarela, K., P. Kalliokoski, and O. Seppanen (eds.). Volume 2. Chemicals in Indoor Air, Material Emissions. Pages 407–412. Indoor Air '93: Proceedings of the 6th International Conference on Indoor Air Quality and Climate. Espoo, Finland. July 4–8, 1993.

Dunn, J.E. and B.A. Tichenor. 1988. Compensating for sink effects in emissions test chambers by mathematical modeling. Atmos. Environ. 22:885–894.

Jorgensen R.B., H.N. Knudsen, and P.O. Fanger. 1993. The influence on indoor air quality of adsorption and desorption of organic compounds on materials. In: Jaakola, J.J.K., R. Ilmarinen, and O. Seppänen (eds.). Indoor Air '93: Proceedings of the 6th International Conference on Indoor Air Quality and Climate. Volume 1. Helsinki University of Technology. Espoo, Finland.

Kjaer, U.D. and P.A. Nielsen. 1993. Adsorption studies on dust samples from the indoor environment. In: Jaakola, J.J.K., R. Ilmarinen, and O. Seppänen (eds.). Indoor Air '93: Proceedings of the 6th International Conference on Indoor Air Quality and Climate. Volume 1. Helsinki University of Technology. Espoo, Finland. July 4–8, 1993.

Levsen, K. and S. Sollinger. 1993. Textile floor coverings as sinks for indoor air pollutants. Pages 395–400 in: Saarela, K., P. Kalliokoski, and O. Seppanen (eds.). Volume 2. Chemicals in Indoor Air, Material Emissions. Indoor Air '93: Proceedings of the 6th International Conference on Indoor Air Quality and Climate. Espoo, Finland. July 4–8, 1993.

Matthews T.G., A.R. Hawthorn, and C.V. Thompson. 1987. Formaldehyde sorption and desorption characteristics of gypsum wallboard. Environ. Sci. Technol. 21(7):629–634.

Meyer, B. 1983. Indoor Air Quality, Addison-Wesley Publishing Company, Inc., Reading, MA.

Murray, D.M. and D.E. Burmaster. 1996. Residential air exchange rates in the United States. Empirical and estimated parametric distributions by season and climactic region. Risk Anal. 15(4):459–465.

National Association of Home Builders. 1996. Characteristics of new single-family homes: 1975–1996. http://www.nahb.com/facts/forecast/sf.hmtl.

Reeves, P. 1996. Characteristics of homes changing, written as a weekly real estate column for Scripps Howard News Service. Las Vegas Sun. October 23.

Renes, S., B.P. Leaderer, L. Schapp, H. Verstraelen, and T. Tosun. 1985. An evaluation of sinks terms in removing NO_2 and SO_2 from outdoor air. CLIMA 2000. 4:221–226.

Seifert, B. and H.R. Schmahl. 1987. Quantification of sorption effects for selected organic substances present in indoor air. In: Seifert, B., H. Esdorn, M. Fischer, H. Rüden, and J. Wegner (eds.). Indoor Air '87. Proceedings of the 4th International Conference on Indoor Air Quality and Climate. August 17–21. Volume 1. Institute for Soil, Water, and Air Hygiene. Berlin (West).

Tichenor, B.A., L.E. Sparks, M.D. Jackson, Z. Guo, and S.A. Rasor. 1990a. The effect of wood finishing products on indoor air quality. In: Gay, B.W. (USEPA) and R.K.M. Jayanty (Research Triangle Institute) (Tech. Prog. Chairmen). Proceedings of the EPA/AWMA international symposium on measurement of toxic and related air pollutants. May 1990. Raleigh, North Carolina. Air and Waste Management Association. Pittsburgh, Pennsylvania. 1096 pages.

Tichenor, B.A., Z. Guo, M.A. Mason, and J.E.Dunn. 1990b. Evaluation of indoor air pollutant sinks for vapor phase organic compounds. In: Walkinshaw, D. (ed.). Indoor Air '90: Proceedings of the 5th international conference on indoor air quality and climate. July 29–August 3. Volume 2. Toronto, Canada. Canada Mortgage and Housing Association. Ottawa, Ontario.

Tichenor, B.A., Z. Guo, J.E. Dunn, L.E. Sparks, and M.A. Mason. 1991. The interaction of vapor phase organic compounds with indoor sinks. Indoor Air. 1:23–35. Danish Technical Press. Copenhagen, Denmark.

Tucker, W.G. 1991. Emission of organic substances from indoor surface materials. Environ. Internat. 17:357–363.

USHUD (U.S. Department of Housing and Urban Development). 1990a. Minimum Property Standards, Code of Federal Regulations. Title 24. Part 200. Subpart S. Sections 200.925 and 200.926.

USHUD (U.S. Department of Housing and Urban Development). 1990b. Manufactured Home Construction and Safety Standards. Code of Federal Regulations. Title 24. Part 3280. Sections 3280.103 and 3280.710.

USEPA (U.S. Environmental Protection Agency). 1987a. Household solvent products. A national usage survey. WESTAT, Inc. Rockville, Maryland. Contract Number 68-02-4243. Exposure Evaluation Division, Office of Pesticides and Toxic Substances. Washington DC. NTIS No. PB88-132881.

USEPA (U.S. Environmental Protection Agency). 1987b. National usage survey of household cleaning products. WESTAT, Inc. Rockville, Maryland. Contract Number 68-02-4243. Exposure Evaluation Division, Office of Pesticides and Toxic Substances. Washington DC.

USEPA (U.S. Environmental Protection Agency). 1987c. National household survey of interior painters. WESTAT, Inc. Rockville, Maryland. Contract Number 68-02-4243. Exposure Evaluation Division, Office of Pesticides and Toxic Substances. Washington DC.

USEPA (U.S. Environmental Protection Agency). 1992. Tests of indoor air quality sinks. Environmental Monitoring Systems Laboratory. Las Vegas, Nevada. EPA contract number EPA-68-DO-0137, EPA report number EPA/600/R-92/094.

USEPA (U.S. Environmental Protection Agency). 1996. Results tables from a detailed analysis of the national human activity pattern survey (NHAPS) response. Draft report. Prepared by: Tsang, A.M. and N.E. Klepeis. Lockheed Martin. Contract Number 68-W6-001. Delivery Order Number 13. National Exposure Research Laboratory, U.S. Environmental Protection Agency. Las Vegas, Nevada. 534 pages.

Wilson, D.J. and I.S Walker. 1992. Feasibility of passive ventilation by constant area vents to maintain indoor air quality in houses. In: IAQ '92. Environment for People. Proceedings of an international conference on indoor air quality and climate. October 19–21, 1992. San Francisco, California. American Society of Heating, Refrigeration, and Air-Conditioning Engineers. Atlanta, Georgia. 1993. 460 pages.

V

Presenting, Interpreting, and Communicating the Results of Residential Exposure Assessments

INTRODUCTION

DISTINGUISHING BETWEEN UNCERTAINTY AND VARIABILITY

It is important, at the outset, for an exposure assessor to consider issues of uncertainty and variability. Uncertainty about the accuracy of exposure and dose estimates and variability in key parameters are critical aspects of every assessment, which must be taken into account when constructing realistic scenarios, selecting appropriate human exposure factors, and interpreting and communicating results. The USEPA recommends that exposure assessors distinguish between uncertainty and variability, where uncertainty represents a lack of knowledge about factors affecting exposure or risk and variability represents true heterogeneity across people, places, and time. Uncertainty is an important issue for assessors because it can lead to inaccurate or biased estimates, whereas variability is important because it can affect the precision of estimates and influence the degree to which they can be generalized. Depending on the circumstances, uncertainty and variability can complement or confound one another (USEPA 1995, 1996).

In the context of exposure assessment, uncertainty and variability have fundamentally different ramifications for risk-related decision making. Uncertainty forces decision makers to judge how probable it is that exposures have been either overestimated or underestimated for members of the exposed population, while variability forces decision makers to confront the fact that there are individuals in the population exposed to concentrations both above and below virtually any choosen reference point (USEPA 1996).

ADDRESSING UNCERTAINTY

According to the National Research Council (NRC 1994) . . . *Uncertainty can be defined as a lack of precise knowledge as to what the truth is, whether qualitative or quantitative. That lack of knowledge creates an intellectual problem—that we do not know what the "scientific truth' is, and a practical problem—we need to determine how to assess and deal with risk in light of that uncertainty.* In the NRC's opinion, pervasive uncertainty is the "dominant analytic difficulty" in risk-based decisions, and has no immediate solution given the many gaps in our knowledge and understanding. The NRC suggests, however, that a powerful solution exists to some aspects of the problem, namely, systematic analysis of the sources, nature, and implications of uncertainty. This view is in accordance with USEPA's Guidance on Risk Characterization, which states . . . *Particularly critical to full characterization of risk is a frank and open discussion of the uncertainty in the overall assessment and in each of its components. Both qualitative and quantitative evaluations of uncertainty provide useful information to users of the assessment* (USEPA 1995).

Residential Exposure Assessment, edited by Baker *et al.*
Kluwer Academic/Plenum Publishers, New York, 2000.

Evaluation of uncertainty-related problems is a key part of exposure assessment, creating a need for assessors to identify, evaluate, and describe consequential uncertainties, and to point out how they affect confidence in the final exposure or dose estimate. There are several reasons why formal consideration and discussion of uncertainty is important (USEPA 1992, 1995, 1996, NRC 1994). First, information from different sources carries different kinds of uncertainty, and knowledge of these differences is important when uncertainties are combined for characterizing exposure and risk. Second, the risk assessment process, with management input, involves decisions regarding the collection of additional information (versus living with uncertainty), and a discussion of the uncertainties will help to identify where additional information could contribute significantly to reducing uncertainties in exposure and risk estimates. Third, emphasis on identifying and elucidating uncertainty serves to focus attention on the most critical human exposure factors, highlighting key areas that are likely to be the subject of debate and disagreement. Fourth, unrecognized uncertainty-induced biases can significantly affect the accuracy of exposure estimates, resulting in "best estimates" that do not accurately reflect actual, real-life exposures. And fifth, a clear and explicit statement of the strengths and limitations of an exposure assessment (or a risk assessment) requires a clear and explicit statement of related uncertainties.

Because uncertainty is a comprehensive yet diffuse concept, which quickly becomes analytically complex, it is helpful to categorize uncertainty into a few major types (Bogen 1990, Finkel 1990, Morgan and Henrion 1990, NRC 1994). The USEPA has classified uncertainty in exposure assessment into three broad categories (USEPA 1992, 1996):

- Scenario Uncertainty—uncertainty regarding missing or incomplete information needed to fully define exposure and dose
- Parameter Uncertainty—uncertainty regarding values for parameters needed to calculate exposure or dose, and
- Model Uncertainty—uncertainty regarding gaps in scientific theory required to make exposure and dose predictions on the basis of causal inferences. This taxonomy provides a useful framework for identifying, evaluating, and communicating uncertainty related to exposure assessment.

The USEPA (1992) Exposure Assessment Guidelines distinguish between two different approaches for examining uncertainty: uncertainty characterization, which generally involves a qualitative discussion of the reasoning behind key decisions, such as rejection of certain data, construction of specific scenarios, and selection of particular values for human exposure factors, and uncertainty assessment, which entails a more quantitative analysis of the magnitude of uncertainty and its effect on exposure and dose estimates. Uncertainty characterization is most appropriate for simple assessments, where not much quantitative information is available. Uncertainty assessment is a phased process, wherein the assessor starts with relatively straightforward measures (*e.g.*, ranges) and simple analytical techniques (*e.g.*, sensitivity analysis), and then decides whether to proceed to increasingly more complex and sophisticated levels of analysis.

The primary source of uncertainty in most exposure assessments is a fundamental lack of knowledge about important human exposure factors for the populations and situations of interest. Both qualitative and quantitative examinations of uncertainty, along with a clear, concise description of decision-related implications, will provide valuable information to decision makers and improve the likelihood that exposure estimates will be used appropriately.

ADDRESSING VARIABILITY

Variability is another important issue for exposure assessors that, like uncertainty, affects interpretation and influences confidence in the final estimates. It is imperative, therefore, that assessors understand and take account of inherent variability in the underlying human exposure factors used for scenario-based calculations of exposure and dose. For purposes of exposure assessment, there are basically three general types of variability related to human exposure factors (USEPA 1996):

- Spatial Variability—systematic differences across geographic locations
- Temporal Variability—fluctuations over time, and
- Human Variability—variability in human characteristics and/or behaviors:
  (a) For one individual over some defined period of time
  (b) Between individuals at a point or interval in time
  (c) Within sociodemographic groups, and
  (d) Between sociodemographic groups.

Spatial variability refers to systematic differences in exposure factors that occur between different geographic locations. Considering residence location, for example, there are likely to be differences in shellfish consumption between those living near the ocean and in the Midwest, in commuting time between those living in urban, suburban, and rural areas, and in time spent outdoors between those living in warmer southern climates and in colder northern climates. On a more local scale, examples of spatial variability include differences in activity patterns between inner city residents and those living in the suburbs, in consumption of homegrown vegetables between urban and rural residents, and in duration and frequency of pesticide exposure between city dwellers and farmers.

Temporal variability refers specifically to changes over time, either short or long-term, in human exposure factors. Seasonal fluctuations in consumption of fruit and vegetables, human activity patterns, use of consumer products, and lawn and gardening activities are examples of longer-term variability, while shorter-term variation includes differences in human activities according to the time of day and in consumption patterns of food and beverage between weekdays and weekends.

Human variability can be thought of as four different types:

- Intraindividual
- Interindividual
- Intragroup, and
- Intergroup.

Intraindividual variability refers to differences in human exposure factors that occur over a specified period of time for the same individual, such as changes in height, weight, and surface area over a lifetime or in activity patterns from workdays to holidays. Interindividual variability refers to differences between individuals at a point or interval in time, as, for example, differences in inhalation rates between a jogger and someone sitting at a desk or in duration and frequency of asbestos exposure between a maintenance worker and a businessman. Intragroup variability refers to differences between individuals who are members of a particular sociodemographic group, while intergroup variability refers to systematic differences between specific sociodemographic groups. Among the groups of people that might be important from an exposure-related perspective are population segments defined by demographic characteristics, such as age, gender, health status, ethnicity, and race, by socioeconomic attributes, such as income, occupation, education, and access to health care, and by behaviors and lifestyles, such as activity patterns, tobacco and alcohol use, fitness level, and nutritional status.

In the context of risk-related decision making, the National Research Council (NRC 1994) has identified four different strategies for dealing with variability:

- Ignore the variability and hope for the best. This strategy tends to be most successful when the variability is small and any estimate that ignores it will not be far from the truth.
- Explicitly disaggregate the variability. Where the quantity seems to change smoothly and predictably over some range, continuous mathematical models may be fitted to the data in place of a discrete step function. In other cases, it is easier to disaggregate the data by considering all the relevant geographic or sociodemographic groups in the population.
- Use the average value of a quantity that varies. This strategy is not the same as ignoring the variability, but rather follows from a decision that the average value can be estimated reliably in light of the variability, and that it is a good surrogate for the variable quantity.
- Use a maximum or minimum value of a quantity that varies. This strategy is the most common way of dealing with variability in risk assessment and tends to be most successful when the measures needed to protect or account for the person or situation with the extreme value will also suffice for the remainder of the distribution.

As pointed out by the National Research Council (NRC 1994), the crucial point to bear in mind about all these strategies for dealing with variability is that unless someone measures, estimates, or at least roughly models the extent and nature of the variability, any strategy will be precarious. Exposure assessors must, therefore, confront the issue of variability head-on, explicitly considering and formally reporting its effect on scenario-based calculations of exposure and dose. Analogous to the situation for uncertainty, variability can be evaluated either qualitatively (variability characterization) or quantitatively at varying levels of sophistication (variability assessment).

MOVING FROM POINT ESTIMATES TO PROBABILISTIC ANALYSES

Issues of uncertainty and variability relate directly to the question of how best to calculate and describe exposure, dose, and related health risk. Ideally, we would like to characterize and describe the statistical distribution of exposure and dose for the population and situation of interest, including evaluation of the salient aspects of uncertainty and variability. In practice, however, probabilistic analysis is often hindered by the scarcity of knowledge and understanding about critical human exposure factors and pollutant concentrations. This paucity of information conflicts with legislative and statutory mandates that require immediate regulatory decisions, and has fostered development of "science policies," such as exposure scenarios, default values, and single-point estimates, which are shortcuts that allow assessors to estimate exposure, dose, or risk despite data deficiencies. Some of these science-policy shortcuts necessarily require heroic assumptions, which leaves regulators with a conundrum. Do the advantages of being able to make quantitative estimates based on limited information outweigh the disadvantages of large and often ill-defined uncertainties in the final exposure estimate? In most instances, regulatory agencies have had to answer this question in the affirmative.

As a result, state and federal health risk assessments historically have produced single-point estimates of risk. Because associated uncertainty and variability are seldom communicated adequately, these point estimates are routinely criticized as artificially precise, overly conservative, and often misleading. Yet notwithstanding their limitations and problems, point estimates have a legitimate, if limited role in exposure assessment (1996). The USEPA's Guidelines for Exposure Assessment (USEPA 1992) suggest that it is useful to separate the question of "how much protection is enough?" from the question of "are we protecting real people?" In the latter case, point estimates can be valuable tools to calculate "bounding, worst-case, and high-end estimates" of exposure and dose.

A bounding estimate is an estimate of individual exposure, dose or risk that is higher than that incurred by the person in the population with the highest actual exposure, dose, or risk. This kind of calculation is helpful in developing statements that exposure, dose or risk is "not greater than" the bounding estimate. A worst-case estimate is an estimate of the maximum possible individual exposure, dose, or risk that can conceivably occur, whether or not this value actually occurs or can be observed in the population. This type of calculation is helpful for developing statements that exposure, dose, or risk "can be as high as" the worst-case estimate. A high-end estimate is a plausible estimate of the individual exposure, dose, or risk for persons at the upper end of the distribution. Conceptually, the high-end is above the 90th percentile, but not higher than the individual in the population who has the highest actual exposure, dose, or risk. This type of calculation is helpful for developing statements that exposure, dose, or risk estimates "represent actual high-end values for real people" or "are above the 90th percentile and less that the highest exposed person."

Because a bounding estimate is constructed specifically to overestimate actual exposure or dose, we can assume that everyone in the population is less exposed. It is not always clear, however, just how much greater the bounding estimate is than the highest exposed person in the population, which is to say, we are not always sure just how "conservative" the estimate actually is. If the calculated bounding estimate is below any level of concern for exposure or dose, then it is reasonable to conclude that no one in the population is exposed at or above the benchmark. The converse, however, is not true. If bounding estimates are above a level of concern, it does not necessarily mean that real people in the population are exposed above the benchmark. More rigorous and indepth analyses is needed to establish whether there is, in fact, any reason for concern. Bounding estimates are useful, inexpensive tools for preliminary evaluations, allowing assessors to screen out levels of exposure or dose that represent negligible health risks, and identify those that require additional work before we can decide whether they are of concern or not (Callahan 1996).

High-end estimates are especially important to assessors and decision makers because they represent actual values that are known to occur at the upper end of the exposure or dose distribution. It can be particularly valuable to use high-end estimates together with worst-case estimates to define a range for the high-end of the distribution. It is helpful, for example, to have a sense of whether the approximate range of exposures between the 90th percentile and the maximum possible value is narrow or wide.

Although bounding estimates, worst-case estimates, and high-end estimates are practical tools that can have beneficial effects if used properly, there is little disagreement that regulatory agencies have placed too much emphasis on single-point estimates and not enough on related issues of uncertainty and variability. Moreover, there is widespread agreement that, to the extent practicable, probabilistic methods should be used to characterize uncertainty and variability in exposure and dose estimates. Many observers advocate using Monte Carlo techniques to represent parameters that greatly affect exposure and dose estimates by distributions rather than point estimates (NRC 1994, Finley *et al.* 1994).

Monte Carlo techniques, which are, by and large, conducted on personal computers using commercially available software, derive an estimate of the distribution of exposure or dose in the population by repeated random sampling from the distribution of values for each parameter in the potential dose equation. Advocates argue that estimating potential dose through numerous iterations using randomly generated values from the defined probability functions offers several advantages over point estimates: makes greater use of the scientific evidence and data without sacrificing conservatism

- It avoids compounding of "worst-case" assumptions and uncertainty, and
- It provides quantitative information on uncertainty (NRC 1994).

On the other hand, critics of the use of Monte Carlo analysis in exposure assessments question the "value added" by this more rigorus analysis when there is a paucity of information on the distributions for many of the exposure factors that drive the exposure assessment. Therefore, it is important to recognize that Monte Carlo techniques also have their limitations and must be used with care and deliberation (Callahan 1996). For example, it is often not possible to determine objective probabilities for important parameters, and subjective probabilities may be either unavailable, inappropriate, or difficult to obtain. Accordingly, Monte Carlo techniques usually require the assessor to use professional judgment in making assumptions and inferences necessary to develop required probability distributions. Moreover, care must be taken to ensure that use of unbounded input distributions, such as lognormal functions, do not lead to unrealistically high values for those at the high end of the distribution, and that correlation between input parameters (*e.g.*, covariance), as, for example, between body weight and surface area, are taken into account in the analysis. As with single-point estimates, realistic Monte Carlo estimates depend on the skill, expertise, and knowledge of the assessor and it is essential that the exposure assessor characterize in a transparent fashion the various assumptions and data sources that underly the Monte Carlo analysis. USEPA is developing guidance on the proper role of Monte Carlo analysis for exposure assessment along with principles that can be applied in constructing and/or evaluating a Monte Carlo exposure assessment. As part of that effort, a workshop was held to address the issues of selecting input data distributions, separating variability from uncertainty, and presenting results to decision makers (USEPA 1996).

Workshop participants emphasized the importance of a tiered assessment approach starting at the screening level with point estimates, followed by a sensitivity analysis to identify the most important pathways and exposure routes, and, finally, followed by a Monte Carlo analysis of only the most important pathways and exposure routes. Criteria were also suggested for reviewing and evaluating the quality of Monte Carlo analyses (*i.e.*, how to tell a "good" analysis from a "bad" one).

The goal is to eventually obtain enough data and scientific understanding so that regulatory agencies, and others, can conduct probabilistic risk assessments that provide decision makers with better and more complete information. With that goal in mind, the National Research Council (NRC 1994) has observed . . . *Once USEPA succeeds in supplanting single point estimates with quantitative descriptions of uncertainty, its risk assessors will still need to summarize these distributions for risk managers (who will continue to use numerical estimates of risk as inputs to decision-making and risk communication). It is therefore crucial to understand that uncertainty analysis is not about replacing "risk numbers" with risk distributions or any other less transparent method. It is about consciously selecting the appropriate numerical estimate(s) from out of an understanding of the uncertainty.*

COMMUNICATING EXPOSURE AND DOSE ESTIMATES

Regardless of whether exposure assessment is based on single-value estimates or probabilistic approaches, the assessor must decide how best to summarize, interpret, and report results so that risk managers have the information they need to make informed and well-reasoned decisions about protecting public health. The NRC and the USEPA have each provided general guidance on how assessors should integrate and communicate findings.

In its 1994 report, Science and Judgment in Risk Assessment, the NRC called attention to the fact that no matter whether statutory mandates require risk managers to balance costs and benefits, determine acceptability of "residual risks," or make other risk-related judgments, a bottom-line summary of the risk is critical. According to the NRC, such risk summaries should include at least three types of information:

- A fractile-based summary statistic, such as the median (50th percentile) or a 95th-percentile upper confidence limit, which denotes the probability that the uncertain quantity will fall an unspecified distance above or below some associated value
- An estimate of the mean and variance of the distribution, which along with the fractile-based statistic provides crucial information about how the probabilities and the absolute magnitudes of errors interrelate, and
- A statement of the potential for errors and biases in these estimates of fractile, mean, and variance.

A more comprehensive discussion of this subject is provided in the USEPA's (1995) Guidance on Risk Characterization, which is aimed at giving direction to agency assessors. The document states three guiding principles for risk characterization, the step wherein assessment results are summarized and communicated:

- Integrate information from the preceding steps in risk assessment (*i.e.*, hazard identification, dose-response assessment, exposure assessment)
- Explicitly discuss both uncertainty and variability, and
- Present conclusions and information regarding the strengths and limitations of the assessment.

Referring specifically to exposure assessment, the USEPA guidelines further state that exposure assessors should address five key issues:

- The basis for values and input parameters used for each exposure scenario
- The confidence in assumptions made about human behavior and characteristics and the relative likelihood of different exposure scenarios
- The major human exposure factor or factors (including concentrations in carrier media) thought to account for the greatest uncertainty in the exposure or dose estimate
- The link between the exposure (and dose) information and important risk descriptors, and
- Other information that may be important for the particular assessment, such as background levels in the environment or other important sources not considered.

A "risk descriptor" is defined as a statement that conveys important information and key insights about the estimated risk values. For example, risk assessors typically make statements about the expected number of excess cancer cases in a population exposed to a specific average daily potential dose, or an individual's lifetime risk of developing cancer from a particular lifetime average daily potential dose. Exposure assessment plays a pivotal role in development of risk descriptors, which are based, in large part, on the estimated exposure distribution for the population of interest. The USEPA's Guidance for Risk Characterization lists three general types of risk descriptors that should be included in agency assessments:

- Descriptors of individual risk, including measures of central tendency (*i.e.*, either arithmetic mean or median) and high-end risks (*i.e.*, 90th percentile and above)
- Descriptors of population risk, including probabilistic estimates of the expected number of cases (for cancer) and estimated percentage of the population with risk greater than some reference value (for noncancer effects), and
- Descriptors of risk for highly exposed (*e.g.*, occupational or hobby-related exposures) and/or highly susceptible groups (*e.g.*, genetic predisposition, preexisting illness) in the population.

REFERENCES

Bogen, K.T. 1990. Uncertainty in environmental health risk assessment. Garland Publishing. New York, New York.

Callahan, M. 1996. Special issue: Commemoration of the 50th anniversary of Monte Carlo. Human and Ecological Risk Assessment. 2(4):627–1038.

Finkel, A.M. 1990. Confronting uncertainty in risk management. Resources for the Future. Washington, D.C.

Finley, B., D. Proctor, P. Scott, N. Harrington, D. Paustenbach, and P. Price. 1994. Recommended distributions for exposure factors frequently used in health risk assessment. Risk Anal. 14(4):533–553.

Morgan, M.G. and M. Henrion. 1990. Uncertainty: A guide to dealing with uncertainty in quantitative risk and policy analysis. Cambridge University Press. New York, New York.

NRC (National Research Council). 1994. Science and judgment in risk assessment. National Academy Press. Washington, D.C. 652 pages.

USEPA (U.S. Environmental Protection Agency). 1992. Guidelines for exposure assessment. Fed. Reg. 57(104):22888–22938. May 29.

USEPA (U.S. Environmental Protection Agency). 1995. Guidance for risk characterization. Science Policy Council. Washington, D.C.

USEPA (U.S. Environmental Protection Agency). 1996. Exposure factors handbook. NCEA-W-0005. Washington, D.C.

12

QUALITATIVE AND QUANTITATIVE UNCERTAINTY ANALYSIS

Chapter 1 defines residential exposure assessment as the process of determining, through direct or indirect means, the doses that the individuals receive from sources of contamination in the residential environment. Measures of exposure and dose and the factors used to determine these measures are subject to both uncertainty and variation. As discussed in the introduction of the present chapter, uncertainty is a measure of the state of knowledge of an investigator. Uncertainty in dose estimates arises from a variety of sources, including: uncertainty in measurements of the factors used in estimating exposures, uncertainty in the accuracy and precision of models, and uncertainty in problem formulation. In addition, residential exposures vary across individuals and across time for an individual. Exposure related factors such as source terms and individual's behaviors vary in time and space (Price *et al.* 1991, 1996a). This variation greatly contributes to the complexity of uncertainty analyses (Morgan and Henrion 1990, USEPA 1992).

Strictly speaking, variation is a separate issue from uncertainty. Variation is function of the event under consideration and is independent of the state of knowledge of the investigator. Characterizing the uncertainty in measures of variation (the 95th percentile of a dose distribution or a standard deviation of a distribution) is not intrinsically different from characterizing the uncertainty of a factor that has no variation (*e.g.*, the Henry's Constant of a agent).

However, there are a number of reasons why the issues and concepts of uncertainty and variation are closely intertwined. First, limitations in the ability to model certain measures of variation with an acceptable level of confidence affect the options for management of variation. Second, techniques used for management of uncertainty (Monte Carlo analysis) lend themselves to measurement of variation. Third, variation can be viewed as a source of uncertainty in certain exposure metrics. Specifically, variation is a source of uncertainty in estimate of the dose to a randomly selected individual in an exposed population. Finally, certain risk management tools, such as the use of "worst case assumptions," are intended to address both variation and uncertainty resulting in policies that manage both issues. As a result, this chapter will address both uncertainty and certain aspects of the characterization of variation.

The chapter begins with a discussion of the various sources of uncertainty in residential exposure assessments. These sources include measurement errors, modeling errors, and scenario uncertainty (USEPA 1992). Discussion then follows on options for management of uncertainty in exposure estimates, including the use of techniques for propagating uncertainty and variability in exposure models. The chapter concludes with a brief discussion on techniques for presenting information on the uncertainty and variation in estimates of residential exposure. Due to limitations in size, this chapter contains only a survey of the issues. For more in-depth information, the reader is referred to Morgan and Henrion (1990).

SOURCES OF UNCERTAINTY IN RESIDENTIAL EXPOSURE ASSESSMENTS

Measured Values

Measurements of the factors used in residential exposure assessment are subject to uncertainty. The sources of uncertainty occur as a result of the processes used in developing the data. These sources include analytical uncertainty, survey limitations, and errors in the understanding of the data.

The factors used in exposure assessment range from structural measurements such as air exchanges or the presence of wood stoves, to behaviors such a time activity patterns, and measurements of contaminant levels. All of these factors are subject to some form of uncertainty.

Measurements of a given quantity of interest, be it radon, NO_x levels from gas stoves, VOC emissions from cleaning products, or surface contamination by pesticide residues, are subject to several types of uncertainties. The first of these is uncertainties is the analytical method used to quantify the concentration of the material of interest. Typically the *lower* limit on this uncertainty is the variability of a set of measurements taken from a set of reference samples containing a single known concentration. More commonly, one will get a set of calibration measurements, where multiple concentrations of the material of interest are analyzed using a particular method of quantification (Caulcutt and Boddy 1983). The level of variability of actual measurements will always be higher than that found in a set of replicate laboratory standards, or variability estimates derived from a calibration curve, both because real samples do have variability, and because other materials in the environment may interfere with the measurement of the concentration of the material of interest.

One specific issue in measurements of factors is the issue of censored data. Most analytical methods have detection limits below which the method cannot discriminate between the measurement and zero. Data sets that contain nondetects are said to be left-censored because all one knows about the low values (on the left side of the axis) are that they are less than the detection limit. The presence of censoring can affect the estimation of value dependent quantities such as the mean and variance as well as percentiles below the detection limit. Also, censoring affects representation of the data by a parametric distribution. The more severe the censoring, the more serious the effect on these estimates. However, censoring does not affect estimates of percentiles for the noncensored data.

If the number of nondetects is low, the easiest approach is to simply assume that concentrations in nondetects are equal to one-half the detection limit. An assumption that is often made is to assume that nondetects follow a triangular distribution between zero and the detection limit. This assumption results in nondetect values being assigned an average value of one third the detection limit. These approaches work acceptably well if the number of nondetect values is less than 10 percent.

Where larger numbers of nondetect values occur, one can fit the data above the detection limit and extrapolate the distribution to the censored data (Gilbert 1987, Helsel 1990). If one can assume a normal or lognormal distribution, one can even make some progress in bounding exposures when all of the observations are nondetects. Ginevan (1994) has shown that through the use of order statistics, and appropriate ancillary information on the lower bound for the standard deviation, one can construct statistical bounds for percentiles of the nondetect distribution.

A second source of uncertainty in measurements is biases that occur from the survey process. Survey design can greatly affect the value of a factor. In general, use of survey data should be restricted to the questions the survey was designed to answer. For example, if a survey is intended to identify the maximum level of contamination in a residence, then the distribution of measurements may be a biased estimate of the levels of contamination randomly contacted by individuals. Unfortunately, the field of exposure assessment is relatively new and, as a result, much of the data used in exposure assessments were originally collected for other purposes.

Surveys that fail to properly consider spatial and temporal variation may result in biased or misleading results. For example, when a pesticide is sprayed in a house, it will volatilize rapidly from smooth surfaces and much more slowly from porous surfaces. In this case, air concentrations will increase with the initial release of pesticide by spraying, then fall rapidly as the pesticide on the smooth surface volatilizes, and finally fall more slowly as the pesticide from the porous compartments volatilizes. If such data are reported only in terms of a first-order decay rate, the results will not be an accurate description of the actual air concentrations. Another example is the current dilemma on characterizing exposures from tap water consumption. The best data on tap water intake were derived from a survey of short-term nutritional habits. As such, distributions of consumption rates from this survey may not be reflective of long-term drinking patterns (Ershal and Cantor 1989).

Spatial structure can also be lost with certain measurement techniques. Combination of data collected in different locations will result in the loss of information on the spatial relationships of the data. The original structure of the data is not always readily apparent from the final data. An example of this is the use of passive dosimetry that gives a measure of average exposure over an 8-hour workday but provides no clue on the relative contribution of specific sources or temporal patterns of exposure.

Uncertainty from use of models. As discussed above, exposures to substances can be characterized by a number of methods including biomonitoring, point-of-contact monitoring, and exposure scenarios (USEPA 1992). All of these approaches require some degree of modeling to derive the dose rate estimates used by toxicologist. However, exposure scenarios are most dependent on models.

Exposure scenarios are descriptions of a behavior or collection of behaviors that define an individual's interactions with a contaminated media. These descriptions provide the basis for dose rate models and the values of the inputs for these models (Ott 1981, 1984, 1990, Duan 1988, McKone and Daniels 1991, USEPA 1985abcd, 1986ab, 1987ab, 1988, 1991ab). The values of these inputs may be the products of fate and transport models or models of behavior. Such dose rate models can range from simple multiplicative formulae to large complex computer models that require extensive computer resources (RSI 1998, Helton *et al.* 1997). Certain exposure models estimate the dose to a particular individual such as the "typical" individual or the "high-end exposed" individual while others model dose rates across a population (Finley *et al.* 1994, Hattis and Burmaster 1994).

The results of all types of models are subject to uncertainty. Modeling uncertainty can be divided into two sources: model uncertainty and uncertainty in the model inputs. Model uncertainty can be defined in terms of the differences between the model results and the actual risks. Model structure may contribute to model uncertainty if there is uncertainty about the composition or completeness of the model (*e.g.*, whether all relevant pathways, activities, and subpopulations have been included) or if there are descriptive errors, aggregation errors, or errors in professional judgment (USEPA 1992).

Model uncertainty will increase when models are extrapolated beyond the context in which they were developed. Here "context" can refer to physical characteristics (*e.g.*, type of soil or geographic location), temporality (*e.g.*, short-term or long-term), time frame (*e.g.*, 1960s or 1990s), database (*e.g.*, adult or child), or range of inputs (*e.g.*, inside or outside the range where the model has been tested). Such uncertainties increase the importance of being able to reflect the different relative degrees of plausibility associated with different models in the quantitative exposure assessment.

MANAGEMENT OF UNCERTAINTY

Uncertainty unlike variation is a function of an investigator's state of knowledge. Additional effort can in many instances reduce uncertainty. The decision on the level of effort necessary to reduce uncertainty is a risk management decision. As the National Research Council of the National Academy of Science discussed risk assessors must seek to balance the cost and time of additional analyses against the value of reducing uncertainty (NRC 1994). The basic strategy for this management is to initially rely on low-cost methods of assessing exposure that have a low potential for underestimating risk. These techniques can be used to efficiently screen out low risk concerns. Concerns that are not screened out by these techniques can be subjected to additional analyses that require additional data or more resource intensive modeling techniques. This process continues in an iterative fashion until the cost of assessment in terms of either resources or time) exceeds the cost of management of the exposure of concern or all concerns are demonstrated to have acceptable risks.

Management of Analytical Sources of Uncertainty

Uncertainty in analyses can be reduced by the collection of additional data. New data can reduce the uncertainty by increasing the number of data points, including data that is more directly relevant, or is free from the biases of prior studies. A second option is the reanalysis of existing data sets to adjust for sampling bias. An example of this is Israeli and Nelson analysis of duration of time lived in a residence to correct for a frequency bias that is inherent in U.S. census data.

Management of Modeling Uncertainty

There are a number of strategies for managing model uncertainty, the evaluation and of the limitations and biases of a particular model using sensitivity analyses, selection of an optimal model or models, and the evaluation all of the plausible models and model validation.

Both qualitative and quantitative sensitivity analyses are an important part of the quantitative characterization of uncertainty and should quantify the impact of different alternatives (*e.g.*, different models, assumptions, and factor values) on the dose characterization (Inman and Helton 1988). Quantitative sensitivity analyses should also provide a quantitative perspective on the robustness of the dose characterization with respect to these alternatives. In addition, quantitative sensitivity analyses should help prioritize and direct research and resource allocation.

Sensitivity analyses can quantify the dependence or association between a dose characterization and its component alternatives. The cases where the dose is characterized by a single number (*e.g.*, a constant or the mean, median, or 95th percentile of a distribution) are considered initially. In these cases, scatter plots or correlation can be used to evaluate the impact of a single alternative that is quantitative (*e.g.*, the mean duration of an activity). Linear regression can be used to evaluate how the mean exposure concentration changes with respect to multiple quantitative alternatives (*e.g.*, drinking water consumption, shower duration, and residential duration). Analysis of variance techniques can be used to quantify the impact on the mean exposure concentration of qualitative factors (*e.g.*, three alternative models). Analysis of covariance techniques and combining linear regression models for quantitative factors with analysis of variance models for qualitative factors can also be used to analyze the behavior of the mean exposure concentration.

Model selection at various stages of an iterative analysis of risk can be based on factors such as use the need for a "default" model, the need for simplicity, convenience, consistency, policy, or legislation, or select one on the basis of the judgement of someone (an expert or a panel of experts). Screening assessments, for example, are generally based on one model that is selected to be reasonably certain of obtaining exposure values at least as large as the anticipated exposures. One alternative to the use of default model is to present the results from all plausible alternative models (NRC 1994).

A second alternative approach to the use of a specific model is to use multiple models and decision trees to appropriately weigh the results. The evaluation of multiple models using a decision tree has been investigated by a number of researchers (Evans *et al.* 1993, 1994, Holland and Sielken 1993, Taylor 1993, Sielken 1996). Under this approach the components of exposure models are defined in terms of exhaustive categories of models where the probability of a category can be specified based on empirical evidence or professional opinion. This approach can be incorporated into Monte Carlo models of total uncertainty.

Models can also be partially or entirely validated using appropriate data sets. Validation is the process of examining the performance of the model compared to actual observations under situations representative of those being assessed. In most situations, only partial validation is possible due to data deficiencies or model complexity. Environmental models can usually be validated only under a specified set or range of conditions, i.e. usually on an ad hoc basis (due to the assumptions that are invariably necessary).

The increasing number of biomonitoring and point of contact monitoring studies provides an opportunity for validating models of exposure. Examples of such studies include monitoring of pesticide residues in children and adults and the comparison of the results to predicted values of exposure. Monitoring studies of PCBs and mercury in the blood of anglers consuming contaminated fish have also been considered for model validation.

Modeling uncertainty has not been appropriately considered in uncertainty assessments. There is an unfortunate tendency to characterize uncertainty in terms of just uncertainty in model inputs rather than both model uncertainty and input uncertainty (McKone and Ryan 1989, Bogen 1990, Slob 1994).

PROPAGATION OF UNCERTAINTY AND VARIABILITY IN MODELS

The preceding sections of this chapter have discussed the major sources of uncertainty and variability in estimates of human exposure. This section describes how this information may be used to characterize the uncertainty and variability in the final estimate of exposure. In particular, a series of techniques are presented for the propagation of uncertainty and variability in exposure factors into exposure estimates.

Propagation of Uncertainty

The basic approach for the analysis of propagation of uncertainty in exposure factors involves the following steps:

- Define the model and its inputs
- Characterize the uncertainty in each factor using distributions
- Select a method for the analysis of propagation of uncertainty, and,
- Analyze the results to determine the overall uncertainty, the key contributors to uncertainty, and the potential benefits of uncertainty reduction.

Much has been written about this topic, but essentially there are two classes of approaches for the analysis of the propagation of uncertainty, analytical approaches (such as Gauss' Law of Error Propagation and lognormal error analysis) and numerical approaches (such as Simple Monte Carlo and/or Monte Carlo with Latin Hypercube simulation). Gauss' Law of Error propagation has limited value where the distributional form of the model output is unknown (Bevington 1969).

Lognormal error analysis is quite useful for analysis of the propagation of uncertainty in powers, products and ratios where the inputs can be adequately described using lognormal distributions. The central proposition of lognormal error analysis is that powers, products and/or ratios of lognormal variables are themselves lognormal.

The two most commonly applied numerical approaches for analysis of the propagation of uncertainty are Monte Carlo analysis (see side bar) and Monte Carlo with Latin Hypercube analysis. Monte Carlo analysis is simply numerical experimentation. Rather than using a closed form analytical expression to assess the propagation of uncertainty, Monte Carlo analysis relies on repeated sampling from the probability distributions of each of the uncertain factors to characterize the uncertainty in estimates of exposure. Monte Carlo analysis involves four steps:

- Drawing a set of possible values of each of the uncertain factors in the model
- Evaluating the model output that corresponds to these particular values of the factors

- Repeating the first two steps N times and generating N corresponding outputs, and
- Statistically analyzing the results from step (iii) to characterize the distribution of model outputs and to determine the key contributors to the uncertainty in the output.

Monte Carlo analysis is relatively simple to perform on current desktop computers and a number of commercial software packages for performing Monte Carlo are available. Monte Carlo analysis can handle both large and small errors, and applies to virtually any function that can be evaluated by the computer. Monte Carlo analysis provides information on the overall uncertainty, both the variance and the shape of the distribution of errors, but does not *directly* provide information on the relative contributions of each uncertain factor to the overall uncertainty. However, the results from a Monte Carlo analysis can itself be analyzed statistically using regression and analysis of variance (ANOVA) approaches to obtain insights about this issue.

In basic Monte Carlo analysis, the values of input variables are selected purely randomly. For many problems, this is quite adequate. However, as the number of variables of interest increases, the chance increases that certain regions of the factor space will not be adequately covered. The problem of inadequate coverage may be addressed by either increasing the sample size (*i.e.*, increasing the number of realizations in each simulation) or by stratified sampling. Latin Hypercube sampling is the most commonly used approach for stratified sampling in numerical analysis of error propagation.

Other types of numerical techniques include discrete probability distribution and factoral design. In general, these approaches can be viewed as approximations of Monte Carlo analysis that were created as a means of avoiding the computational demands of the approach. With the advent of low-cost computers, these techniques have declined in use.

Propagation of Uncertainty and Variability

The same approaches for modeling uncertainty in a model can be used to model variability. Thus, Monte Carlo and related modeling approaches have been widely used to characterize interindividual variation in the distribution of dose rates in exposed population (Finley *et al.* 1994).

As Morgan and Henrion (1990) have discussed, a factor that varies across the individuals in a population is also subject to uncertainty. In this case, the investigator must be careful to decompose differences in measurements into two elements, one representing uncertainty and another reflecting variability (Frey 1992, Bogen and Spear 1987). Frey (1992) discusses the simultaneous analysis of uncertainty and variability in the context of Monte Carlo analysis, and uses the term two-dimensional Monte Carlo analysis to describe the procedure. A number of publications using this technique have appeared in the peer review literature (Frey and Rhodes 1996, Cohen *et al.* 1996).

Application of Propagation Techniques to Modeling Uncertainty. Modeling uncertainty can be incorporated into a Monte Carlo framework. This is done by incorporating multiple alternative models into one "super model" the choice of which alternative model is selected in each iteration of a Monte Carlo analysis. This approach requires that the set of models exhaustively represent of options for modeling and the relatively likelihood if applicability is known for each model.

Developing Distributions for Exposure Factors. Developing distributions for a scenario-based exposure assessment has been the subject of a number of publications (Taylor 1993, Finley *et al.* 1994, AIHC 1994, USEPA 1996). The choice of a distribution for an assessment must be based on the analysis that is being performed. The assessor must address the following issues:

- Is the assessment attempting to evaluate uncertainty, variation, or both
- What is assessment evaluating a population or an individual
- What is the population being evaluated
- What is the time frame for the exposures, and
- Will intra-individual variation in exposure be considered.

Assessments of variation of exposure across a population will want to use distribution of interindividual variation in the factors used in estimating exposure. In contrast, if the goal is to estimate the dose received by a specific individual then the distribution should reflect the uncertainty in the exposure factors for this individual. If the goal is to characterize the risk to a random member of a population then both uncertainty and variability should be considered (USEPA 1992, Morgan and Henrion 1990).

It is important to define whether a distribution is characterizing variability, uncertainty, or a combination of the two (Bogen and Spear 1987). Variability in a factor may occur as the result of interindividual variation (*e.g.*, one adult is taller than the next), age-related variation (*e.g.*, adults eat more food than children), or spatial variation (*e.g.*, pollutant concentration is greater at one location than another) (Taylor 1993). Actual data on exposure factors will reflect both variability and uncertainty. It is often useful to attempt to separate the two.

The population under consideration can have a dramatic effect on the choice of distributions. For example, children have different tap water consumption rates than adults. Farm families have much longer residence time than urban home owners. Typical air exchange rates are differ for homes in the northeast versus the south (Murray and Burmaster 1996).

Time frames of exposures have a dramatic effect on exposure distributions. Consider the interindividual variation in exposure duration. If the time scale of a residential exposure is a single day then the 80th percentile will be exposed for the entire period of concern (Robinson and Thomas 1991). If the time period for exposure is a lifetime then the fraction of time exposed falls to less than 0.15 for 80th percentile of the population.

Time-scale related variability can be particularly troublesome when attempting to characterize rare events, such as consumption of certain foods, particular behaviors, or use of certain consumer products in the home. In these instances, short-term measurements may suggest, for example, that the vast majority of individuals do not participate in the activity during the measurement period and that a very small number of individuals participate on a daily basis. Use of such findings to characterize behavior by the population will poorly characterize variability among the general population. In general, the interpersonal variability will decrease with the length of time over which the factor is measured (Buck *et al.* 1995, Ebert *et al.* 1996).

Finally, exposures that occur over long periods of time tend to have significant intra-individual variation. For example, life time exposure to a pesticide residue in food will be a function of the average of doses that vary greatly from day to day. Accurate modeling of long term exposures may require that this day-to-day variation be explicitly modeled (Price *et al.* 1996ab).

Because of the relationship between the purpose of the assessment and the distributions the concept of a single "standard distribution" for an exposure factor is generally not appropriate. Instead, the researcher is directed to determine the most appropriate distribution from libraries of distributions or to derive distributions from population specific data. However, certain factors such as age and gender specific body weights generally play a minor role in estimating dose and difference in the distributions are not significant in different populations. For these factors, it is possible to define a single distribution that can be widely used.

Characterizing Exposure Factors Using Parametric and Nonparametric Techniques. It is possible to develop both parametric and nonparametric characterizations of exposure factors. In order for the variability in an exposure factor to be described by a parametric distribution such as the normal or lognormal, data must meet statistical testing and physical appropriateness requirements. There are a number of techniques that can be used to determine the distribution that fits a data set. The Kolmogrov-Smirnov (KS) tests discussed in Conover (1980) is an excellent general test.

There are a variety of ways of testing for fit to a normal distribution (remember that for a lognormal the same tests apply to the log-transformed data). One good method is to standardize the data by subtracting the sample mean and dividing by the sample standard deviation, and applying Lillifors test for normality. This test is a modification of the KS goodness of fit test which takes into account the fact that the data have been standardized (Wilkinson *et al.* 1994).

A second, graphical, method is the normal quantile plot, as depicted in Figure 12-1. The figure shows the values of the data (here the data are just 100 random normal numbers plotted against their expected normal scores). The expected normal scores are the Z scores predicted from an observation s rank and the total number of observations. For example, the largest value in a sample of 50 has an expected normal score of about 2.2. The Wilk-Shapiro statistic referred to in the footnote of Figure 12-1 is a goodness-of-fit statistic often included with a normal quantile plot (Gilbert 1987). It represents the correlation between the observed data and their expected normal scores. Note that, since the expected value of this statistic is much greater than zero, one cannot use the usual tests to assess its significance.

If one has a limited amount of data, a statistical fit only says that one cannot reject the candidate distribution. If one has a large amount of data, the fit may offer fairly strong evidence that the data are in fact from a particular candidate distribution. However in this case one can use a nonparametric density function to describe the data (Silverman 1986). Such nonparametric density estimates have the added advantage that, given adequate data, they are always appropriate.

Distributions Based on Surrogate Data. Production of an exposure assessment often requires that dozens of factors be evaluated, including exposure concentrations, intake rates, exposure times, and frequencies. A combination of monitoring, survey, and experimental data, fate and transport modeling, and professional judgment is used to develop distributions for these factors. Often the only available data are not completely representative of the population being assessed. Some examples are the use of activity pattern data collected in one geographic region to evaluate the duration of activities at a Superfund site in another region; use of national intake data on consumption of a particular

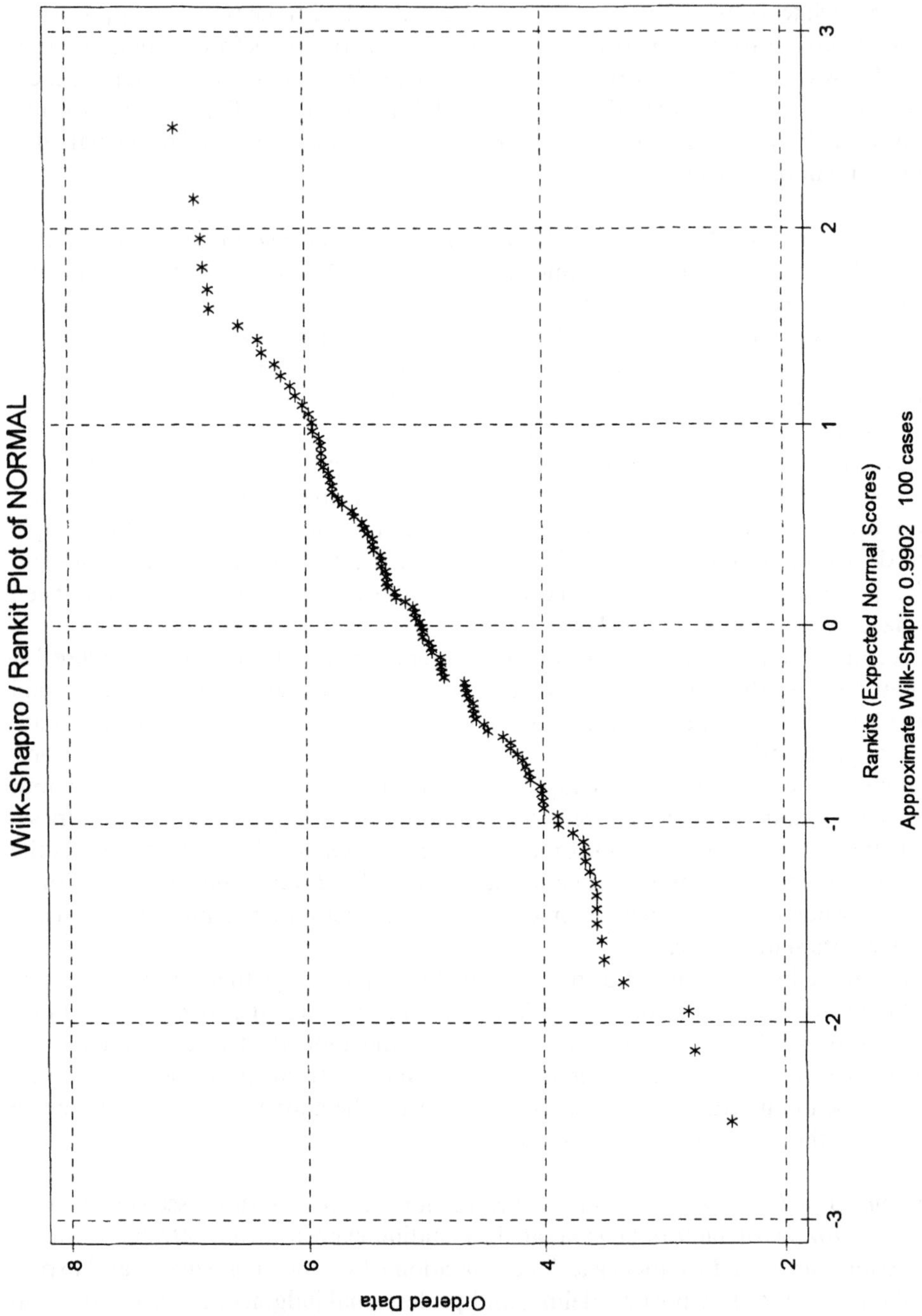

Figure 12-1. A normal scores plot of 100 random normal numbers. Note that the points lie approximately on a straight line.

food item to estimate regional intake; and use of data collected from volunteers to represent the general population.

Characterizing Extreme Values for Exposure Factors

A major challenge in developing distributions for exposure factors is characterizing extreme values. Because characterizing the upper end of the distribution is a common goal of exposure assessment, there is a desire on the part of assessors to accurately characterize the values that result in the highest exposures.

This poses a problem for developing distributions. A central percentile of the range of variation for an exposure factor can always be estimated with less uncertainty than an extreme percentile. Thus, for many exposure factors, there is little disagreement on appropriate values between the 10th and 90th percentiles. If only limited data are available, estimates of extreme percentiles can be quite unstable. Additionally, technical constraints on certain measurement techniques can limit our knowledge of and our ability to characterize extreme behavior. Finally, extreme percentiles are also sensitive to the averaging time of the measurement technique and the presence of nonhomogeneous subpopulations within the sample population.

PRESENTATION TECHNIQUES FOR QUANTITATIVE UNCERTAINTY ANALYSES

There are two objectives in presenting the findings of a quantitative risk assessment. The first is to present the audience with the key insights of the assessment and present them in a balanced context. To meet this objective, the assessor should balance the strengths, weaknesses, and limitations of the assessment and express his/her confidence in the analysis. The second objective is to provide sufficient information so that re-viewers can make an independent evaluation of the assessment.

When developing presentations of uncertainty an assessor must consider the audience. It is important that the risk assessor not overestimate the sophistication of the audience, but at the same time, not underestimate their ability and willingness to learn. While the insights the audience will develop will ultimately be qualitative they will be derived from both quantitative and descriptive information presented.

There is a range of insights and perspectives the risk assessor should seek to convey to his/her audience, which include:

- An appreciation of the overall degree of confidence in the analysis
- An understanding of the key sources of uncertainty and their impacts;
- An understanding of the critical assumptions and their importance to the analysis, and
- An understanding of any uncertainties which have not been addressed and how they affect confidence in the analysis.

In conjunction with any quantitative risk assessment, it is important to provide a well thought out descriptive component in which the risk assessor interprets the findings in clear, easy to understand language in a context meaningful for the audience. This may require several levels of presentation and narration depending on the needs and sophistication of the audience.

Presentation Techniques

There are several presentation techniques that have proven useful for presenting the findings from quantitative risk assessments (USEPA, 1997). For some audiences, presenting key summary statistics and their ranges or credible intervals can be useful. Important summary statistics can include the arithmetic or geometric mean, the mode, the median, and important percentiles (*e.g.*, 90th, 95th, or 99th percentiles). For each key summary statistic presented to an audience, an appropriate credible interval should be presented in parentheses immediately following the statistic. Graphical techniques can play an indispensable role in communicating the findings of a quantitative risk assessment.

It is important that the risk assessor select a clear and uncluttered graphical style in an easily understood format. Displaying too much data or inappropriate data will weaken the effectiveness of the effort. Having decided which information to display, the risk assessor should tailor a graphical presentation to the informational needs of specific audiences. Table 12-1 summarizes useful graphic construction principles proposed by Cleveland (1994).

Three graphical styles have proven to be particularly effective in communicating quantitative findings:

- Boxplots (Tukey boxplots, box and whisker plots) as explained in Tukey (1977) are a very effective graphic display for summarizing either uncertainty or variability. Boxplots provide easily comprehended visual summaries of the median, interquartile range, and skewness (see Figure 12-2).
- Probability density function plots (PDF plots, histograms) plot the absolute frequency or cumulative frequency of a random variable. Histograms are best for displaying the relative probability of values, the most likely values (modes), and the shape of the distribution (skewness, kurtosis).
- Cumulative probability plot (CDF plot, quantile plot) plots the sample-derived probability or expert-solicited probability that the value of a variable or uncertain statistic is less than a specific value. CDF plots are generally considered best for displaying fractiles, including the median, probability intervals and confidence intervals. CDF plots of variation can also express uncertainty in the form of confidence intervals around the median distribution of variation.

In general, assessors should seek to avoid excessively complicated graphs. When possible in publications and reports, graphs should be accompanied by a table of the relevant data. PDF and CDF plots can be presented on the same graph with the location of the mean clearly indicated on both curves with a solid point. Do not depend on the audience to correctly interpret any visual display of data. Always provide a narrative in the report interpreting the important aspects of the graph. Descriptive statistics and box plots generally serve the less technically-oriented audience well. Probability density and cumulative probability plots are generally more meaningful to risk assessors and uncertain.

Descriptors of Variation in Exposure

In presenting the results of an exposure assessment it is often useful to use exposure descriptors to characterize interindividual variation (USEPA 1992). There are three descriptors used to describe individual exposures: (1) measures of central tendency or

Table 12-1. Basic principles for good graphic construction (Cleveland 1994)

CLEAR VISION

- ▲ Make the data stand out. Avoid superfluity.
- ▲ Use visually prominent graphical elements to show the data.
- ▲ Use a pair of scale lines for each variable. Make the data rectangle slightly smaller than the scale-line rectangle. Tick marks should point outward.
- ▲ Do not clutter the interior of the scale-line rectangle.
- Do not overdo the number of tick marks
- Use a reference line when there is an important value that must be seen across the entire graph, but do not let the line interfere with the data.
- Do not allow the data labels in the interior of the scale-line rectangle to interfere with the quantitative data or to clutter the graph.
- Avoid putting notes and keys inside the scale-line rectangle. Put a key outside, and put notes in the caption or in the text.
- Overlapping plotting symbols must be visually distinguishable.
- Superposed data sets must be readily visually assembled.
- Visual clarity must be preserved under reduction and reproduction.

❑

CLEAR UNDERSTANDING

- Put major conclusions into graphical form. Describe everything that is graphed. Draw attention to the important features of the data. Describe the conclusions that are drawn from the data on the graph. Error bars should be clearly explained.
- When logarithms of a variable are graphed, the scale label should correspond to the tick mark labels.
- Proofread graphs.
- Strive for clarity.
- The aspect ratio of a graph is an important factor for judging rate of change. When the orientations of line segments are judged to decode information about rate of change, bank the segments to 45°.

❑

SCALES

- Choose the range of tick marks to include or nearly include the range of data.
- Subject to the constraints that scales have, choose the scales so that the data rectangle fills up as much of the scale-line rectangle as possible.
- It is sometimes helpful to use the pair of scale lines for a variable to show two different scales.
- Choose appropriate scales when data on different panels are compared.
- Do not insist that zero always be included on a scale showing magnitude.
- Use a logarithmic scale when it is important to understand percentage change or multiplicative factors.
- Showing data on a logarithmic scale can cure skewness toward large values.
- Use scale breaks only when necessary. If a break cannot be avoided, use a full scale break. Do not connect values on two sides of a break. Taking logs can cure the need for a break.

❑

GENERAL STRATEGY

- Graphing data should be an iterative, experimental process.
- A large amount of quantitative information can be packed into a small region.
- Many useful graphs require careful, detailed study.
- Graph data two or more times when it is needed.

❑

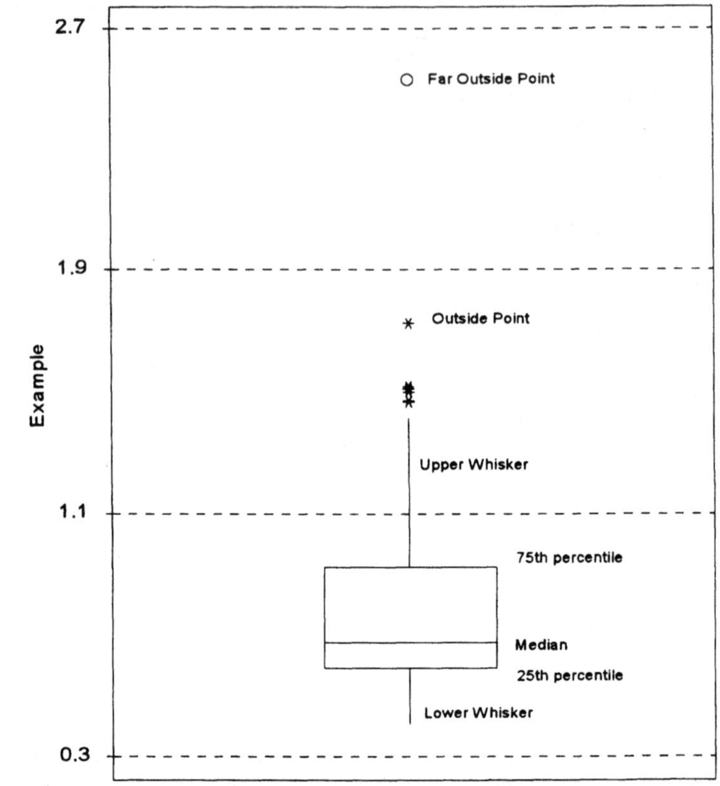

Figure 12-2. A sample box plot. The median is the 50% point of the data; the upper hinge (UH) is the 75% point of the data. The lower hinge (LH) is the 25% point of the data; the upper whisker extends from the UH to the largest data value less than the UH plus 1.5 times the difference between the UH and the LH hinges (the interquartile range (IQ)); the lower whisker extends from the LH to the smallest data value greater than the LH minus 1.5 times the IQ; Outside points are either between the UH plus 1.5 IQ and UH plus 3 IQ or LH minus 1.5 IQ and LH minus 3 IQ. Far outside points are beyond UH plus 3 IQ or LH minus 3 IQ.

most likely values, (2) descriptors of upper-end or high-end percentiles, and (3) descriptors of subpopulations.

Presenting estimates of central tendency is an attempt to convey information about the most likely exposures in a specified population. Measures of central tendency include estimates of the mode, median, and for moderately-skewed distributions, the mean. For *lognormal-like* distributions (positively bounded, *uni*modal distributions with positive skewness) the mean will not, except for the case of chronic exposure, be a valid descriptor of central tendency. For quantitative risk assessments involving short-term exposures, the median is a frequently used measure of central tendency. This measurement corresponds to the general public's impression of what a typical individual.

The arithmetic mean provides an estimate of the expected value of the exposure or risk distribution in a population. This, as noted, is a correct measure of chronic exposure. However, the mean is heavily influenced by high-end quantiles in positively skewed distributions. When the arithmetic mean is substantially greater than the median, indicating high skewness, the risk assessor should try to minimize any unintentional high value bias or other possible misinterpretations by the audience by providing a careful discussion of the selected statistics.

High-end descriptors refer to estimates of an upper-end percentile from the cumulative distribution of exposure, selected by the risk assessor and from percentiles above the 90th percentile but below the highest true exposure or risk thought to occur in the exposed population. The objective of this description is to provide a plausible estimate of the more highly exposed individuals in the defined population.

Within an exposed population, there are often population segments that experience significantly different exposures because the individual falls into an objectively definable subpopulation. For example, pica behavior results in a population with a distinctive soil ingestion rate distribution. Exposure assessments for these populations may not be predicted by the distribution of dose in the general population and exposure descriptors of the central tendency and high-end should be assessed and presented separately.

Descriptors of Uncertainty

Every summary statistic of exposure is inherently uncertain. It is incumbent upon the risk assessor to provide quantitative estimates of this uncertainty whenever possible. Because uncertainty always has subjective elements, it is recommended that the term *credible interval* be used to describe the plausible range of uncertainty around any exposure or risk statistic instead of the more statistically-grounded term *confidence interval.* When reference to an uncertain exposure or risk statistic is made in a report or presented to a lay audience, an estimate of its median value and its credible range should be given. For example, *the median estimate of the population risk attributable to radon in drinking water is 114 excess cancer deaths per year (credible range 37 to 415).* The reported credible range should be based on either the central 90 or 95 percentiles. For technically-oriented audiences truncated box plots showing the selected credible interval, inter-quartile range, and median are recommended. The boxplot should be clearly labeled as to which percentiles are being used. For scientific or peer review audiences, the full exposure or risk distribution should be shown and should include the lower and upper bound distributions.

ACKNOWLEDGMENTS

The following individuals made significant contributions to this chapter—Dr. John Evans, Dr. Robert Sielkin, and Ms. Natalie Wilson.

REFERENCES

AIHC (American Industrial Health Council). 1994. Exposure factors sourcebook. American Industrial Health Council. Washington DC. 118 pages.

Bevington, P.R. 1969 Data reduction and error analysis for the physical sciences. McGraw-Hill Book Company. New York, New York.

Bogen, K.T. 1990. Uncertainty in environmental health risk assessment. Garland Publishing. New York, New York. 195 pages.

Bogen, K.T. and R.C. Spear. 1987. Integrating uncertainty and interindividual variability in environmental risk assessment. Risk Analysis 7(4):427–436.

Buck, R.J., K.A. Hammerstrom, and P.B. Ryan. 1995. Estimating long-term exposures from short-term measurements. J. Exposure Anal. Environ. Epidemiol. 5(3):359–373.

Caulcutt, R. and R. Boddy. 1983. Statistics for analytical chemists. Chapman and Hall, London.

Cleveland, W.S. 1994. The elements of graphing data. Revised edition. Hobart Press. Summit, New Jersey.

Cohen, J.T., M.A. Lampson, and T.S. Bowers. 1996. The use of two-stage Monte Carlo simulation techniques to characterize variability and uncertainty in risk analysis. Human Ecol. Risk Assess. 2(4):939–971.

Conover, W.J. 1980. Practical non-parametric statistics. John Wiley, New York. 507 pages.

Duan, N. 1988. Estimating microenvironment concentration distributions using integrated exposure measurements. Pages 15–114 in: Starks, T.H. (ed.). Proceedings of the research planning conference on human activity patterns. EPA-600/4-89/004. U.S. Environmental Protection Agency, Office of Research and Development. Washington DC.

Ebert, E.S., T.J. Barry, M.N. Gray, and N.W. Harrington. 1996. Estimated rates of fish consumption by anglers participating in the Connecticut Housatonic River creel survey. North. Amer. J. Fish. Management 16:81–89.

Ershal, A.G. and K.P. Cantor. 1989. Total water and tapwater intake in the United States: Population-based estimates of quantities and sources. National Cancer Institute, Bethesda, Maryland. 263-MD-810264. May.

Evans, J.S., J.D. Graham, G.M. Gray, and R.L. Sielken, Jr. 1993. A distributional approach to characterizing low-dose cancer risk. Risk Anal. 14(1):25–34.

Evans, J.S., G.M. Gray, R.L. Sielken, Jr., A.E. Smith, C. Valdez-Flores, and J.D. Graham. 1994. Use of probabilistic expert judgment in uncertainty analysis of carcinogenic potency. Reg. Toxicol. Pharmacol. 20(1):15–36.

Finley, B., D. Proctor, P. Scott, N. Harrington, D. Paustenbach, and P. Price. 1994. Recommended distributions for exposure factors frequently used in health risk assessment. Risk Anal. 14(4):533–553.

Frey, H.C. 1992. Quantitative analysis of uncertainty and variability in environmental policy making. Directorate for Science and Policy Programs. American Association for Advancement of Science. Washington DC.

Frey, H.C. and D.S. Rhodes. 1996. Characterizing, simulating, and analyzing variability and uncertainty: An illustration of methods using an air toxics emissions approach. Human Ecol. Risk Assess. 2:762–797.

Gilbert, R.O. 1987. Statistical methods for environmental pollution monitoring. Van Nostrand Reinhold. New York, New York. 320 pages.

Ginevan, M.E. 1994. Bounding the mean concentration for environmental contaminants when all observations are below the limit of detection. Pages 123–128 in: Proceedings of the annual meeting of the American Statistical Association. Section on statistics and the environment. American Statistical Association. Washington DC.

Helsel, D.R. 1990. Less than obvious: Statistical treatment of data below the detection limit. Environ. Sci. Technol. 24:1766–1774.

Holland, C.D. and R.L. Sielken, Jr. 1993. Quantitative cancer modeling and risk assessment. PTR Prentice Hall, Inc. Englewood Cliffs, New Jersey.

Hattis, D. and D.E. Burmaster. 1994. Assessment of variability and uncertainty distributions for practical risk analyses. Risk Anal. 14(5):713–730.

Helton, J.C., D.R. Anderson, B.L. Baker, J.E. Bean, J.W. Berglund, W. Beyeler, R. Blaine, K. Economy, J.W. Garner, S.C. Hora, R.C. Lincoln, M.G. Marietta, F.T. Mendenhall, N.H. Prindle, D.K. Rudeen, J.D. Schreiber, A.W. Shiver, L.N. Smith, P.N. Swift, and P. Vaughn. 1997. Computational implementation of a systems prioritization methodology for the waste isolation pilot plant: A preliminary example. Reliability Engineer. System Safety 57:223–266.

Inman, R.L. and J.C. Helton. 1988. An investigation of uncertainty and sensitivity analysis techniques for computer models. Risk Anal. 8(1):71–90.

McKone, T.E. and P.B. Ryan. 1989. Human exposures to chemicals through food chains: An uncertainty analysis. Environ. Sci. Technol. 23(9):1154–1163.

McKone, T.E. and J.I. Daniels. 1991. Estimating human exposure through multiple pathways from air, water, and soil. Reg. Toxicol. Pharmacol. 13(1):36–61.

Morgan, M.G. and M. Henrion. 1990. Uncertainty: A guide to dealing with uncertainty in quantitative risk and policy analysis. Cambridge University Press. New York, New York. 332 pages.

Murray, D.M. and D.E. Burmaster. 1996. Residential air exchange rates in the United States. Empirical and estimated parametric distributions by season and climactic region. Risk Anal. 15(4):459–465.

NRC (National Research Council). 1994. Science and judgment in risk assessment. National Research Council. National Academy Press. Washington DC. 652 pages.

Ott, W. 1981. Computer simulation of human air pollution exposures to carbon monoxide. Paper 81–57.6 presented at the 74th annual meeting of the Air Pollution Control Association. Philadelphia, Pennsylvania.

Ott, W. 1984. Exposure estimates based on computer generated activity patterns. J. Toxicol. Clin. Toxicol. 21:97–128.

Ott, W.R. 1990. A physical explanation of lognormality of pollutant concentrations. JAWMA 40:1378–1383.

Price, P., J. Sample, and R. Strieter. 1991. Determination of less-than-lifetime exposures to point source emissions. Risk Anal. 12(3):367–382.

Price, P.S., C.L. Curry, P.E. Goodrum, M.N. Gray, J.I. McCrodden, N.W. Harrington, H. Carlson-Lynch, and R.E. Keenan. 1996a. Monte Carlo modeling of time-dependent exposures using a microexposure event approach. Risk Anal. 16(3):339–348.

Price, P.S., S.H. Su, J.R. Harrington, and R.E. Keenan. 1996b. Uncertainty and variation in indirect exposure assessments: An analysis of exposure to tetrachlorodibenzo-*p*-dioxin from a beef consumption pathway. Risk Anal. 16(2):263–277.

Robinson, J.P. and J. Thomas. 1991. Time spent in activities, locations, and microenvironments: A California-National Comparison Project report. U.S. Environmental Protection Agency, Environmental Monitoring Systems Laboratory. Las Vegas, Nevada.

RSI (Risk Science Institute). 1998. Aggregate exposure assessment: Risk Science Institute workshop report. International Life Sciences Institute. Washington DC.

Sielken, R.L. Jr. 1996. Comprehensive realism's use of trees and distributions to reflect uncertainty and variability and to overcome the limitations of default assumptions. In: Ricci, P.F. (ed.). Journal of Energy Engineering: New methods to deal with uncertainty and applications in assessing energy systems. In press.

Silverman, B.W. 1986. Density estimation for statistics and data analysis. Chapman and Hall. London, United Kingdom.

Slob, W. 1994. Uncertainty analysis in multiplicative models. Risk Anal. 14(4):571–576.

Taylor, A.C. 1993. Using objective and subjective information to develop distributions for probabilistic exposure assessment. J. Exposure Anal. Environ. Epidemiol. 3(3):285–298.

Tukey, J.W. 1977. Exploratory data analysis. Addison-Wesley. Boston, Massachusetts.

USEPA (U.S. Environmental Protection Agency). 1985a. Methods for assessing exposure to chemical substances. Volume 2: Methods for assessing exposure to chemicals in the ambient environment. EPA-560/5-85/002, NTIS PB86-107067. Office of Toxic Substances, Office of Pesticides and Toxic Substances. Washington DC.

USEPA (U.S. Environmental Protection Agency). 1985b. Methods for assessing exposure to chemical substances. Volume 4: Methods for enumerating and characterizing populations exposed to chemical substances. EPA-560/5-85/004, NTIS PB86-107042. Office of Toxic Substances, Office of Pesticides and Toxic Substances. Washington DC.

USEPA (U.S. Environmental Protection Agency). 1985c. Methods for assessing exposure to chemical substances. Volume 5: Methods for assessing exposure to chemical substances in drinking water. EPA-560/5-85/005, NTIS PB86-1232156. Office of Toxic Substances, Office of Pesticides and Toxic Substances. Washington DC.

USEPA (U.S. Environmental Protection Agency) 1985d. Methodology for characterization of uncertainty in exposure assessments. EPA/600/8-86/009. Office of Toxic Substances, Office of Pesticides and Toxic Substances. Washington DC.

USEPA (U.S. Environmental Protection Agency). 1986a. Methods for assessing exposure to chemical substances. Volume 8: Methods for assessing environmental pathways of food contamination. EPA-560/5-85/008. Office of Toxic Substances, Office of Pesticides and Toxic Substances. Washington DC.

USEPA (U.S. Environmental Protection Agency). 1986b. Guideline on air quality models (revised). EPA-450/2-78/027R. Office of Air Quality Planning and Standards, Office of Air and Radiation. Research Triangle Park, North Carolina.

USEPA (U.S. Environmental Protection Agency). 1987a. Supplement A to the guideline on air quality models (revised). EPA-450/2-78/027R. Office of Air Quality Planning and Standards, Office of Air and Radiation. Research Triangle Park, North Carolina.

USEPA (U.S. Environmental Protection Agency). 1987b. Selection criteria for mathematical models used in exposure assessments: Surface water models. EPA-600/8-87/042. Office of Health and Environmental Assessment, Office of Research and Development. Washington DC.

USEPA (U.S. Environmental Protection Agency). 1988. Selection criteria for mathematical models used in exposure assessments: Groundwater models. EPA-600/8-88/075. Office of Health and Environmental Assessment, Office of Research and Development. Washington DC.

USEPA (U.S. Environmental Protection Agency). 1991a. Selection criteria for mathematical models used in exposure assessment: Atmospheric dispersion models. DRAFT. EPA-600/8-91/038. Office of Health and Environmental Assessment. Office of Research and Development. Washington DC.

USEPA (U.S. Environmental Protection Agency). 1991b. Dermal exposure assessment: Principles and applications. EPA-600/8-91/011A. Office of Health and Environmental Assessment. Office of Research and Development. Washington DC.

USEPA (U.S. Environmental Protection Agency). 1992. Guidelines for exposure assessment. Fed. Reg. 57(104):22888–22938. May 29.

USEPA (U.S. Environmental Protection Agency). 1996. Exposure factors handbook. EPA/600/P-95/002F(a–c), Washington DC.

USEPA (U.S. Environmental Protection Agency). 1997. Guiding Principles for Monte Carlo Analysis. EPA/630/R-97/001. Risk Assessment Forum, Washington, DC.

Wilkinson, L.M. Hill, P. Howe, G. Birkenbeuel, J. Beck, and J. Liu. 1994. SYSTAT for DOS: Using SYSTAT, Version 6. SYSTAT. Inc. Evanston, Illinois.

13

COMMUNICATING ABOUT RESIDENTIAL EXPOSURES

Residential exposures are unique in that they are influenced by residents preferences and practices as well as the characteristics of products and their delivery systems. For these products the traditional risk management team of state and local government officials and industry product specialists should be augmented to include intermediaries such as lawn care professionals and the end use consumer.

This volume is devoted to exposure assessment. Most residents are ultimately concerned with risk hence risk is usually the focus of communication efforts. Separately characterizing exposure has several advantages. Citizens tend to equate exposure to risk rather than considering dose response issues. When asked to agree or disagree with the statement "It's not how much of a chemical I'm exposed to but whether I'm exposed at all," about 80 percent strongly agreed (baseline study). By communicating each component separately in the risk message separately it may impart a clearer understanding of the situation. Another major reason for focusing attention on exposure is that particularly in residential exposures is that improving personal practices may be a major factor in reducing exposure and hence risk.

Because of the broad audience that needs to be involved in information and action about residential exposures, communication is perhaps even more important here than in other environmental issues. Good communication will provide advantages from a public health, economic, and public policy point of view.

RISK MANAGEMENT

Better risk management is the end goal for all the components of environmental management. Risk assessment, risk communication and risk management are linked so that no one component can be considered without considering the others. However, each component has specific attributes and responsibilities. Risk assessment should fully characterize exposures and the toxicological characteristics of the issue. The risk assessor must be careful not to bury policy decisions (*e.g.*, safety factors) in away that their effects on risk estimates are not readily apparent.

The risk manager should carefully evaluate the risk estimates, the distribution of exposures, the policy and public values issues and consider the full range of risk management options. The major risk management tools are education, economic and social incentives or disincentives, and regulatory and legal restrictions. Selecting among the various risk management options, from least coercive to most coercive, depends on the

Residential Exposure Assessment, edited by Baker *et al.*
Kluwer Academic/Plenum Publishers, New York, 2000.

level of risk (risk assessment) and the political and public acceptability of the management options (risk communication).

Communication is pivotal in the process. Individuals cannot change without the knowledge of alternatives and consequences, and without the motivation, skills, and support systems to do so. Even the most stringent laws and regulations cannot function without broad-based public support.

PRINCIPLES AND KEY ELEMENTS OF RISK COMMUNICATION

Establish Clear Goals and Objectives

Understanding the reasons why a risk communication program is needed and setting attainable goals are the preeminent concerns in achieving changes in knowledge, attitudes, and behaviors. The process of setting the objectives will help to identify gaps in the available information about the problem being addressed. Without clear objectives, success cannot be measured. But these objectives should form a road map, not a tunnel, and should not become static. They should be adjusted, based on information from evaluation research through the planning and implementation process.

The goals and objectives should emphasize the comprehensive nature of most successful public health campaigns, including (Flay and Burton 1988):

- An integrated series of communication activities
- Multiple activities and channels
- Targeting specific audience segments
- Long duration, and
- A clear purpose.

An empirical, iterative process is one of the best ways to achieve success. It includes development, implementation, evaluation, and any necessary midcourse corrections of the program (USEPA 1989).

The purpose may be to make people aware of a particular exposure and provide options for personal actions. As the potential public health significance and public consensus increases the goal may include trying to promote protective behaviors. Trying to promote measurement of radon levels and encourage mitigation efforts is an example. These programs have met with mixed success. Another example is the USEPA's Consumer Labeling initiative. The stated goal of this initiative is to enable consumers "to make informed choices and to use these products safely as directed."

The program plan must provide an operational framework for the goals and objectives. The steps to success are easy to enumerate, but can be difficult to achieve:

- Understand the Exposure and the Risk
- Understand audience segment characteristics, including:
  - Demographics
  - Psychographics
  - Information utilization characteristics
  - Perceptions of risks
  - Perceptions of information sources
- Develop and test messages

- Motivate audiences to attend to the messages
- Motivate audiences to want to change behavior
- Transmit skills necessary to achieve behavior change
- Deliver messages effectively
  - Time messages appropriately
  - Select appropriate channels
  - Select appropriate spokespersons
  - Deliver proper dosage
- Reinforce messages and behaviors
  - Encourage interpersonal communication among target audience
  - Encourage broader social change to support individual behavior change
- Measure the effects of the campaign

Understand the Risk

The receptivity of populations and individuals to any message, and ultimately behavior change, is, in part, related to their perception of risk. Individuals and groups must feel at risk before they will commit to or demand change. Many years of research have produced a long list of characteristics that are associated with audiences' perceptions of risks (see Table 13-1). Some examples of characteristics are level of internal control, whether the risk affects children in particular, and whether the risk is voluntary or imposed. Many of the characteristics relate to whether personal action or sacrifice is needed to reduce the risk. Individuals frequently deny risks that require personal action to manage.

The implications for behavior change of this research are complex. The complexity comes from the inability to determine whether the perception reflects a reluctance to make personal changes to reduce risk, or if the failure to act reflects the way the risk is perceived. Regardless of the sequence of events, understanding public perception of risks provides useful information for risk professionals who need to predict the barriers and opportunities for individual and societal change.

For example, cigarette smoking was changed from an individual behavior change issue to a social problem by identifying environmental tobacco smoke as a risk to others. This provided anti-smoking advocates with a reason and a means to put in place strong social deterrents such as no-smoking regulations. Social acceptance of interference with personal freedom is enhanced by a present health risk to others as opposed to a simple nuisance or personal risk imposed voluntarily, even when that risky behavior has economic and social consequences for the community.

Other examples of risks that present unique communication problems are drugs such as Accutane®, which is highly effective in the treatment of severe acne but which causes severe birth defects. Therefore, a strategy to educate health professionals to avoid treating women who might be or become pregnant required a major communications effort. Vaccines present a case where the risk to the individual from the rare but severe side effects has to be traded against the risks to the community in the event of an outbreak of various infectious diseases. Again, communication strategies are key to the public's understanding of the issues involved.

In the environmental arena, problems that Americans believe they, as individuals, can help solve are seen as much less serious. For example, litter and indoor air pollution ranked 16th and 17th in terms of environmental anxieties, with more than seven out of ten people saying that they can "do a lot" about these problems. Solid waste and auto

Table 13-1. Factors affecting the public's evaluation of risk

Catastrophic potential—Fatalities and injuries that are grouped in time and space (*e.g.*, airplane crashes) cause more concern than fatalities and injuries that are scattered or random in time and space (*e.g.*, automobile accidents).	*Attributability*—People are more concerned about risks that are perceived to be due to human actions (*e.g.*, industry accidents) than about risks that are perceived to be natural in origin (*e.g.*, acts of God).
Familiarity—People are more concerned about risks that are unfamiliar (*e.g.*, ozone depletion due to emissions of fluorocarbons) than about risks that are familiar (*e.g.*, household accidents).	*Impact on children*—Activities that are perceived as putting children specifically at risk (*e.g.*, school bus accidents) cause more concern than activities not generally so perceived (*e.g.*, adult smoking).
Understanding—Activities characterized by poorly understood exposure mechanisms or processes (*e.g.*, exposure to radiation) cause more anxiety than those characterized by apparently well understood exposure mechanisms or processes (*e.g.*, pedestrian accidents or slipping on ice).	*Dread*—The public is more concerned about certain dreaded risks that evoke a response of fear, terror, or anxiety (*e.g.*, exposure to potential carcinogens from toxic waste dumps or to nuclear radiation) than about risks that are not especially dreaded (*e.g.*, common colds and household accidents).
Uncertainty—Individuals are more concerned about risks that are scientifically unknown or uncertain (*e.g.*, recombinant DNA experimentation) than about risks that are relatively well-known to science (*e.g.*, actuarially documented automobile accidents).	*Institutional trust*—Situations where the responsible risk management institution is perceived to lack trust and credibility (*e.g.*, criticisms of the Nuclear Regulatory Commission for its perceived close ties to industry) cause more concern than those where the responsible risk management institution is perceived to be trustworthy and credible (*e.g.*, trust in the management of recombinant DNA risks by universities and the National Institutes of Health).
Controllability—Risks that are perceived to be not under individuals' personal control (*e.g.*, traveling as a passenger in an airplane or automobile) engender more concern than risks that are perceived to be under their personal control (*e.g.*, driving an automobile).	*Accident history*—People are more concerned about activities that have a history of major and sometimes minor accidents (*e.g.*, nuclear power plant accidents such as the accident at Three Mile Island) than about activities that have a history of no major or minor accidents (*e.g.*, recombinant DNA experimentation).
Voluntariness of exposure—People are more concerned about risks that they perceive to be involuntary (*e.g.*, exposure to unlabeled food additives or to air or water pollution) than about risks that they perceive to be voluntary (*e.g.*, smoking, sunbathing, or mountain climbing).	*Equity*—People are more concerned about activities that are characterized by a perceived inequitable distribution of risks and benefits (*e.g.*, offshore oil exploration) than about those characterized by a perceived equitable distribution of risks and benefits (*e.g.*, vaccination).
Effects on future generations—People are more concerned about activities that pose risks to future generations (*e.g.*, genetic effects due to exposure to radiation) than about risks that do not pose risks to future generations (*e.g.*, skiing accidents).	*Clarity of benefits*—People are more concerned about hazardous activities that are perceived to have unclear benefits (*e.g.*, nuclear power generation) than about hazardous activities that are perceived to have clear benefits (automobile driving).
Victim identity—People are more concerned about risks to identifiable victims (*e.g.*, a yachtsman lost at sea or a trapped coal miner) than about risks to statistical victims (*e.g.*, statistical profiles of automobile accident victims).	*Personal stake*—People are more concerned about activities that they believe place them (or their families) personally and directly at risk (*e.g.*, drinking water contamination due to local dumping of hazardous waste) than about activities that do not place them (or their families) personally and directly at risk (*e.g.*, dumping of hazardous waste at sea or in other remote sites).
Reversibility—People are more concerned about activities characterized by potentially irreversible adverse effects (*e.g.*, acid rain) than about those characterized by reversible adverse effects (*e.g.*, injuries from sports or household accidents).	

Adapted from Covello and Merkhofer (1992). Based on Slovic *et al.* (1980), Vlek and Stallen (1981), Litai, *et al.* (1983), Renn (1981).

exhaust, which are largely caused by consumer behavior, ranked eighth and tenth, respectively. However, fewer than half the people felt they could do "a lot" about these problems (Roper Organization 1990). Denial of personal risk is a key element in noncompliance with drug therapy and failure to modify lifestyles to reduce risk factors for chronic or infectious diseases. One conclusion to draw from his is that people adjust their perception of risk and personal control to reflect their willingness to act.

Understand Target Audiences

Target audiences need to be characterized in terms of both perception of risks and the specific leverage points that may be effective in motivating changes in attitudes or behavior. In addition, credible sources and effective channels need to be identified. Some of these may vary by the type of message as well as by the type of audience. Standard audience segmentation techniques such as focus groups, central intercept interviews, risk perception surveys, and psychometric studies can be used to investigate audience characteristics (USEPA 1989). Empirical descriptive studies, both qualitative and quantitative, can guide strategy development in the same way that epidemiology guides clinical research and practice. But as with pharmaceuticals, the effects of information on individuals and populations need to be tested and monitored for effectiveness and side effects.

One technique that has been particularly effective in identifying the kind of information that will be most influential has been the use of indepth interviews to characterize how subjects are making decisions (Bastrom *et al.* 1992). By comparing the lay decision model with the expert model for evaluating the same situation, the missing information and the timing of the decision can be identified. At least one study on radon exposure suggests that communications interventions using this approach are more effective in influencing decisions. This approach has been dubbed by some as "mental models," and obviously assumes cognitive processing of the information.

Determining whether audiences are processing information cognitively (central processing) or simply responding to external cues, such as source credibility (peripheral processing), has a strong influence on the message approach. Personal relevance is a major determinant of central versus peripheral processing (Hammond 1990).

The amount and kind of information in which a particular target audience is interested is an important variable. For example, the amount of information patients desire about prescription medications varies; some only want basic instruction, while others want a full discussion of various alternatives and effects (McCallum 1989a). Hence, characterizing audiences as to their receptivity for information is important in planning communication strategies.

Finally, some audience segments are subject specific. Careful evaluation of the differences in a target audience's response by subject area for both messages and spokespersons is necessary. For example, target audience response to messages about marijuana can be quite different from those about cocaine. The decision of where to target generic messages, and where to target specific messages, can influence the program's effectiveness.

Develop Effective, Tested Messages

Due to the two-way nature of risk communication, effective messages about risks must contain the information the audience wants, as well as the information the program designer wishes to convey. In addition, the message should clarify information that audiences might otherwise misunderstand and show sensitivity to the emotions, concerns, values, *etc.*, of the audiences. Also, messages must be tested to determine how effective they are likely to be. Standard message-testing techniques include focus groups, reading level evaluation, and central intercept interviews, as well as theater testing (USEPA 1989). Choosing a testing method depends on the size of the investment in the communication effort and judgement about the consequences of delivering an ineffective message or the wrong message.

The language in the message has emotional and political impact, as well as cognitive value. Using "dump" rather than "landfill," or "medicine" as opposed to "drug," elicits very different responses. The relationship between the perception of illicit versus ethical drugs is an interesting example. In calling for clarity in messages about psychoactive substances, some experts have suggested that there is a continuum of drugs that society uses, ranging from prescription drugs to over-the-counter drugs to legal and socially accepted drugs, such as alcohol, to illicit drugs such as marijuana, cocaine, and others including illicit prescription drugs. What is the interaction between benefit messages about prescription and over-the-counter medicines (for example, take a pill to stop the pain), and risk messages about illicit drugs (as in "This is your brain on drugs . . ." as the egg fries in the skillet)? (McCallum 1989a). While there is little evidence that the public is confused about these differences, and illicit and ethical drugs may not be perceived as a continuum, the selection of words does affect perceptions. In one focus group study, risks were evaluated as greater when prescription products were referred to as "drugs" than when the word "medicine" was used. When prescription products were referred to as medicines, the benefits were rated as greater. There may be merit in developing more clarity in the words used in these arenas (McCallum 1989b).

Motivational appeals, both to gain attention and to change behavior, range from humor to fear to role models of people we would like to be and, hence, the appeal of slice-of-life commercials. Fear has been one of the more controversial options. Can we scare people into healthful behavior or into avoiding risks? Fear is an emotion which also has a cognitive component. Fear arousal does create a state that can lead to behavior change. Whether the message is processed or denied, or whether the potential thrill of the activity is enhanced, depends on how it is incorporated.

Some guidance for the use of fear appeals have been developed (Table 13-2) (NHLBI 1987).

In addition to the motivational component of the message, it is necessary to transfer to the audience the skills needed to act in response to the message. Depending on the particular situation, this may be a straightforward instruction (such as where to seek services) or, at a later stage in the process, it may be a complex set of behaviors for self-care.

Linking the motivation to an action in a *timely* manner is where programs can fail or succeed. In an asbestos notification campaign for shipyard workers, awareness of this hazardous exposure was linked to the need to take a series of protective action steps (*e.g.*, quit smoking) to reduce a synergistic risk (Freimuth and Van Nevel 1981). Linking these messages provided a risk reduction strategy for those whose anxiety level was raised by information that they had been exposed to a cancer-causing substance. In another case, the lack of access to drug treatment services has deterred the effectiveness of substance abuse communications programs (The White House 1989). Timing and consumer convenience are the most critical issues. At a diabetes control meeting, one hospital reported that it was able to eliminate emergency admissions due to improper insulin dosages by implementing a 24-hour hotline so that patients could get their questions answered when *they* needed the information.

Table 13-2. Use of fear appeals

✓ Realize that individuals may react differently	➪ Research among target audiences is essential
✓ Choose a credible spokesperson	➪ Audiences may discount messages from noncredible sources
✓ Make the message as personally relevant as possible	➪ Audiences must feel vulnerable
✓ Encourage a sense of personal empowerment	➪ Audiences must believe that they can protect themselves
✓ Aim for immediacy in the message	➪ Audiences respond more to "here and now" consequences
✓ Be aware of the overall effect of the message	➪ Fear of the problem (*e.g.*, AIDS) should not overpower the message about the solution (*e.g.*, safe sex)
✓ Use appropriate channels for fear messages	➪ Interpersonal channels (*e.g.*, doctor-patient) may be better than the media
✓ Consider the effect of fear appeals on other health influentials (health practitioners and policy makers)	➪ Fear appeals may drive the policy process
✓ Realize that fear messages have a significant wear-out factor	➪ Fear appeals may drive the policy process
✓ Fear may be more effective in motivating avoidance of risk	➪ Fear may not be best for reinforcing positive behavior

Source: NHLBI 1987.

Spokespersons

Credible spokespersons are important to the delivery of any message. Perception of credibility is a complex commodity that includes caring, competence, and dedication. In a tracking study of six communities across the United States, respondents were asked about how much they used various sources of information, how knowledgeable these sources were, and how much they trusted the sources. These data(see Table 13-3) show the independence of trust and perceived competence. Doctors were the most trusted and were considered somewhat knowledgeable, while industry officials were the least trusted but were considered the most knowledgeable (USEPA 1993). Between 1988 and 1992, there was a decrease in trust of authoritative sources overall, but their relative positions remained the same.

Validation of the importance of health or environmental information by a physician or other credible authority can help to gain the attention of an individual, even if the actual information is given by a nurse or health educator. For example, in programs to increase compliance with drug therapy, having the physician say "Tom (or Pam) is going to give you some important information about your medicine," improves the effectiveness of the education.

Health professionals are particularly trusted sources of information on health and environmental issues. However, the physician may miscommunicate and reinforce misconception. An environmental example illustrates a possible effect of this discrepancy in perception (McCallum 1992). In 1988, a survey of the public questioned respondents about five common health problems they might have experienced in the previous month: headaches, nausea, irritation of the eye, nose, or throat, shortness of breath, and skin rashes. They were then queried about whether they thought the symptoms were related to chemicals in the environment, and whether they went to a doctor for treatment. After

Table 13-3. Perception of information sources: comparison of information, trust, and knowledge

	Percent indicating					
	Amount of information received is a lot		A lot of trust		Very knowledgeable	
	1988	1992	1988	1992	1988	1992
News reporters	27	25	27	18	17	12
Environmental groups	21	16	40	28	53	44
Friends/relatives	7	9	34	32	9	11
LEPC	6	6	28	18	33	24
State government	6	6	12	8	29	32
Local government	5	5	11	5	22	12
Federal government	4	6	12	8	36	38
Chemical industry officials	3	4	8	6	58	62
Doctors	3	8	46	28	27	23
USEPA	—	9	—	22	—	52

the respondents' symptoms were reviewed, a strong relationship was found between consulting a doctor and viewing the environment as the cause of the problem ($X^2 = 141$, df = 1, $p < .001$). Approximately 41 percent of respondents who consulted a doctor about symptoms believed the environment was the cause of the problem, compared with 19 percept of those who experienced symptoms but did not consult a doctor.

In 1989, a survey looked at the physician's view. Physicians were queried about patients seen in the previous week with headaches, nausea, skin rashes, eye, nose, and throat irritation, and shortness of breath, and about the relationship of these symptoms to chemicals in the air, water, or soil. With skin rashes, for example, the physicians had seen 1,741 patients with skin rashes in the preceding week, an average of three patients per doctor. The physicians independently attributed five percent to chemical exposure and reported that an additional 11 percent of the patients had told them the symptoms were environmentally induced. The doctors reported concurring with the patients' attribution 80 percent of the time (McCallum 1992). Each of the other symptoms produced about six patients per week. Shortness of breath, and eye, nose, and throat irritation were attributed to the environment for about 14 percent of cases—ten percent by patients and about four percent independently by doctors. Again, the physicians concurred with the patients' attribution to chemicals in over 80 percent of cases. Nausea and headaches yielded lower chemical attribution rates by doctors and lower concurrence of doctor with patient. In focus groups conducted in conducted in conjunction with the survey, results suggest that the high rate of concurrence with the patient's attribution to environment is not based on a careful review of the exposure history, but is a casual response to the patient's concern.

Because doctors are viewed as the most credible authorities on health risks, including environmental risks, a physician's attribution of a symptom to the environment can have a great impact on the patient. The research suggests that many physicians do not understand the importance that the public gives to the physician's role in these matters. They seem to treat a patient's questions and comments about environmental health risks as general conversation, and do not see their own comments as the authoritative medical statements the patient sees them to be. So, in many cases, physicians allow their patients

Table 13-4. Conflicting priorities between media and public health objectives

Mass media objective	Public health objectives
To entertain or inform	To educate
To cover short-term events	To conduct long-term campaigns
To deliver salient pieces of information	To create understanding of complex information
To reflect society	To change society
To address personal concerns	To address societal concerns
To make a profit	To improve public health

Source: Adapted from USDHHS 1991.

to go on blaming the environment for symptoms that are probably linked to tobacco or other lifestyle choices. In focus groups physicians were not interested in being educated. They wanted information about local problems about which their patients might ask specific questions (Santos and McCallum 1993). In programs to promote healthier lifestyles, it has been important to give physicians easy-to-use tools. It is important to give them clinically useful environmental information, as well.

Channels

Mass media is the primary channel through which health risk information gets to most members of the general public. The media and public health professionals share the same clients—the public—but they have conflicting priorities (see Table 13-4).

The mass media transmit health information through news coverage and features such as talk shows. The messages may convey information that is life-saving or misleading or alarmist. Health education and health reporting have conflicting goals. The need for repetition for education is the antithesis of news. A particular health issue can graduate from occasional to frequent coverage, in part because the public wants it, but it has to be packaged as news (see Table 13-5).

Entertainment programs display stories involving health and environmental issues. The behavior of actors in films and television can provide role models for behavior. Television has become our major storyteller and a prime agent of socialization. The images are sometimes positive, but are more frequently negative.

Advertising, both commercial and public service, also influences our attitudes and behaviors about health and the environment. The placement of public service advertising has become more difficult, and commercial messages have increasing numbers of health or environmental claims. Two major problems with commercial advertising, and in some cases with nonprofit groups, are:

Table 13-5. How health becomes news

- Human Interest (*e.g.*, Rock Hudson's suffering from AIDS)
- Different events that are related to some problem (*e.g.*, Betty Ford's drug dependency, death of River Phoenix)
- Patience to keep the media's attention
- Personal contacts with journalists
- Local news has more time to fill

Source: USDHHS 1991.

Table 13-6. Health images in entertainment programming

• Physical illness is romanticized
• Mental illness is villainized
• Physicians are powerful
• Most health professionals are absent
• Violence is frequently without consequences
• Snacking and drinking alcohol are responses to stress
• Overweight people are rare
• Sex is adolescent and titillating. Little is done to prevent pregnancy or sexually transmitted diseases

Source: USDHHS 1991.

- Misleading claims, and
- Advertising to vulnerable populations such as teenagers who may be more influenced by cigarettes, beer, and other advertising.

Better consumer acceptance of companies who take environmentally positive actions and market "green" products can be an incentive for risk reduction. Standards for claims and ways to avoid manipulative messages is necessary to make these incentives work and to increase positive public response.

Pointing out the ways in which some popular media reinforce negative images or behaviors frequently leads advocates of social change to call for "guidelines" (see Table 13-6). The media in general mirror rather than map society (Klaidman 1991). The media may be an indicator of more than an influence on society's norms. The process of encouraging "positive" behavior must recognize the need to bypass the media, as well as use it. Also, as society becomes more and more multimedia, the definition of literature needs to be expanded to include all forms. Schools and families need to help future generations understand the symbols and become active rather than passive users of these information sources.

Reinforce Messages and Behaviors

Behavioral change is difficult. To be successful, a campaign has to repeat the message over and over. Further, the person making the behavior change has to get positive feedback both from himself and from society.

Using interpersonal channels in addition to the media is usually more effective than using either channel alone. As more and more people adopt the behavior, it becomes the normative behavior for that group. The change in attitude towards cigarette smoking took a long time—almost 30 years—but it happened. One major factor was the commitment to keeping the health implications of smoking in front of the public over the years. Another was identifying smoking as a hazard to others.

The need for sustained effort is further illustrated by the recent increase in drug use by high school seniors (Washington Post 1994). The report on a survey by the University of Michigan shows an increase in marijuana and tobacco and a decrease in alcohol use. One research report suggests that each new generation overcomes its naivete about drug use. Also, recent campaigns against underage drinking may have resulted in some substitution of other drugs. Other campaigns that have faltered because of complacency about the gains being made include tuberculosis and cigarette smoking among women.

EVALUATION

Evaluation should be part of any communication effort; without it, there is no way to measure progress or to identify the need to make timely changes. Evaluation can be an effective tool, making programs more efficient and effective.

Problems and difficulties arise in the evaluation of risk communication activities due to conflicts and misunderstandings of values, goals, resources, and usefulness. Because the process of evaluation is a value-laden endeavor, significant practical and policy implications are likely to occur (USEPA 1991).

Goals and Values

A difficulty in evaluating risk communication programs is the challenge of identifying appropriate goals (*i.e.*, to raise awareness of an issue, to bring about action, to increase participation). Only when a program's goals and expected effects are clearly defined can there be significant evaluation. However, for many programs this level of specificity is difficult, if not impossible. In addition, evaluators and program implementors often disagree on what the goals of the program are to begin with, as well as which goals should be evaluated.

Communication goals should be considered and built into the risk management program from the beginning. Without clear goals, it is impossible to measure the success or failure of any program.

Any risk communication program has many stakeholders—including government agencies, public interest groups, industry groups, and individuals—who are interested in the conduct and effectiveness of the program. Different audiences have different perspectives, different goals and values, need different types of information, and different risk communication activities require different types of evaluation studies.

For example, designated driver campaigns have been successful in reducing alcohol-related accidents, but some feel that it implies acceptance of alcohol abuse or underage drinking. In the quest to stop the spread of AIDS, needle exchange programs and explicit campaigns have sparked similar values-related controversy. Thus, communication efforts could be called successful or unsuccessful depending on the goals. When the results of the evaluation are used to inform future program plans and dissemination efforts, planners need to ask questions such as: What was the original goal of the project? Has a particular ideology shaped the project or the evaluation? What values framework will be used in the evaluation and interpretation of the results? Are there competing points of view that need to be considered?

RESOURCES

Ongoing evaluation of a program is important. With early, and continuing, evaluation, implementors are able to get feedback on the effectiveness of the program. If the activities are not working as intended, there is time to reevaluate the program and make corrections which, in the long run, may save time and money.

In spite of the benefits of ongoing evaluation of risk communication activities, it is often disregarded, especially if no evaluation has been planned or budgeted for in advance.

Managers are often reluctant to evaluate programs. Many managers believe that it is an expensive process that calls for skills beyond those represented in the organization. Some are unwilling to support research that might show that the program is not working as intended. In addition, program managers often allocate all the available resources to the program activities, without considering if the messages are being communicated and are accomplishing the desired purposes.

Some strategies to bring about increased attention to, and usage of, evaluation research in risk communication programs are:

- Integrate evaluation activities into the program from the beginning
- Use quick, simple, and informal evaluation methods, when appropriate
- Develop incentives for planners to fund evaluation activities, in order to better understand the effectiveness of the program
- Encourage managers to develop clear evaluation plans, and
- Encourage implementors to record and share successes.

❑

COMMUNICATING ABOUT RESIDENTIAL HAZARDS A CASE STUDY OF LAWN CARE CHEMICALS

Understanding consumers is key to estimating exposures and developing strategies for communicating information about potential hazards and protective strategies. A series of focus groups was held in Baltimore, Maryland and Boston, Massachusetts by FOCUS GROUP—Risk Communication Consultants. Three groups were conducted in each city: one with homeowners using professional lawn care services and two with homeowners who take care of their lawns themselves.

The study was funded by the Professional Lawn Care Associate of America (PLCAA) to examine the effectiveness of posting requirements for the application of lawn care pesticides and chemicals. PLCAA was interested in determining whether proposed additional homeowner notifications were needed. Specifically, the study was conducted to determine: (1) whether residential homeowners took notice of the posting signs, (2) perceptions, attitudes and feelings associated with the use of lawn care pesticides and chemicals, (3) what knowledge and perceptions they had towards the posting and other informational materials, and (4) whether homeowners had any specific information needs.

The findings from these groups provide insight into consumer/homeowner awareness, perceptions, and attitudes regarding potential residential hazards. This case study sheds insight into the effectiveness of labeling and communication materials on homeowner behavior and knowledge regarding specific lawn-care pesticides and chemicals as a residential hazard.

AWARENESS OF SIGNS

Posted signs about pesticide applications were widely recalled. Participants knew that the signs signified the use of chemicals and meant that people and pets should stay off the lawn. Customers and noncustomers both seemed to be equally aware of the signs, with noncustomers tending to mention the signs earlier in their discussions than did customers. While people said that the signs "are small," they could generally recall that the signs contains a company name and a date. There was little awareness about how long the signs are supposed to stay up. Customers indicated that they left signs up until "the next time I mow" or "until it blows away."

Most people were also aware that "a requirement" exists for posting the signs. There was little evidence that anyone had been alarmed by the signs, or had sought more information from a homeowner whose yard was posted or from a service after posting took place. People who do not use a service voiced more concern about what risks the signs imply than people who used a services, but no one said he or she had ever asked a question or objected to a neighbor about signs or service. Among noncustomers, some participants were concerned about the fact that "children and animals don't read" or that "kids don't exercise caution." Some noncustomers thought the postings were required only for "sprayed" chemicals. Neither customers nor noncustomers had ever actively sought additional information on lawn care products, application safety, health risks or effects, or any other aspects of product use.

ATTITUDES AND PERCEPTIONS REGARDING EXPOSURE, HEALTH, AND SAFETY

Most participants, including both customers of lawn care services and noncustomers, seemed tolerant of the risks associated with lawn care (*e.g.*, recall of the signs and awareness of their meaning had not prompted inquiries or complaints). For customers, the benefits of using a lawn care service, (*e.g.*, convenience and appearance results) far outweighed risks.

When concerns were expressed about lawn care products, the primary concern was health effects. Environmental impacts (*e.g.*, effects on groundwater did not seem to be a concern among customers or noncustomers. Homeowners who do not use professional services were more likely to express concern about the safety of the lawn care chemicals than those who used the service. It was not possible to determine from this study whether the difference in concern prompted using the service or whether, once a decision to use a service is made, the voluntary nature of the decision reduces risk perception.

Many of the nonusers of lawn care services indicated the primary reasons were concern over the use of chemicals (health and safety concerns) and economic reasons. "Chemicals" were often cited as a reason for not using a service or for discontinuing a service previously used. People with young children and/or pets were more likely to raise concerns about the effect of pesticides on children and pets or to indicate that they had stopped using, or decided against using, a service due to perceived health risks and safety of exposure for children and pets.

People who used lawn care services were more likely (than noncustomers) to believe professional products are different and can only be purchased and use by "trained professionals" with a license. They also tended to believe that the services' chemicals are safer if applied by professionals. There was also a general perception that granular products are safer than chemicals that are sprayed.

Interestingly, homeowners who do not use services but did apply materials themselves to their lawns thought that the chemicals they buy are safer than those the services use, primarily because they are "probably less concentrated." It never occurred to them to post a sign after they personally apply something to their lawn—or to ask their neighbors to post such a sign. Some homeowners also seemed unaware that the products they use are chemicals. For example, one homeowner was adamant about not using chemicals on her lawn and was amazed that the 10-10-10 she applies contains chemicals. Another common perception was that only weed control products are chemicals, not lime and fertilizers.

Some customers seemed to think that fertilizers are better than pesticides because fertilizers "make things grow" while pesticides "kill things". Others seemed aware that "fertilizers are damaging the water." There was less familiarity with "herbicides" than other pesticides. Initially, homeowners believed organic pesticides are safer. However, very few people thought that they were worth paying more for. People given definitions of organic chemicals and natural chemicals to read regarded most often that this information was interesting and new to them. People assumed that "organic" meant "natural."

There was a perception among some that professionally applied chemicals are safer "because they are regulated." This belief was not widespread. Some participants questioned the value or credibility of regulation. People were concerned about the training and expertise of the people who apply lawn chemicals. Homeowners who do not use professional services were more likely to doubt the credibility of the professionals' training and expertise, citing incidents of "burned" shrubs and other mistakes by professionals, the young age of the workers they see ("just college kids"), or the transient nature of the workers. They were also more likely to mention that they regularly see workers without gloves or any other apparent protective equipment, which suggested "poor training" to them.

ADEQUACY OF INFORMATION

In general, homeowners expressed little desire for information or any motivation for seeking information about the products used in lawn care. Most were not knowledgeable about the chemical content of specific products. Homeowners seemed generally satisfied with the amount of information they have, or feel they could obtain about the chemicals used on their lawns or in their neighborhoods. When queried during the discussion, participants *said* that they wanted more information, but their reported behavior suggested otherwise.

Participants appreciated and were reassured by the information left behind by their services about what had been done to the lawn after a service call. However, there was little evidence that homeowners actually read the information or had concerns about what had been applied to their lawn. Customers of lawn care services were generally satisfied that the services supply adequate information and that more is available if they want to call. None of the law care service customers reported ever being questioned by neighbors or anyone else about what posting signs mean or what had been applied to the lawn.

While some people mentioned an interest in having more information, their behaviors seem ed to indicate otherwise. Only a few people took the PLCAA brochure or the state information with them when offered to do so. None of the participants indicated any intent to seek more information or to take other action. There was some general interest in having a toll-free telephone number to call for information. People with lawn care services thought they would just call their services if they wanted information.

SOURCES OF INFORMATION

In some States there is also state-mandated information. In the Baltimore customer group, participants were asked to review the Maryland state-mandated information. Participants seemed to resist reading it and felt it was too complicated. In one of the Boston noncustomer groups, participants felt the state information was "more credible" and easier to read. Interestingly, participants' comments of liking the brevity of the supplemental state information were inconsistent with earlier statements claiming a desire for more in-depth information. The USEPA's review of the voluntary PLCAA brochure did not improve credibility of the materials provide. In fact, the USEPA's disclaimer of endorsement raised concerns about inadequate regulation of chemical usage.

CONCLUSIONS AND IMPLICATIONS

These qualitative studies provide some insights into the effectiveness of existing communication strategies, some opportunities for new ones, and valuable information for guiding data collection for future exposure assessments.

Posting has been an effective communication strategy for achieving the desired action of avoiding treated lawns. Both customers of professional lawn care services and noncustomers are familiar with posted signs about pesticide applications. People have observed the signs and understand that they mean people and pets should stay off the treated lawn for a period of time. It is important to note that posting has not caused undue alarm by either customers or noncustomers. It may be appropriate to consider literature for customers explaining the purpose of posting and ensure that customers are provided with a contact number for obtaining additional information.

In general, there are inconsistent perceptions regarding the residential risk of exposure to lawn care chemicals. Customers understate the risk of professionally applied chemicals while noncustomers underestimate the risk of products that they purchase and apply themselves. Participants (including both noncustomers and customers who personally apply products) were not very knowledgeable about or attentive to safe practices with the products they apply themselves. This suggests that there is a decrement in exposure to be achieved through better consumer education.

The results of this study demonstrate that use patterns vary considerably across individuals, suggesting the appropriateness of distributional analysis and the need for more quantitative data. The misunderstanding about terms such as organic and what a chemical is argues for very careful wording and test questions for any survey to avoid misclassification of responses.

REFERENCES

Bastrom, A., B. Fischoff, and M.G. Morgan. 1992. Characterizing mental models of hazardous processes. J. Social Issues. 48:85–100.

Flay, B.R. and D. Burton. 1988. Effective mass communication campaigns for public health. In: Mass media and health: Complexities and conflicts. Rancho Mirage, California. September 17–19, 1988.

Freimuth, V.S. and J.P. Van Nevel. 1981. Reaching the public: The asbestos awareness campaign. J. Communications 31:155–167.

Hammond, S. 1990. Issue involvement and the elaboration likelihood model: A field experiment using environmental risk messages. University of Maryland. College Park, Maryland.

Klaidman, S. 1991. Health in the headlines: The stories behind the stories. Oxford University Press. New York, New York.

McCallum, D.B. 1989a. Semantics of prescription drugs. Videotape program on risk communication. Institute for Health Policy Analysis. Georgetown University Medical Center. Washington DC.

McCallum, D.B. 1989b. Communicating the benefits and risks of prescription drugs. With S. Yenney. Institute for Health Policy Analysis. Georgetown University Medical Center. Washington DC.

McCallum, D.B. 1992. Physicians and environmental risk communications. Health and Environmental Digest April 1992. Pages 3–5.

NHLBI (National Heart, Lung, and Blood Institute). 1987. Seminar results: Fear as a persuasion technique. Unpublished proceedings of a conference on Fear Communications. December 1987.

Roper Organization. 1990. The environment: Public attitudes and individual behaviors. Report commissioned by S.C. Johnson & Son, Inc. July 1990.

Santos, S.L. and D.B. McCallum. 1993. Kanawha Valley health effects study: Risk communication research project. Focus group and key interview project. Focus Group. Medford, Massachusetts. April.

The White House. 1989. National Drug Control Strategy. Washington DC.

USEPA (U.S. Environmental Protection Agency). 1989. Toxic and hazardous substances, Title III and communities. An outreach manual for community groups. C. McNeil, E.B. Arkin, and D.B. McCallum.

USEPA (U.S. Environmental Protection Agency). 1991. Evaluation of effective risk communications. In: A. Fisher, M. Pavlova, and V. Covello (eds.). Proceedings of a workshop. January 1991. Interagency Task Force on Environmental Cancer and Heart and Lung Disease. Report number EPA/600/9-90/054. Washington D.

USEPA (U.S. Environmental Protection Agency). 1993. Public knowledge and perceptions of chemical risks in six communities: Follow up survey results. Report by D.B. McCallum and S.L. Santos.

Washington Post. 1994. Illicit drug use rises among U.S. teenagers. Pierre Thomas. February 1, 1994. Page A-1.

Appendix A

GLOSSARY

Definitions for many of the terms in this glossary are drawn from CMA Exposure Glossary, USEPA Exposure Factors Handbook of 1997, and the USEPA Exposure Assessment Guidelines of 1992.

Absorption—The penetration of a substance through a barrier (*e.g.*, the skin, gut, or lungs).

Absorbed Dose—See "Internal Dose".

Absorption Factor—The fraction of an agent (*e.g.*, a chemical) making contact with an organism that is systemically distributed in the organism.

Acceptable Daily Intake (ADI)—Same as "Reference Dose". Estimate of largest amount of exposure to an agent on a daily basis not expected to result in adverse effects.

Acute Exposure—An exposure of short duration and/or rapid onset. An acute toxic effect is one that develops during or shortly after an acute exposure to a toxic substance.

Administered Dose—The amount of substance given to a test subject (human or animal) in determining dose-response relationships, especially through ingestion or inhalation. In exposure assessment, since exposure to agents is usually inadvertent, this quantity is called "Potential Dose".

Agent—A chemical, physical, mineralogical, or biological entity that may cause deleterious effects in an organism after the organism is exposed to it.

Applied Dose—The amount of a substance in contact with the primary absorption boundaries of an organism (*e.g.*, skin, lung, gastrointestinal tract) and available for absorption.

Average Daily Dose (ADD)—The average dose received on any given day during a period of exposure, expressed in mg/kg body weight/day. Ordinarily used in assessing noncancer risks.

Bioavailability—The state of being capable of being absorbed and available to interact with the metabolic processes of an organism. Bioavailability is typically a function of properties of agents, physical state of the material to which an organism is exposed, and the ability of the individual organism to physiologically take up the agent.

Biologically Effective Dose—The amount of a deposited or absorbed agent that reaches the cells or target site where an adverse effect occurs, or where that agent interacts with a membrane surface.

Body Burden—The amount of a substance stored in the body a particular time, in the body as a result of exposure. Body burdens can be the result of long term or short term storage, for example, the amount of metal in bone, the amount of a lipophilic substance such as PCB is adipose tissue, or the amount of carbon monoxide (as carboxyhemoglobin) in the blood.

Bootstrap—Bootstrap analysis is a technique of replicated resampling (usually by computer) or an original data set for estimating standard errors, biases, confidence intervals, and other measures of statistical accuracy. It can automatically produce accuracy estimates in almost any situation, including very complicated ones, without requiring the statistician to make doubtful assumptions about the original distribution.

Bounding Estimate—An estimate of exposure, dose, or risk that is higher than that incurred by the person in the population with the highest exposure, dose or risk. Bounding estimates are useful in developing statements that exposures, doses, or risks are "not greater than" the estimated value.

Chronic Exposure—A persistent, recurring, or long-term exposure, as distinguished from acute. Chronic exposure may result in health effects (such as cancer) that are delayed in onset, occurring long after exposure has ceased.

Delta Method—The delta method, or infinitesimal jackknife, is an approach for assessing standard errors and influence functions that can be applied to bootstrap statistics. In some situations, it may be more efficient than jackknife.

Dimensionality—Along with temporal pattern release, one of the two terms used to describe emission sources qualitatively. Dimensionality defines the sources of air pollution as particular points in space. There are three basic types: point sources, such as smokestacks; line sources, such as cars moving along a road; or area sources, such as a group or residences burning wood for heat.

Direct Exposure—Exposure to a subject who comes into contact with an agent via the medium in which it was initially released to the environment. Examples include exposures mediated by cosmetics, other consumer products, some food and beverage additives, medical devices, over-the-counter drugs, and single-medium environmental exposures.

Dose—The amount of substance available for interaction with metabolic processes or biologically significant receptors after crossing the outer boundary of an organism (see potential dose, applied dose, and internal dose definitions). The amount of the agent available for interaction by any particular organ or cell is termed the delivered dose for that organ or cell.

Dose Rate—Dose per unit time, or example in mg/day, sometimes also called dosage. Dose rates are often expressed on a per-unit-body-weight basis, yielding units such as mg/kg/day (mg/kg-day). They are also expressed as averages over some time period (for example, a lifetime).

Dose Response Assessment—The determination of the relationship between the magnitude of potential, applied, or internal dose and a specific biological response. Response can be expressed as measured or observed incidence, percent response in groups of subjects (or populations), or the probability of occurrence or a response in a population.

Dose Response Curve—A graphical representation of the quantitative relationship between potential, applied or internal dose of an agent, and a specific biological response to that agent.

Duration of Exposure—Toxicologically, there are three categories describing duration of exposure: acute (one-time), subchronic (repeated, for a fraction of a lifetime), and chronic (repeated, for nearly a lifetime).

Exposure—Contact of a chemical, physical, or biological agent with the outer boundary of an organism. Exposure is quantified as the concentration of the agent in the medium in contact integrated over the time duration of that contact.

Exposure Assessment—The determination or estimation (qualitative or quantitative) or the magnitude, frequency, duration, and route of exposure.

Exposure Limits—Suggested or mandatory restrictions placed upon exposure typically established to assure that exposure below these levels will result in no or minimal adverse effects.

Exposure Monitoring—To keep track of, regulate, or control exposure (see Exposure). Frequently conducted relative to a reference exposure limit. See Exposure Limits).

Exposure Pathway—The physical course an agent takes from the source to the organism exposed.

Exposure Route—The way an agent enters an organism after contact (*e.g.*, by ingestion, or dermal absorption).

Exposure Scenario—A set of facts, assumptions, and inferences about how exposure takes place that aids the exposure assessor in evaluating, estimating, or quantifying exposures.

Extrapolation—Estimation of unknown values by extending or projecting from known values.

Frequency of Exposure—The number of times an exposure occurs in a given period; exposure may be continuous, discontinuous but regular (*e.g.*, once daily), or intermittent (*e.g.*, less than daily, with no standard quantitative definition).

Hand-To-Mouth Transfer—Transfer of agent from a surface to hand to mouth. This pathway is significant for infants and toddlers who have frequent hand-to-mouth contact.

High-End Exposure (Dose) Estimate—A plausible estimate of individual exposure or dose for those persons at the upper end of an exposure or dose distribution, conceptually above the 90th percentile, but not higher than the individual in the population who has the highest exposure or dose.

Human Equivalent Dose—A dose that, when administered to humans, produces an effect equal to that produced by a dose in animals.

Incidental Ingestion—Unintentional intake of small amounts of agents, particularly associated with children's from hand-to-mouth activity.

Indirect Exposure—Often defined as an exposure involving multimedia transport of agents from source to exposed individual. Examples include exposures to chemicals deposited onto soils from the air, chemicals released into the ground water beneath a hazardous waste site, or consumption of fruits or vegetables with pesticide residues.

Infectious Agent—An agent (always a living organism or virus) that can cause infection in an appropriate host.

Infectious Dose—The amount of an infectious agent that must reach an appropriate site in a susceptible host before infection is likely to occur.

Intake—The amount of contact with a medium containing an agent. Used for estimating dose received from a particular medium.

Internal Dose—The amount of a substance penetrating across the absorption barriers (the exchange boundaries) of an organism, via either physical or biological processes. This term is often use synonymously with absorbed dose.

Interpolation—The estimation of values of a function (*e.g.*, risk, effect) between known values or entities based upon existing information.

Jackknife—As applied to bootstrapping, jackknife is a statistical technique to compute the standard error of the bootstrap estimates of statistical parameters. Further, jackknife can provide influence functions with allow, for example, the influence of any one of the original data points on the length and shape of a bootstrap confidence interval to be estimated. The jackknife estimates of the bootstrap error can usually be computed by rearrangement of the original bootstrap calculations without further resampling.

Latin Hypercube—As applied to Monte Carlo analysis, a technique for increasing the number of samples taken from the tails of the distribution and thus reducing the number of iterations required to get adequate definitions of the extreme of the distribution.

Less-Than-Lifetime Exposure—For exposures of limited duration, the time-in-life during which exposure took place may be important in an exposure assessment. For example, for a teratogenic agent (one causing birth defects) to exert its effects, exposure must occur while the subject is *in utero*.

Lifetime Average Daily Dose (LADD)—Total dose received over a lifetime multiplied by fraction of lifetime during which exposure occurs, expressed in mg/kg body weight/day. Ordinarily used for assessing cancer risk.

Lowest Observed Adverse Effect Level (LOAEL)—Lowest exposure level in an experiment at which any adverse effects are observed.

Lowest Observed Effect Level (LOEL)—The lowest exposure level in an experiment at which any effects are observed.

Margin of Exposure (MOE)—The ratio of the NO(A)EL or NOEL to an estimated dose/exposure level.

Margin of Safety (MOS)—The maximum amount of exposure producing no measurable effect in animals (or studied humans) divided by the actual amount of human exposure in a population. Used in assessing noncancer risks, the margin of safety is the ratio between the no-observed-(adverse)-effect level (NO(A)EL) in a toxicity study and actual or estimated intake of an agent experienced by humans. Regulatory decisions are often made based on the perceived adequacy of the MOS.

Matrix—A specific type of medium (*e.g.*, surface water, drinking water) in which the analyte of interest may be contained.

Maximum Daily Dose (MDD)—Maximum dose received on any given day during a period of exposure, expressed in mg/kg body weight/day. Ordinarily used for assessing noncancer risks.

Maximally Exposed Individual (MEI)—The single individual with the highest exposure in a given population. Also "maximum exposed individual."

Media—The air, water, soil, biota (plants and animals that transport agents on or in their tissues) that make up the physical pathway between the source of an agent and the exposed individual.

Microenvironment—Well-defined surroundings such as the home, office, automobile, kitchen, store, etc. that can be treated as homogeneous (or well-characterized) in the concentrations of a chemical or other agent.

Modeling—Use of mathematical equations to simulate and predict real events and processes.

Monte Carlo Simulation—A technique that can provide a probability function of estimated exposure using distributed values of exposure factors in an exposure scenario. The Monte Carlo simulation involves assigning a joint probability distribution to the input variables (*i.e.*, exposure factors) of an exposure scenario. Next, a large number of independent samples from the assigned joint distribution are taken and the corresponding outputs calculated. This is accomplished by repeated computer runs (*i.e.*, >=1,000 iterations) using random numbers to assign values to the exposure factors. The simulated output represents a sample from the true output distribution. Methods of statistical inference are used to estimate, from the exposure output sample, some parameters of the exposure distribution, such as percentiles, mean, variance, and confidence intervals. The Monte Carlo simulation can also be used to test the effect that an input parameter has on the output distribution.

No Observed Adverse Effect Level (NOAEL)—Highest exposure level in an experiment that does not produce an observable adverse effect.

No Observed Effect Level (NOEL)—That level of exposure which, over a lifetime, is not expected to result in any adverse effects.

No Significant Risk Level (NSRL)—That level of exposure which, over a lifetime, is not expected to result in any adverse effects.

Pathway—The physical course an agent takes from the source to the organism exposed.

PB-PK Modeling—Physiologically-based pharmacokinetic modeling. Broadly: utilization of known anatomical and physiological functions as a basis for pharmacokinetic models to permit the rational prediction of the events occurring during the disposition of a drug (or other compound) throughout the body. Commonly used in a more restricted sense to refer to multicompartmental models in which certain parts of the body (*e.g.*, organs) are lumped together and described by a single concentration level, and linked by blood flow. Differs from "classical" pharmacokinetics by using a compartment as an actual local tissue region, rather than an abstract mathematical construct.

Permissible Exposure Levels (PELs)—The legal limit for occupational exposure to airborne concentrations of several hundred agents. Established by OSHA (U.S. Occupational Safety and Health Administration).

Personal Measurement—A measurement collected from an individual's immediate environment using active or passive devices to collect samples.

Pharmacokinetics—The study of the time course of absorption, distribution, metabolism, and excretion of a foreign substance (*e.g.*, drug or pollutant) in an organism's body.

Point-of-contact Exposure—Exposure expressed as the product of the concentration of an agent in the medium of exposure and the duration and surface area of contact with the body surface (*e.g.* mg/cm2-hours). Some agents do not need to be absorbed into the body but rather produce toxicity directly at the point of contact (*e.g.*, the skin, mouth, GI tract, nose, bronchial tubes, or lungs). In such cases the absorbed dose is not the relevant measure of exposure; rather it is the amount of toxic agents coming directly into contact with the body surface.

Point-of-contact Measurement of Exposure—An approach to quantifying exposure by taking measurements of concentration over time at or near the point of contact between the agent and an organism while the exposure is taking place.

Population at Risk—A group of subjects with the opportunity for exposure to an agent.

Potential Dose—The amount of agent contained in material ingested, air breathed, or bulk material applied to the skin.

Range—The difference between the largest and smallest values in a measurement data set.

Reasonable Worst Case Exposure or Risk Range—The lower portion of the "high end" of the exposure, dose, or risk distribution. The reasonable worst case conceptually should be targeted at or above the 90th percentile in the distribution, but below the 98th percentile.

Reconstruction of Dose—An approach to quantifying dose after exposure has occurred from evidence within an organism such as levels of agents in tissues or fluids, or from evidence of other biomarkers of exposure.

R_fD—EPA reference dose.

Risk—The probability of deleterious health or environmental effects.

Risk Assessment—The characterization of the potential adverse effects on human life or health, or on the environment. According to the National Research Council's Committee on the Institutional Means for Assessment of Health Risk, human health risk assessment includes: description of the potential adverse health effects based on an evaluation of results of epidemiologic, clinical, toxicologic, and environmental research (hazard identification); extrapolation from those results to predict the type and estimate the extent of health effects in humans under given conditions of exposure (dose-response assessment); judgments as to the number and characteristics of persons exposed at various intensities and durations (exposure assessment); summary judgments on the existence and overall magnitude of the public-health problem and characterization of the uncertainties inherent in the process of inferring risk (risk characterization).

Route—The way an agent enters an organism after contact (*e.g.*, by ingestion, inhalation, or dermal absorption).

Safety Factor—The ratio between exposure to an element or situation that would be expected to cause loss, injury or peril, and anticipated actual exposure to the causative element/agent.

Scenario Evaluation—An approach to quantifying exposure by measurement or estimation of both the amount of a substance contacted, and the frequency/duration of contact, and subsequently linking these together to estimate exposure or dose.

Sensitivity Analysis—A technique that tests the sensitivity of an output variable to the possible variation in the input variables of a given model. The purpose of sensitivity analysis is to quantify the influence of input variables on the output variable and develop bounds on the model output. The sensitivity of the output variable of a given mathematical model depends on the nature of the mathematical relationship of the model (and plausible values of its input variables). For a given model, the sensitivity of the output variable with respect to each input variable is computed, and the sensitivities of all input variables are compared. When computing the sensitivity with respect to a given input variable, all other input variables are held fixed at their nominal vales. Sensitivity can be calculated for a point estimate of an input variable or over a range of an input variable. Varying several input parameters at the same time will often highlight interaction effects in the model which are not obvious during "one at a time" variation.

Setting—The place or situation in which a person is exposed to the agent. Setting is often modified by the activity that a person is undertaking, for example, occupational or in-home exposures.

Source—The activity or entity from which an agent is released for potential human exposure.

Source Characterization Measurements—Measurements made to characterize the rate of release of agents into the environment from a source of emission such as an incinerator, landfill, industrial facility, or consumer product.

Subchronic Exposure—An exposure of intermediate duration between acute and chronic.

Subject—An exposed individual, whether human or an exposed animal or organism in the environment. An exposed individual is sometimes called a receptor.

Surrogate Data—Substitute data or measurements on one substance used to estimate analogous or corresponding values of another substance.

Systemic Dose—The dose of agent within the body (*i.e.*, not localized at the point of contact). Thus skin irritation cause by contact with an agent is not a systemic effect, but liver damage due to absorption of the agent through the skin is.

Temporal Distribution of Release—Along with dimensionality, one of the two terms used to qualitatively describe emission sources. Release rates over time can be quite diverse, and are usually described as continuous, cyclic, intermittent, random, or concentrated.

Threshold Limit Value (TLV)—The concentrations for several hundred agents recommended by the American Conference of Governmental Industrial Hygienists (ACGIH) as safe occupational exposures for the average, healthy worker who is exposed 8 hours a day during a 5-day work week.

Time-Activity Pattern—Information on activities in which various individuals engage, length of time spent performing various activities, locations in which individuals spend time, and length of time spent by individuals within various environments.

Time Weighted Average (TWA)—The average value of a parameter (such as concentration of an agent in air) that varies over time.

Uncertainty Factors—Factors used to adjust for multiple sources of uncertainty encountered in using experimental animal data for predicting effects on humans, such as intraspecies variation, interspecies variation, synergism, and different route of exposure (*i.e.* oral versus inhalation).

VOC—Volatile Organic Compound.

Worst Case—A semiquantitative term referring to the maximum possible exposure, dose, or risk, that can conceivably occur, whether or not this exposure, dose, or risk actually occurs or is observed in a specific population. Historically, this term has been loosely defined in an ad hoc way in the literature, so assessors are cautioned to look for contextual definitions when encountering this term. It should refer to a hypothetical situation in which everything that can plausibly happen to maximize exposure, dose, or risk does in fact happen. This worst case may occur (or even be observed) in a given population, but since it is usually a very unlikely set of circumstances, in most cases, a worst-case-estimate will be somewhat higher than occurs in a specific population. As in other fields, the worst-case scenario is a useful device when low probability events may result in a catastrophe that must be avoided even at great cost, but in most health risk assessments, a worst-case scenario is essentially a type of bounding estimate.

Appendix B

EXPOSURE MODELS AND SOFTWARE

The more prominent exposure-related models and software products are briefly described in this Appendix. It must be noted that this is not a comprehensive list and that the mention of model and software names is not an endorsement of any kind. It is left to the reader to decide the appropriate model and software of choice for the exposure related problem at hand. Additional software systems, exposure related databases, and online systems are addressed in Hakkinen (1997). An overview and discussion of most human exposure models of the National Institute of Public Health and Environmental Protection (RIVM) in the Netherlands is given by Vermeire *et al.* (1997). Van de Meent *et al.* (1995) discuss screening models for exposure assessment, including exposure through food and water and through consumer products.

Table B-1 lists 27 models/software products discussed in this Appendix. The subsequent paragraphs briefly describe their capabilities, computer requirements, and available contacts. Other relevant models, not described in this Appendix, include ASTM's RBCA, DRES, MAVRIQ, and MCCEM.

AERIS (AID FOR EVALUATING THE REDEVELOPMENT OF INDUSTRIAL SITES)

AERIS is a soil toxicity model developed by SENES Consultants Ltd., Ontario, Canada. The model, which is specifically designed to meet the soil testing requirements for the Decommis-sioning Steering Committee in Canada, provides estimates of concentrations of chemicals in contaminated soil that are used to help establish cleanup guidelines or to indicate when remedial actions are required. Model inputs include site specific data, physical and chemical properties, transport calculation methods, and exposure parameters. The model also estimates human exposures resulting from contaminated soil. The PC-based model includes a lookup database that contains physical and chemical properties of certain chemicals.

AMEM

AMEM (Arthur D. Little Migration Estimation Model, USEPA 1990) is a computer program that facilitates the rapid estimation of the fraction of additives originally in polymer sheets that will migrate under various conditions. With AMEM, one can

Table B-1. List of models and software products available for quantitating residential exposures

Model name		Model description
1	AERIS	Aid for Evaluating the Redevelopment of Industrial Sites
2	AMEM	Arthur D. Little Migration Estimation Model
3	API DSS	American Petroleum Institute Decision Support System
4	BEAD	Benzene Exposure and Absorbed Dose
5	BEAM	Benzene Exposure Assessment Model
6	CalTOX	California Toxicity Model
7	CHEMCAN	
8	CONSEXPO	Consumer Exposure Model
9	CONTAMx	
10	DERMAL	Dermal Model
11	DEPM	Dietary Exposure Potential Model
12	EASE	Estimation and Assessment of Substance Exposure
13	EML/IMES	Exposure Models Library/Integrated Model Evaluation System
14	INDOOR/EXPOSURE/RISK	IAQ for Windows (USEPA)
15	MEPAS	Multimedia Environmental Pollutant Assessment System
16	MMSOILS	Multiple Media Soil Model
17	NEM/pNEM/HAPEM-MS	NAAQS Exposure Model
18	PCGEMS	Personal Computer Graphical Exposure Modeling System
19	RESRAD	Residual Radiation
20	RISKPRO	
21	RISK*ASSISTANT	
22	SCIES	Screening Consumer Inhalation Exposure Software
23	SHAPE	Simulation of Human Air Pollution Exposures
24	SmartRISK	
25	THERdbASE	Total Human Exposure Risk database and Advanced Simulation Environment
26	TOXLT/TOXST	Toxic Modeling System Long-Term/Short-Term
27	USES (and EUSES)	Uniform System for the Evaluation of Substances (European USES)

estimate migration of additives to fluid and solid external phases. The USEPA Office of Toxic Substances, Exposure Evaluation Division (EED) assesses the potential for exposure to chemicals—used as additives in polymeric materials or which are the monomers or low molecular weight oligomers contained in polymers—by using AMEM or similar models. AMEM, which also has subroutines that estimate many physical properties (partition coefficients, mass ttransfer coefficients, and diffusion coefficients of additives in polymers, air, and water), is coded in FORTRAN for operation on PCs.

API DSS (AMERICAN PETROLEUM INSTITUTE DECISION SUPPORT SYSTEM)

API DSS estimates exposure concentrations and subsequent risk from a suite of fate and transport models coupled with toxicity parameters. Relevant exposure pathways evaluated by the models include vapor intrusion (from soil) into homes, volatilized chemicals from bathroom showers, soil ingestion, and ingestion of food and water.

BEAD (BENZENE EXPOSURE AND ABSORBED DOSE)

BEAD (MacIntosh *et al.* 1995) is a population-based, multiple-exposure pathway, microenvironmental model of 24-hour average inhalation exposures and total absorbed doses of benzene. The model was developed to:

- Provide a tool for estimating the distribution of benzene personal air concentrations and total absorbed doses for a large population
- Examine the determinants of interindividual variability of exposures and absorbed doses of benzene, and
- Explore the accuracy and precision of predictions of population exposures and absorbed doses of benzene made with monitoring results from past field studies.

A two-dimensional Monte Carlo simulation approach is used in the model to estimate the uncertainty about the predicted population exposure and absorbed dose distributions. A principal advantage of this approach to uncertainty analysis is that the relative contribution of the input variables to prediction uncertainty can be easily identified. Decisions can then be made regarding the appropriate measures to be taken to reduce the parameter uncertainty, where the overall goal is to minimize prediction uncertainty.

BEAM (BENZENE EXPOSURE ASSESSMENT MODEL)

BEAM (USEPA 1993) generates benzene inhalation exposure profiles for different human subgroups. It considers exposure to benzene in air through the inhalation pathway only. The model has three major input components: (1) human location patterns, (2) ambient (background) benzene concentrations, and (3) microenvironmental benzene concentrations. It matches available location and activity data with environmental concentration data to obtain exposure profiles for 24-hour periods. Exposure concentrations are obtained by applying a superposition principle to contributions from the ambient and different microenvironments. Inhalation dose is then obtained by applying inhalation rate to exposure concentrations.

BEAM is currently a product under the data-and-model based-system, THERdbASE. The minimum computer requirements of THERdbASE include an Intel-based 486 or higher PC, at least 8 Mb of RAM, at least 40 Mb of available disk space, at least a color VGA monitor, and Microsoft Windows 3.1. The most recent installable version of THERdbASE underwent beta-testing and a beta version is on the Internet (WWW site: http://hrcweb.lv-hrc.nevada.edu).

CalTOX

The Department of Toxic Substances Control (DTSC), within the California Environmental Protection Agency, has the responsibility for managing the State's hazardous-waste program. As part of this program, The DTSC funded the development of the CalTOX program (McKone 1993a, 1993b, 1993c). CalTOX has been developed as a set of spreadsheet models and spreadsheet datasets to assist in assessing human exposures and defining soil clean-up levels at uncontrolled hazardous wastes sites. CalTOX addresses contaminated soils and the contamination of adjacent air, surface water, sediments, and ground water. The modeling components of CalTOX include a multimedia transport and transformation model, exposure scenario models, and add-ins to quantify uncertainty and variability. The multimedia transport and transformation model is a dynamic process that can be used to assess time-varying concentrations of contaminants introduced initially to soil layers or for contaminants released continuously to air, soil, or water. This model assists the user in examining how chemical and landscape properties impact both the ultimate route and quantity of human contact. Multimedia, multiple-pathway exposure models are used in CalTOX to estimate average daily doses within a human population. The exposure models encompass twenty-three exposure pathways. The exposure assessment process consists of relating contaminant concentrations in the multimedia model compartments to contaminant concentrations in the media with which a human population has contact (personal air, tap water, foods, household dusts soils, *etc.*).

CHEMCAN

CHEMCAN, similar to CALTOX and USES, is a multiple-media fugacity model designed to estimate the distribution of single chemicals in multiple media. Given emission rates and chemical properties, it estimates chemical concentrations in air, food, soil, and water. CHEMCAN was developed for Environment Canada.

CONSEXPO (CONSUMER EXPOSURE MODEL)

CONSEXPO (Van Veen 1995) is a modeling approach based on simple exposure and uptake models. In order to cope with the diversity in consumer products, it is based on a general model framework that provides a general setting for widely differing exposure situations and, secondly, it offers a number of predefined exposure and uptake models, which users can link to build a complete exposure and uptake model. The starting points are the inhalation, dermal, and ingestion exposure pathways. For each of these pathways, a limited number of models are available to model exposure and uptake. The program reports several important exposure variables, namely, the per-event concentration, the yearly averaged concentration, the fraction taken up, the amount taken up during a year (per year and cumulative), and the uptake per kilogram body weight per day. The program also allows for stochastic parameters, in order to propagate the effects of variable and/or uncertain parameters to the final exposure and uptake estimates. If there are stochastic parameters, the resultant distributions can be displayed and studied.

The CONSEXPO program is related to the USES program discussed elswhere in this Appendix. CONSEXPO can be executed as a stand-alone program or as part of the framework of USES. The minimum computer system requirements for CONSEXPO are an Intel-based PC, Microsoft Windows 3.1, and at least 1 Mb of available hard drive space. A math coprocessor is not mandatory, but highly recommended.

CONTAMx

The National Institute of Standards and Technology (NIST) has, over the past several years, developed a series of public domain computer programs (CONTAM86,

CONTAM87 . . . *etc.* . . . CONTAM94) for calculating air flow and contaminant dispersal in multiple zone buildings. These programs take a multiple-zone network approach to airflow analysis. Airflow paths include doorways, small cracks in the building envelope, and a simple model of the air handling system.

The most recent version of CONTAMx, CONTAM94, works on a Intel-based PC in the DOS environment. A graphical interface is used to create and edit building descriptions. Future version of this program will include exposure assessment.

CSOIL (RIVM)

A model to calculate exposures to soil contamination. Parts of it model residential exposure in a simplified way. The model is described in Van den Berg (1994).

DERMAL

DERMAL was developed to assist the Economics, Exposure, and Technology Division of the USEPA Office of Pollution Prevention and Toxic Substances, in performing screening-level assessments of the potential dose rates resulting from dermal contact with consumer products containing new and existing chemicals in consumer products. The model calculates screening-level estimates of annual individual dermal potential dose rates to components of 16 consumer product categories. Exposures are calculated based on the weight fraction of the chemical of interest in the consumer product and assuming deposition of a film of liquid on the dermal surface from contact with the product. Conservative default values are provided for most of the input parameters required to run the model for each of the 16 consumer product categories.

The model is programmed in Turbo C++ and is designed to run on a PC using a DOS environment.

DEPM (DIETARY EXPOSURE POTENTIAL MODEL)

DEPM is a model and database system developed by the USEPA Office of Research and Development to correlate food information in a format for dietary exposure modeling. Currently, the database system includes information from several, national, government-sponsored surveys and monitoring programs. In the model, food consumption is based on 11 food groups, containing approximately 800 core food types, established from over 6,500 common food items. A unique feature of the DEPM is the use of recipes, developed for exposure analysis, that link consumption survey data to the contaminant residue information. The summary databases are aggregated in a fashion that allows analyst selection of demographic factors, such as age and sex groups, geographical regions, ethnic groups, and economic status. The model was developed for personal computers with the data files designed in dBASE IV with FoxPro for Windows applications programs for queries and reporting.

DEPM Beta Version (2.2) may now be downloaded by qualified individuals. These individuals should have an interest in evaluating dietary exposure. They should also be interested in testing the model and providing meaningful feedback to the USEPA.

Minimum equipment includes an Intel-based 80486 CPU and Windows 3.1x. The DEPM was developed by Technical Assessment Systems, Inc., Washington, DC.

EASE

The EASE system was developed by the United Kingdom Health and Safety Executive in conjunction with the Artificial Intelligence Applications Institute. The system is a knowledge-based one that stores knowledge on specialized subjects along with the reasons for that knowledge. EASE estimates exposure of workers to notified substances. In particular, it estimates inhalation exposure to dust and vapor, and dermal exposure.

EML/IMES (EXPOSURE MODELS LIBRARY/INTEGRATED MODEL EVALUATION SYSTEM)

EML, developed by the USEPA Office of Research and Development, is a collection of exposure models distributed in a CD-ROM medium (USEPA 1996), to provide a compact and efficient means to distribute exposure models, documentation, and the IMES. The disc contains over 120 models that may be used for transport modeling and exposure assessments. The model files may contain source and/or executable code, sample input files, and other data files, sample output files, and in many cases, model documentation in Wordperfect, ASCII text, or other similar formats. IMES assists in selecting appropriate models, provides literature citations on model validations, and demonstrates model uncertainty protocols. The IMES software is an MS-DOS application, can be used on an Intel-based PC, and is capable of running on a network. Model codes and documentation can be downloaded from the CD-ROM to a hard drive. The most recent version, which is the third edition, has an HTML interface to view model directories and Internet source for some models.

Table B-2 lists the models included in the EML/IMES CD-ROM. The CD-ROM and technical assistance can be obtained from USEPA (1998).

INDOOR EXPOSURE/RISK (IAQ MODEL FOR WINDOWS, RISK VERSION 1.0)

RISK is the third in a series of indoor air quality models developed by the Indoor Air/Radon Mitigation Branch of the USEPA National Risk Management Research Laboratory. The first model, INDOOR, was designed to calculate the indoor pollutant concentrations from indoor sources. The second model, EXPOSURE, extended INDOOR to allow calculation of individual exposure. RISK extends EXPOSURE to allow analysis of individual risk to indoor pollutant sources. Risk estimates generated by models such as this one are useful mainly for the purpose of comparing scenarios rather than for determining absolute risks to individuals or populations.

RISK uses data on source emissions, room-to-room air flows, air exchange with the outdoors, and indoor sinks to predict concentration-time profiles for all the rooms. The

Table B-2. Models included in the exposure models library/integrated model evaluation system

Medium	Models
Nonpoint	AGNPS, ANSWERS, BASINS, CREAMS, DR3M-QUAL, EPIC, FHWA, GLEAMS, GWLF, HELP, HSPF, MRI,
Source	NPSGUIDE, NPSMAP, P8-UCM, PREWET, PRZM, Q-ILLUDAS, SIMPLE, SLAMM, SLOSSPHOS, STORM, STREAM, SWAT, SWMM, SWRRBWQ, USGSREGR, WATERSHED, WMM, WQAM
Air	APRAC3, AQDM, ARRPA, AREALDOC, BLP, BPIP, CAL3QHC, CAL3QHCR, CALINE, CALINE3, CALPUFF, CDM2, CMB7, COMPDEP, COMPLEX1, CRSTER, CTDMPLUS, CTSCREEN, DEGADIS, EDMS, EKMA, ERT, FDM, GUIDANCE, HIWAY2, IMPACT, INPUFF, ISC3, LONGZ, MESOPUFF, MPRM, MPTER, MULTIMAX, OCD, PAL-DS, PCRAMMET, PLUMES5, PLUVUE2, PPSP, PTPLU, RAM, RPM, RPM2, RTDM, RTM2, RVD2, SCREEN2, SCREEN3, SCSTER, SDM, SHORTZ, STAR, TCM2, TEM8, TOXLT, TOXST, TSCREEN, UAM, VALLEY, VISCREEN, WINDROSE, WYNDVALLEY
Surface Water	AIDE, CEQICM, CEQR1, CEQRIV1, CEQUALICM, CEQUALR1, CEQUALRIV1, CEQUALW2, CEQW2, CLC, CORMIX, DYNHYD5, DYNTOX, EUTRO5, EXAMS, FATE, GCSOLAR, HEC-5Q, HEC-6, HECDSS, HSPF, HYDRO2D-V, YDRO3D, LC50, MEXAMS, MICHRIV, MINTEQA2, OXYREF, PCPROUTE, PLUMES, QUAL2E, REACHSCAN, RECOVERY, RESTMP, RIVMOD, SEDDEP, SLSA, SMPTOX, THERMS, TOXI5, TWQM, WASP5, WQAM, WQRRS, WQSTAT
Ground water	AGU10, AQUIFEM, ASM, AT123D, BALANCE, BIOPLUM2, BTEX, CATTI, CFEST, CHAIN, CHAINT, CHEMFLO, CONMIG, DBAPE, DPCT, FE3DGW, FEMWASTE, FEMWATER, FLUMP, GEOPACK, GETOUT, GLEAMS, GWCOMPEN, GWFL3D, GWTR3D, HELP, HSSM, HST3D, ICE1, IGWMC, MMT, MOC, MOCDENSE, MODFLOW, MOFAT, MT3D, NWFT/DVM, OASIS, PATHS, PESTAN, PHREEQE, PHREEQEP, PLUME, PLUME2D, PRINCETON, PRZM, RANDWALK, RETC, RITZ, RUSTIC, RWH, SAFTMOD, SESOIL, SHALT, SOLUTE, STF, SUTRA, SWANFLOW, SWENT, SWIFT, SWIFT2, SWIP2, TETRANS, TRAFRAP-WT, TRANS, TRUST, TRIPLOT, UNSAT2, USGS2D, USGS2D-MOC, V3, VADOFT, VHS, VLEACH, VTT, WALTON, WATEQF, WELFLO, WHAEM, WHPA, WORM, WQAM
Multimedia	BASINS, GLEAMS, HELP, HSPF, MEPAS, MMSOILS, MULTIMED, PATRIOT, PCGEMS, PIRANHA, PRZM, RUSTIC, SWAT, SWRRBWQ, WQAM

MODEL—model included in the Exposure Models Library (EML); MODEL—model discussed in the Integrated Model Evaluation System (IMES); MODEL—model included in the EML and discussed in the IMES.

concentration-time profiles are then combined with individual activity patterns to estimate exposure. Risk is calculated using a risk-calculation framework. The model allows analysis of the effects of air cleaners, located in either/or both the central air circulating system and individual rooms, on IAQ and exposure. The model allows simulation of a wide range of sources including long-term steady state source, on/off sources, and decaying sources. Several sources are allowed in each room. The model allows the analysis of the effects of sinks and sink re-emissions on IAQ. The results of test house experiments have been compared with model predictions. The agreement between predicted concentration-time profiles and the test house data were good. The model is designed to run in the Windows operating environment.

RISK can be obtained from the National Technical Information Service, 5285 Port Royal Road, Springfield, VA 22161 (Telephone: 703-487-4650). The complete report is entitled "IAQ Model for Windows, RISK Version 1.0: User Manual," (Order No. PB96-501 929).

MEPAS (MULTIMEDIA ENVIRONMENTAL POLLUTANT ASSESSMENT SYSTEM)

MEPAS (Strenge and Chamberlain 1995, Droppo *et al.* 1993) was developed by Battelle Pacific Northwest Laboratory (PNL) for the U.S. Department of Energy (USDOE). The system was developed to rank USDOE sites having potential hazardous chemical and radioactive releases. The key objective of MEPAS is to rank sites by calculating human health risk to the population surrounding the site. MEPAS calculates "hazard potential index" (HPI) values for a site by summing up risk factors associated with various exposure scenarios. This system has wide applicability to a range of environmental problems using air, ground water, surface water, overland, and exposure models.

MEPAS integrates source, transport, and exposure models into a single system. The algorithms in MEPAS accommodate the following ten components:

- Source term
- Overland pathway
- Ground water (vadose and saturated zones) pathway
- Surface water pathway
- Atmospheric pathway
- Exposure routes
- Hazard assessment (chemical carcinogens and noncarcinogens, radionuclides)
- Contaminant transport and exposure scenarios
- User-friendly PC shell, and
- Chemical database.

Contaminant transport media are ground water, overland flow, surface water, and atmosphere. Human uptake occurs through ingestion (of contaminated water, soil, crops, animal products, and aquatic foods), inhalation (of airborne contaminants), and dermal contact (with chemicals and radionuclides). The hydrologic media consist of the hydrologic source term, unsaturated and saturated ground water zones, and surface and runoff water. The source term can be computed internally or specified at receptor locations or by specified flow. The source term geometries include point, line, and area sources. Limitations of the hydrologic pathway include negligible leaching of the source by ground water, and flow in the virtual direction only.

The atmospheric pathway consists of the atmospheric source term and atmospheric transport processor. Source terms consist of point sources and area sources. Atmospheric transport of contaminants utilizes an enhanced Gaussian plume model, and computes long-term exposure for a sixteen-sector grid using average stack parameters. Enhancements to the plume model include deflection of wind speed to account for variability in local surface roughness, and can consider radioactive decay depletion and first order chemical reactions. Only simple sources can be modeled, and particulate contaminants originate from area sources only.

MEPAS calculates an average dose over 70-years time increments for a number of user specified receptor locations. Dose is calculated for each transported contaminant. For radioactive contaminants, the dose is expressed as the effective dose-equivalent from each contaminant. MEPAS uses the ICRP dose-conversion factors to convert the rate of exposure to dose.

MMSOILS: MULTIMEDIA CONTAMINANT FATE, TRANSPORT, AND EXPOSURE MODEL

MMSOILS estimates the human exposure and health risk associated with releases of contamination from hazardous waste sites. It is a multimedia model addressing the transport of a chemical in ground water, surface water, soil erosion, the atmosphere, and accumulation in food. The human exposure pathways considered in the methodology include: soil ingestion, air inhalation of volatile organic compounds and particulates, dermal contact, ingestion of drinking water, consumption of fish, consumption of plants grown on contaminated soil, and consumption of animals grazing on contaminated pasture. For multimedia exposures, the methodology provides estimates of human exposure through individual pathways and combined exposure through all pathways considered. The risk associated with the total exposure is calculated based on chemical-specific toxicity data.

The methodology is intended for use as a screening tool. It is critical that the results are interpreted in the appropriate framework. The intended use of the exposure assessment tool is for screening and relative comparison of different waste sites, remediation activities, and hazard evaluation. The methodology can be used to provide an estimate of health risk for a specific site, but the uncertainty of the estimated risk may be quite large (depending on the site characteristics and available data). This uncertainty must be considered in any decision making process.

MMSOILS was developed to operate on an Intel-based PC. Other minimum system requirements are 512 KB available random access memory, one $3\frac{1}{2}''$ or $5\frac{1}{4}''$ floppy disk drive, one hard disk drive (2.0 MB storage available), and DOS 3.x or higher. Because of the extensive computations involved in MMSOILS, it will operate most effectively on a 386 or 486 series computer equipped with a math coprocessor.

NEM, pNEM, AND HAPEM-MS (NATIONAL AMBIENT AIR QUALITY STANDARD EXPOSURE MODEL, PROBABILISTIC NATIONAL AMBIENT AIR QUALITY STANDARD EXPOSURE MODEL, AND HAZARDOUS AIR POLLUTANT EXPOSURE MODEL FOR MOBILE SOURCES)

NEM and pNEM are a series of models developed by the USEPA Office of Air Quality Planning and Standards to estimate human exposure to selected motor vehicle pollutants and residing in the USEPA National Computing Center (NCC) in Research Triangle Park, NC. The models simulate the movements of specific population groups (called cohorts) through zones of varying air quality. Each zone is typically defined by a geographic location and a microenvironment. The movements of each cohort are determined by the use of activity diary data specific to the demographic characteristics of the cohort. The activity data are also specific to day of the week, season, and temperature. Depending on the application, cohort movements may account for trips to work places or to schools.

Between 1984 and 1994, 13 different versions of NEM and pNEM have been developed for four NAAQS (National Ambient Air Quality Standard) pollutants—CO, NO_2, O_3, Pb. HAPEM-MS was developed to estimate exposures to toxic pollutants (*e.g.*, benzene) emitted by mobile sources (Johnson *et al.* 1993). Early versions (NEM) were deterministic in nature and later versions (pNEM and HAPEM-MS) include Monte Carlo sampling from input distributions. Most of these models use U.S. Census data directly so that modeling results can be related to the population as a whole or to specific segments within it. Estimates of outdoor air quality can be derived from diffusion models, ambient monitors, or personal exposure monitors. Thus, the models are flexible with respect to air quality inputs.

PC—GEMS (PERSONAL COMPUTER—GRAPHICAL EXPOSURE MODELING SYSTEM)

GEMS supports exposure and risk assessments by providing access to single-medium and multimedia fate and exposure models, physical and chemical properties estimation techniques, statistical analysis, graphics and mapping programs with related data on environments, sources, receptors, and populations. Under development since 1981, GEMS provides analysts with an interactive, easily learned interface to various models, programs, and data needed for exposure and risk assessments on the USEPA's National Computing Center (NCC) VAX cluster. PC-GEMS is a stand-alone version of GEMS that can be run on a PC.

The environmental models in GEMS are atmospheric, surface water, land unsaturated (soil) and saturated (ground water) zones, and multimedia in nature. Methods for estimating octanol-water partition and adsorption coefficients, bioconcentration factor, water solubility, melting and boiling point, vapor pressure, Henry's constant, acid dissociation constant, lake/stream volatilization rate, and atmospheric half-life are available. Data sets are related to environmental characteristics (climate, soil, rivers, ground water, vegetation), source releases (POTWs and industrial water discharges, Census business patterns, RCRA permit sites), and receptors (population and household estimates for 1970, 1980, 1990, and 1995 by small area census district and drinking water facilities). Mapping routines in GEMS support Tektronix 4014 and color 4105/06/25/35, and ASCII terminals or PCs. PC-GEMS supports Hercules, CGA, and EGA graphics.

PC-GEMS is available with air, surface water and ground water models and associated environmental and 1980 population data encompassing most of the U.S. Minimum system requirements include IBM AT or compatible, Math Coprocessor (8087/80287/80387), 640K RAM, and 20 Mb of dedicated disk space or Bernoulli cartridge.

RESRAD (RESIDUAL RADIATION)

RESRAD was developed by the Argonne National Laboratory to evaluate residual concentrations of radionuclides in soil, concentrations of airborne radon decay products, external gamma radiation levels, surface contamination levels, and radionuclide concentrations in air and water, and to determine radiation dose and excess lifetime cancer risks to an onsite resident (a maximally-exposed individual or a member of a critical population group).

The computer code RESRAD, developed for the USDOE and accepted as a primary code for remedial action activities (DOE Order 5400.5), determines site-specific residual radioactive material cleanup guidelines based on calculations of the radiation dose to hypothetical residents or workers on the site. The nine environmental pathways considered in RESRAD are direct exposure, dust inhalation, radon, and ingestion of plant foods, meat, milk, aquatic foods, water, and soil. Currently, the RESRAD code is being adapted to integrate chemical and radiological health risks. Other RESRAD developments are incorporating uncertainty analysis and decontamination and decommissioning analysis capabilities. The development of the code is funded by the USDOE.

RISKPRO

RISKPRO provides an integrated system that includes environmental modeling programs, site, and chemical tracking and supporting tools. Six programs are provided to simulate the transport and transformation of a chemical through air, water, and soil. The models interact with each other. Chemical property estimation and environmental databases are included.

RISKPRO is a compartmentalized system consisting of:

- AUTOEST—a chemical property estimating program
- ISCLT (Industrial Source Complex Long Term Model)—an air dispersion modeling program
- PTPLU (Point Plume)—a single source Gaussian dispersion modeling program
- SESOIL (Seasonal Soil Compartment Model)—a soil transport modeling program
- AT123D—a ground water transport modeling program, and
- EXAMS (Exposure Analysis Modeling System)—a surface water modeling program.

Also included within the system are several databases that contain climate information for the U.S., geographic information, census data, and a catalog of SMILES notations. RISKPRO was produced by General Sciences Corporation (GSC) as an updated, user-friendly version of PCGEMS.

RISK*ASSISTANT

Developed by Hampshire Research Institute, RISK*ASSISTANT for Windows, a commercial product, provides a straightforward tool for evaluating complex risks, useful for both novice and professional risk assessors. The software provides information on hazard, exposure, and risk, and allows flexibility to tailor the assessment to local sites. Risks from air, surface water, ground water, soil, sediment, and food can be assessed. The software allows the choice of chemical and concentration data sets, alternative populations, exposure scenario factors, and toxicity data sets. Risks can be calculated for single scenarios (*e.g.*, showering) or multiple scenarios (drinking water, showering, and eating local fish).

Sensitivity analyses can be performed to determine the effect of exposure variables on risk estimates. Transport models included are ISCLT2 (air dispersion model to predict chemical concentrations in the air surrounding points where toxic chemicals are released as air emissions) and STREAM (surface water model to predict the downstream concentrations of contaminants in rivers and streams). Over 15 Mb of hydrology data on U.S. water bodies is part of the software.

Databases in the software supplied include hazard data for 650 chemicals. They provide USEPA data on carcinogenic and noncarcinogenic toxicity, from the Integrated Risk Information System (IRIS) or Health Effects Assessment Summary Tables (HEAST). Other toxic hazard databases can be easily incorporated.

SCIES (SCREENING CONSUMER INHALATION EXPOSURE SOFTWARE)

SCIES was developed to assist the Economics, Exposure, and Technology Division of USEPA's Office of Pollution Prevention and Toxic Substances in performing screening-level assessments of the potential dose rates resulting from inhalation of new and existing chemicals in consumer products. The model calculates screening-level estimates of average individual inhalation potential dose rates to components of consumer products that can be classified into eleven different product categories. The model estimates potential dose rates for both actively-exposed users of the product and passively-exposed nonusers. Default values are suggested for each parameter required to run the model for each product category. These values are based on exposure scenarios, volatility classifications, and residence occupancy patterns. The model combines results of an effort to measure ventilation flows within residences with a two-zone mass balance model to allow estimation of potential dose rates to both consumer product users and nonusers.

The model is programmed in Turbo Pascal and is designed to run on a personal computer using a DOS environment.

SHAPE (SIMULATION OF HUMAN AIR POLLUTION EXPOSURES)

SHAPE (Ott *et al.* 1988) generates carbon monoxide inhalation exposure profiles for different human subgroups. It considers exposure to carbon monoxide in air through the inhalation pathway only. The model has two major input components: (1) human location patterns, and (2) microenvironmental carbon monoxide concentrations. It matches available location and activity data with environmental concentration data to obtain exposure profiles for 24-hour periods. Exposure concentrations are obtained by applying a superposition principle to contributions from the ambient and different microenvironments.

SmartRISK

SmartRisk is a stand-alone commercial application designed for conducting human health risk assessments. Calculations follow the USEPA paradigm: Hazard Identification, Exposure Assessment, Dose-Response Assessment, and Risk Characterization. Exposure is characterized through scenarios and multiple pathways are considered. Exposure scenarios require exposure media contact, exposure durations, and exposure frequencies. All calculations are deterministic. Risk reports can be exported to a spreadsheet format in order to perform Monte Carlo Simulation using Crystal Ball or @RISK.

For additional information, contact Pioneer Systems Development (Telephone: 206-822-4461, email: pioneer@halcyon.com).

THERdbASE (TOTAL HUMAN EXPOSURE RISK DATABASE AND ADVANCED SIMULATION ENVIRONMENT

THERdbASE is the result of a cooperative agreement between USEPA's Office of Research and Development (National Exposure Research Laboratory, Las Vegas). THERdbASE is being developed as a PC-based computer modeling and database system that contains exposure and risk related information. The system provides an optimal framework for the construction of a suite of exposure- and risk-related models within the Modeling Engine by using information available in data files within the Database Engine.

Data can be viewed as a table, coded fields can be viewed as decoded fields, fields can be set to "show" or "hide" mode, and multiple data files can be viewed at the same time. In the "advanced" mode, user files can be edited. Data records can be queried and simple statistics (summary statistics—mean, standard deviation, minimum and maximum; percentile values at desired intervals; and linear regression on two numerical data fields) can be performed. Data can be printed, saved, or exported. New user files can be created and data can be imported.

Input to models is achieved through a standardized procedure. Inputs can be provided as single values, custom distributions (normal, lognormal, *etc.*), distributions based on data files present in THERdbASE, or specific percentile values. Efficient algorithms are provided to optimally access input data, to perform the numerical simulations, and to generate appropriate output data. Multiple model runs can be done through a batch process. Outputs from models is done in the following two ways: as THERdbASE data files or as preset graphs.

In the Internet version of THERdbASE, the following data files are included:

- 1990 Bureau of Census Population Information
- California Adult Activity Pattern Study (1987 and 1988)
- AT&T-sponsored National Activity Pattern Study (1985)
- 1992–1994 National Human Activity Patterns Study (NHAPS)
- Chemical Agents from Sources
- Chemical Agent Properties
- Air Exchange Rates
- Information from the USEPA TEAM (Total Exposure Assessment Methodology) studies
- Information from the USEPA NOPES (Nonoccupational Pesticide Exposure Study), and
- Human Physiological Parameters.

In the internet version of THERdbASE, the following models are included:

- Location Patterns
- Chemical Source Release—Instantaneous Emission
- Chemical Source Release—Timed Application
- Indoor Air (2-Zone)

- Indoor Air (N-Zone)
- Exposure Patterns for Chemical Agents
- Benzene Exposure Assessment Model (BEAM)
- Source Based Exposure Scenario (Inhalation and Dermal)
- Film Thickness Based Dermal Dose
- PBPK-Based Dermal Dose

The computer requirements of THERdbASE are an Intel-based 486 or higher PC, at least 8 Mb of RAM, at least 40 Mb of available disk space, at least a color VGA monitor, and Microsoft Windows 3.1. The most recent installable version of THERdbASE underwent beta-testing and a beta version available from the Internet (WWW site: http://hrcweb.lv-hrc.nevada.edu).

TOXLT/TOXST

The Toxic Modeling System Long-Term (TOXLT) is a PC-based model that has been developed in conjunction with the release of the new version of the EPA's Industrial Source Complex (ISC2) Dispersion Models and the development of a tiered modeling approach for assessing the risks due to sources of hazardous air pollutants. The purpose of TOXLT is to assist in the evaluation of the lifetime cancer risks and chronic noncancer hazards that may result from long-term exposure to toxic air pollutants. The ISCLT2 model is used to simulate annual average pollutant concentrations, which are then used to estimate cancer risk levels or hazard index values at each user-specified receptor. These outputs presume: (1) a hypothetical individual exists at each receptor, (2) no contribution from "background" sources (*i.e.*, sources not specifically included in the simulation), and (3) pollutant contributions in a mixture are additive (*i.e.*, there are no synergistic or antagonistic interactions between pollutants).

The Toxic Modeling System Short-Term (TOXST) addresses the problem of estimating expected exceedances of specified short-term health effects thresholds in the vicinity of continuous and intermittent toxic releases. Certain industrial facilities emit airborne toxic chemicals known to be harmful when their concentrations exceed a specified health effect threshold value for a specified length of time. However, releases of such chemicals often occur intermittently. This random emission pattern makes it difficult to predict the frequency with which ambient concentrations will exceed the health effect threshold. An analysis in which worst case dispersion conditions are assumed to coincide with each release would tend to overestimate the exceedance frequency, since a certain proportion of release is likely to occur under more favorable dispersion conditions. On the other hand, assuming that each release occurs in conjunction with some more favorable dispersion condition, or that the source operates continuously at some long-term average emission rate, would likely underestimate the exceedance rate.

TOXST attempts to avoid the problems of underestimation and overestimation of exceedance rates resulting from random emission patterns by using the Monte Carlo simulation of source emissions of user-specified durations and rates at randomly selected points in time over a simulated long period of time. In addition, TOXST maintains the capability of simulating continuous emission sources along with intermittent emission, thereby providing a more realistic simulation of actual industrial operations.

The former Integrated Toxic Expected Exceedance Model (INTOXX), which was based on the superseded version of the Industrial Source Complex Short-Term (ISCST)

Model, was revised to TOXST. TOXST allows for Monte Carlo post-processing output from the EPA model ISCST2 to account for the variability in emissions. The model has been expanded to allow for emissions distributions (both daytime and night time), bath operations, mutually-exclusive source groups, and multiple hourly averaging.

Both TOXLT and TOXST can be accessed via the Exposure Models Library (EML), discussed elsewhere in this Appendix.

USES—Uniform System for the Evaluation of Substances & EUSES (Europeon USES)

The Uniform System for the Evaluation of Substances (USES) (RIVM, VROM, WVC 1994), was developed in the Netherlands by the National Institute of Public Health and Environmental Protection (RIVM), the Ministry of Housing, Spatial Planning and Environment (VROM), and the Ministry of Welfare, Health, and Cultural Affairs (WVC). USES provides a single framework for comparing the potential risks of difference chemical substances released to multiple media of the environment. It is an integrated modeling system that includes multiple environmental media and multiple human exposure pathways. The exposure assessment in USES starts with an estimate of substance emissions to water, soil, and air during the various life-cycle stages of a substance and follows its subsequent distribution in the total environment. The result of this type of multimedia assessment are the Predicted Environmental Concentrations (PECs) and an estimate of the daily intake by human receptors. In general, PECs are compared to "no-effect" levels for organisms in the environment, which are derived by extrapolating single-species toxicity tests to field situations. The estimated daily intake by humans is compared to the no-observed-adverse-effect-level (NOAEL) for mammals or to the no-observed-effect-level (NOEL) for humans.

REFERENCES

Droppo, J.G. Jr., J.W. Buck, D.L. Strenge, and B.L. Hoopes. 1993. Risk Computation for environmental restoration activities. J. Hazardous Materials. 35:341–352.

Hakkinen, P.J. 1997. Computer software programs, databases, and the use of the Internet, World Wide Web, and other online systems. In: Vlasta Molak (ed.). Fundamentals of Risk Analysis and Risk Management. Lewis Publishers. CRC Press, Inc.

Johnson, T., M. McCoy, and J. Capel. 1993. Enhancements to the hazardous air pollutant exposure model (HAPEM) as applied to mobile source pollutants. Report for the Atmospheric Research and Exposure Assessment Laboratory, United States Environmental Protection Agency. Research Triangle Park, North Carolina.

MacIntosh, D.L., J. Xue, H. Ozkaynak, J.D. Spengler, and B.P. Ryan. 1995. A population-based exposure model for benzene. J. Exposure Anal. Environ. Epidemiol. 5(3):375–403.

McKone, T.E. 1993a. CalTOX, a multimedia total-exposure model for hazardous-waste sites. Part I: Executive summary. UCRL-CR-111456Pt I. Lawrence Livermore National Laboratory. Livermore, California.

McKone, T.E. 1993b. CalTOX, a multimedia total-exposure model for hazardous-waste sites. Part II: The dynamic multimedia transport and transformation model. UCRL-CR-111456Pt II. Lawrence Livermore National Laboratory. Livermore, California.

McKone, T.E. 1993c. CalTOX, a multimedia total-exposure model for hazardous-waste sites. Part III: The multiple-pathway exposure model. UCRL-CR-111456Pt III. Lawrence Livermore National Laboratory. Livermore, California.

Ott, W.R., D. Mage, and L. Wallace. 1988. Validation of the simulation of human activity and pollutant exposure (SHAPE) model using paired days from the Denver, Colorado, Carbon Monoxide Field Study. Atmos. Environ. 22:2101–2113.

RIVM, VROM, WVC. 1994. Uniform system for the evaluation of substances (USES), version 1.0. National Institute of Public Health and Environmental Protection (RIVM). Ministry of Housing, Spatial Planning and Environment (VROM). Ministry of Welfare, Health, and Cultural Affairs (WVC). The Hague, The Netherlands, VROM Distribution No. 11144/150.

Strenge, D.L. and P.J. Chamberlain II. 1995. Multimedia environmental pollutant assessment system (MEPAS): Exposure pathway and human health impact assessment models. PNL-10523. Pacific Northwest Laboratory. Richland, Washington.

USEPA (U.S. Environmental Protection Agency). 1990. Methods for assessing exposure to chemical substances. Volume II. Methodology for estimating the migration of additives and Impurities from Polymeric Materials. EPA Report No. 560/5-85-015. Exposure Evaluation Division. Office of Toxic Substances. Washington DC. September.

USEPA (U.S. Environmental Protection Agency). 1993. Estimation of the Exposure to Benzene of Selected Populations in the State of Texas Using the Benzene Exposure Assessment Model (BEAM). Environmental Monitoring Systems Laboratory. Las Vegas, Nevada.

USEPA (U.S. Environmental Protection Agency). 1996. Exposure models library/IMES CD-ROM. Office of Health and Environmental Assessment, Exposure Assessment Group. Washington DC. Hhttp://www.epa.gov/ncea/www1/imes.htm.

USEPA (U.S. Environmental Protection Agency). 1998. Richard Walentowicz. Office of Research and Development. Telephone: 202-260-8922. E-mail: walentowicz.rich@epamail.epa.gov). Washington DC.

Van de Meent D., J.H.M. De Bruijn, F.A.A.M. De Leeuw, A.C.M. De Nijs, J.T. Jager, and T.G. Vermeire. 1995. H. 4 Exposure modelling. Pages 103–145 in: C.J. Van Leeuwen and J.L.M. Hermens (eds.). Risk assessment of chemicals: An introduction. Kluwer Academic Publishers, Dordrecht.

Van den Berg, R. 1994. Blootstelling van de mens aan bodemverontreiniging. Een kwalitatieve en kwantitatieve analyse, leidend to voorstellen voor humaan toxicologische C-toestingswaarden (beperkt herziene versie). RIVM report 725201/006.

Van Veen, M.P. 1995. CONSEXPO—A Program to Estimate Consumer Product Exposure and Uptake. Report No. 612810.002. RIVM. Bilthoven, The Netherlands.

Vermeire, T.G., M.P. Van Veen, M.P.M. Janssen, en R.C.G.M. Smetsers. 1997. De schatting van de blootstelling van de mens aan stoffen en straling: Het RIVM-onderzoek. RIVM report 601132/002.

Appendix C

EUROPEAN DATA SOURCES AND LEGISLATION

This Appendix summarizes European data sources and legislation on new and existing substances with respect to residential exposure. The compilation of data sources given here is only a starting point. The intention is to give an overview, and to point to relevant data and literature sources in order to provide the reader with leads to more detailed information.

The European Union (EU) Directives on the assessment of new and existing substances affect the notification (new substances), data gathering, prioritization (existing substances), classification, risk assessment and risk reduction of substances and products containing these substances. The guidance accompanying the directives sets the standard for risk assessment, including exposure assessment. The guidance specifically addresses human exposure to or via consumer products as well as through ambient air (following industrial emissions) and via the food chain—topics relevant to residential exposure.

LEGISLATION IN THE EUROPEAN UNION

The legal context of risk assessments for new and existing substances can be summarized by two directives on the assessment of new and existing substances and the Technical Guidance Document (TGD) accompanying the directives. The EU legislation on new and existing substances is summarized in Table C-1.

Existing substances are listed in the European Inventory of Existing Commercial Chemical Substances (EINECS), which includes all 100,000 or so substances which were on the market in the European Commission (EC) on 18 September 1981. New substances are those launched after that date and therefore do not appear in EINECS. Directives are to be translated into the legislation of each Member State. Regulations are to be enforced as such. The following is an excerpt from the introduction of the TGD for New and Existing Substances, explaining the directives.

EXISTING EU DIRECTIVES

Council Directive 67/548/EEC (as amended for the seventh time by Directive 92/32/EEC) on the approximation of the laws, regulations, administrative provisions

Table C-1. European Union regulations for new and existing substances

	New substances	Existing substances
Definition	Not on the EU-market for the 10 years prior to 18 September 1981 and not appearing in EINECS[1]	Appearing in EINECS[1]
Basic regulations	Directive 92/32/EC (EC 1992)	Council Regulation (EC) 793/93 (EC 1993c)
Risk assessment legislation	Directive 93/67/EC (EC 1993b)	Commission Regulation (EC) 1488/94 (EC 1994b)
TGDs supporting risk assessment legislation	EC, 1993b and combined TGDs (EC 1996)	EC (1994b) and combined TGDs (EC 1996)

[1]European Inventory of Existing Commercial Chemical Substances.

relating to the classifications, packaging and labeling of dangerous substances requires the manufacturer or importer of a new substance, before they place it on the market, to notify it to the competent authority of the Member State in which it is manufactured or into which it will be imported. Having received the notification, the competent authority is required to carry out an assessment of the risks of the substance to man and the environment in accordance with the principles set out in Commission Directive 93/67/EEC.

The Council Regulation (EEC) number 793/93 on the evaluation and control of existing substances requires under Article 10 the real or potential risk for man and the environment of priority substances to be assessed using principles which have been laid down in the Council regulation (EC) No. 1488/94 on risk assessment for existing substances. The risk assessments are carried out by competent authorities designated by the responsible Member States to act as rapporteurs.

A number of other EU Directives exist that also deal with residential exposure issues. The EU Directive 76/769 on restrictions on marketing and the use of chemicals provides a general framework for bans or restrictions on the marketing and use of dangerous substances and preparations. For example trichloroethylene, tetrachloroethylene and carbon tetrachloride may not be used in ornamental objects, *tris*(2,3-dibromophenyl) phosphate may not be used in textile articles (Directive 79/663), and the use of benzene is banned in toys (Directive 82/806). Other directives include:

- 91/442, which establishes a scheme for requiring the packaging of certain dangerous preparations to be fitted with child-resistant fastenings
- 75/440, which classifies sources of surface water for the abstraction of drinking water
- 76/160, which sets parameters for which Member States must set imperative or guidance values to protect bathing water
- 76/678, which regulates cosmetic products, and
- 88/378, which places restrictions on chemicals in toys.

TECHNICAL GUIDANCE DOCUMENTS (TGDs)

The general principles for the risk assessment of new and existing substances laid down in Directive 93/67/EEC and regulation 1488/94 do not include extensive technical detail for conducting hazard identification, dose-response (concentration and effect) assessment, exposure assessment, and risk characterization in relation to human health and the environment. The TGD for New and Existing Substances provides supplementary technical detail. The guidance is not legally binding, and competent authorities may use other methods or approaches if they are more appropriate, provided that they are scientifically justified and compatible with the general principles laid down in Directive 93/67/EEC or Regulation 1488/94.

RESIDENTIAL EXPOSURE ASSESSMENT IN THE TGD FOR NEW AND EXISTING SUBSTANCES

The TGD discusses human exposure assessment as a function of occupational exposure, consumer product exposure, and indirect exposure via the environment. Dietary exposure and exposure to air, water, and soil are described as a part of indirect exposure via the environment. Consumer products are the most important direct source of residential exposure. Exposure assessment of substances in consumer products is initiated with the following sequence:

(1) Identification of all sources of consumer exposure and gathering information about these sources and the way they might cause exposure. This means that there must be a wide knowledge of the sources of information available about products in each country. If the information that is gathered indicates that the substance is *not* used in or as consumer products, it is not necessary to continue the assessment. If the substance is present in consumer products, further quantification is necessary and the sequence is continued.
(2) Quantification of the potential exposure in the form of concentrations in the exposure media that may contact the human body. Quantification may follow from exposure data as reported in one of the data sources. Otherwise, simple models may be applied (Appendix IV of the TGD) to obtain a reasonable worst case prediction of the exposure.
(3) Quantification of the contact with the exposure media containing the substance by defining exposure routes and exposure patterns, including contact durations, contact frequencies, and site of contact.
(4) Summation of the relevant exposures to each subject group. Both chronic an acute exposure measures may be needed, depending on the expected effects and exposure patterns.
(5) An assessment of the risks associated with consumer product exposure. Usually, the risk assessment takes into account (1) short-term exposure during the event, using acute LD_{50}s or NOAELs derived from short-term experiments, (2) repeated, long-term exposure, using NOAELs derived from (sub)chronic experiments, (3) sensitization and irritation, (4) effects indicated by classification of the substance into such categories as "danger of serious health by prolonged exposure," "possible risk of irreversible effects," "may cause cancer," and "may impair fertility."

In cases where Steps 2 through 5 lead to the conclusion that a substance is "of concern," the assessment is refined by the collection of conclusive data or by using more complex exposure models to obtain a more realistic exposure prediction. Exposure levels based on measured data and on modeling may be considered to be in agreement if they are within the same order of magnitude.

DATA AND MODEL SOURCES

As a companion to the models presented in Appendix B of this book, this Appendix summarizes data sources in the European Union. Many exposure factors will differ from those of the United States. For instance, the construction of houses is different, implying different sizes and means of ventilation. As another example, the dietary composition in the United States differs from that in Europe, and even within the European Union, differences between the south and the north are present. First, a number of data sources are described that cover multiple exposure pathways. Then, data sources specific for food and drinking water, indoor air, and consumer products are described. From these data sources, a gross overview of sources—rather than a complete listing of sources—will be presented within the European Union.

Data Sources Covering Multiple Pathways of Exposure

A number of European data sources aim at summarizing existing data from multiple pathways of exposure. An important source of residential exposure data focusing on Germany is contained in the standards of Arbeitsgemeinschaft der Leitenden Medizinalbeamtinnen und—beamten der Länder and Sagunski *et al.* (1998). The standards summarize anthropometric data, time budgets, food consumption, drink water consumption, and oral soil intake. Based primarily on German data and supplemented with other European and American data sources, this source proposes a number of default values. Also, a Task Force of the European Centre for Ecotoxicology and Toxicology of Chemicals has presented a description of nonoccupational exposures through food, drinking water, and air in the EU (ECETOC 1994).

The above-mentioned TGD for new and existing substances has a broader scope, not only including human exposure, but also including guidance to the assessment of ecosystem exposures and guidance to application of toxicity tests. A summary with respect to the exposure and risk assessment of consumer products has been presented above. In addition, the TGDs contain appendices of simple screening models (Appendices 1, 4, and 7) and use data (Appendix 6). The use data indicate interesting differences in use patterns in Northern and Southern Europe. According to the Appendix, people in Southern Europe use more grams of laundry and cleaning products per wash per household than those in Northern Europe, slightly compensated by a somewhat lower use frequency. However, the source and investigative methods used to generate the data are not documented in the TGDs.

The EUSES program (Vermeire *et al.* 1994, ECB 1996) is directly related to the TGDs in offering a program that performs the screening calculations as suggested in the Technical Guidance Documents. It produces a risk characterization for a particular agent, for both the ecosystem and humans. Residential exposure in EUSES includes exposure to consumer products and to agents in the environment via food, drinking water, and air.

Food and Drinking Water

Exposure to compounds in food and drinking water can be calculated by direct and indirect methods (Parmar *et al.* 1997). Findings on nitrate and nitrite in a number of European states using various methods has been reported in the Council of Europe (1995). These proceedings provide a useful entry with respect to available data sources, and present a case study on nitrate and nitrite exposure via food and drinking water.

Intake by direct methods is determined from duplicate diet studies, where each food item eaten is duplicated for analysis. Vaessen *et al.* (1996) have reported the results of a duplicate diet study in the Netherlands and analyzed, among other compounds, nitrate and nitrite, PCBs, several pesticides, and agents migrating from packaging material. Similar studies have been performed by Vahter *et al.* (1990) in Sweden and Sherlock *et al.* (1983) in the United Kingdom.

Intake by indirect methods is calculated by merging concentrations of agents in every possible food item with the composition of the diet. Slob developed a statistical method to calculate age-specific food exposure from concentration data and diet composition, which reveals intra- and interindividual variation (Slob 1993a,b, Slob and Krajnc 1994, Van Veen *et al.* 1995). An overview of average food consumption data for all EU member states has been prepared by ECETOC (1994), using data from Euromonitor (1992). Food consumption data for specific states are reported. Data sources include:

- For the Netherlands—WVC and LNV (1990), Voorlichtingsbureau voor de Voeding (1993), Heisterkamp and Olling (1996)
- For Germany—Adolf *et al.* (1994a,b)
- For France—Combris *et al.* (1995)
- For United Kingdom—Gregory *et al.* (1990), and
- For the Basque country—Jalón González-Moreno *et al.* (1997).

Food concentration data are reported by various sources from most European countries. In the Netherlands, DLO-RIKILT maintains a database of residues of several agents in food. Results are reported annually in such publications as Van Klaveren (1995). Schuddeboom (1993) has summarized data sources on the nitrate and nitrite contents of food, and has discussed processing and storing effects. In the United Kingdom, the MAFF (1987) has provided, among others, an overview of the presence of plasticisers in food. The Danish Levnedmiddelstyrelsen published results from a monitoring system covering 1988 to 1992 (LST 1995), including nutrients and contaminants. Jalón González-Moreno *et al.* (1997) have reported on a monitoring program in the Basque countries.

Indoor Air

Indoor air quality is set by sources inside and outside the house. Sources outside the house emit agents into the air, which then may be carried into the house by ventilation (Ekberg 1994, 1996, Hoffman *et al.* 1996). Lebret (1985) studied the impact of air pollution on indoor air quality and data sources in the Netherlands with regard to monitoring are reported by Lebret *et al.* (1996). Bluyssen *et al.* (1996) reported on a European study of indoor air quality in office buildings, resulting from concerns about the "sick building syndrome." A second source of agents penetrating the house are those evaporating from contaminated soil and penetrating the house from below. Waitz *et al.* (1996) model this source and provide default values applicable to the Netherlands.

Ventilation will add outdoor contaminants to indoor air, and will remove contaminants from indoor sources. It is thereby an important and sensitive parameter determining indoor concentrations of agents. Van der Wal *et al.* (1991) reported ventilation volumes in Dutch houses, showing that they are generally between 0.3 and 9 air changes per hour. Stymne *et al.* (1994) reported values at the lower bound for Swedish houses, with a median of 0.3 to 0.4 air exchanges per hour. Unpublished results of the National Institute for Public Health and the Environment (RIVM) in The Netherlands show that ventilation in well isolated rooms may drop to 0.1 air changes per hour. De Vries and Bartholomeus (1973) determined an upper bound of 4.6 air exchanges per hour in Dutch houses. Mechanical ventilation provides ventilation volumes of 1.05 to 1.35 air changes per hour (Van der Wal *et al.* 1991). Note that mechanical ventilation does not imply air conditioning—typically not present in most European houses. Many (northern) European mechanical ventilation systems cause a slightly negative static indoor pressure, whereas American ventilation systems are often designed for a slightly positive indoor pressure (Ekberg 1996).

Many sources of indoor air pollution (Brussels conference on the present and future of indoor air quality 1989) include such items as carpets (Clausen *et al.* 1993), and paint and lacquer (Clausen *et al.* 1991, Hansen *et al.* 1991) emitting volatile organic compounds or chlorodifluoromethane released from sprays (for instance, see Hartop and Adams 1989). The European collaborative action on indoor air quality and its impact on man have produced 3 main reports on testing emissions from products and materials (ECA-IAQ, 1994, 1995, 1997). Other reports provide additional information on indoor air quality and the role of product and materials emissions. Many consumer products also emit volatile organic compounds to indoor air (see the following section for discussion). Seifert and Ullrich (1987) provide methodologies for measuring and classifying indoor sources, as well as a number of examples of indoor air pollution in Germany. Hoffman *et al.* (1996) and Berry (1996) have reported recent measurements of indoor VOC concentrations in Germany and the United Kingdom, respectively, with the latter also reporting nitrogen dioxide and biological particulates. These studies included correlations among VOCs and correlations among household characteristics, occupant behavior, and presence of industries in the vicinity. In the Netherlands, overviews of agents and sources have been provided by Van Dongen *et al.* (1989) and Fraanje *et al.* (1993), both using national and international sources of information.

A special class of indoor pollutants are agents formed during combustion while cooking and heating. Of note is carbon monoxide, which is formed in low amounts during combustion. Carbon monoxide production rapidly increases when combustors are not clean and when the air used for combustion contains over 1 to 2 percent carbon dioxide (De Vries and Bartholomeus 1973). Accidents then happen when the ventilation is insufficient to remove carbon monoxide and provide oxygen-rich and carbon dioxide-depleated air. Given its large stores in northwestern Europe, natural gas is used extensively for cooking and heating, leading to a common proliferation of gas-fueled devices in households.

Not only are the *concentrations* in indoor air important for exposure, but also the *contact duration* for determining exposure and possible *health effects*. Time-activity patterns, or time budgets, have been investigated to reveal the time spend indoors in various environments (*e.g.*, school, work, and home). Time activity patterns have been generated for:

- The Netherlands—Janssen *et al.* (1993) and Freijer *et al.* (1997)
- France—Grimler (1992)
- Germany—Dörre (1997), and
- Croatia—Croatian Radiation Protection Association (http://mimi.ini.hr:80/~ksega/index.html).

Hoffman *et al.* (1996) used personal sampling in combination with recorded time activity patterns to find behaviors that correlated with significant exposures to VOCs. For example, reading a newspaper correlated with ethylbenzol, xylol, propyl benzol, and ethyl toluol exposures.

Consumer Products

Consumer products have a diversity of composition, packaging, usage, *etc.* This diversity implies that exposure to agents in or from consumer product may follow any route of exposure, or any combination of routes. Assessing exposure to agents in or from consumer products requires knowledge about (1) product composition, (2) the concentrations that occur in any medium that can be contacted (which will depend on product usage), (3) contact patterns with the product and all media containing agents from the product, and (4) the intake and uptake of agents (Van Veen 1996).

Product composition data can be gathered from a number of sources, among which labels, poison information centers (*e.g.*, Hofstee *et al.* 1990, Velvart 1993, Flyvholm and Andersen 1993), and product registers (*e.g.*, at the Swedish Kemikalien Inspection, KEMI) are the most important ones. Environmental health groups may also provide product composition information (*e.g.*, Annema 1988, Reijnders *et al.* 1994). Both models and measurements are used to estimate concentrations in media containing chemical agents from the product. Screening models have been provided in the TGD, and by Vermeire *et al.* (1993), ECETOC (1994), and Van de Meent *et al.* (1995). Mechanistic exposure models have been provided by Van Veen (1997). Van Veen (1995, 1996, 1997) has also provided CONSEXPO as a program to combine contact, exposure, and uptake models for the inhalation, dermal, and oral routes of exposure for a single compound. Measurements of exposure concentrations have been reported by Clausen *et al.* (1991), Hansen *et al.* (1991), Hartop and Adams (1989), Seifert and Ullrich (1987), Brown *et al.* (1993), and Crump *et al.* (1996). Van Schooten *et al.* (1994) have assessed exposure to polycyclic aromatic hydrocarbons in coal tar shampoo, using 1-hydroxypyrene as a biomarker.

Two weaknesses in the data used to assess exposure to consumer products are (1) product usage and (2) product contact. Some information is present in Appendix 6 of the TGD. Data on extreme uses can be gathered from poison information centres (Velvart 1993, and others), but these data often represent accidental exposure scenarios. Weegels (1997) has investigated consumer product use—comparing questionnaire, diary, and videotape methods—and provides data on dish washing products, cleaning products, hair styling products, paints and pesticides. The author concluded that questionnaires tend to overestimate product usage. Variables like amount used and duration of use exhibited a large variation (coefficient of variation of 1 was commonly encountered) and variables were barely correlated. For example, the amount of dish washing fluid used did not correlate with the amount of water used.

For toxicological assessments, an estimate of intake or uptake is necessary. Some consumer products cause local effects, such as dishwasher detergents (Madarikan 1990), or alkali cleaner (Rider and Tarar 1995). Other agents may cause systemic effects after their uptake in the body. Measurements of dermal exposure and uptake of octopyrox in shampoo have been provided by Black and Kamat (1988). Ten Berge (in Wilschut *et al.* 1995) developed a dermal uptake model that not only models uptake from a aqueous solution, but also from air. Wilschut *et al.* (1995) have provided an overview of empirical relations to estimate dermal permeability for agents dissolved in aqueous solutions. The CONSEXPO program (Van Veen 1996, 1997) contains a number of simple diffusive uptake models for inhalation, dermal, and oral routes of exposure.

ACKNOWLEDGMENTS

The authors would like to thank W. Mennes, J. van Koten-Vermeulen, K. van Velze, E. Lebret (RIVM, Rijksinstituut voor Volksgezondheid en het Milieu, The Netherlands), D.R. Crump (BRE, United Kingdom), B. Seifert (UBA, Germany), H. Sagunski (BAGS, Germany), and G. Moore (KEMI, Sweden) for their assistance in preparing this Appendix and its references.

REFERENCES

Adolf, T., W.R. Eberhardt, S. Hartmann, A. Herwig, H. Heseker, B. Matiaske, K.J. Moch, and J. Rosenbauer. 1994a. Ergebinsse der nationalen verzehrstudie (1985–1988) über die lebensmittel- und nährstofaufnahme in de bundesrepubliek Deutchland. Wissenschaftlicher fachverlag Dr. Fleck. Niederkleen.

Adolf, T., W.R. Eberhardt, H. Heseker, S. Hartmann, A. Herwig, B. Matiaske, K.J. Moch, R. Schneider, and W. Kübler. 1994b. Ergänzungsband zum ernährungsbericht 1992 (Oktober 1985 bis Januar 1989) auf der basis der nationalen verzehrsstudie. Wissenschaftlicher fachverlag Dr. Fleck. Niederkleen.

Annema, J.A. 1988. Mooi is anders. Pages 1–133 in: Stichting natuur en milieu. Utrecht, The Netherlands.

Arbeitsgemeinschaft der leitende medizinalbeamtinnen and beamten der länder. 1995. Standards zur expositionsabschätzung. Bericht des Ausschusses für Umwelthygiene. Gehörde für Arbeit, Gesundheit und Soziales. Hamburg, Germany. 163 pages.

Berry, R.W. 1996. Indoor air quality in homes. The Building Research Establishment Indoor Environment Study. BRE Reports BR299 and BR300. CRC Ltd. Watford, Hertsfordshire, United Kingdom.

Black, J.G. and V.B. Kamat. 1988. Percutaneous absorption of octopyrox. Food Chem. Toxicol. 26:53–58.

Bluyssen, P.M., E. De Oliveira Fernandez, L. Groes, G. Clausen, P.O. Fanger, O. Valbjørn, C.A. Bernhard, and C.A. Roulet. 1996. European indoor air quality audit project in 56 office buildings. Indoor Air 6:221–238.

Brown, V.M., D.R. Crump, and C. Yu. 1993. The use of chamber methods to measure the emission of volatile organic compounds from building and consumer products. Pages 27–28 in: Proceedings of the International Conference on VOCs in the Environment. London, United Kingdom.

Brussels conference on the present and future of indoor air quality. 1989. Proceedings. C.J. Bieva, Y. Courtois, and M. Govaerts (eds.). February 14–16, 1989. Excerpta Medica. ISBN 0 444 81134 6. Amsterdam, The Netherlands. 462 pages.

Clausen, P.A., P. Wolkoff, E. Holst, and P.A. Nielsen. 1991. Long-term emission of volatile organic compounds from waterborne paints—methods of comparison. Indoor Air 4:562–576.

Clausen, P.A., B. Laursen, P. Wolkoff, E. Rasmusen, and P.A. Nielsen. 1993. Emission of volatile organic compounds from a vinyl floor covering. In N. Nagda (ed.). Modeling of Indoor Air Quality and Exposure. Pages 3–13. ASTM STP 1205. Philadelphia, Pennsylvania.

Combris, P., P. Bertail, C. Boizot, and J.-C. Poupa. 1995. La conformation alimentaire en 1991: Distribution des quantites consommees a domicile. Laboratoire de Recherche sur la Consummation. Brandenburg, France. 256 pages.

Council of Europe. 1995. Health aspects of nitrates and its metabolites (particularly nitrite). Council of Europe. Strasbourg, France. 336 pages.

Crump, D., R. Squire, C. Yu, and M. Atkinson. 1996. Small chamber methods for characterizing formaldehyde emission from particleboard. Pages 211–224 in: Tichenor, B.A. (ed.) Characterizing sources of indoor air pollution and related sink effects. American Society for Testing and Materials (ASTM). STP 1287. Philadelphia, Pennsylvania.

De Vries, H. and P.H.J. Bartholomeus. 1973. Gas en milieuhygiene in huis. GAS 8:1–15.

Dörre, W.H. 1997. Time-activity patterns of some selected small groups as a basis for exposure estimation: a methodological study. J. Exposure Anal. Environ. Epidemiol. 7:471–492.

ECA-IAQ (European Collaborative Action "Indoor Air Quality and its Impact on Man"). 1994. Formaldehyde emission from wood-based materials: guideline for the determination of steady-state concentrations in test chambers. Report 2, EUR 12196 EN, OOP. Luxembourg.

ECA-IAQ (European Collaborative Action "Indoor Air Quality and its Impact on Man"). 1995. Determination of VOCs emitted from floor materials and products. Interlaboratory comparison of small chamber measurements. Report 13, EUR 15054 EN, OOP. Luxembourg.

ECA-IAQ (European Collaborative Action "Indoor Air Quality and its Impact on Man"). 1997. Evaluation of VOC emissions from building products—Solid flooring materials. Report 18, EUR 17334 EN, OOP. Luxembourg.

ECB (European Chemical Bureau). 1996. EUSES (European Union System for the Evaluation of Substances). EUR 17308 EN. Ispra, Italy.

ECETOC (European Centre for Ecotoxicology and Toxicology of Chemicals). 1994. Assessment of non-occupational exposure to chemicals. Technical Report 58. Brussels, Belgium. 96 pages.

Ekberg, L.E. 1994. Outdoor air contaminants and indoor air quality under transient conditions. Indoor Air 4:189–196.

Ekberg, L.E. 1996. Relationships between indoor and outdoor contaminants in mechanically ventilated buildings. Indoor Air 6:41–47.

Euromonitor. 1992. European marketing data and statistics 1992. ISBN 0 86338 403 X.

Flyvholm, M., and P. Andersen. 1993. Identification of formaldehyde releasers and occurrence of formaldehyde and formaldehyde releasers in registered chemical products. Am. J. Industrial Med. 24:533–552.

Fraanje, P., F. Steenhuisen, K. Van Velze, H. Laan, and J. Janus. 1993. Vluchtige organische stoffen in het binnenmilieu van woningen. RIVM (Rijksinstituut voor Volksgezondheid en het Milieu). Rapport Number 222302/002. 113 pages.

Freijer, J.I., H.J. Th. Bloemen, S. De Loos, M. Marra, P.J.A. Rombout, G.M. Steentjes, and M.P. Van Veen. 1997. AIRPEX: a computer model to assess exposure of humans to air pollution. RIVM (Rijksinstituut voor Volksgezondheid en het Milieu). Report 650010.005. 77 pages.

Gregory, J., K. Foster, H. Tyler, and M. Wiseman. 1990. Dietary and Nutritional Survey of British Adults. Her Majesty Stationary Office. London, United Kingdom.

Grimler, G. 1992. Les rythmes quotidiens en France. Resultats de l'enquete "Les emplois du temps des francais en 1985–1986". INSEE Resultats n 34–35. INSEE. Grenoble, France.

Hansen, L.F., G.D. Nielsen, J. Tøttrup, A. Abildgaard, O.F.D. Jensen, M.K. Hansen, and O. Nielsen. 1991. Biological determination of emission of irritants from paint and lacquer. Indoor Air 2:95–110.

Hartop, P.J. and M.G. Adams. 1989. Simulated consumer exposure to propellant HCFC22 (chlorodifluoromethane) in aerosol personal products. Internat. J. Cosmetic Sci. 11:27–34.

Heisterkamp, S.H. and M. Olling. 1996. Haalbaarheid human blootstelling voeding. Report No. 604502.001. RIVM (Rijksinstituut voor Volksgezondheid en het Milieu) Bilthoven, The Netherlands. 25 pages.

Hoffman, K., R. Schwabe, C. Krause, C. Schultz, B. Seifert, and D. Ullrich. 1996. Umwelt-Survey 1990/91 Band IV: Personengebundene Exposition gegenüber flüchtigen organischen Verbindungen in den alten Bundesländern. WaBoLu-Hefte 4/96. Umwelt Bundes Amt. Berlin, Germany. 135 pages.

Hofstee, A.W.M., J. Meulenbelt, and T.J.F. Savelkoul. 1990. Informatieverstrekking bij vergiftigingen door het Nationaal Vergiftigingen Informatie Centrum. Gebaseerd op gegevens van 1985–1989. RIVM (Rijksinstituut voor Volksgezondheid en het Milieu) Report Number 348802011. 48 pages.

Jalón González-Moreno, M., I. Urieta Guijarro, L. Macho Eiras, and M. Azpiri Luzar. 1997. Vigilancia de la contaminacion quimica de los alimentos en la Communidad Autonoma del Pais Vasco, 1990–1995. Gobierno Vasco. Bilbao, Spain. 187 pages.

Janssen, N.A.H., B. Brunefreef, J.P. Zock, P. Fischer, and J. Van Scheindelen. 1993. Activiteitenpatronen in verband met populatie-expositie aan luchtverontreiniging, Verslag van een orienterend onderzoek. Report 1993-466. Vakgroep humane epidemiologie en gezondheidsleer. Lanfbouw Universiteit Wageningen. Wageningen, The Netherlands. 80 pages.

Lebret, E. 1985. Air Pollution in Dutch homes. An exploratory study in environmental epidemiology. Wageningen Agricultural University. Report R-138/Report 1985-221. Wageningen, The Netherlands. 142 pages.

Lebret, E., P.H. Fischer, B.A.M. Staatsen, E.A.M. Franssen, A.E.M. Den Hollander, and D.J.M. Houthuijs. 1996. Monitoring of exposures, body burdens and health effects of environmental pollutants in The Netherlands. Position paper from the perspective of environmental epidemiology. Report 529104 001. RIVM (Rijksinstituut voor Volksgezondheid en het Milieu). Bilthoven, The Netherlands. 66 pages.

LST (Levnedmiddelstyrelsen). 1995. Food Monitoring 1988–1992. Publication 232 (English version). Ministry of Health. Copenhagen, Denmark. 222 pages.

Madarikan, B.A. 1990. Ingestion of dishwasher detergent by children. Brit. J. Clin. Practice 44:35–36.

MAFF (Ministry of Agriculture, Fisheries, and Food). 1987. Survey of plasticiser levels in food contact material and in food. Food Surveillance Paper No. 21. Her Majesty's Stationary Office. London, United Kingdom.

Parmar, B., P.F. Miller, and R. Burt. 1997. Stepwise approaches for estimating the intakes of chemicals in food. Regulatory Toxicol. Pharmacol. 26:44–51.

Reijnders, L., A. Tosserams, H. Van Wieringen, and G. Den Blaauwen-Kerst. 1994. Stop de Plaag. De Geus, Breda. 175 pages.

Rider, M.A. and M.N. Tarar. 1995. Burns caused by domestic alkalis. J. Accident Emergency Med. 12:130–131.

Sagunski, H., H. Neus, I. Ollroge, and A. Kappos. 1998. Public health needs for exposure assessment and its realisation in Germany. J. Amer. College Toxicol. In press.

Schuddeboom, L.J. 1993. Nitrates and nitrites in foodstuffs. Health protection of consumers. Council of Europe Press. Strasbourg, France. 125 pages.

Seifert, B. and D. Ullrich. 1987. Methodologies for evaluating sources of volatile organic chemicals (VOC) in homes. Atmos. Environ. 21:395–404.

Sherlock, J.C., G.A. Smart, B. Walters, W.H. Evans, D.J. McWeeny, and W. Cassidy. 1983. Dietary surveys on a population at Shipham, Somerset, United Kingdom. Sci. Total Environ. 29:121–142.

Slob, W. 1993a. Modeling human exposure to chemicals in food. RIVM (Rijksinstituut voor Volksgezondheid en het Milieu). Report Number 639102002. 20 pages.

Slob, W. 1993b. Modeling long-term exposure of the whole population to chemical in food. Risk Anal. 13:525–530.

Slob, W. and E.I. Krajnc. 1994. Interindividual variability in modeling exposure and toxicokinetics: A case study on cadmium. Environ. Health Perspect. 102:78–81.

Stymne, H., C.A. Boman, and J. Kronvall. 1994. Measuring ventilation rates in the Swedish housing stock. Building and Environment. 29:373–379.

Vaessen, H.A.M.G., A. Van Ooik, J.W. Van Loon, J.J.H. Van Vliet, A.J. Kieft, G. Van den Burg, and R. Ritsema. 1996. Duplicaat 24-uurs voedingonderzoek 1994—verzamelen en bewerken van het monstermateriaal. RIVM (Rijksinstituut voor Volksgezondheid en het Milieu). Report Number 389101.008. Bilthoven, The Netherlands. 37 pages.

Vahter, M., M. Berglund, L. Friberg, L. Jorhem, B. Lind, S. Slorach, and A. Akesson. 1990. Dietary intake of lead and cadmium in Sweden. Var Foda 44(suppl 2). National Food and Drug Administration. Uppsala, Sweden.

Van de Meent, D., J.H.M. De Bruijn, F.A.A.M. De Leeuw, A.C.M. De Nijs, J.T. Jager, and T.G. Vermeire. 1995. Exposure modeling. Pages 103–145 in: Van Leeuwen, C.J. and J.L.M. Hermens (eds.). Risk assessment of chemicals: An introduction. Kluwer Academic Publishers. Dordrecht, The Netherlands.

Van der Wal, J.F., A.M.M. Moons, and H.J.M. Cornelissen. 1991. Indoor air quality in renovated Dutch homes. Indoor Air 4:621–633.

Van Dongen, J.E.F., J.F.L. Gerbrandy, R.G. de Jong, R.W. Lanting, M.J. Leupen, and G. de Mik. 1989. Gezondheidskundige aspecten en kwaliteit van het binnenmilieu, vooronderzoek. TNO-NIPG. Leiden, the Netherlands. 118 pages.

Van Klaveren, J.D. 1995. KAP verslag 1995 resultaten residuebewaking in Nederland. RIKILT-DLO. Wageningen, The Netherlands.

Van Schooten, F., E.J.C. Moonen, E. Rhijnsburger, B. Van Agen, H.H.W. Thijssen, and Kleinjans. 1994. Dermal uptake of polycyclic aromatic hydrocarbons after hairwash with coal-tar shampoo. Lancet 344:1505–1506.

Van Veen, M.P. 1995. CONSEXPO, a program to estimate consumer product exposure and uptake. Report Number 612810.002 RIVM (Rijksinstituut voor Volksgezondheid en het Milieu). Bilthoven, The Netherlands. 46 pages.

Van Veen, M.P. 1996. A general model for exposure and uptake from consumer products. Risk Anal. 16:331–338.

Van Veen, M.P. 1997. CONSEXPO 2.0, consumer exposure and uptake models. Report No. 612810 005. RIVM (Rijksinstituut voor Volksgezondheid en het Milieu). Bilthoven, The Netherlands. 46 pages.

Van Veen, M.P., R. Van der Berg, and W. Slob. 1995. A statistical exposure model (STEM) applied to nitrate intake in the Dutch population. Pages 75–81 in: Health aspects of nitrate and its metabolites (particularly nitrite). Council of Europe. Strasbourg, France.

Velvart, J. 1993. Toxicologie der haushaltsproducte. H. Huber verlag. Bern, Switzerland. 463 pages.

Vermeire, T.G., P. Van der Poel, R.T.H. Van der Laar, and H. Roelfzema. 1993. Estimation of consumer exposure to chemicals: Application of simple models. Sci. Total Environ. 136:155–176.

Vermeire, T.G., P.T.J. van der Zandt, H. Roelfzema, and C.J. van Leeuwen. 1994. Uniform system for the evaluation of substances. I. Principles and structure. Chemosphere 29:23–38.

Voorlichtingsbureau voor de Voeding. 1993. Zo eet Nederland, de resultaten van de Voedselconsumptiepeiling 1992. Voorlichtingsbureau voor de Voeding. Den Haag, The Netherlands. 196 pages.

Waitz, M.F.W., J.I. Freijer, P. Kreule, and F.A. Swartjes. 1996. The VOLASOIL risk assessment model based on CSOIL for soils contaminated with volatile compounds. RIVM (Rijksinstituut voor Volksgezondheid en het Milieu). Report Number 715810.014. Bilthoven, The Netherlands. 189 pages.

Weegels, M.F. 1997. Exposure to chemicals in consumer products. Technical University. Delft, The Netherlands. 159 pages.

Wilschut, A., W.F. Ten Berge, P.J. Robinson, and T.E. McKone. 1995. Estimating skin permeation. The validation of five mathematical skin permeation models. Chemosphere 30:1275–1296.

WVC (Ministerie van Welzijn, Volksgezondheid en Cultuur) and LNV (Ministerie van Landbouw, Natuur en Visserij). 1990. Wat eet Nederland. Resultaten van de voedselconsumptiepeiling 1987–1988. Rijswijk.

INDEX

MIX
Papier aus verantwortungsvollen Quellen
Paper from responsible sources
FSC® C105338

If you have any concerns about our products,
you can contact us on
ProductSafety@springernature.com

In case Publisher is established outside the EU,
the EU authorized representative is:
Springer Nature Customer Service Center GmbH
Europaplatz 3, 69115 Heidelberg, Germany

Printed by Libri Plureos GmbH
in Hamburg, Germany